Phenolic Resins

Springer
*Berlin
Heidelberg
New York
Barcelona
Hong Kong
London
Milan
Paris
Singapore
Tokyo*

A. Gardziella · L. A. Pilato · A. Knop

Phenolic Resins

Chemistry, Applications, Standardization, Safety and Ecology

2nd completely revised edition

With 309 Figures and 166 Tables

Springer

Dr. Arno Gardziella (ehem. Bakelite AG)
Rüdinghausener Berg 4
D-58454 Witten
Germany

Dr. Louis A. Pilato
Phenolic Resin Technology
Advanced Composite Systems
598 Watchung Road
Bound Brook, NJ 08805
USA

Dr. Andre Knop
Schillerstraße 31
D-61350 Bad Homburg
Germany

ISBN 3-540-65517-4 Springer-Verlag Berlin Heidelberg New York

Cataloging-in-Publication Data applied for

Die Deutsche Bibliothek – CIP-Einheitsaufnahme

Gardziella, Arno: Phenolic resins : chemistry, applications, standardization, safety and ecology ; with 166 tables / A. Gardziella ; L. A. Pilato ; A. Knop. – 2., completely rev. ed. – Berlin ; Heidelberg ; New York ; Barcelona ; Hong Kong ; London ; Milan ; Paris ; Singapore ; Tokyo : Springer, 1999
   1. Aufl. u. d. T.: Knop, André: Phenoloc resins
   ISBN 3-540-65517-4

This work is subject to copyright. All rights are reserved, whether the whole or part of the material is concerned, specifically the rights of translation, reprinting reuse of illustrations, recitation, broadcasting, reproduction on microfilms or in any other way, and storage in data banks. Duplication of this publication or parts thereof is permitted only under the provisions of the German Copyright Law of September 9, 1965, in its current version, and permission for use must always be obtained from Springer-Verlag. Violations are liable for Prosecution under the German Copyright Law.

© Springer-Verlag Berlin Heidelberg 2000
Printed in Germany

The use of general descriptive names, registered names, etc. in this publication does not imply, even in the absence of a specific statement, that such names are exempt from the relevant protective laws and regulations and free for general use.

Product liability: The publisher cannot guarantee the accuracy of any information about dosage and application contained in this book. In every individual case the user must check such information by consulting the relevant literature.

Production Editor: Christiane Messerschmidt, Rheinau
Typesetting: Fotosatz-Service Köhler GmbH, Würzburg
Coverdesign: Design & Production, Heidelberg

SPIN: 10489866   2/3020-5 4 3 2 1 0 – printed on acid-free paper

Preface

A backward glance of the many new industries that emerged in the 20th century would surely recognize communications, automobile, aircraft, computer and several others that have had a global impact on world economy. Yet another industry, and an often neglected industry, made its debut early in the 20th century – the Plastics Industry. The Plastics Industry owes its identity to the brilliance of Dr. Leo Hendrik Baekeland (1863–1944).

He discovered the technique, currently in use to this day, to manufacture highly crosslinked plastics by transforming monomeric and/or oligomeric phenolic materials into attractive phenolic products.

Today phenolics represent one of the many different types of commercially available plastics (thermoset and thermoplastic). Phenolics are distinguished by a broad array of application areas that utilize phenolics as compared to other thermoset or thermoplastic resins. Thermoplastic resins transformed into molded products, films or synthetic fibers (polypropylene as an example) are rapidly recognized as "plastics" whereas the phenolic resin is a component in a material system and the identity of the phenolic resin within the system is not easily identified as "plastic". These systems consist of fiber reinforced composites, honeycomb paneling, electrical laminates, acid resistant coatings, wood panels, glass fiber or rock wool insulation. Phenolic resin identity is hidden and has little consumer recognition or identity. The apparent hidden but security/comfort aspect of phenolic resins is best illustrated by considering the use of phenolic resin as binder in friction linings or automobile brakes. Although there are as many as 30 components including phenolic resin binder in the brake formulation, phenolic resin is the preferred binder in this "life or death" application. During the braking operation, the kinetic energy of the vehicle is largely converted into heat; peak temperatures of greater that 800 °C occur at the surface of the brake lining, depending on the stress and type of vehicle. Through the analyses of various bonding functions (temporary, complementary, carbon forming and chemically reactive) the performance of phenolic resin as the preeminent binder for the automobile brake is best described. It maintains the integrity of the brake system.

The main objective of this publication is to familiarize the reader with phenolic resins by describing in more detail the subject of phenolic resins than the previous publications: A. Knop and Scheib, "Chemistry and Application of Phenolic Resins" (1979); and A. Knop and L. A. Pilato, "Phenolic Resins, Chemistry, Applications and Performance – Future Directions" (1985). The title of this current publication is deliberately long and is due to the authors' attempt to broaden the scope of phenolics beyond the achievements of the earlier books.

For those acquainted with phenolic resins, they will appreciate the novel approach to identify the role of phenolic resins as adhesives by considering SIX functions to illustrate the binding versatility of phenolic resins. For those less familiar readers, their introduction to phenolic resins will be rewarding since they will encounter raw materials, chemistry, reactions, mechanisms and resin production in the beginning of the book. The chemical portion is based on current scientific publications. The applications areas and the phenolic resins used in these market areas are extensively illustrated and described.

Once both neophyte and veteran complete the chemistry and six bonding functions segments, comparison of phenolic resins with other thermosetting resins as well as thermoplastic resins are described in which both filled and unfilled resin systems are compared. These comparisons provide ample evidence for the superiority of phenolic resins with many of the competitive thermosetting and thermoplastic resins. Prominent features of phenolic resins are:

1. Excellent thermal behavior
2. High strength level
3. Long term thermal and mechanical stability
4. Excellent fire, smoke, and low toxicity characteristics
5. Excellent electrical and thermal insulating capabilities
6. Excellent cost performance characteristics

These criteria provide an entry into many application areas which utilize phenolic resins along with the appropriate bonding function. In each application area the final product which combines phenolic resin with other components exhibits the necessary performance characteristics of the intended application. By considering the type of bonding function (permanent, temporary, complementary, etc.) and the role of the phenolic resin in the commercial product, it allows one to examine critically the fundamental aspects that are responsible in the successful design and performance of the resulting product. It is possible that the bonding function(s) coupled to new phenolic resin mechanistic perceptions and new analytical techniques may stimulate or challenge the reader to develop improved products/processes.

The timing of the section related to standardization and certification is opportune. Currently standardization of phenolic based electrical laminates and molding compounds is encountering considerable change and is becoming even more international or global in nature. An extensive ISO and CEN system of test methods for phenolic resins now exists and of great importance for purposes of

ISO 9001 ff certification, European standardization and globalization of plastics production and application. It is discussed in detail.

The chemical industry and resin manufacturers remain vigilant in their endeavors to provide environmental and worker safety. Enhancement of worker safety and related environment is a continuing activity of both resin manufacturer and resin processor. Continued reduction of residual monomers to very low levels, accelerated changes of solvent based systems to aqueous types, environmentally friendly resins, melt processing... are evergreen activities ensuring collaborative efforts between resin manufacturer and resin processor. Recycling, renewable raw materials, and future development guidelines for phenolic resins are discussed.

With technical advice, brochures, photos and drawings, numerous experts and business associates in the USA and FRG have greatly aided us in expanding our level of knowledge. Corporate management of Bakelite AG, Iserlohn, and company employees in the development and application research, analytical and environmental technology departments as well as in the business areas of phenolic resins and molding compounds have joined in the development of this new reference work with many comments and suggestions. We would like to extend our warmest thanks to all of them.

August 1999   Arno Gardziella, Louis A. Pilato, Andre Knop

Table of Contents

Abbreviations XXI

Conversion Factors XXV

Part A	**Chemistry/Production**	1
Chapter 1	**Raw Materials**	3
1.1	Phenols	3
1.1.1	Physical Properties of Phenol	4
1.1.2	Supply and Use of Phenol	4
1.1.3	Phenol Production Processes	5
1.1.3.1	Cumene Oxidation (Hock Process)	5
1.1.3.2	Toluene Oxidation Process	8
1.1.3.3	Nitrous Oxide Oxidation of Benzene	9
1.1.3.4	Other Synthesis Processes	9
1.1.4	Alkylphenols	9
1.1.4.1	Methylphenols	10
1.1.4.2	Higher Alkylphenols	11
1.1.5	Phenols from Coal and Mineral Oil	12
1.1.6	Resorcinol	12
1.1.7	Bisphenol-A	13
1.2	Aldehydes	14
1.2.1	Formaldehyde	14
1.2.2	Production and Economics of Formaldehyde	16
1.2.3	Paraformaldehyde	20
1.2.4	Trioxane and Cyclic Formals	20
1.2.5	Hexamethylenetetramine	21
1.2.6	Other Aldehydes	21
	References	22
Chapter 2	**Phenolic Resins: Chemistry, Reactions, Mechanism** . . .	24
2.1	Introduction	24
2.2	Chemistry	24
2.2.1	Resole	25
2.2.1.1	Methylol Phenol(s)	25

2.2.1.2	Oligomerization	27
2.2.2	Novolak	36
2.2.2.1	Bisphenol F	36
2.2.2.2	Oligomerization	39
2.2.2.2.1	Random	39
2.2.2.2.2	High Ortho	46
2.2.2.2.3	Modified Novolaks	48
	Bismethylol Para Substituted Phenols 48 – Diethers 49 – Dialcohols 49 – Diolefins 50 – Enzymes 50	
2.3	Reactions	51
2.3.1	Functional Group Appendage to Phenolics	51
2.3.1.1	Epoxy	51
2.3.1.2	Allyl, Benzyl Group	55
2.3.1.3	Cyanate Ester	56
2.3.2	Ring Formation with Phenolic	57
2.3.2.1	Benzoxazine	57
2.3.3	Alkylation	60
2.3.3.1	Friedel-Crafts	60
2.3.3.2	Hydroxymethylation	60
2.3.4	Miscellaneous	60
2.3.4.1	Polyimide	60
2.3.5	Natural Products	61
2.4.	Mechanism	61
2.4.1	Cure	61
2.4.1.1	Resole	62
2.4.1.1.1	Thermal/Viscoelastic Analyses	62
2.4.1.1.2	NMR Analyses	64
2.4.1.2	Novolak	66
2.4.1.3	Non-Hexa Cure	72
2.4.1.4	Level of Cure	74
2.4.2	Toughness/Alloys/IPN	75
2.4.3	Hazardous Occurrences	77
2.5	Summary/Trends	78
	References	79

Chapter 3 Production of Phenolic Resins 83

3.1	Production of Novolak	87
3.2	Production of Resole	89
3.3	Continuous Process	89
	References	90

Chapter 4 Structure (Methods of Analysis) 91

4.1	Introduction	91
4.2	Structural Characteristics	91
4.2.1	Hydrodynamic Volume Characterization	91
4.2.1.1	Chromatography	92

4.2.2	MALDI	92
4.2.3	Compositional Structure	96
4.2.3.1	Oligomer, Uncured Resin	96
4.2.3.2	Cured Resin	96
4.3	Solution Properties	97
4.3.1	Novolaks	97
4.3.2	Resoles	97
4.3.3	Amine Catalyzed Resoles	98
4.3.4	"All ortho" Novolaks	101
4.4	Complexation	101
4.4.1	"All ortho" Novolaks	101
4.4.2	Ethyl Acetate Appendage	101
4.4.3	Trioxyethylene Ether Appendage	102
4.5	Mechanical Properties	103
4.6	Fire Properties	104
4.7	Summary/Trends	105
	References	106

Part B Applications of Phenolic Resins ... 107

Chapter 5 Thermosets: Overview, Definitions, and Comparisons ... 109

References ... 120

Chapter 6 Economic Significance of Phenolic Resins, Survey of Applications, and Six Bonding Functions ... 122

6.1	Permanent Bonding	126
6.1.1	Wood Composites	127
6.1.1.1	Introduction	127
6.1.1.2	Wood Structure/Surface	128
6.1.1.3	Wood Sources	129
6.1.1.4	Wood Composite Materials	130
6.1.1.4.1	Wood Panel and Engineered Lumber Products	130
6.1.1.4.2	Adhesives	130

Phenolic Resins 131 – Panels 131 – Plywood 131 – Resins, Additives, and Formulations 133 – Production of Plywood 134 – Co-Reaction of PF with UF 136 – Color Improvement 137 – Oriented Strand Board (OSB)/Waferboard 137 – Core and Surface Resins 138 – Resin Distribution by Imaging 140 – MDI 141 – Particleboard 141 – Classification of PB 141 – Emission Studies 142 – Medium Density Fiberboard (MDF) 142 – Hardboard 142

6.1.1.4.3	Engineered Lumber Products	143

Laminated Veneer Lumber (LVL) 143 – Parallel Strand Lumber (Parallam) 143 – Parallam® Features 145

6.1.1.5	New Process and Equipment Developments in Panel Manufacture	145

6.1.1.5.1	Steam Injection Pressing (SIP)	145
6.1.1.5.2	Wider Presses	146
6.1.1.5.3	Increased Length – OSB Flakes	146
6.1.1.5.4	Uniform Mats	146
6.1.1.6	Natural Resins	146
6.1.1.6.1	Introduction	146
6.1.1.6.2	Lignin	147
6.1.1.6.3	Tannin	148
6.1.1.7	Non-Wood Materials	149
6.1.1.8	Structural Wood Gluing	149
6.1.1.8.1	Introduction	149
6.1.1.8.2	Resorcinol Adhesives	150
6.1.1.8.3	New Areas	151
	FRP Glulam Adhesive 151 – Coupling Agent 152	
6.1.1.9	Summary/Future Directions	152
6.1.2	Insulation	153
6.1.2.1	Inorganic Fibers	153
6.1.2.1.1	Introduction	153
6.1.2.1.2	Types of Inorganic Fibers	154
6.1.2.1.3	Resins for Inorganic Fibers	156
6.1.2.1.4	Resin Formulation	157
6.1.2.2	Phenolic Foam	158
6.1.2.2.1	Introduction	158
6.1.2.2.2	Resins	159
6.1.2.2.3	Catalysts	160
6.1.2.2.4	Blowing Agents	160
6.1.2.2.5	Surfactants	161
6.1.2.2.6	Wetting Agents	162
6.1.2.2.7	Fillers	162
6.1.2.2.8	Foam Stabilizers	162
6.1.2.2.9	Foam Applications	162
	Thermal Insulation 162 – Low Density Foam (<50 kg/m^3) 162 – Corrosion 163 – Closed Cell 163 – Medium to High Density Foam 163 – Floral Foam 163 – Orthopedic Foam 164 – Mine/Tunnel Foam 164 – Hybrid Phenolic Foams 165 – Polyurethane 165 – Urea Formaldehyde 165 – Furfuryl Alcohol 165	
6.1.2.2.10	Foaming Equipment	165
	Batch 166 – Continuous 166	
6.1.2.2.11	Properties of Various Phenolic Foam Products	166
6.1.2.2.12	Foam Testing	166
	Fire Conditions 166 – Thermal Conductivity 168	
6.1.2.2.13	Aerogels	168
6.1.2.2.14	Summary/Trends	169
6.1.2.3	Phenolic Resin-Bonded Textile Felts (DIN 61 210)	170
6.1.2.3.1	Introduction (Textile Recycling, Raw Materials, Applications)	170
6.1.2.3.2	Definitions	170

6.1.2.3.3	Composition of Textile Felts	171
6.1.2.3.4	Types of Resins for Textile Felts	172
6.1.2.3.5	Manufacturing Processes	176
6.1.2.3.6	Properties and Applications	179
6.1.2.3.7	Property and Quality Testing	180
6.1.2.3.8	Recycling	184
6.1.3	Phenolic Molding Compounds	187
6.1.3.1	Introduction: Economic Aspects	187
6.1.3.2	Composition of Phenolic Molding Compounds (Resins, Fillers, and Reinforcing Agents)	188
6.1.3.3	Phenolic Molding Compound Standardization (Current Status)	194
6.1.3.4	Production of Molding Compounds	197
6.1.3.5	Processing of Molding Compounds	199
6.1.3.6	Test Methods (Application of ISO Standards)	203
6.1.3.6.1	Bulk Density (ISO 60)	205
6.1.3.6.2	Process Shrinkage (ISO 2577)	205
6.1.3.6.3	Tensile Strength (ISO 527)	205
6.1.3.6.4	Bending Strength (ISO 178)	205
6.1.3.6.5	Charpy Impact Resistance (ISO 179/1eU) and Charpy Notched Impact Strength (ISO 179/1eA)	206
6.1.3.6.6	Heat Distortion Temperature (ISO 75)	206
6.1.3.6.7	Maximum Application Temperature (IEC 60216, Part 1)	206
6.1.3.6.8	Flammability Test (UL 94)	207
6.1.3.6.9	BH Incandescent Rod Flammability Method (IEC 60707)	207
6.1.3.6.10	Water Absorption (ISO 62)	207
6.1.3.6.11	Specific Surface Resistance (IEC 60093)	207
6.1.3.6.12	Specific Volume Resistance (IEC 60093)	208
6.1.3.6.13	Dielectric Loss Factor and Dielectric Number (IEC 60250)	208
6.1.3.6.14	Dielectric Strength (IEC 60243, Part 1)	208
6.1.3.6.15	Tracking Resistance (CTI and PTI Comparative Tracking Index, IEC 60112)	209
6.1.3.7	Flow Behavior of Phenolic Molding Compounds	209
6.1.3.8	Application of Phenolic Molding Compounds	212
6.1.3.9	Standardized and Non-Standardized Molding Compounds (as Examples for Applications)	220
6.1.3.10	Properties of Phenolic Molding Compounds Compared to Those of Other Materials	224
6.1.3.11	Phenolic Novolaks and Epoxidized Phenolic Novolaks for Production of Epoxy Molding Compounds	230
6.1.4	Impregnation of Paper and Fabric (Overview)	231
6.1.4.1	Molded Laminates (Introduction)	232
6.1.4.2	Molded Laminates (Survey of Technologies and Diversification)	234

6.1.4.3	Economic Considerations and Background (Electrical Laminates and Printed Circuit Boards)	237
6.1.4.4	Standardization of Molded Laminates	244
6.1.4.5	Raw Materials and Impregnation	248
6.1.4.6	Fabric Laminates	250
6.1.4.7	Applications of Paper- and Fabric-Based Laminates	251
6.1.4.8	Industrial Filter Inserts	251
6.1.4.9	Miscellaneous Impregnation Applications	254
6.1.5	High Performance and Advanced Composites	259
6.1.5.1	Introduction	259
6.1.5.2	Composite Market Segments	260
6.1.5.3	Fiber Reinforced Plastics	263
6.1.5.4	Phenolic Resin Developments	263
6.1.5.5	Fiber Developments	264
6.1.5.6	Composite Fabrication Processes	264
6.1.5.6.1	Impregnation	265
	Prepreg 265 – Cargo Liner 266 – Ballistics 266 – Carbon-Carbon Composites 267 – Honeycomb Core Sandwich Construction 268	
6.1.5.6.2	Filament Winding	272
6.1.5.6.3	Pultrusion	275
6.1.5.6.4	Sheet Molding Compound	279
6.1.5.6.5	Resin Transfer Molding	280
6.1.5.6.6	Hand Lay Up	281
6.1.5.7	Summary/Trends	281
6.1.6	Miscellaneous	282
6.1.6.1	Chemically Resistant Putties and Chemical Equipment Construction	282
6.1.6.2	Lampbase Cements ("Socket Putties")	285
6.1.6.3	Various Applications (Brush Cements, Casting Resins, and Concrete Flow Promoters)	287
6.1.6.4	Phenolic Resin Fibers	288
6.2	Temporary Bonding	291
6.2.1	Foundry	292
6.2.1.1	Introduction (Economic and Technical Survey)	292
6.2.1.2	Hot Curing Processes	300
6.2.1.3	The Shell Molding Process	300
6.2.1.3.1	Sand Conditioning	301
	Hot Coating 301 – Warm Coating 301	
6.2.1.3.2	Fabrication of Cores and Shells	301
6.2.1.4	The Hot Box Process	302
6.2.1.5	The Warm Box Process	304
6.2.1.6	Cold Curing with Direct Addition of a Curing Agent (No Bake Process)	305
6.2.1.6.1	Acid Curing	305
6.2.1.6.2	No Bake Curing with Added Esters	307
6.2.1.6.3	No Bake Process with Isocyanate Curing	307

6.2.1.7	Gas Curing Processes Using Phenolic Resin Binders	308
6.2.1.7.1	The Polyurethane Cold Box Process	309
6.2.1.7.2	The Methyl Formate Process	311
6.2.1.7.3	The CO_2-Resole Process	312
6.2.1.7.4	The Acetal Process	312
6.2.1.7.5	Other Gassing Processes	313
6.2.1.8	General Remarks on Core/Mold Fabrication Processes Using Phenolic Resins	313
6.2.2	Abrasives	314
6.2.2.1	Introduction (Grinding, Abrasives)	314
6.2.2.2	Economical Significance	315
6.2.2.3	Grinding Wheels (Classification, Definitions)	317
6.2.2.4	Grinding Wheel Design, Bonded Abrasives (Composition and Stresses)	319
6.2.2.5	Liquid and Powdered Binders (Phenolic Resins)	323
6.2.2.6	Cold Pressed Cutting and Roughing Wheels	328
6.2.2.7	Production and Use of Glass Fabric Inserts	333
6.2.2.8	Binders for High Wet Strength Grinding Wheels and Segments	336
6.2.2.9	Hot Pressed, Highly Compacted Grinding Wheels (HP Wheels)	337
6.2.2.10	Diamond Wheels	339
6.2.2.11	Phenolic Resins as Binders for Coated Abrasives	340
6.2.3	Friction Linings	353
6.2.3.1	Introduction (General Information)	353
6.2.3.2	Demands on Friction Linings	355
6.2.3.3	Composition of Friction Linings	358
6.2.3.4	Phenolic Resins and Properties of Friction Linings	361
6.2.3.5	Manufacture of Friction Linings	365
6.2.3.5.1	Process 1 (Dry Mixes for Hot Pressing or Warm Shaping with Subsequent Oven Curing)	365
6.2.3.5.2	Process 2 (Wet Mixes for the Calender and Extrusion Processes)	366
6.2.3.5.3	Process 3 (Impregnation of Fiber Textiles and Yarns)	367
6.2.4	Auxiliaries for Petroleum and Natural Gas Production (Proppant Sands and Tensides)	370
6.3	Complementary Bonding	373
6.3.1	Phenolic Resins for Coatings and Surface Protection	374
6.3.1.1	Introduction: History and Possible Applications	374
6.3.1.2	Phenolic Resins as Binders in Coatings (Types of Resins and Modifications)	375
6.3.1.3	Phenolic Resin Coatings Composition	377
6.3.1.4	Packaging Coatings (Protective Interior Coatings)	377
6.3.1.5	Anticorrosion Primers	380
6.3.1.6	Electrically Insulating Coatings	383

6.3.1.7	Miscellaneous Coatings Applications (Printing Inks, Alkyd Coatings, Photosensitive Coatings, Powder Coatings)	384
6.3.1.8	Photoresist/Imaging	385
6.3.1.8.1	Introduction	385
6.3.1.8.2	Photoresist	386
	Positive Resists 388 – Negative Resists 388 – Cresol Novolak Microstructure 389 – Molecular Weight and Molecular Weight Distribution 390 – Cresol Novolak Preparation 391 – Chemically Amplified Resists (CAR) 395	
6.3.1.8.3	Summary/Trends	397
6.3.2	Phenolic Resins as Additives in the Rubber Industry	398
6.3.2.1	Vulcanization Resins	399
6.3.2.2	Tackifying Resins: Resins Used to Increase the Tack of Rubber Mixes	401
6.3.2.3	Reinforcing Resins: Phenolic Resins Used for Reinforcement of Rubber Mixes	403
6.3.3	Binders for the Adhesives Industry	405
6.3.3.1	Polychloroprene-Based Adhesives	406
6.3.3.2	Nitrile Rubber Adhesives	408
6.3.3.3	Polyurethane-Based Contact Adhesives	409
6.3.3.4	Rubber/Metal Bonding	410
6.3.3.5	Friction Lining Adhesives	412
6.3.3.6	Other Applications of Phenolic Resins in Adhesives Formulations and Sealants	414
6.3.3.6.1	Metal Bonding	414
6.3.3.6.2	Cladding Adhesives	414
6.3.3.6.3	Natural Rubber-Based Bonding Adhesives	414
6.3.3.6.4	Hot Melts	415
6.3.3.6.5	Gaskets	415
6.3.3.6.6	Sealants	415
6.4	Intermediate and Carbon-Forming Bonding	416
6.4.1	Pyrolysis of Phenolics	417
6.4.2	Phenolic Resins as Binders for Refractories	425
6.4.2.1	Definitions	425
6.4.2.2	Economic Importance and Applications	427
6.4.2.3	Environmental Impact in Use of Phenolic Resins	430
6.4.2.4	Shaped Products	432
6.4.2.5	Manufacture of Bricks	434
6.4.2.6	Isostatically Pressed Products for Continuous Casting	438
6.4.2.7	Slide Gates, Graphite Crucibles, Insulating Plates	441
6.4.2.8	Unshaped Refractory Mixes	442
6.4.2.9	Impregnation of Refractories	442
6.4.3	Carbon and Graphite Materials	444
6.4.3.1	Introduction (General Information)	444

6.4.3.2	Phenolic Resins as Binders for Carbon and Graphite Engineering Materials (Including CFC)	446
6.4.3.3	Phenolic Resins as Impregnating Agents for Porous Carbon or Graphite Articles	450
6.4.3.4	Electrode Production	450
6.4.4	Glassy Carbon (Polymeric Carbon)	452
6.5	The Chemically Reactive Bonding Function	455
6.5.1	Summary: Epoxidation, Alkoxylation, Polyurethane	455
6.5.2	Alkylphenolic Resins as Dye Developers for Carbonless Copy Paper	456
6.5.2.1	Thermography	459
6.5.3	Antioxidants/Stabilizers	460
6.5.3.1	Introduction	460
6.5.3.2	Types of Antioxidants/Stabilizers	460
6.5.3.2.1	Natural Antioxidants	460
6.5.3.2.2	Synthetic Antioxidants	460
6.5.3.3	Classification of Antioxidants/Stabilizers	461
6.5.3.3.1	Primary Antioxidants	461
6.5.3.3.2	Secondary Antioxidants	461
6.5.3.4	Mechanism of Oxidation	461
6.5.3.5	Effect of Antioxidants/Stabilizers	462
6.5.3.5.1	Hindered Phenols and Secondary Aromatic Amines	462
6.5.3.5.2	Trivalent Phosphorus Compounds	462
6.5.3.5.3	Hindered Amines (HALS)	462
6.5.3.6	Market Areas	463
6.5.3.6.1	Food/Beverages	463
6.5.3.6.2	Petroleum Products	464
	Fuels 464 – Lubricating Oils 464	
6.5.3.6.3	Rubber Compounds	464
6.5.3.6.4	Polymeric Materials	464
6.5.3.7	New Developments	466
6.5.3.7.1	Polymer Bound	466
6.5.3.7.2	Benzotriazole Types	466
6.5.3.7.3	Triazine	467
6.5.3.7.4	Lactone/Hydroxylamine	467
6.5.3.8	Summary/Trends	467
	References	468

Chapter 7 Chemical, Physical and Application Technology Parameters of Phenolic Resins 488

7.1	Introduction (Purpose and Objective)	488
7.2	Standard Chemical and Physical Tests	490
7.3	Description of Physical and Chemical Test Methods (ISO 10082) and Their Significance [1, 2]	491
7.3.1	ISO 3146, Melting Range, Melting Behavior	492

7.3.2	ISO 8620, Screen Analysis with the Air Jet Screen and Particle Size Analysis as Specified by ISO 13 320 (Laser Method)	492
7.3.3	ISO 60, Bulk Density	494
7.3.4	ISO 2811 and ISO 3675, Determination of Density	495
7.3.5	ISO 2555, ISO 3219, and ISO 12 058, Determination of Viscosity	495
7.3.6	ISO 8975, Determination of pH	496
7.3.7	ISO 9944, Determination of Electrical Conductivity	497
7.3.8	ISO 8989, Water Miscibility	497
7.3.9	ISO 8819, Determination of the Flow Distance	497
7.3.10	ISO 8987, ISO 9396, and ISO 11 409, Measurement of the Curing Characteristics	498
7.3.11	ISO 9771, Acid Reactivity of Liquid Phenolic Resins	500
7.3.12	ISO 8618, Determination of Nonvolatile Components	500
7.3.13	ISO 8974, Residual Phenol, Gas Chromatographic Determination	501
7.3.13.1	DIN 16 916-02-L 1 and DIN 38 409-16, Conventional Methods for Determination of Residual Phenol	501
7.3.14	ISO 11 402, Determination of Free Formaldehyde in Phenolic Resins and Co-Condensates	502
7.3.15	ISO 8988, Hexamethylene Tetramine (HMTA) Content	503
7.4	Miscellaneous Chemical Test Methods Used for Analysis of Phenolic Resins	504
7.4.1	ISO 11 401, Liquid Chromatography for Separation of Phenolic Resins	504
7.4.2	Thin-Layer Chromatography (Company-Specific Method)	505
7.5	Application Technology Testing	507
7.5.1	Refractories	509
7.5.2	Foundry Binders	509
7.5.3	Abrasives (Grinding Wheels, Abrasive Shapes, and Coated Abrasives)	510
7.5.4	Friction Linings	510
7.5.5	Impregnating Resins for Industrial and Paper Base Electrical Laminates	510
7.5.6	Resins for the Coatings Industry	511
7.5.7	Resins for the Adhesives Area	512
7.5.8	Resins for the Rubber Industry	512
7.5.9	Resins for Textile Felt Production and Related Applications	512
7.5.10	Plywood and Resorcinol Resin Adhesives	512
7.5.11	Resins for Floral Foam	512
	References	513

**Chapter 8 Industrial Safety and Ecological Questions
(Raw Materials, Recycling, Environment)** 514

8.1	Toxicological Properties, Hazards Labeling [2–5] . .	515
8.2	Workplace, Exhaust Air, Low-Monomer Resins	516
8.3	Flue Gas Treatment	520
8.4	Analytical Determinations at the Workplace and in Flue Gas	522
8.4.1	Example of a Measurement Procedure	523
8.5	Waste, Recycling	524
8.6	Renewable Raw Materials (Furfuryl Alcohol, Lignin, Tannin)	527
	References	529

**Chapter 9 Conclusion: Guidelines for Future Developments
of Phenolic Resins and Related Technologies** 532

Subject Index 535

Abbreviations

ABS	automotive braking system
ABS	acrylonitrile butadiene styrene
ACGIH	American Conference of Governmental Industrial Hygienists
AHMT	4-amino-3-hydrazino-5-mercapto-1,2,4-triazole
AK	acrylate resin
AMS	alpha-methylstyrene
ASTM	American Society for Testing and Materials-West Conshohocken, PA
ATBN	amine terminated butadiene nitrile rubber
ATP	α-tocopherol
BHA	butylated hydroxy anisole
BHT	butylated hydroxy toluene
BMC	bulk molding compounds
BMI	bismaleimide
BPA	bisphenol A
BR	butadiene rubber
BS	British Standard Institute, London, UK
BT	bismaleimide triazine
C-C	carbon-carbon composites
CAR	chemically amplified resists
CB	coated back
CBN	corundum boron nitride
CCA	chromated copper arsenate
CCFC	chlorinated fluorocarbon
CFB	coated front and back
CEN	European Committee for Standardization, Brussels, Belgium
CF	coated front
CFC	carbon fiber composite; also chlorofluorocarbons
CHP	cumene hydroperoxide
CNSL	cashew nutshell oil
CP/MAS	cross polarization magic-angle spinning
CP	cold punching
CR	polychloroprene
CTBN	carboxyl terminated butadiene nitrile rubber

CVD	chemical vapor deposition
CVI	chemical vapor infiltration
DDT	p,p'-dichloro diphenyl [trichloromethyl] methane
DEA	dielectric analysis
DFG	Deutsche Forschungsgemeinschaft/ German Research Society
DIN	Deutsche Institut für Normung (Berlin, Germany) Institute for Standardization
DIPAC	distributed parameter continuum
DMA (DMTA)	dynamic mechanical (thermal) analysis
DMBA	dimethyl benzyl alcohol
DMSO	dimethyl sulfoxide
DNQ	diazonaphthoquinone
DRAM	dynamic random access memory
DSC	differential scanning calorimetry
DUV	deep ultra violet
ECN	cresol novolak
EDC	electrophoretic dip coat
EEP	ethoxy ethyl propionate
EP	epoxy resins
EPA	Environmental Protection Agency, USA
EPDM	ethylene propylene diene terpolymers
FF	furan resin
FRP	fiber reinforced polymers (plastics)
FST	fire-smoke-toxicity
FTIR	Fourier transform infrared spectroscopy
FW	filament winding
GPC	gel permeation chromatography
HALS	hindered amine light stabilizers
HCFC	hydrochlorofluorocarbon
HMR	hydroxy methylated resorcinol
HMTA	hexamethylene tetramine
HPLC	high performance liquid chromatography
HRP	horse radish peroxidase
IC	integrated circuits
IEC	International Electrotechnical Commission (Geneve, Switzerland)
IFWI	instrumental falling weight impact
ILS	interlaminar shear strength
IMO	International Maritime Organization (UK)
IPN	interpenetrating network
ISO	International Organization for Standardization (Genf, Switzerland)
LEL	lower explosive limit
LIFT	lateral ignition and flame spread test
LOI	limiting oxygen index
LPA	low profile additive

LSL	laminated strand lumber
LVL	laminated veneer lumber
MAK	occupational exposure limit (Germany)
MBS	methacrylate butadiene styrene
MC	moisture content
MDF	medium density fiberboard
MDI	methylene diisocyanate
MF	melamine formaldehyde
MF	methyl formate
MMA	methyl methacrylate
MP	melamine phenolic
MTBE	methyl *tert*-butyl ether
MUF	melamine urea formaldehyde
MW	molecular weight
MWD	molecular weight distribution
NBR	acrylonitrile butadiene copolymers
NBS	National Bureau of Standards
NEMA	National Electrical Manufacturers Association (USA)
NIST	National Institute of Standards & Technology (USA)
NMP	N-methyl pyrolidone
NMR	nuclear magnetic resonance
NR	natural rubber
OSB	oriented strand board
PB	particleboard
PBOCST	poly[4-(ter-butoxycarbonyl)oxy] styrene
PBOX	phenylene bisoxazoline
PBW	parts by weight
PEAR	polyester amide resin
PEI	polyetherimide
PES	polyether sulfone
PET	polyethylene terephthalate
PF	phenol formaldehyde resin
PG	propyl gallate
PGME	propylene glycol monomethyl ether
PGMEA	propylene glycol monomethyl ether acetate
PHS	polyhydroxy styrene
PI	polyimide
PIR	polyisocyanurate
PLB	Bakelite AG internal resin designation
PMF	phenol melamine formaldehyde
PPO	polyphenylene oxide
PSF	polysulfone
PSL	parallel strand lumber
PT	phenolic triazine
PUR	polyurethane
QA	quality assurance
QC	quality control

R	thermal resistivity
RF	resorcinol formaldehyde
RH	relative humidity
RPF	resorcinol phenol formaldehyde
RSST	reactive system screening tool
RTM	resin transfer molding
S	silicone
SBP	soybean peroxidase
SBR	styrene butadiene rubber
SIA	Semiconductor Industry Association (USA)
SIP	steam injection pressing
SMC	sheet molding compound
TBA	torsional braid analysis
TBHQ	*tert*-butyl hydroquinone
TEC	thermal expansion coefficient
TI	temperature index
TLC	thin layer chromatography
TMA	thermomechanical analysis
TOC	total organic carbon
UEL	upper explosive limit
UF	urea formaldehyde resin
UFP	urea formaldehyde phenol resin
UHMWPE	ultra high molecular weight polyethylene
UKOOA	UK Offshore Operators Association
UL	Underwriters Laboratory, Northbrook, IL 60062
UP	unsaturated polyesters
UV	ultra violet
VDE	Verband Deutscher Elektrotechniker (Frankfurt, Germany); German Society of Electrical Engineers
VDG	Verein Deutscher Giessereifachleute (Düsseldorf, Germany) German Foundry Men's Society
VOC	volatile organic compound
WB	waferboard
XPS	X-ray photoelectron spectroscopy

Conversion Factors: Metric/SI Units to U.S. Customary and Commonly Used Units[a]

Length	1 mm	0.039370 inches
	1 cm	0.393701 inches
	1 m	39.3701 inches; 3.28084 feet; 1.093613 yards
	Yard	0.9144 m
	Inch	2.54 cm.
	Foot	0.3084 m
Area	1 cm^2	0.15500 square inches
	1 m^2	10.76391 square feet
	1 sq. inch	6.4516 cm^2
Volume	1 cm^3	0.0328084 cubic inches
	Gallon U.S.	0.0037854 m^3 ; 3.7853 liters
Mass	1 kg	2.2046 pound
	1 pound	0.45359 kg
Density	1 g/cm^3	62.41 lb/ft^3
Pressure	1 at	1 kp/cm^2; 0.980665 bar; 1.01325 × 10^5 Newton/m^2
Temperature	°C	(°F − 32) × 5/9
	°F	(°F × 5/9)+32
Viscosity	1 cp	1 mPa s; 0.001 kg/m s; 0.000672 lb m/ft s
Thermal conductivity	1 W/m · K	0.86 kcal/ m h °C; 0.579 BTU/ft h °F
Heat quantity	1 kcal	4.187 kJ

[a] From: R.H. Perry et al. (1997) Perry's chemical engineers' handbook, 7th edn. McGraw-Hill, New York.

Part A
Chemistry/Production

CHAPTER 1

Raw Materials

According to ISO 10082 phenolic resins are obtained by the reaction of phenols with aldehydes. Both parent compounds, phenol and formaldehyde, are by far the most important components in commercial phenolic resin production.

1.1 Phenols

Phenols are a family of aromatic compounds with the hydroxyl group bonded directly to the aromatic nucleus. They differ from alcohols in that they behave like weak acids and dissolve readily in aqueous sodium hydroxide, but are insoluble in aqueous sodium carbonate. Phenols are colorless solids with the exception of some liquid alkylphenols. Selected physical properties of phenols are listed in Table 1.1.

Table 1.1. Physical properties of phenols [1–6]

Name		MW	MP °C	BP °C	pK$_a$ 25 °C
Phenol	hydroxybenzene	94.1	40.9	181.8	10.00
o-Cresol	1-methyl-2-hydroxybenzene	108.1	30.9	191.0	10.33
m-Cresol	1-methyl-3-hydroxybenzene	108.1	12.2	202.2	10.10
p-Cresol	1-methyl-4-hydroxybenzene	108.1	34.7	201.9	10.28
p-tert Butylphenol	1-tert-butyl-4-hydroxybenzene	150.2	98.4	239.7	10.25
p-tert Octylphenol	1-tert-octyl-4-hydroxybenzene	206.3	85	290	–
p-tert Nonylphenol	1-nonyl-4-hydroxybenzene	220.2	–	295	–
2,3-Xylenol	1,2-dimethyl-3-hydroxybenzene	122.2	75.0	218.0	10.51
2,4-Xylenol	1,3-dimethyl-4-hydroxybenzene	122.2	27.0	211.5	10.60
2,5-Xylenol	1,4-dimethyl-2-hydroxybenzene	122.2	74.5	211.5	10.40
2,6-Xylenol	1,3-dimethyl-2-hydroxybenzene	122.2	49.0	212.0	10.62
3,4-Xylenol	1,2-dimethyl-4-hydroxybenzene	122.2	62.5	226.0	10.36
3,5-Xylenol	1,3-dimethyl-5-hydroxybenzene	122.2	63.2	219.5	10.20
Resorcinol	1,3-dihydroxybenzene	110.1	110.8	281.0	–
Bisphenol-A	2,2-bis(4-hydroxyphenyl)propane	228.3	157.3	–	–

1 Raw Materials

Table 1.2. Physical properties of phenol [2, 5, 7]

CAS registration number	108–95–2
EG registration number	604–001–00–2
MW	94.11
MP/BP	40.9/181.7 °C
Relative density 20 °C	1.071
Dissociation constant in water 20 °C	1.28×10^{-10}
Flash point (DIN 51758)	81 °C
Ignition temperature (DIN 51794)	595 °C
Explosion limits LEL/UEL	1.7%/8.6%
Vapor pressure 20 °C	0.02 kPa
MAK limit	19 mg/m^3/5 ppm
OSHA PEL	TWA 5 ppm

1.1.1
Physical Properties of Phenol

The melting point of pure phenol (40.9 °C) is lowered considerably by traces of water – approximately 0.4 °C per 0.1 % of water content; over 6 % renders it liquid at room temperature. A mixture with 10 % of water is called phenolum liquefactum which is mostly used in industrial resin production. Phenol forms azeotropic mixtures with water. In the temperature range up to 68.4 °C its miscibility with water is limited; above this temperature it is completely miscible.

In the solid state phenol is colorless. When exposed to air it rapidly turns pink if it contains impurities, in particular iron or copper, from the production process or storage. Phenol is highly toxic and exposure limits have to be strictly controlled. Personnel who handle phenol should wear protective clothing, safety glasses and rubber gloves. Selected properties of phenol are presented in Table 1.2; more information on toxicology and risk assessment [7–11] is presented in Chap. 8.

1.1.2
Supply and Use of Phenol

The largest use of phenol is the production of phenol-formaldehyde resins and bisphenol-A (polycarbonate and epoxide resins), followed by cyclohexanone (caprolactam), aniline (MDI), 2,6-xylenol (PPO) and alkylphenols (Table 1.3).

In 1997, only about 2% of the world production of phenol was derived from coal. Among the synthetic processes, the cumene process is by far the most prevalent.

In 1997, the world-wide phenol capacities amounted to 5.4 million tonnes (USA 2.0 million tonnes, West Europe 1.8 million tonnes, Asia 1.4 million tonnes) (Table 1.4). Actual world production was about 5.0 million tonnes. Capacity and demand are predicted to be in balance by around 1999 (operation rate 100%).

Table 1.3. Breakdown of phenol consumption 1997

%	USA	West-Europe	Japan
Phenolic resins	35	28	30
Bisphenol-A	35	26	38
Caprolactam	17	31	–
Aniline	3	–	15
Alkylphenols	5	4	3
Others	5	11	14

Table 1.4. Major phenol producers in the world

Company	Capacity in thousands of tonnes
Phenolchemie	775
Allied Signal	405
Enichem	390
Shell Chemical	320
General Electric/Mt. Vernon	315
Aristech	290
Georgia Gulf	275
Dow Chemical	250
Mitsui Toatsu	200
Chiba Phenol	200
Mitsui Petrochemical	190
Mitsubishi Petrochemical	180
Ertisa	160
Rhone-Poulenc	150
Nippon Steell Chemical	120
Rhodia	120
Taiwan Prosperity Chem.	100
Kumho Shell	100
Dsm	100
Worldwide total	*5400*

1.1.3
Phenol Production Processes

The cumene process (Figs. 1.1 and 1.2) is by far the most important industrial process for the production of phenol and accounts for more than 95% of the production capacity. The phenol synthesis based on cumene was discovered by Hock and Lang [12] in Germany where the first pilot plant was constructed jointly by Rütgerswerke and Bergwersksgesellschaft Hibernia with Hock's assistance soon after World War II.

1.1.3.1
Cumene Oxidation (Hock Process)

Cumene [13] (isopropylbenzene), produced by alkylation of benzene with propylene according to the reaction at Eq. (1.1) over a solid phosphoric acid

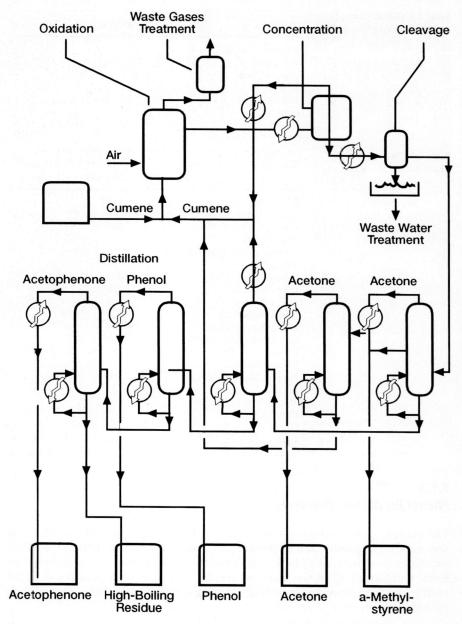

Fig. 1.1. Cumene process flow diagram. (Drawing: Phenolchemie, D-45966 Gladbeck)

Fig. 1.2. Phenolchemie's new 200,000 tonnes per year phenol plant in Antwerp, which came on-stream in 1993

catalyst (UOP Process), is oxidized in the liquid phase to cumene hydroperoxide (CHP) according to the reaction at Eq. (1.2). CHP decomposes to phenol and acetone very rapidly under acidic conditions and elevated temperatures in a mechanism shown at Eq. (1.3) as postulated by Seubold and Vaugham [14].

$$\text{Benzene} + H_3C-CH=CH_2 \xrightarrow{\text{Phosphoric acid}} \text{cumene} \quad (1.1)$$

$$\text{Propylene}$$

$$\text{cumene} + O_2 \text{ (Air)} \xrightarrow{\text{Catalyst}} \text{Cumene hydroperoxide CHP} \quad (1.2)$$

$$\text{CHP} \xrightarrow[-H_2O]{H^{\oplus}} \left[\underset{\substack{H_3C-\overset{|\overset{\oplus}{O}}{\underset{|}{C}}-CH_3 \\ \text{Ph}}}{} \xrightarrow{\text{Rearrangement}} \underset{H_3C-\overset{\overset{\oplus}{C}}{\underset{\|}{C}}-CH_3}{\underset{O-\text{Ph}}{}} \xrightarrow{+H_2O} \underset{H_3C-\overset{\overset{\oplus}{OH_2}}{\underset{|}{C}}-CH_3}{\underset{\text{Ph}}{}} \right]$$

$$\longrightarrow \underset{\text{Phenol}}{\text{PhOH}} + \underset{\text{Acetone}}{H_3C-\overset{O}{\underset{\|}{C}}-CH_3} + H^{\oplus} \tag{1.3}$$

In the commercial process, cumene is oxidized with air to CHP (95% yield) which is concentrated and cleaved in presence of an acidic catalyst at high efficiency (99%) to phenol and acetone. The catalyst is removed and the reaction mixture fractionated. By-products of the oxidation are acetophenone and dimethyl benzyl alcohol (DMBA). DMBA is dehydrated in the cleavage reaction to alpha- methylstyrene (AMS), which may be hydrogenated to cumene and recycled for oxidation or optionally recovered as a pure product (Table 1.5). Acetone and phenol are purified by distillation. With AMS hydrogenation, 1.31 tonnes of cumene will produce 1 tonne of phenol and 0.615 tonnes of acetone.

Safety is a critical aspect in designing and operating a phenol plant. The oxidation occurs close to the flammability limit. CHP is an unstable material which can violently decompose under certain conditions.

1.1.3.2
Toluene Oxidation Process

A two-step oxidation process based on toluene was developed by Dow Chemical. In the first stage, toluene is oxidized in the liquid phase with atmospheric oxygen in the presence of a cobalt catalyst to benzoic acid and a variety of by-products. In the second step, the benzoic acid is decarboxylated in the

Table 1.5. Commercial phenol quality

Phenol	%	99.9
Cresols	ppm	20–300
Benzofurane and methylbenzofurane	ppm	50
Methylstyrene and cumene	ppm	5
Acetophenone	ppm	3
Magnesium	ppm	0.05
Copper	ppm	0.01
Cobalt	ppm	0.01

presence of air and a copper catalyst to produce phenol (Eq. 1.4). The overall yield of phenol from benzoic acid is around 88%.

$$\text{Toluene} + 1.5\, O_2 \xrightarrow[\text{(Co)}]{140°C\,/\,3\text{ bar}} \text{Benzoic acid} \;(+ H_2O) \longrightarrow \text{Phenol} + CO_2 \qquad (1.4)$$

This process has been carried out in four industrial plants (about 4% of the total world wide phenol production), the latest being constructed in Japan by Nippon Steel Chemical in 1991 (120,000 tonnes per year).

1.1.3.3
Nitrous Oxide Oxidation of Benzene

A new one-step route that produces phenol directly from benzene has been jointly developed by Monsanto and Boreskov Institute of Catalysis (BIC, Novoibirsk, Siberia). The method involves nitrous oxide wastestream (from Monsanto's adipic acid production via oxidation of cyclohexane) as the oxidant for phenol preparation. Instead of disposing of the nitrous oxide by incineration, the novel catalysis scheme developed by BIC eliminates costly nitrous oxide disposal and generates a feedstock, phenol, at little cost. Successful pilot plant testing of the process has prompted Monsanto to commercialize the process by the year 2000 in their Pensacola, FL facility.

1.1.3.4
Other Synthesis Processes

Several other synthesis processes (listed below) have no commercial importance:

- Chlorination of benzene and alkaline hydrolysis of chlorobenzene
- Chlorination of benzene and steam hydrolysis of chlorobenzene (Raschig process, Raschig-Hooker and Gulf oxychlorination process)
- Sulfonation of benzene and benzenesulfonate decomposition in molten sodium hydroxide to sodium phenolate (A. Wurtz and A. Kekule)
- Cyclohexene conversion to cyclohexanol-cyclohexanone mixtures followed by dehydrogenation to phenol (Mitsui Petrochemical 1989)

1.1.4
Alkylphenols

Alkylphenols are phenol derivatives wherein one or more of the ring hydrogens have been replaced by an alkyl group. Monomethyl derivatives of phenol (hydroxy derivatives of toluene) are commonly designated as cresols, dimethyl derivatives as xylenols. Like phenol, they are typically solids at 25 °C (Table 1.1). The solubility of alkylphenols in water decreases with the number of carbons attached to the ring; on the other hand, the solubility in hydrocarbons increases. The solubility of alkali phenolates depends on the position of the alkyl group. The position of the alkyl group also strongly influences the reactivity towards formaldehyde. The main synthetic pathway to alkylphenols is the alkylation of phenols with alkenes by use of an acidic catalyst, generally a sulfonic acid or aluminum catalyst. Judicious choice of catalyst and reaction conditions allows some control over selectivity and the amount of by-products and waste.

In general, pure compounds are not necessary for resin production. Thus for higher alkylphenols simple batch reactors similar to those for resin production, preferably operating under pressure, are satisfactory. Phenol is loaded first, then the catalyst is added. The alkene is added at such a rate that the reactor's heat removal capability is not exceeded and the desired reaction temperature is maintained as the selectivity is greatly affected by temperature. Monoalkylation is favored by high phenol excess.

Purification, if necessary, is performed by multiple distillation. Raw material costs make up 50–80% of the total manufacturing costs.

In general, the use of alkylphenols for resin application is decreasing because of their higher costs and stronger odor compared to phenol.

1.1.4.1
Methylphenols

Cresols exist as three isomers depending on the position of the methyl group in relation to the hydroxyl group, designated as *ortho*, *meta* or *para*. Locants or substituents located in positions relative to the hydroxyl group, i.e., methyl groups, are used for designation of dimethyl as phenols xylenols. In the past, coal tar was the main source of cresols and xylenols. Today, synthesis processes [15] based on toluene and/or phenol predominate for cresols. The use of xylenols is very limited and decreasing. The main industrial routes to cresols are the alkylation with propylene followed by splitting of the hydroperoxide, and the chlorination process.

The chlorination of toluene with Cl_2 is performed at about 30 °C with $FeCl_3/S_2Cl_2$ as catalyst. The isomeric mixture obtained is hydrolyzed with NaOH at 280–300 bar pressure and 390 °C yielding mainly the *meta* isomer (around 50%) with an approximately equal *ortho* and *para* content (2:1:1). A difficult to separate *m-/p-* mixture is obtained after *o*-cresol is removed by distillation. If pure *m*-cresol is desired, the chlorotoluene fraction is separated first and the *o*-chlorotoluene hydrolyzed later.

Chemistry and technology of toluene alkylation are similar to the cumene process. Toluene is reacted with propylene in the presence of $AlCl_3$ or other Friedel-Crafts catalysts at 60–80 °C to obtain a mixture of cymenes (Eq. 1.5) with a *m-/p-* ratio of about 2:1 and less then 5% of *o*-cymene. About a 63% *m-*, 32% *p-* and, 5% *o*-cymene mixture is then oxidized with air to the hydroperoxide (20% conversion only) and then split in acidic medium to

$$\text{Toluene} + \text{Propylene} \xrightarrow{\text{Catalyst}} \text{Cymene} \xrightarrow{\text{Oxidation}} \quad (1.5)$$

$$\text{Cymene hydroperoxide} \xrightarrow{H^{\oplus}} \text{m-Cresol} + \text{Acetone} \quad (1.6)$$

cresols and acetone (Eq. 1.6) and a variety of by-products. Two plants in Japan are operated by this process (Mitsui Petrochemical and Sumitomo Chemical).

The phenol route is based on alkylation of phenol with methanol either in the gas phase (Koppers, Pitt-Consol, Croda) or in the liquid phase (Chemische Werke Lowi, UK-Wesseling). In the gas phase process, methanol and phenol vapors pass over an aluminum oxide catalyst at approximately 350 °C under moderate pressure. Mainly o-cresol and 2,6-xylenol are obtained. If 2,6-xylenol is desired as the main product (for PPO production), magnesium oxide is employed as catalyst. In the liquid phase reaction, performed at 300–350 °C and 40–60 bar pressure, aluminum methylate or zinc bromide is used as catalyst. Main product is o-cresol. A synthesis process for 3,5-xylenol based on isophorone was developed by Rütgers AG.

For the separation of m- and p-cresols, special methods are required due to their similar boiling points. By heating a m-/p-cresol mixture with urea and cooling afterwards, a crystalline addition compound of m-cresol and urea is obtained (urea process).

Furthermore, p-cresol when gently heated to 90 °C forms a crystalline addition compound with anhydrous oxalic acid. Chemical processes are based on the different reaction rates during sulfonation (sulfuric acid process) or alkylation (isobutylene process).

In the phenolic resin area, cresols are used for the production of coatings, resins for electrical laminates, and antioxidants. High purity novolaks based on o-cresol and m- and p-cresol are used in photoresist areas and for the production of electronic grade epoxy resins.

1.1.4.2
Higher Alkylphenols

The important members among the higher alkylphenol family for resin production are the *para*-derivatives, *p-tert*-butylphenol, *p-tert*-octylphenol and *p*-nonylphenol. They are used for antioxidants, rubber crosslinking agents, adhesives, coating resins, printing inks, carbonless paper chemicals, and oil field demulsifiers among others.

p-tert-Butylphenol is prepared from phenol and butylene, which is mainly produced by dehydration of *tert*-butyl alcohol or cracking of methyl *tert*-butyl ether (MTBE). Technical grade *p-tert*-butylphenol is used for resin application which accounts for 60–70% of the total consumption. Disobutylene is used for *p*-octylphenol production and nonene for *p*-nonylphenol.

1.1.5
Phenols from Coal and Mineral Oil

An average of approximately 1.5% crude phenols, mainly phenol (0.5%) as well as *o*-, *m*-, and *p*-cresol, 2,3-, 2,4-, 2,5-, 2,6- and 3,5-dimethylphenol, is found in coal tar. Their extraction from a coal tar fraction [16] is performed with dilute sodium hydroxide (8–12%). Pyridine bases and hydrocarbons are removed from the phenoxide solution by steam distillation and the phenols precipitated with carbon dioxide. They are extracted with diisopropylether and separated by distillation and crystallization.

Phenols are further obtained from condensates of coke oven gases and waste waters of coal gasification plants. The extraction is performed with benzene/sodium hydroxide (Pott-Hilgenstock process [17]) or with diisopropylether (Lurgi's Phenosolvan process [18]).

Hydrocarbon Research Inc. developed a two-step process [19] to convert lignin to phenols and benzene. Lignin is first depolymerized by hydrocracking in a fluidized bed reactor, and the resulting aromatic mixture is dealkylated to phenol and benzene.

Low rank coal can be depolymerized [20] to phenols using acidic catalysts (phenol/BF_3-complex or *p*-toluenesulfonic acid).

A further source of phenols is the petroleum industry. During the catalytic cracking process, various phenolic compounds are formed. They are extracted from the bottoms with aqueous sodium hydroxide and recovered by precipitation with carbon dioxide. Separation and purification is conducted by distillation and other methods.

1.1.6
Resorcinol

Resorcinol (1,3-dihydroxybenzene) as a dihydric benzene exhibits a very high reaction rate towards formaldehyde compared to phenol. The primary market area is the rubber industry (over 50%). Resorcinol-formaldehyde condensates enhance the adhesion between tire cord and rubber. In structural wood adhesives resorcinol increases the curing rate and allows cure at ambient temperature. On the other hand, its use is limited because of the high cost.

Resorcinol [21] is obtained from benzene by sulfonation (Indspec Chemical Corp.) or alkylation with propene (Mitsui Petrochemical, Sumitomo Chemical). It is produced only in a few specialized plants. World production is in the range of 40,000 tonnes per year.

$$\text{Benzene} + 2\,SO_3 \xrightarrow{H_2SO_4} \text{Benzene disulfonic acid} \xrightarrow[-H_2O]{NaOH} \text{Resorcinol} \qquad (1.7)$$

In the sulfonation process, benzene is converted to 1,3-benzene disulfonic acid with sulfur trioxide at 150 °C. The sulfonation product, benzene-1,3-disulfonate, is neutralized with sodium hydroxide or sodium sulfite, dried and fused with sodium hydroxide at 350 °C (Eq. 1.7). Disodium resorcinate is dissolved in water and neutralized with sulfuric acid to give resorcinol. Finally, resorcinol is extracted with diisopropylether and purified by distillation.

$$\text{diisopropylbenzene} \xrightarrow{O_2} \text{dihydroperoxide} \xrightarrow{H^{\oplus}} \text{resorcinol} + 2\,H_3C\text{-}\overset{O}{\underset{\|}{C}}\text{-}CH_3 \qquad (1.8)$$

The alkylation process of benzene with propene is performed in two plants in Japan. Chemically, the process (Eq. 1.8) corresponds to the Hock process for the production of phenol. The alkylate, which is made in a liquid phase using an $AlCl_3$-HCl complex as catalyst, is subjected to isomerization/transalkylation. The mixture is separated in three fractions, the *m*-diisopropylbenzene fraction is oxidized in an aqueous alkaline medium with compressed air at about 100 °C in a cascade reactor. The conversion is in the range of 20%. The heterogeneous oxidate is subjected to phase separation, the hydroperoxide crystallized, dissolved in acetone and fed to the cleavage reactor. Final purification is performed by distillation. The overall process yield is 75% based on benzene. Very high safety standards are necessary in the plant.

1.1.7
Bisphenol-A

Bisphenol-A, 2,2-bis(4-hydroxyphenyl) propane (BPA), is used for the production of colorless resins, e.g., for coating resins. In the newer commercial BPA production processes [22], acetone and phenol in excess are reacted in presence of a cation exchange resin catalyst. The reaction mixture is separated by distillation and the crude product containing BPA, phenol, and impurities transferred to a crystallizer. Separated crystals are washed with pure phenol, then melted and sent to the prilling tower. Polycarbonate and epoxy grade BPA can be produced in this way.

Older processes use sulfuric acid as catalyst and thioglycolic acid or mercaptans as promoters. Iron and 2.4-BPA isomer content are critical for the color of the resins obtained.

1.2 Aldehydes

Formaldehyde is practically the only carbonyl component used in the synthesis of industrial phenolic resins. Special resins are made with other aldehydes, e.g., acetaldehyde, butyraldehyde, furfural, glyoxal, or benzaldehyde, but these have not achieved much commercial importance. Ketones are also rarely used. Physical properties of some aldehydes [23–25] are compiled in Table 1.6.

1.2.1 Formaldehyde

Formaldehyde, the first of the series of aliphatic aldehydes, was discovered in 1859 by Butlerov and has been manufactured on a commercial scale since the beginning of the twentieth century. Because of its variety of chemical reactions [26, 27] and relatively low cost (basically reflecting the cost of methanol) it has become one of the most important industrial chemicals. Formaldehyde is a hazardous chemical [7–11]. It causes eye, upper respiratory tract, and skin irritation. Significant eye, nose, and throat irritation does not generally occur until concentration levels of 1 ppm. EPA has classified formaldehyde as a "probable human carcinogen" (group B) under its Guidelines for Carcinogen Risk Assessment [28]. However, no evidence exists for identifying formaldehyde as a "probable human carcinogen" (group B). Facilities that manufacture or consume formaldehyde must strictly control workers exposure following workplace exposure limits. Further information on toxicology and risk assessment of formaldehyde is covered in Chap. 8.

Table 1.6. Physical properties of some aldehydes [23–25]

Type	Formula	MP °C	BP °C
Formaldehyde	$CH_2=O$	−92	−21
Acetaldehyde	$CH_3CH=O$	−123	20.8
Propionaldehyde	$CH_3-CH_2-CH=O$	−81	48.8
n-Butyraldehyde	$CH_3(CH_2)_2-CH=O$	−97	74.7
Isobutyraldehyde	$(CH_3)_2CH-CH=O$	−65	61
Glyoxal	$O=CH-CH=O$	15	50.4
Furfural	$\begin{array}{c}HC-CH\\ \diagup\quad\diagdown\\ HC\quad\;\;C-CH=O\\ \diagdown\;\diagup\\ O\end{array}$	−31	162

Table 1.7. Physical properties of formaldehyde [8, 23, 24]

CAS registration number	50-00-0
EG registration number	605-002-01-2
MAK limit	0.5 ppm
OSHA STEL	2 ppm
MW	30.03
MP/BP	−19.2/118 °C
Dissociation constant, H_2O at 0/50 °C	$1.4 \times 10^{-14}/3.3 \times 10^{-13}$
Flash point, 40% solution	60 °C
Explosion limits LEL/UEL	7%/73%

At ordinary temperatures, formaldehyde is a colorless gas with a pungent odor. Physical properties are shown in Table 1.7. It is highly reactive and commonly handled in aqueous solutions containing variable amounts of methanol where it forms predominantly adducts with the solvent [26, 27], i.e., equilibrium mixtures of methylene glycol (Eq. 1.9), polyoxymethylene glycols (Eq. 1.10), and hemiformals of these glycols with methanol (Eq. 1.11). Even in concentrated solutions the content of nonhydrated HCHO is very small (less then 0.04%). MWD of the oligomers was investigated by NMR [29] and GPC techniques. Chemical equilibria of polyoxymethylene glycol formation in formaldehyde solutions in water (and D_2O) and of polyoxymethylene hemiformal formation in methanolic formaldehyde solution was examined by ^1H and ^{13}C NMR spectroscopy [29]. For similar formaldehyde concentration, the chain length of formaldehyde oligomeric products in aqueous solution is larger than in methanol solution. This fact explains why precipitation of solids (long chain oligomer) occurs at a lower overall formaldehyde concentration in aqueous solutions than in methanol solution and why solid precipitation from aqueous formaldehyde solution can be avoided by adding methanol. Furthermore with increasing temperature, the average chain length decreases in solution whereas in methanolic formaldehyde solution temperature has a negligible effect on chain length.

In acidic solutions, hemiacetals react to acetals by elimination of water.

$$HO-CH_2-OH \tag{1.9}$$

$$HO-(CH_2-O)_n-CH_2-OH \tag{1.10}$$

$$CH_3-OH + HO-CH_2-OH \leftrightarrow CH_3-O-CH_2-OH + H_2O \tag{1.11}$$

With the action of basic catalysts, formaldehyde easily undergoes a disproportionation reaction to methanol and formic acid, known as the Cannizzaro reaction (Eq. 1.12).

A surprising development related to the Cannizzaro reaction was reported by Ashby and coworkers [30]. The Ashby group showed that formaldehyde reacts with NaOH (Cannizzaro conditions) to yield formic acid and hydrogen quantitatively. Typically, conditions exist in which both hydrogen formation, and classical Cannizzaro reaction proceed simultaneously. This is far different to the report of Loew in 1887 suggesting a very small quantity of hydrogen as being formed in the Cannizzaro reaction. Ashby showed that the reaction of 4 mol l^{-1} NaOH with 15 mmol l^{-1} formaldehyde resulted in the formation of hydrogen from 31% of formaldehyde, the remainder going to normal Cannizzaro products. Chrisope and Rogers [31] point out that with only 0.5 mol l^{-1} NaOH, 20 mmol l^{-1} formaldehyde solution can be converted into 1–2% hydrogen and the remainder into Cannizzaro products. Depending on ratio of NaOH to formaldehyde, it is possible to generate some hydrogen resulting in potentially hazardous conditions.

Depending on manufacture and storage, formaldehyde solutions always contain trace amounts of formic acid (around 0.05%).

$$2\ H_2C=O\ +\ NaOH\ \rightleftharpoons\ CH_3OH\ +\ H_3C-C\underset{ONa}{\overset{O}{\diagup\!\!\!\diagdown}} \quad (1.12)$$

With concentrations ranging from 37 to 56, wt%, aqueous formaldehyde solutions used normally in resin production are unstable. Both formic acid and paraformaldehyde formation depend on temperature, time, and metallic ions (iron) concentration. Low temperature storage lowers the formic acid formation rate, but increases the tendency towards paraformaldehyde precipitation. Commercial formaldehyde grades are therefore stabilized with various amounts of methanol up to 15%. Other stabilizers include methyl- and ethylcelluloses, polyvinylalcohol or isophthalobisguanamine at concentrations from 10 to 1000 ppm. Storage containers should be of stainless steel; iron containers are not suitable. Containers with plastic linings or made from reinforced plastics may also be used. Higher than 50% solutions of formaldehyde are obtained by addition of paraformaldehyde.

1.2.2
Production and Economics of Formaldehyde

Worldwide, most phenolic resin producers without a captive phenol source very often have their own formaldehyde production. Access to formaldehyde and low cost methanol is a very important competitive factor. About 90% of the methanol produced is based on natural gas as feedstock. Others include naphtha, heavier oil fractions and coal. Natural gas cost is the most significant factor besides plant size (capital charges) in siting a plant at a specific location [32, 33]. For example comparing a 200,000 tonnes per year methanol plant in the Middle East and Germany, gas and energy costs in the Middle East account for about 10% of the total methanol production cost, in Germany about 40%.

Competitive sizes of newer plants are in the range of 700,000 tonnes per year leading to a cost decrease of about 20% compared to a 200,000 tonnes per year plant (German cost situation 1995).

Methanol process [34] consists of three main sections, synthesis gas preparation, methanol synthesis and methanol purification. Because of extensive heat recovery and recycle streams, there is a considerable interaction between the three. The principal, highly exothermic synthesis reactions are as follows:

$$CO + 2H_2 \rightarrow CH_3OH \qquad (1.13)$$

$$CO_2 + 3H_2 \rightarrow CH_3OH + H_2O \qquad (1.14)$$

All methanol synthesis processes employ a synthesis loop to overcome equilibrium conversion limitations at typical catalyst operating conditions. The high pressure process was rendered obsolete when ICI in the United Kingdom and Lurgi in Germany developed a highly active and selective copper-zinc-alumina catalyst. The resulting low pressure process [35] dominates the methanol production. Four basic reactor types are in use, quench converter, multiple adiabatic, tube-cooled, and steam-raising converter. Crude methanol is purified by conventional distillation with one to three towers depending on purity required.

Sufficient quantities of methanol should be available according to a market study [36] shown in Table 1.8. Over the next years, the aggregate average demand growth is expected to be 3.7%. Due to new plant additions underway for Trinidad, Chile, and Qatar, lower methanol plant utilization rates are forecast. On the other hand, consolidation has occurred in the market. One global supplier has access to almost 25% of the total production and handles nearly 50% of the merchant market. Other sources [37] are more critical concerning the realization of announced new plants and estimate operation rates over 90% with the continued strong growth over 20% predicted for methyl *tert*-butyl ether (MTBE) as federal clean air regulations in the US mandate the use of oxygenates in reformulated gasoline.

The global formaldehyde production is expected to increase at a rate of 2–3% per year. Formaldehyde production in the U.S. is more or less stable and the production in Southeast Asia is growing and offsets decreases in other parts of the world. The production of formaldehyde-based thermosetting

Table 1.8. CMAI 1997 world methanol analysis (thousand metric tonnes) [36]

Demand (in 1000 t)	1996	2000 Forecast
Formaldehyde	9000	10,200
MTBE/TAME	6700	7900
Acetic acid	1620	2350
Solvents	1000	1120
Methyl methacrylate	770	840
Gasoline	690	760
DMT	340	380
Others	4530	5370
Total demand	24,650	28,920
Total capacity	29,300	37,800

resins (phenol-, urea-, and melamine resins) accounts for about 60% of the total formaldehyde consumption. Other uses are for acetal resins, 1,4-butanediol, pentaerythritol, ethylene glycol, and methylene diisocyanate.

There are two basic processes to produce aqueous formaldehyde [38] from methanol using either a silver or an iron oxide-molybdenum oxide catalyst.

The silver catalyzed reaction occurs at essentially atmospheric pressure and 600–650 °C with two parallel reactions:

$$CH_3OH + 1/2O_2 \rightarrow HCHO + H_2O \qquad (1.15)$$

$$CH_3OH \rightarrow HCHO + H_2 \qquad (1.16)$$

In a typical silver catalyst process, the feed mixture is prepared below the flammability limit by sparging air into heated methanol and combining the vapor with steam. The mixture is sent through a superheater to the reactor equipped with a bed of silver crystals or layers of silver gauze. The product is rapidly cooled and fed to an absorption tower. The bulk of the methanol, formaldehyde, and water is condensed in the lower water cooled section of the absorber; the nearly complete removal of the remaining methanol and formaldehyde occurs in the top of the tower by countercurrent contact with clean process water. Absorber bottoms go to distillation where methanol is recovered and recycled to the reactor. The base stream from distillation, an aqueous solution of formaldehyde, is usually sent to an anion exchange unit where the formic acid is reduced to specification. The product contains up to 55% of formaldehyde and less then 1.5% of methanol.

In new plants, however, the choice is mostly the metal oxide catalyzed process. A typical flow scheme [39] is shown in Fig. 1.3. The reaction is performed at essentially atmospheric pressure (0–6 psig) and around 300 °C in a recirculation loop. Vaporized methanol is mixed with air and recycle gas to be on the methanol lean side of a flammable mixture. It is preheated to about 250 °C and sent to the reactor. Methanol and oxygen react to formaldehyde in the tubes filled with an iron oxide-molybdenum oxide catalyst. Reaction heat is removed by an oil heat transfer medium. The reacted gas which exits the reactor at about 290 °C is cooled to 130 °C before entering the absorber. In the absorber, the formaldehyde is absorbed in water or urea solution. The methanol conversion can be greater then 99% yielding a product with up to 55% formaldehyde and less then 1% methanol. Formic acid is removed by ion exchange. The tail gas, essentially nitrogen and oxygen, with small amounts of methanol, formaldehyde, dimethyl ether, and carbon monoxide, is incinerated.

The absence of a methanol recovery tower is an obvious advantage over the silver process. However, large capacity equipment is used to handle a greater flow of gas (three times greater). The metal oxide catalyst is more tolerant to trace contamination and requires less frequent changes compared to silver.

There has been significant research activity to develop alternative routes to formaldehyde, but without commercial success. The direct oxidation of methane with air at 450 °C and 10–20 bar pressure on an aluminum phosphate catalyst should be mentioned among these processes.

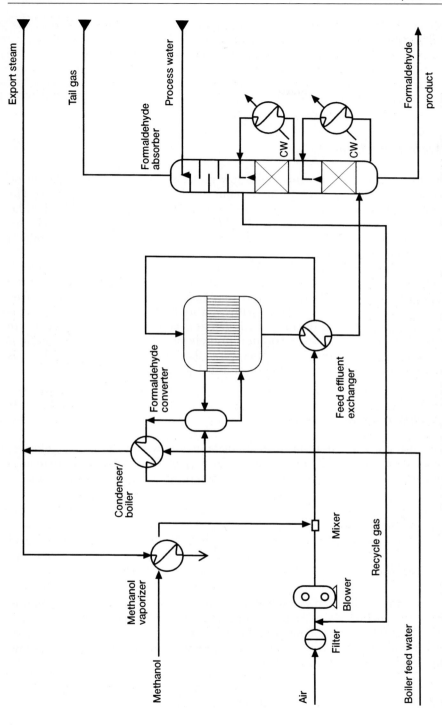

Fig. 1.3. Flow scheme of the iron oxide-molybdenum oxide formaldehyde process. CW = cooling water. (Drawing:Haldor Topsoe A/S)

Table 1.9. Properties of paraformaldehyde [40]

Content of formaldehyde	90–97%
Content of free water	0.2–4%
Specific weight	1.2–1.3 g/cm^3
Flash point	71 °C
Melting range	120–170 °C

1.2.3
Paraformaldehyde

Paraformaldehyde is a white, solid low molecular weight polycondensation product of methylene glycol with the characteristic odor of formaldehyde. The degree of polymerization ranges between 10 and 100. It is obtained by distillation of formaldehyde solutions under different conditions depending on temperature, time, and pressure during distillation. Commercial grades contain up to 6% water (Table 1.9).

Paraformaldehyde is seldom used in resin production because of high costs compared to aqueous formaldehyde solutions and severe problems with the high heat evolution during the start-up phase. Some possible reasons for its use in special instances are resins with high solids content or to avoid distillation of waste water. Paraformaldehyde can be used with acidic catalysts to cure novolak resins or to cure resorcinol prepolymers in cold setting structural wood adhesives. At low cure temperatures the separation of formaldehyde is tolerable, the reaction rate depends on the degree of polymerization.

1.2.4
Trioxane and Cyclic Formals

Trioxane, a cyclic trimer of formaldehyde or methylene glycol, is a colorless solid (MP 62–64 °C, BP 115 °C) and can be prepared by heating paraformaldehyde or a 60–65% formaldehyde solution in presence of 2% sulfuric acid.

$$\text{Trioxane} \xrightarrow[H_2O]{BF_3} HO-CH_2-[O-CH_2]_n-OH \longleftrightarrow CH_2O + H_2O \quad (1.17)$$

Trioxane   Poly-oxymethylene   Formaldehyde

Trioxane can be used as a source of formaldehyde in prepolymer formation or as resin curing agent. Cyclic formals, 1,3-dioxolane, 4-phenyl-1,3-dioxolane, and 4-methyl-1,3-dioxolane, have been recommended [41] as curing agents for novolak resins and high-solids, low-pressure laminating systems, taking into account their action as solvent.

Table 1.10. Physical properties of HMTA [42]

MW	140.2
Relative density	1.39 g/cm^3
Behavior at heating	sublimation at 270–280 °C
Solubility in 100 g H$_2$O, 20/60 °C	87.4/84.4 g

1.2.5
Hexamethylenetetramine

Hexamethylenetetramine, HMTA, is by far the most important compound to cure novolak type of resins. It is prepared from formaldehyde and ammonia according to Eq. (1.18):

$$6\ CH_2O + 4\ NH_3 \rightleftharpoons (CH_2)_6N_4 + 6\ H_2O \qquad (1.18)$$

In reverse, HMTA decomposes at elevated temperatures. In aqueous solution, HMTA is easily hydrolyzed to aminomethylated compounds and is often used as catalyst in the resole synthesis instead of ammonia.

HMTA is quite soluble in water (Table 1.10), but less soluble in methanol or ethanol. The aqueous solution exhibits a weak alkaline action with a pH in the range 7–10. In powder form, HMTA is prone to dust explosions. It is rated as a severe explosion hazard.

1.2.6
Other Aldehydes

Like formaldehyde, acetaldehyde is a gas at ambient temperature (BP 20.8 °C). Acetaldehyde is miscible with water in all proportions and most common organic solvents. The liquid phase oxidation of ethylene is the process of choice for the manufacture of acetaldehyde (Wacker-Hoechst Process) using palladium and cupric chloride catalyst at 130 °C. It reacts with phenol at considerably lower reaction rates as compared to formaldehyde.

C_3 through C_{11} aldehydes [23, 25] are highly flammable and explosive, colorless liquids with pungent odor. They are miscible with most organic solvents, e.g., acetone, ethanol, or toluene, but are only slightly soluble in water. Butyraldehyde is produced by oxo-reaction of propylene (Eq. 1.19).

$$H_3CHC{=}CH_2 + CO \xrightleftharpoons[Rh]{Co\ or} CH_3CH_2CH_2CHO + (CH_3)_2CHCHO \qquad (1.19)$$

Higher aldehydes react with phenol at significantly slower rates compared to formaldehyde. The base catalyzed resole formation is not practical as higher aldehydes easily undergo aldol and Tischenko condensation and self-resinification reactions.

Acidic catalysts are preferred. In particular, non-transition metal phenates (Mg, Zn, or Al) are recommended [43] for the preparation of acetaldehyde-phenol novolak resins in an aprotic solvent medium (12 h/120 °C). Yields in excess of 90% with exclusive ortho substitution (Eq. 1.20) are reported.

$$\text{PhOMgBr} + CH_3CHO \longrightarrow \left[\text{ortho-linked phenol-CH(CH}_3\text{) novolak} \right]_n \tag{1.20}$$

The phenol reaction with higher aldehydes is in general performed under strong acidic conditions, preferably in a water-free system, by continuous aldehyde addition to the phenol melt. However, only acetaldehyde and butyraldehyde have gained limited commercial significance, e.g., in resins for rubber modification, wood fiber binders, and antioxidants. The structure of novolak resins from acetaldehyde (Eq. 1.21) correspond to those obtained by the reaction of acetylene with phenol and cyclohexylamine as catalyst:

$$\text{4-tert-butylphenol} + CH_3CHO \longrightarrow \left[\text{CH(CH}_3\text{)-bridged 4-tert-butylphenol} \right]_n \longleftarrow HC\equiv CH + \text{4-tert-butylphenol} \tag{1.21}$$

The reaction of unsaturated aldehydes, acrolein and crotonaldehyde, with phenol in acidic medium was studied [44] by ^{13}C NMR and GPC. The alkylation reaction via the double bond seems to be the dominant pathway.

References

1. Weast RC (ed) (1988) Handbook of chemistry and physics. CRC Press
2. Jordan W, Van Barneveld H, Gerlich O, Kleine-Boymann M, Ullich J (1991) Phenol. In: Ullmann's encyclopedia of industrial chemistry, 5th edn, vol A 19. VCH, Weinheim, p 299
3. Fiege H, Voges HW, Hamamoto T, Umemura S, Wata T, Miki H, Fujita, Buysch H-J, Garbe D, Paulus W (1991) Phenol derivatives. In: Ullmann's encyclopedia of industrial chemistry, 5th edn, vol A 19. VCH, Weinheim, p 313
4. Pujado PR, Sifniades S (1990) Phenol. In: McKetta, Cunningham (eds) Encyclopedia of chemical processing and design, vol 35. Marcel Dekker, New York, p 372
5. Wallace J (1996) Phenol. In: Kroschwitz JI (ed) Kirk-Othmer, Encyclopedia of chemical technology, vol 18. Wiley, New York, p 592
6. Lorenc JF (1992) Alkylphenols. In: Kroschwitz JI (ed) Kirk-Othmer, Encyclopedia of chemical technology, vol 2. Wiley, New York, p 113
7. Noyes Data Corp. (1988) Extremely hazardous substances, Superfund Chemical Profiles. EPA. Noyes Data Corp., Park Ridge
8. OSHA (1991) Regulated hazardous substances. Occupational Safety and Health Administration. Noyes Data, Park Ridge
9. Rippen (1995) Handbuch Umweltchemikalien. Ecomed, Landsberg
10. Roth, Weller (1995) Sicherheitsfibel Chemie, 5 edn. Ecomed, Landsberg
11. Cook WA (1987) Occupational exposure limits. Worldwide. American Industrial Hygiene Association

12. Hock H, Lang S (1944) Chem Ber 77:257
13. Schulz RC (1993) Cumene. In: Kroschwitz JI (ed) Kirk-Othmer (eds) Encyclopedia of chemical technology, vol 7. Wiley, New York, p 730
14. Seubold FH, Vaugham WE (1953) J Amer Chem Soc 75:3790
15. Franck H-G, Stadelhofer JW (1987) Industrielle Aromatenchemie. Springer, Berlin Heidelberg New York
16. Franck H-G, Collin G (1968) Steinkohlenteer. Springer, Berlin Heidelberg New York
17. Wurm HJ (1976) Chem Ing-Techn 48:840
18. Process Bulletin: Dephenolization of effluents by the phenosolvan process. Lurgi GmbH
19. Hydrocarbon Research Inc. (1980) Chem Eng News, p 35
20. Heredy LM (1981) The chemistry of acid-catalyzed coal depolymerization. Coal Structure, ACS no 192, p 179
21. Schmiedel KW, Decker D (1993) Resorcinol. In: Ullmann's encyclopedia of industrial chemistry, 5th edn, vol A 23. VCH, Weinheim, p 111
22. Chyoda Corp. (1995) Petrochemical processes. Hydrocarbon Processing, March 1995, p 98
23. Miller DJ (1991) Aldehydes. In: Kroschwitz JI (ed) Kirk-Othmer (eds) Encyclopedia of chemical technology, vol 1. Wiley, New York, p 926
24. Gerberich HR, Seaman GC (1994) Formaldehyde. In: Ullmann's encyclopedia of industrial chemistry, 5th edn, vol 11. VCH, Weinheim, p 929
25. Hagemeier HJ (1991) Acetaldehyde. In: Encyclopedia of chemical technology, vol 1. Wiley, New York, p 94
26. Walker JF (1964) Formaldehyde. ACS Monograph 159, 3rd edn
27. Zabicka J (ed) (1970) The chemistry of the carbonyl group. Interscience
28. Hernandez O, Homberg LR, Hogan K, Siegel-Scott C, Lai D, Grindstaff G (1994) Risk assessment of formaldehyde. J Hazardous Materials 39:161
29. Hahnenstein I, Hasse H, Kreiter CG, Maurer G (1994) Ind Eng Chem Res 33(4):1022
30. Ashby EC, Doctorovich F, Neilmann HM, Barefield EK, Konda A, Zhang K, Hurley J, Siemer DD (1993) J Am Chem Soc 115:1171
31. Chrisope DR, Rogers PE (1995) C & EN p2 January 9, 1995
32. LeBlanc JR (1992) Economic considerations for new methanol projects. World Methanol Conference, Monte Carlo
33. Hymas R (1993) The economic rationale for the location of new methanol investment. World Methanol Conference, Atlanta
34. ICI Katalco, Lurgi GmbH (1997) Öl Gas Chemie, The MW Kellog Technology Co.: Hydrocarbon Processing, March 1997, pp 139, 140
35. Lurgi GmbH (1993) Company Bulletin: Integrated low pressure methanol process
36. CMAI (1997) World Methanol Analysis: Hydrocarbon Processing, October 97, p 25
37. Crocco JR (1996) World Methanol Conference, Monte Carlo, December 1996
38. English A, Rovner J, Bown J (1995) Methanol. In: Kroschwitz JI (ed) Kirk-Othmer (eds) Encyclopedia of chemical technology, vol 16. Wiley, New York, p 537
39. Haldor Topsoe A/S (1997) „Petrochemical Processes 97", Hydrocarbon Processing, March 97, p 134
40. Degussa AG, Product Bulletin: Paraformaldehyde
41. Heslinga A, Schors A (1964) J Appl Polymer Sci 8:1921
42. Degussa AG, Product Bulletin: Hexamethylentetramine
43. Casiraghi G, Cornia M, Ricci G, Balduzzi G, Casnati G, Andreeti GD (1983) Polymer Preprints 24(2):183; (1984) Macromolecules 17:19
44. Sebenik A, Osredkar U (1983) ACS Polymer Preprints 24/2:185

CHAPTER 2
Phenolic Resins: Chemistry, Reactions, Mechanism

2.1
Introduction

The chemistry of phenolic resins involves a variety of key factors which are critical in the design of the desired phenolic resin. These include:
- Molar ratio of F to P
- Mode of catalysis: acid, base, metal salt, enzyme
- Liquid, solid, dispersion
- Thermoplastic or thermosetting resin

These low to medium M_w materials can be viewed as "reactive intermediates" which can be cured or can undergo many transformation reactions by appending new reactive groups to the phenolic hydroxyl substituent such as epoxy, allyl, cyanate, or form a new ring structure.

The mechanism of resin preparation and cure are discussed by considering different cure conditions to arrive at a highly crosslinked, infusible structure.

The perennial problem of phenolic brittleness and methods to alleviate or improve resin ductility will be discussed in conjunction with blend/alloy/IPN studies which suggest some improvement in ductility or reduced brittleness.

The exothermic nature of reacting formaldehyde with phenol under acidic or basic conditions has resulted in "rare" uncontrollable reactions due to unknown or unforeseen circumstances. Conditions that may lead to hazardous reactions by both modes of catalysis are mentioned and are to be avoided under all circumstances.

2.2
Chemistry

The reaction of formaldehyde with phenol can lead to either a heat reactive resole or a stable novolak and is dependent upon the mode of catalysis and molar ratio of F to P. The chemistry of phenolic resins has been reviewed extensively [1–9].

A comparison of different catalysts that can be utilized in phenolic resin preparation is shown in Table 2.1.

In describing the preparation of phenolic resins the mode of catalysis of the resulting resin dictates its overall property characteristics. These are summarized in Table 2.1. Phenolic resins are obtained by step-growth polymerization. Since the functionality of P and F varies according to stoichiometry and the latter is related to the type of resin (resole or novolak) it is important to determine functionality. Recently Shipp and Solomon [10] reported that the

$$\text{Phenol} + CH_2O \xrightarrow[F/p \geq 1]{\text{Base}} \text{Resole} \quad (2.1)$$

functionality of P in novolak is in the range of 1.49–1.72 and *not* 2.31 as reported by Drumm and us in our earlier book [1]. Functionality of F is 2. However functionality of P and F in resole is more complex where functionality can be ≤ 3 and formaldehyde never approaches a functionality of 2.

2.2.1
Resole

2.2.1.1
Methylol Phenol(s)

The simplest phenolic component that can cure into a phenolic resin is *o*- or *p*-methylol phenol. The reaction of P with F under basic conditions is initially the addition of F to phenolate (Eq. 2.2) leading to *o*- and *p*-methylol phenol.

$$\text{PhOH} + OH^- \rightleftharpoons \text{PhO}^- + H_2O \xrightarrow{CH_2O} \text{o-methylol phenolate} + \text{p-methylol phenolate} \quad (2.2)$$

Besides monomethylol phenols, some dimethylol phenols, and trimethylol phenol are formed. It is commonly referred to as the Lederer-Manasse reaction.

Table 2.1. Different catalysts for various phenolic resins

Catalyst	Resin Type	F/P	F	P	Physical state	Product stability (RT)	Functional groups	Mode of cure
Base	Resole	≥1	<2	≤3	Liquid, solid, solution	Limited	Methylol, Phenolic	Acid, base, thermal
Acid	Novolak	<1	2	1.49–1.72	Solid	Stable	Phenolic	Hexa
Metal salt	Resole or Novolak	≥1	"High ortho"	"High ortho"	Liquid, solid	Liquid-limited Solid-stable	Methylol, Phenolic	Same as Resole or Novolak
Enzyme	Pseudo Novolak	No (CH$_2$O)	–	–	Solid	Stable	Phenolic	Resin transformation

Preparation of *o* and *p*-methylol phenol have been examined using complexation via cyclodextrin [11, 12] or crown ether [13]. Komiyama investigated the base catalyzed reaction of P + F and sought to increase the amount of *p*-methylol phenol. The reaction usually leads to a 2:1 ratio of *p*:*o* isomers (Eq. 2.3):

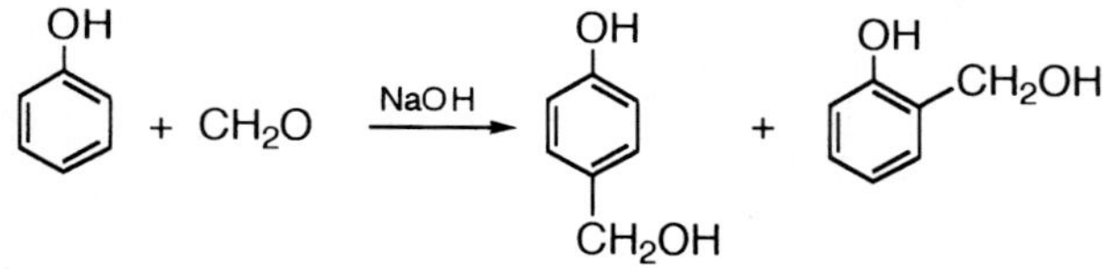

$$\text{Phenol} + CH_2O \xrightarrow{NaOH} \text{p-methylol phenol} + \text{o-methylol phenol} \quad (2.3)$$

Product selectivity is 60% but isolation of *para* isomer free of *ortho* is troublesome. Using β-cyclodextrin, selectivity increased to 3:1 with a 74% yield by introducing a concentration of 0.3 mol l^{-1} β-cyclodextrin. Mechanism of selective synthesis of *para* isomer is shown in Fig. 2.1 which locates phenolate anion in the β-cyclodextrin cavity and a more facile approach of F from the bottom of the cavity to react with the *para* position of the phenolate. Approach of F to the *ortho* position is hindered by complexation of phenolate and β-cyclodextrin. Increased selectivity of *para* over *ortho* was obtained with 2-hydroxy propyl group on the bottom side of the cavity (Fig. 2.2). Use of 0.6 equivalents of HP β-cyclodextrin boosted the selectivity to 94%.

Use of 18 crown 6 ether (Fig. 2.3) with P, paraform and NaOH leads to high selectivity of *p*-methylol phenol (to 88%). Presumably an analogous complexed intermediate is formed with phenolate anion residing inside the crown ether cavity to facilitate the reaction of F to the *para* position.

2.2.1.2
Oligomerization

Resoles are a mixture of methylol phenols, oligomers of varying -mer units and residual amounts of free phenol and formaldehyde. Depending on reaction conditions (molar ratio of F/P, catalyst concentration/type, and temperatures)

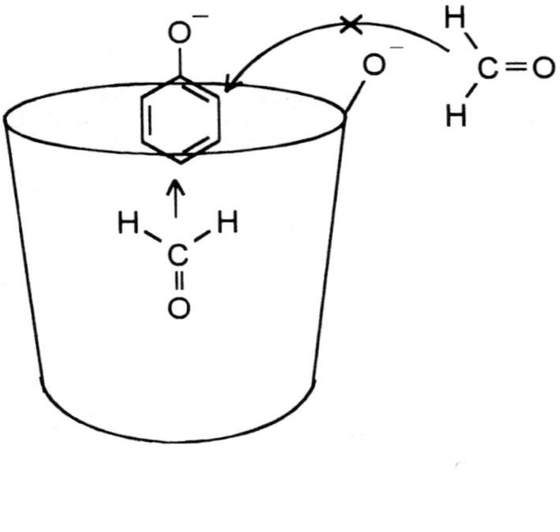

Fig. 2.1. P + F in β-cyclodextrin cavity

Fig. 2.2. Use of 2-hydroxy propyl-β-cyclodextrin

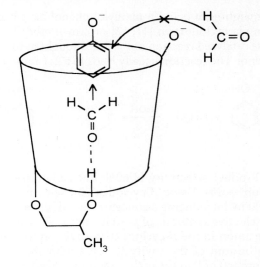

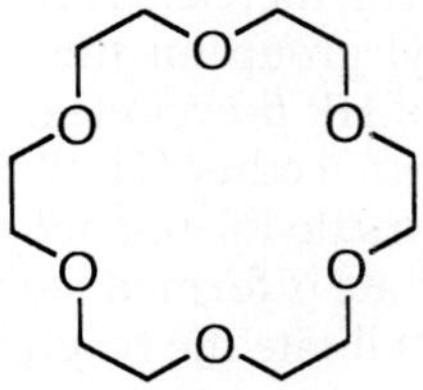

Fig. 2.3. 18 Crown 6 ether

(18 Crown 6 Ether)

at low temperatures (60 °C) only addition of formaldehyde to phenol occurs. In a temperature range above 60 °C, condensation reactions of methylol phenols with phenol and/or methylol phenols occur and lead to prepolymer or the desired resole resin (Eqs. 2.4, 2.5).

Addition (<60 °C)

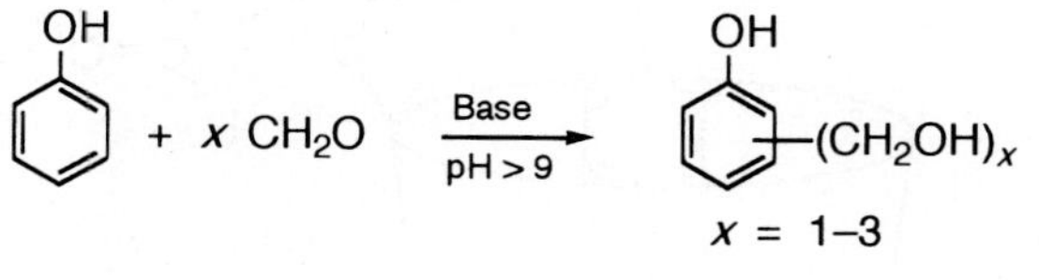

$$\text{PhOH} + x\,CH_2O \xrightarrow[pH>9]{\text{Base}} \text{HO-C}_6H_4\text{-}(CH_2OH)_x \quad x = 1\text{–}3 \tag{2.4}$$

Condensation (60–100 °C)

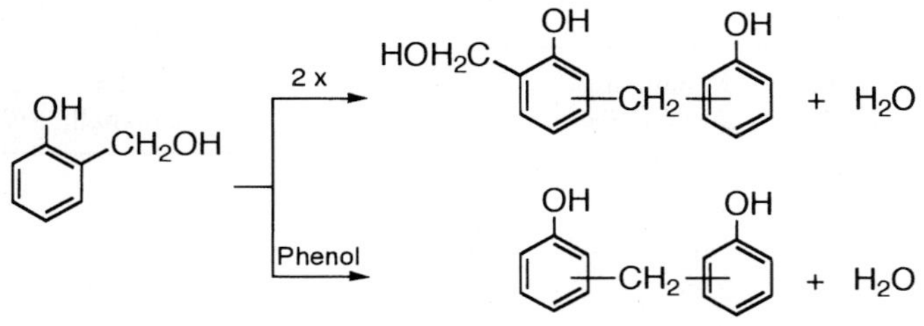

$$\text{(2.5)}$$

The addition of formaldehyde to phenol results in the formation of several methylol phenol monomers (Fig. 2.4). Earlier studies [1] provided kinetic rate data for the formation of various methylolated monomers. Kinetic data was reported with either equimolar ratio of NaOH to P or NaOH/P ratio less than unity. Although these conditions established kinetic rate data for the appearance of the various methylolated phenols (Table 2.2), some authors neglected to consider the role of deficient NaOH in their kinetic analyses since there are differences in acidity among the phenolic compounds emerging during reaction. This, in fact, is why there are differences in the relative values. In an attempt to remedy this, Higuchi et al. [14, 15] have developed a rate equation involving the concentration of hydroxide ion which changes with the change in reaction composition. By computer simulation and accounting for hydroxide ion concentration, fate of reactants (P and F) and products (five methylolated products, 2–6) as function of time can be described. Computer simulated reactions are shown in Figs. 2.5 and 2.6. Figure 2.5 shows F/P molar ratio of 2 and NaOH/P molar ratio of 0.05 with an initial P concentration of

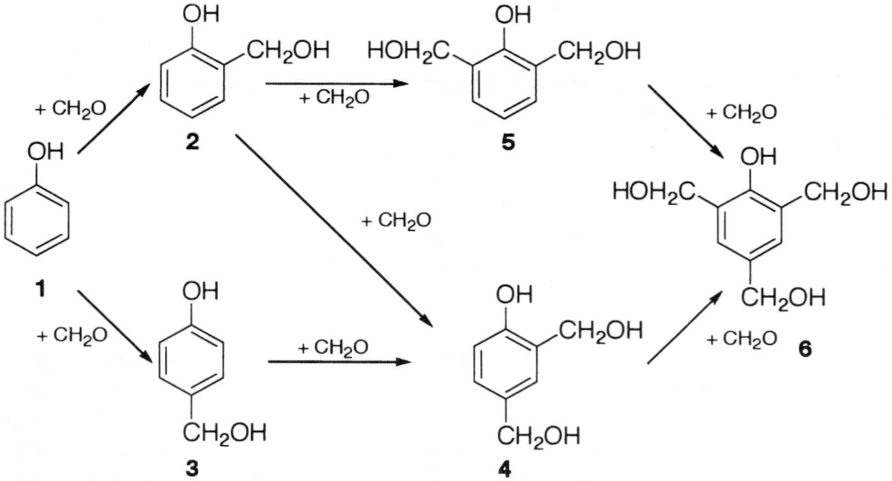

Fig. 2.4. Addition of formaldehyde to phenol

Table 2.2. Relative values of rate constants of methylolated products from reaction of P with F

Reactions	Researchers			
	Freeman and Lewis (1954)	Minami and Ando (1956)	Zsavitsas and Beaulieu (1967)	Eapen and Yeddanapalli (1968)
P+F → 2	1.00	1.00	1.00	1.00
P+F → 3	1.18	2.08	1.09	1.46
2+F → 5	1.66	1.08	1.98	1.70
2+F → 4	1.39	2.58	1.80	3.80
3+F → 4	0.71	0.83	0.79	1.02
5+F → 6	7.94	3.25	3.33	4.54
4+F → 6	1.73	1.25	1.67	1.76

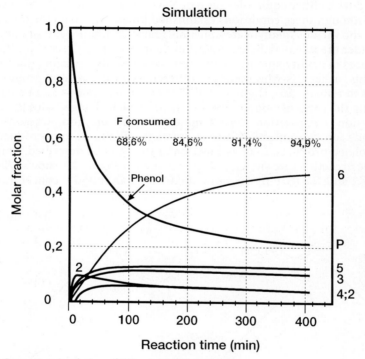

Fig. 2.5. Computer simulation of F/P = 2, NaOH/P = 0.05

3 mol/l and conducted at 70 °C. After 400 min 95 % F is consumed and yet over 0.2 molar fraction of P remains with **6** being the major product (0.5 molar fraction). With similar F/P molar ratio of 2 and initial concentration of P (3 mol/l) but higher NaOH/P molar ratio of 0.4 and lower temperature of 60 °C within 150 min (Fig. 2.6), about 98% of F is consumed, free phenol is lower (0.175 molar fraction) and **6** is lower (0.426 molar fraction). Higuchi's initial

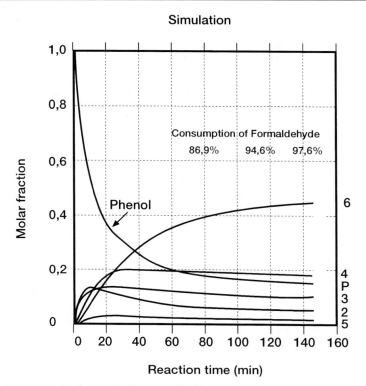

Fig. 2.6. Computer simulation of F/P = 2, NaOH/P = 0.4

study only considered methylolated products for comparison with early work and attempted to generate an "ideal resole" by computer simulation. Temperatures above 60 °C facilitate dimer, trimer formation. Higuchi is currently expanding his studies to include these oligomeric materials. Nevertheless computer simulation provides a method to simulate an "ideal resole" composition without many time-consuming resin preparations.

In a series of publications, Grenier-Loustalot and coworkers [16–19] have examined resole resin preparation under controlled conditions: temperature, stoichiometry, catalyst and pH using a range of techniques such as HPLC, ^{13}C NMR, FTIR, UV, and chemical assay. By monitoring F disappearance and observing the appearance of methylolated monomers, dimers, trimers by using both HPLC and ^{13}C NMR, they showed resole resin structure depends on reaction conditions. All known methylolated phenols (Fig. 2.4, structures 2–6) as well as many dimers and trimers were prepared and fully characterized unambiguously. The primary reaction sequence (Fig. 2.4) was examined by conducting the reaction with a pH of 8 (NaOH) at 60 °C and increasing molar ratio of F/P from 1 to 1.5 [16]. Appearance of various intermediate products is shown by HPLC characterization (Fig. 2.7) from 2 to 48 h reaction time. All the peaks for methylolated phenols (structures 2–6) are readily observed after

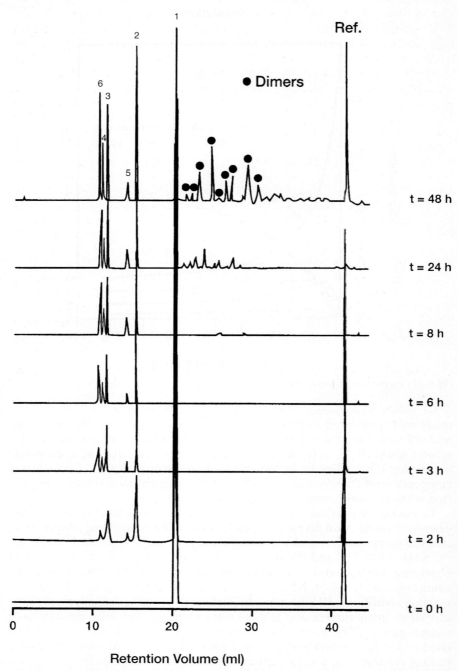

Fig. 2.7. HPLC characterization F/P = 1.5; column spherisorb ODS-2 (Bishoff ICS+), 25 cm long, 5 μm particle size, mobile phase: methanol/water 10/90 to 100/0 in 40 min

2–4 h; after 8 h trace amounts of condensation products (dimers) are barely detectable. After 48 h many dimer peaks are prominent besides monomers (structures 2–6). Using a low temperature of 60 °C addition products are obtained; even at low molar ratio of F/P some trimethylol phenol (structure 6) is obtained. Increasing the pH of the reaction medium at 60 °C with F/P at 1.5, the amount of dimers increased significantly (HPLC) from pH 8 to 9.25 after 8 h; also residual phenol decreased from 42% (pH 8) to 16% (pH 9.25). As dimers are formed, only o,p' and p,p' linkages are observed with no o,o' bridge. The methylene region is conveniently monitored by ^{13}C NMR and distinguishes all three isomeric linkages.

Alkali and alkaline metal hydroxides were evaluated [17] and included Group IA hydroxides: LiOH, NaOH, and KOH as well as Group IIA hydroxides: $Mg(OH)_2$, $Ca(OH)_2$, and $Ba(OH)_2$ under similar reaction conditions (F/P = 1.5, pH 8, 60 °C). The rate of F disappearance was related to ionic radius within family groups: Li > Na > K and Mg > Ca > Ba. Comparing all the catalysts with the disappearance of F resulted in the following ranking: Mg > Ca > Ba > Li > Na > K. Authors propose that the ionic radius and hydrated cation account for the observed trend.

Reacting separately each methylolated phenolic compound (structures 2–6 in Fig 2.4) with an equimolar amount of F at 60 °C and a pH of 8 the reaction of each methylol phenol was followed by HPLC and ^{13}C NMR [18]. Compound 2 with equimolar F led to competition between addition and condensation. A similar trend was noted for 3 (Eq. 2.6):

$$2 \text{ or } 3 + F \xrightarrow[\text{pH 8}]{60\,°C} 4,5,6\,(\text{addition}) + \text{dimers (condensation)} \tag{2.6}$$

Analyses of the dimers indicated that dimers were common to both reactions and were linked by p,p' methylene bridge. Dimer structures are:

This was a surprising observation especially since compound 2 is *ortho* methylolated and should result in some *ortho* linkage. Thus regardless of initial substitution pattern of starting material (2 = *ortho*, 3 = *para*), the majority of condensation dimeric products exhibited p,p' linkage. Kinetic data indicated that the rate of disappearance of 2 or 3 was identical.

Similarly reacting separately 4 and 5 with equimolar F at 60 °C and 8 pH are shown in Fig. 2.8 (compound 4) and Fig. 2.9 (compound 5). Both 4 and 5 form 6 as an intermediate with subsequent condensation reactions involving loss of water and/or F leading to various dimers joined by either o,p' or p,p' methylene linkages. Compound 5 is more reactive than 4 with F and difference in reactivity is attributed to quinoid form:

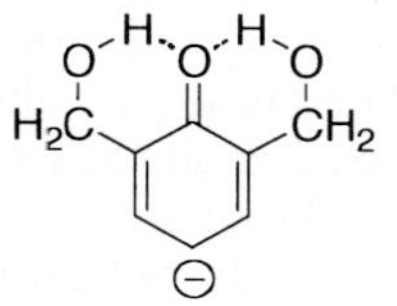

Intramolecular stabilization is due to H bonds and leads to a higher negative charge on the *para* position.

Thus rate data indicate the following ranking:

- $k_{phenol} < k_4 < k_2 < k_3 < k_5$.

Compound 5 disappears rapidly in the reaction medium whereas phenol and compound 4 accumulate. The *para* position is more reactive than the *ortho* position with regard to F or methylol and leads to o,p' and p,p' methylene linkages; no o,o' is detected. Phenol must be consumed in the early stages of the reaction or it will remain unreacted because functional materials (2–5) are more reactive (more acidic) than phenol in reacting with F or methylol.

Hence, it is very unlikely that the following reaction occurs:

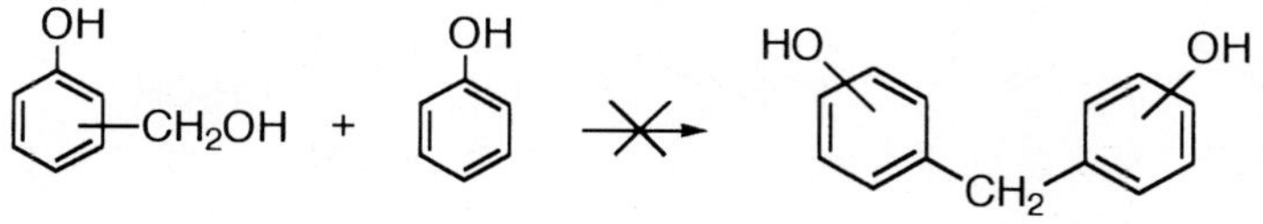

(2.7)

A closely related HPLC study of the kinetics of resole resin catalyzed by either NaOH or triethylamine was recently reported by Mondragon and coworkers [20]. Using a molar ratio of 1.8 for F:P, a temperature of 80 °C, and a pH of 8, the rate of disappearance of both F and P was faster for the triethylamine catalyzed system as compared to NaOH catalyst. The amine catalyzed reaction favored the formation of 2-hydroxymethyl phenol 2 and suggests it is "*ortho* directing". A higher amount of 4-hydroxymethyl phenol 3 was obtained with NaOH. Similarly higher amounts of methylolated phenols occurred with triethylamine as well as oligomeric materials. Data suggest that a much faster reaction occurs with amine when compared with NaOH.

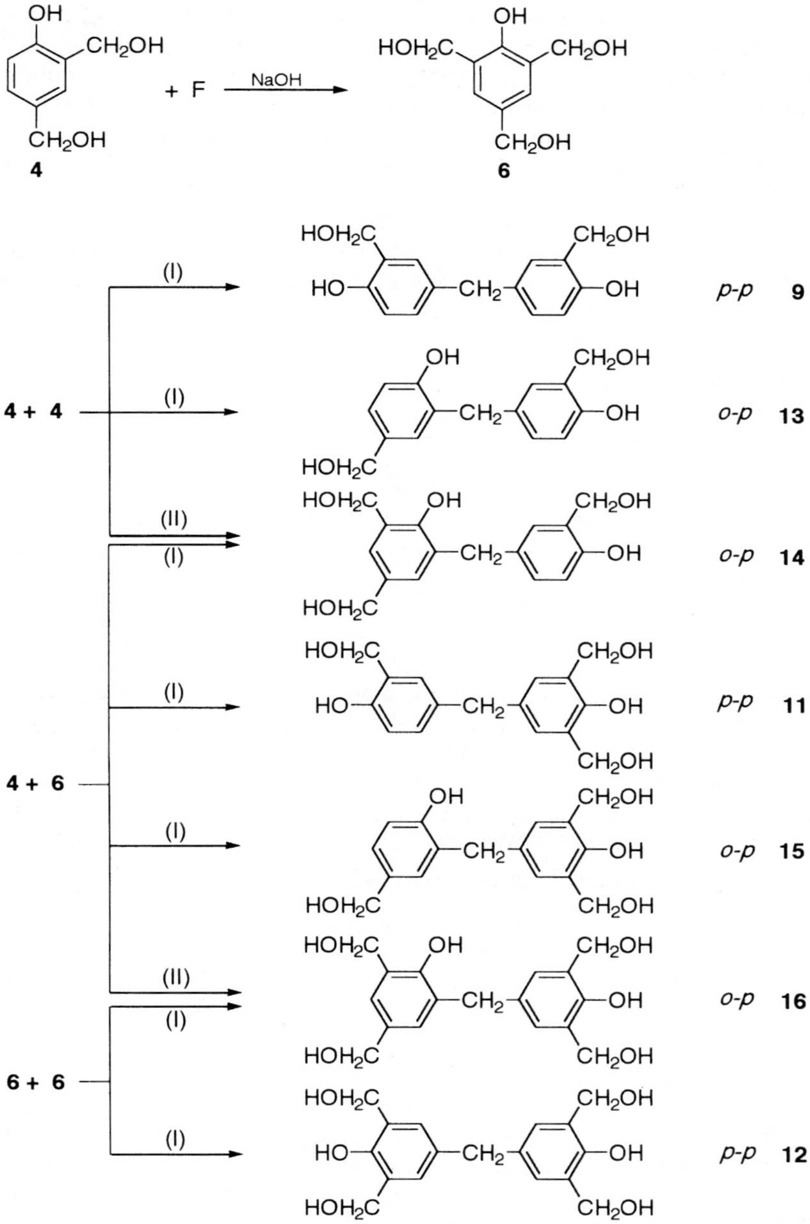

(I) Condensation with loss of one molecule of water and of formaldehyde

(II) Condensation with loss of one molecule of water

Fig. 2.8. Reaction of compound 4 with equimolar F at 60 °C and pH 8

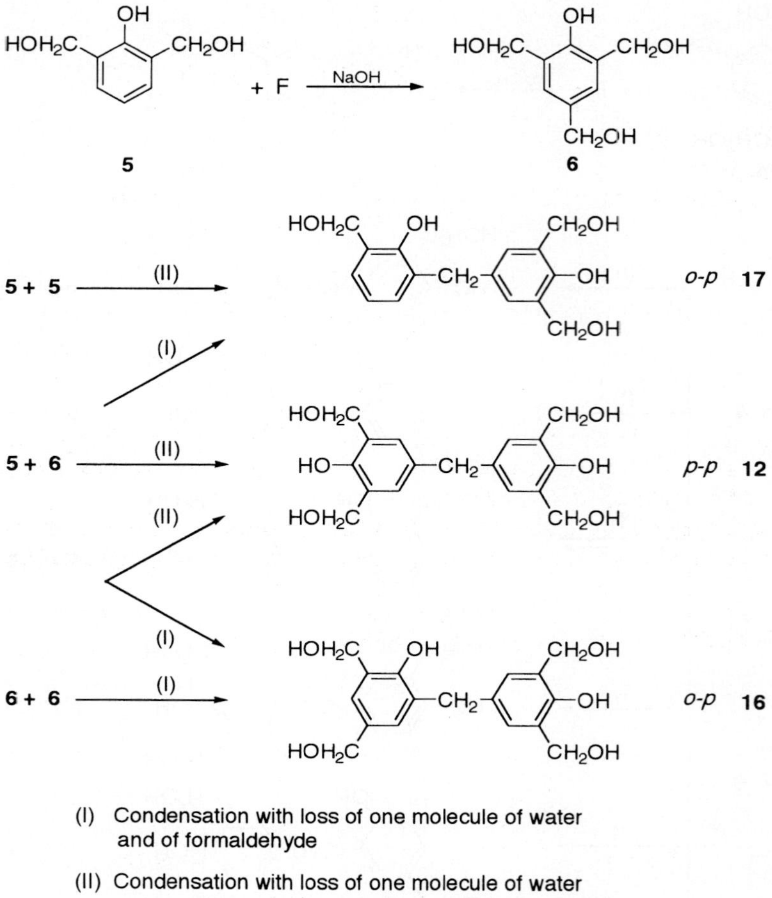
OH
HOH2C
CH2OH
+ F
NaOH
5
6
5 + 5
(II)
(I)
o-p 17
5 + 6
p-p 12
6 + 6
o-p 16
CH2
HO
(I) Condensation with loss of one molecule of water and of formaldehyde
(II) Condensation with loss of one molecule of water

Fig. 2.9. Reaction of compound 5 with equimolar F at 60 °C and pH 8

2.2.2 Novolak

2.2.2.1 Bisphenol F

Bisphenol F, the simplest novolak, is prepared by conducting the reaction of P and F with a large excess of phenol under acidic conditions. A mixture of isomers is obtained: *o,p′* isomer predominates, followed by *p,p′* and *o,o′*, the lowest (Eq. 2.8):

$$\text{Phenol} + CH_2O \xrightarrow{H^+} \begin{cases} \text{o,p'-isomer} \\ \text{p,p'-isomer} \\ \text{o,o'-isomer} \end{cases} \quad (2.8)$$

It is the simplest but most difficult bisphenol to obtain (compared with Bisphenol A, others) because of its propensity to undergo oligomerization to higher MW materials (novolaks).

The mechanism of formation involves initial protonation of methylene glycol (hydrated F) which reacts with phenol at the *ortho* and *para* positions (Eq. 2.9):

$$HOCH_2OH + H^\oplus \rightleftharpoons HOCH_2\overset{\oplus}{O}H_2 \quad (2.9)$$

The intermediate benzyl alcohols (carbonium ions) are transient intermediates (detectable by NMR) as they are transformed into Bisphenol F. Newer processes identify continuous production of Bisphenol F using oxalic acid catalyses [21, 22]. A small scale continuous (400 ml/h) operation is described [21] utilizing a multistage reactor in which the first stage is a stirred vessel with short residence time. The oxalic acid catalyzed reaction of 37% F and P is conducted in the first stage at 80 °C/5 min and then a flow reactor with a 10 min residence time. Yields of 92% Bisphenol F are reported. A large scale continuous process consists of reacting excess phenol with F and oxalic acid in multiple reactors with the first group of reactors in a tube configuration. The process involves introducing 360 kg/h of 20/1 P:F with 0.046% oxalic acid (based on P) and reacting the composition at 70 °C. Continuous removal of 30 kg/h of reaction mixture and continually feeding it into a packed distillation column (water and phenol removal) yielded crude Bisphenol F consisting of 29% p,p'; 38% o,p' and 11% o,o' or 78% Bisphenol F by weight. Besides preparing Bisphenol F, the process yields novolak resin (residue from continuous distillation of Bisphenol F) with very low amounts of phenol and dimer (Bisphenol F).

A novel process involving a heterogeneous liquid/liquid reaction medium via a hemiformal intermediate was recently reported [23]. An equilibrated mixture of F and hemiformal in isobutanol is added dropwise to aqueous solution of oxalic acid in phenol over 3 h (Eq. 2.10):

$$\text{(Eq. 2.10)} \tag{2.10}$$

The reaction mixture is subsequently stirred at 50 °C for 6 h to yield a 95:5 mixture of Bisphenol F isomers and novolaks.

Bisphenol F is a desirable material with many attractive applications paralleling Bisphenol A in use but significantly lower in volume. Bisphenol F epoxy resin possesses a low viscosity (2000–3000 mPa) and is used in solvent free or high solids coatings, floorings, linings, impregnation, moldings, and laminates [24]. Polycarbonates based on Bisphenol F/Bisphenol A exhibit high T_gs, high impact strengths, low brittle-ductile transition temperatures, and flame retarding properties [25]. The latter property is attributable to the Bisphenol F segment. High purity Bisphenol F is used in liquid crystal production [26]. Bisphenol F is combined in a novolak/hexa inorganic filled molding material composition for low pressure injection molded parts with very low flash [27, 28]. Bisphenol F epoxy with high o,o' isomer ($\geq 46\%$) is reported [29] to exhibit lower viscosity (melt and solution viscosity) than typical Bisphenol F epoxy resin containing 10–15% o,o' isomer.

Methods for preparing o,o'-Bisphenol F are known and require special conditions. Casiraghi et al. [30] reported the preparation of o,o'-Bisphenol F among higher p,o' oligomers using phenol and paraformaldehyde in xylene in a pressurized reactor.

Katritzky et al. [31] demonstrated a general procedure for the synthesis of o,o'-Bisphenol F via the reaction of benzotriazolylmethyl phenol with phenol and base (Eq. 2.11):

$$\text{(Eq. 2.11)} \tag{2.11}$$

40 % yield

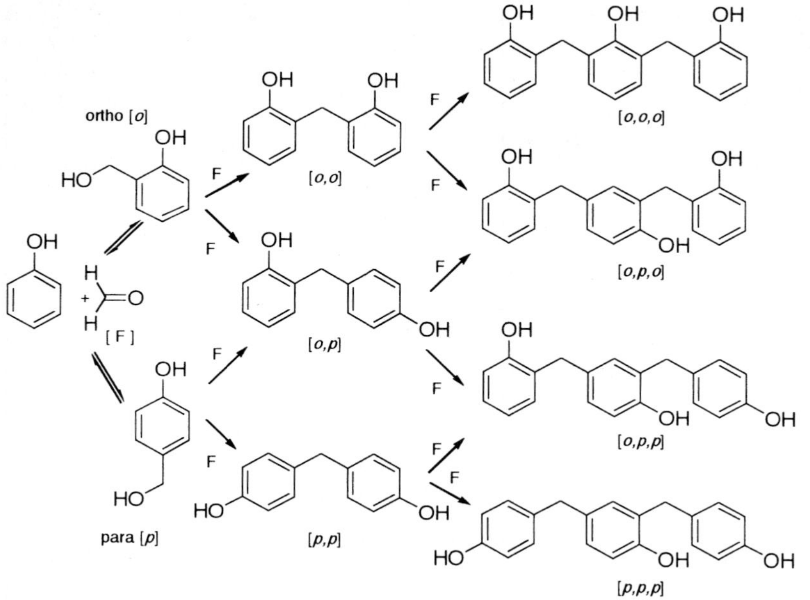

Fig. 2.10. Isomeric structures during oligomerization

2.2.2.2
Oligomerization

2.2.2.2.1
Random

Using lower amounts of phenol (than in the Bisphenol F preparation) but maintaining less than equimolar amount relative to F to avoid gelation, novolaks with varying MWs can be prepared. Depending on the molar ratio of F to P and residual free phenol constant, low melting solids to materials with softening points above 100 °C can be obtained. The overall reaction scheme showing isomeric changes that occur during oligomerization is shown in Fig. 2.10. Beyond Bisphenol F, isomeric changes and number of structural oligomers increases significantly.

Pethrick and Thomson [32] examined novolak synthesis by conducting the oxalic acid catalyzed reaction in an NMR tube. Figure 2.11 shows in situ novolak preparation monitored by ^{13}C NMR. Peak assignments are presented in Table 2.3. The in situ experiments were performed at 80 °C and spectra were obtained at intervals after 1, 3.25, and 5.5 h. Evidence for methylolphenol and its corresponding hemiformal was noted (61–66 ppm) but no identity of *p*-methylolphenol (64 ppm.). Yet indirect evidence for the existence of *p*-methylolphenol exists since *p,p'* and *o,p'* methylene linkages are observed in the latter stages of reaction. Authors propose the absence of NMR peak being

Fig. 2.11. ^{13}C NMR of in situ novolak preparation

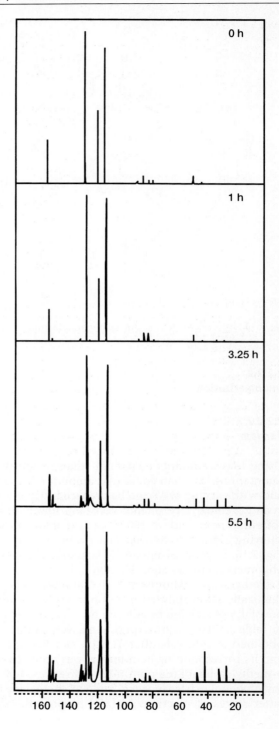

Table 2.3. General assignments of novolak ^{13}C NMR spectra

ppm	Description
150–156	C_1 bearing phenolic hydroxyl
151–152	Trisubstituted (branched)
131–135	C_{4s}
128–131	Free phenol, mono and disubstituted phenolic rings
125–131	$C_{2s,6s}$ and trisubstituted (branched) phenolic
120	C_4 free phenol
115	$C_{2,6}$ free phenol
65	p-methylol phenol
61	o-methylol phenol
40	p,p'-methylene
35	o,p'-methylene
30	o,o'-methylene

observed for p-CH$_2$OH is probably due to its high instability. o-Methylol formation is more stable than p- due to intramolecular H bonding between methylol group and adjacent phenolic group. The first bridging methylene carbons which are characteristic of the resin microstructure were those of p,p' linkage, followed by o,p', with o,o' being the last to appear. It should be noted that although the o,p' methylene group follows the initially formed p,p' linkage, the o,p' linkage will increase and be greater than either p,p' or o,o' bridges. More detailed spectra peaks are assigned to monosubstituted rings (free phenol), disubstituted, and trisubstituted rings (Table 2.3).

Thus in the preparation of novolaks, branching, free phenol content, substitution patterns, and methylene bridging structures can be ascertained by ^{13}C NMR. Depending on type of catalyst, strong acid or a weakly acidic metal salt medium, a statistical distribution of p,p':o,p':o,o' 1:2:1 is obtained for strong acid while the weakly acidic conditions (4–6 pH) leads to high *ortho* resin exhibiting large amounts of o,o' and o,p' and low amounts of p,p'. (Weak acid conditions will be described in the section under High *ortho*). Table 2.4 describes different novolaks.

Table 2.4. Isomeric content of various novolaks

Novolak	% Methylene bridges		
	p-p	o-p	o-o
High *ortho*	2.3	40.2	57.5
Random	25.8	49.1	25.1
Oxalic acid	27.6	49.1	23.3
Phosphoric acid	25.4	51.1	23.5
Sulfuric acid	25.6	48.5	25.9

Table 2.5. Relationship of F/P ratio to novolak softening point

F/P Ratio	Softening Point (°C)
1:0.65	60–70
1:0.75	70–75
1:0.85	80–100

As the molar ratio of F/P is increased, higher melting novolaks are obtained (Table 2.5).

GPC traces of novolak resins (Table 2.5) are shown in Fig. 2.12 where 0 = phenol, 1 = dimer, and 2 = trimer [34]. A shift to higher MW is noted in 0.85 F/P ratio GPC trace; higher MW materials elute early at about 16 min vs little high MW material for 0.65 F/P ratio GPC trace. By comparison, greater amounts of dimer and trimer are observed in 0.65 ratio as compared to 0.85 ratio. GPC traces of 0.75 F/P ratio is intermediate in high MW peak and dimer, trimer peaks.

GPC (ISO 11401, see chapter 7) has been widely used for MWD of novolak resins with MW values being determined using monodisperse polystyrene as standard. In many instances reported MW values for novolaks are approximate and relate to polystyrene values. Solomon and coworkers [35, 36] have attempted to correct this by synthesizing several pure oligomeric phenolic compounds linked by o,o', o,p', and p,p' type linkages (Fig. 2.13, oligomers of 4,

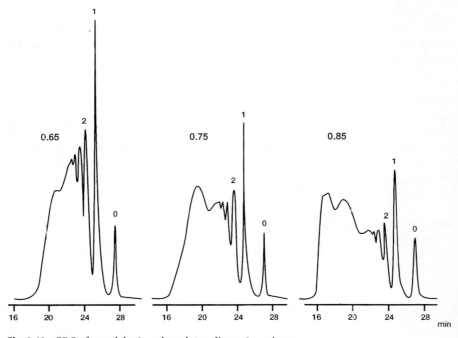

Fig. 2.12. GPC of novolaks 0 = phenol, 1 = dimer, 2 = trimer

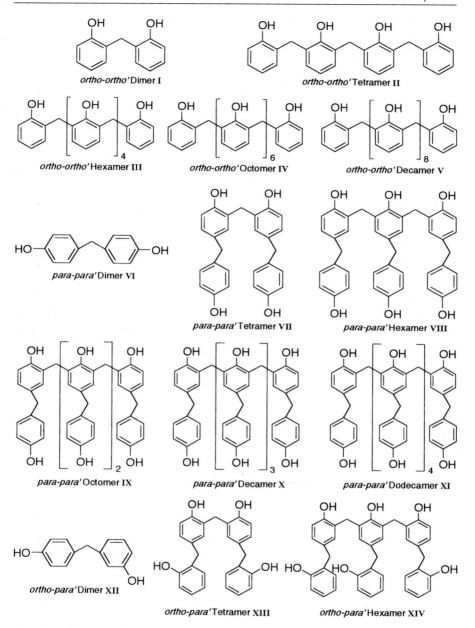

Fig. 2.13. Oligomers of 4, 6, 8, 10, 12 phenol units

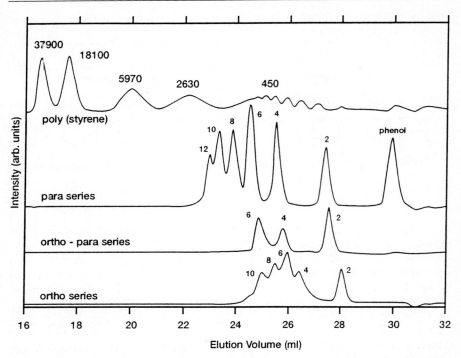

Fig. 2.14. Comparison of polystyrene, and *ortho* series, *ortho para* series and *para, para* series standards for GPC

6, 8, 10, 12 phenol units) and develop calibration curves for novolak resins (Fig. 2.14). Use of three calibration curves (curves of *o,o′*, *o,p′*, and *p,p′*) with GPC analyses of commercial novolaks resulted in better MW values in the low to medium MW range; and more accurate than the polystyrene method.

All three calibration curves must be used due to different novolak isomeric structural units causing variation during elution especially in the high MW region. The authors showed that M_n and M_w values approaching average MW from ^{1}H NMR can be obtained by using the three calibration curves and weighted amounts of isomeric methylene linkages (^{13}C NMR).

Average MW of novolaks can be determined by obtaining the number of phenolic units (n) in the novolak resin from the ratio of methylene to aromatic protons of the ^{1}H NMR spectrum using:

$$\frac{[CH_2]}{[AR]} = \frac{2n-2}{3n+2}$$

The method for determining the value of "n" is fairly accurate up to 7–8 units and typical of novolak resins.

Comparison of commercial random and *ortho* novolaks [35] using the newly developed method (trio of calibration curves combined with weighted fractions

Table 2.6. Molecular weight comparisons of polystyrene standards method, ^{1}H NMR method and newly developed method

Novolak	Molecular Weight Averages						
	New method			Polystyrene			
	M_n	M_w	PD	M_n	M_w	PD	^{1}H NMR
Random	599	1157	1.93	496	1132	2.28	893
Random	477	740	1.66	370	715	1.93	526
ortho	595	1561	2.62	423	1116	2.64	850
ortho	435	1728	2.97	345	1225	3.55	526

PD = polydispersity.

of methylene linkages (^{13}C NMR), polystyrene standards method, and ^{1}H NMR method are summarized in Table 2.6 (see Sect. 7.4.1, ISO 11401, p. 504).

Gelan and coworkers [33] report an improved ^{13}C NMR quantification method for analyzing random and high *ortho* novolaks. Full quantitative spectra of novolaks with high signal-to-noise ratios are obtained by using chromium acetylacetonate during a few hours of spectrometer time. Using proton decoupled ^{13}C NMR, proper peak assignments were established particularly in regions where significant overlap occurs. Both random and high *ortho* novolaks were characterized by determining the degree of polymerization, degree of branching, M_n, isomeric distribution and number of unreacted *ortho* and *para* ring positions per phenolic ring. Table 2.7 contains a summary of Gelan's ^{13}C NMR data.

Gelan's novolaks are relatively low M_n novolaks (M_n from 288 to 400) with higher branching observed in random novolaks relative to the high *ortho* novolaks. Unreacted *para* and *ortho* position per phenolic ring parallel high *ortho* and random isomeric methylene distributions. A higher amount of unreacted *para* positions is highly desirable to facilitate a rapid cure of novolak and directly relates to high *ortho* novolak resins.

Preparation of novolaks is usually conducted in bulk with a molar ratio of < 1 for F to P to avoid gelation. These conditions lead to novolaks with MWs of several thousand daltons. In an attempt to prepare higher MW novolaks,

Table 2.7. ^{13}C NMR quantification data for novolaks

Novolak	n	M_n	α branching[a]	o,o'%	o,p'%	p,p'%	Pu[b]	Ou[c]
High *ortho*	3.88	400	0.15	66.8	30.7	2.47	0.74	0.78
High *ortho*	3.65	375	0.14	58.9	39.4	1.70	0.69	0.86
Random	2.83	288	0.19	22.4	51.3	26.3	0.33	1.38
Random	3.45	353	0.19	24.4	50.7	24.9	0.29	1.29

[a] α obtained by $\alpha = (2-[CH_m])/2$ where CH_m is number of meta carbons originating from free phenol and mono and disubstituted phenol rings. When $\alpha = 0.5$ gelation occurs.
[b] Unreacted *para* positions per phenolic ring.
[c] Unreacted *ortho* positions per phenolic ring.

Table 2.8. ^{13}C NMR analysis of novolak resins

Novolak	Methylene linkage (%)			Phenolic component (%)			Methylene linkage/ phenolic unit
	o-o'	o-p'	p-p'	Terminal	Linear	Branch	–
Conventional	32	43	25	47	44	9	0.81
In acetic acid	33	45	22	26	47	27	1.00
In dioxane	37	45	18	29	45	26	1.00

Yamagishi/Ishida group [37, 38] conducted the acid catalyzed reaction of F and P in polar solvents. (see Structure chapter). Polar solvents are used in the preparation of high MW cresol novolaks (see Photoresist chapter). Using equimolar amounts of F (paraform) to P and HCl, reactions were carried out in solvents such as isopropanol, ethyl propionate, 2-methoxy ethanol, 4-methyl-2-pentanone, dioxane, and acetic acid. Qualitative MW (GPC based on polystyrene) indicate that relatively high MW novolaks are obtained with MW greater than 1×10^6. ^{13}C NMR data (Table 2.8) indicate that solution-based novolaks are similar to conventional novolaks as they relate to methylene linkages but exhibit higher branching.

2.2.2.2.2
High Ortho

High o,o' resins are obtained under weak acidic conditions (pH 4–6) with an excess of P to F and divalent metal salts such as Zn, Mg, Cd, Pb, Cu, Co, and Ni, preferably acetates. The initial reaction is proposed to occur through chelation of phenol and F through metal carboxylate (Eq. 2.12):

$$\text{Phenol} + M(Ac)_2 \longrightarrow \text{chelated intermediate} \xrightarrow{CH_2O} \text{intermediate} \longrightarrow \text{o-methylol phenol} + M(Ac)_2 + H_2O \quad (2.12)$$

$M(Ac)_2$ = Metal Acetate

The chelated intermediate is then transformed into o-methylol phenol. Evidence for o-methylol phenol and its accompanying hemiformal was observed by Tüdos and coworkers [39] via ^1H NMR and GPC. Concentration of both materials increased as pH was increased from 4.1 to 6.5. Oligomerization occurs to high o,o' resin whose amount of methylene bridging for o,o' and o,p' is reasonably similar (47–49%) with less than 5% for *p,p'*. Some methylene ether is also noted.

Solution properties of all *ortho* linear oligomer were examined by the Yamagishi/Ishida group [40]. All the oligomers (n = 2 to 7) were highly solvat-

$$\left(\begin{array}{c}\text{OH}\\ \text{CH}_2\end{array}\right)_n$$

ed and molecularly dispersed in acetone while in $CHCl_3$ strong H bonding was observed. Both inter- and intramolecular hydrogen bonds facilitated bimolecular aggregation of dimer and trimer into tetramer and hexamer like cyclic (calixarene) structures (Fig. 2.15). Tetramer (n = 4) and hexamer (n = 6) oligomers formed pseudo calixarene structures by intramolecular hydrogen bonds. All *ortho* linear oligomers with *p-tert*-butyl substituent and derivatized with ethyl acetate or trioxyethylene ether groups appended to the phenolic hydroxyl were compared in metal cation extraction with 18 crown 6 ether [41]. Metal cation extraction data suggested linear all *ortho* oligomers/ethyl acetate derivatives were similar to ethyl acetate attached to calixarene but lower in percentage extraction. Similarly the complex structure of trioxyethylene ether derivative of the linear all *ortho* oligomers suggested a structure like 18 crown ether (Fig. 2.16). Thus the linear all *ortho* oligomers with appropriate derivatives develop "pseudocyclic" conformations as either "calixarene-like"

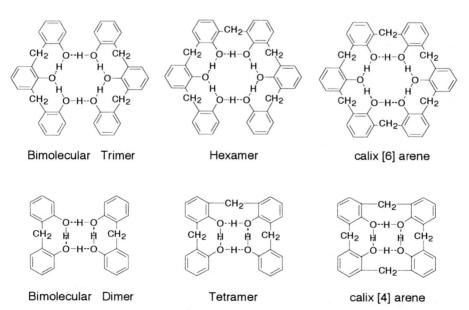

Fig. 2.15. Molecular conformations of calixarene type materials

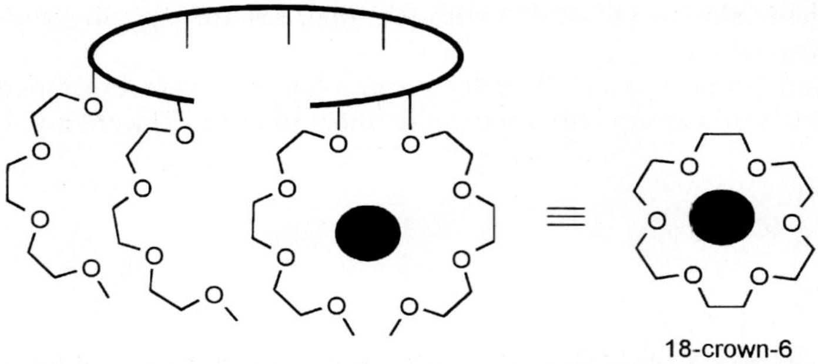

Fig. 2.16. "Crown ether like/calixarene like" structure with alkali metal cation

or "crown ether-like" to exhibit favorable cation metal extraction characteristics (see Structure Chapter 4).

2.2.2.2.3
Modified Novolaks

A variety of modified novolaks can be prepared under acid catalyzed conditions involving various reactants such as bismethylol *para* substituted phenols and similar materials, dialcohols, diethers, and novolaks. The latter (novolaks) can be modified by reaction with olefins, rosin, and others. Besides these basic reactants, other phenols and aldehydes have been used, particularly in the preparation of high T_g novolaks for photoresist area (see Photoresist) as well as epoxy hardeners for electronic potting products. In some cases the modified novolak is obtained without formaldehyde or any aldehyde.

Bismethylol Para-Substituted Phenol

The reaction of bismethylolated phenolics (particularly cresol) provides a facile method for preparing random and block copolymers with various phenolic reactants (Eq. 2 13):

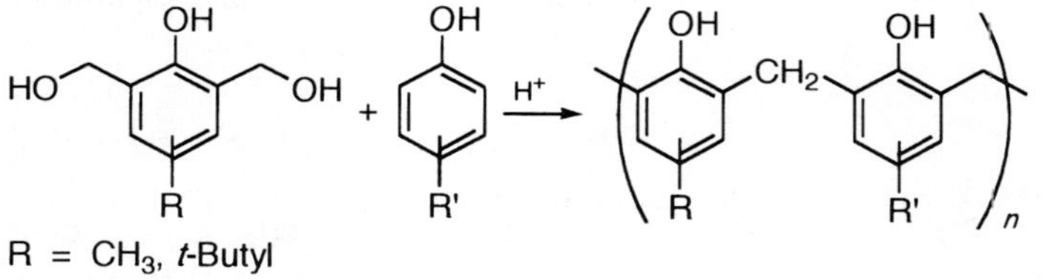

$$R = CH_3, \textit{t-Butyl} \quad (2.13)$$

When R, R' = CH_3 Zampini et al. [42, 43] and Jeffries et al. [44] have prepared random and block copolymers of *p*- and *m*-cresol novolaks. The Yamagishi/Ishida group [40, 41] used *t*-butyl substituted (R = R' = *t*-butyl) materials to obtain all *ortho* linear oligomers.

Naphthol type novolaks have also been reported [45] (Eq. 2.14):

$$\text{(structure)} \quad (2.14)$$

Diethers

Materials similar to Xylok® resin (**1**) continue to be described in the literature [46, 47]. These are "formaldehyde free" novolaks which exhibit increased high temperature resistance compared with conventional novolaks (Eqs. 2.15 and 2.16):

$$\text{(reaction scheme)} \quad (2.15)$$

$$\text{(reaction scheme)} \quad (2.16)$$

Dialcohols [48]

$$\text{(reaction scheme)} \quad (2.17)$$

Diolefins

Another "formaldehyde free" novolak that is an intermediate material and transformed into epoxy or cyanate ester products (see later) is the novolak from phenol and dicyclopentadiene [49] (Eq. 2.18):

$$\text{PhOH} + \text{dicyclopentadiene} \xrightarrow{CF_3SO_3H} \text{novolak} \quad (2.18)$$

Enzymes

The enzymatic oxidative polymerization of phenolic monomers has received considerable attention due to mild reaction conditions and is shown to occur in the absence of formaldehyde. Several groups notably Pokora of Enzymol, Kobayashi of Tohoku University, and Akkara and Kaplan of US Army RD & E Center have been engaged in enzymatic biooxidation of phenolic materials. Early work by Pokora at the Mead Corp. with subsequent acquisition of the enzyme technology by Enzymol identified a relatively inexpensive enzyme system to polymerize phenols. Soybean peroxidase (SBP), readily available from soybeans, in conjunction with hydrogen peroxide easily oxidizes substituted phenols [50] (Eq. 2.19):

$$\text{4-R-phenol} + H_2O_2 \xrightarrow{SBP} \text{poly(4-R-phenol)} \quad (2.19)$$

The reaction can be carried out in a mixed aqueous solvent system (acetone, ethanol, or isopropanol) at 50–60 °C with insoluble product removed by filtration and solvents recycled. More costly horse radish peroxidase (HRP) enzyme with H_2O_2 has been evaluated by Kobayashi et al. [51] and Akkara et al. [52] leading to similar oxidatively polymerized materials.

Use of SBP or HRP requires hydrogen peroxide co-reactant and an aqueous polar solvent environment. Although mild reaction conditions (<60 °C) and formaldehyde free system are beneficial factors, overall biocatalytic system is more expensive than conventional novolak catalysts. In addition these substituted novolaks prepared by enzymatic oxidative polymerization have limited applications. Enzymol claims active programs in coatings, adhesives, antioxidants, and micro-electronics.

2.3
Reactions

Selective reactions can be directed to the hydroxyl functionality of the phenolic group whereby the transformed product has new utility in applications that are non-attainable by the phenolic resin. Moreover in many instances these appendages eliminate water by-product evolution and minimize void content. These transformations are preferably conducted on the stable novolak system. These new functional groups or appendages vary from epoxy, allyl, cyanate groups to the formation of an oxazine ring structure, Friedel-Crafts alkylation and imidization. Similar reactions can be conducted with natural products containing phenolic functionality such as lignin, tannin, and cashew.

2.3.1
Functional Group Appendage to Phenolics

2.3.1.1
Epoxy

The most important transformation of phenolics is the introduction of an epoxy group onto the hydroxyl group leading to a variety of epoxy compounds which can be difunctional or polyfunctional (Eq. 2.20):

$$HO-\langle\bigcirc\rangle-X-\langle\bigcirc\rangle-OH \xrightarrow[NaOH]{H_2C-CHCH_2Cl \atop \diagdown O \diagup}$$

$$X = CH_2, C(CH_3)_2$$

$$H_2C-CHCH_2O-\langle\bigcirc\rangle-X-\langle\bigcirc\rangle-OCH_2CH-CH_2 \qquad (2.20)$$

The simplest novolak, Bisphenol F (X = CH$_2$), is transformed into the corresponding epoxy resin whose value as a low viscosity epoxy resin system was mentioned earlier. Bisphenol A [X=C(CH$_3$)$_2$] epoxy resin is a large volume commodity epoxy resin used in electrical laminates (Sect. 6.1.4.3), coatings, adhesives, composites, etc. [53, 54].

The use of high purity epoxy o-cresol novolak (ECN) combined with phenolic novolak hardener represents general purpose epoxy potting system for encapsulation via transfer molding of solid-state devices such as diodes, transistors, and integrated circuits (Eq. 2.21):

$$\left[\begin{array}{c} OH \\ \\ -CH_2- \end{array}\right]_n + \left[\begin{array}{c} O-CH_2 \\ OCH_2CH \\ CH_3 \\ -CH_2- \end{array}\right]_n \longrightarrow \text{Crosslinked Product} \qquad (2.21)$$

These silica filled compositions are characterized by high modulus and high T_g and low thermal expansion coefficients (TEC). These three characteristics determine to some extent the internal stress in the epoxy molding material and are important considerations in the design of new epoxy/hardener systems. Volume of ECN molding compounds is about 90,000 tonnes per year (1996) and is expected to grow to 120,000 tonnes per year in the year 2000. With large chip size and increasing DRAM density, new characteristics such as reducing thermal stress and improving solder crack resistance have become important due to high packaging density and surface mount configuration. The latter requires infrared lamp and vapor phase solder baths (215–260 °C).

Soldering deterioration of various epoxy molded IC packages after moisture absorption has been evaluated by examining many different epoxy resins (23 types) in conjunction with 5 different novolak hardeners [55]. Cured systems exhibiting low modulus at high temperatures and good adhesion to the die pad showed good durability. Difunctional epoxies based on limonene cresol novolak epoxy, tetramethyl biphenyl epoxy, and naphthalene epoxy showed excellent durability. Similarly studies conducted by the Hitachi Ltd. group [56] identified ways to reduce thermal stress by considering shape and filler (silica) content, combined with lower viscosity resin systems. Introduction of siloxane polymer in the epoxy also reduced thermal stress without sacrificing T_g. Water absorption behavior of the encapsulated materials is equally critical due to the high temperature solder conditions (infrared lamp and vapor solder reflow). The Hitachi group showed that water absorption is closely related to crosslink density of matrix resin (ϱ), and it follows that moisture increased as ϱ increased. Ordinarily a low ϱ would indicate a low T_g. The authors developed resin conditions with low ϱ and high T_g to reduce moisture absorption of the resulting resin and yet maintain high adhesion and low stress. Several candidate epoxy resins and novolak hardeners were identified and are shown in Fig. 2.17.

Thus a new resin system (structure not disclosed) was developed with both low crosslink density and high T_g due to the enhancement of intermolecular interaction of the crosslinked network resin system. DMA data of new compositions (identity not revealed) were improved compared with conventional ECN system. Thus high T_g, low moisture/stress and high adhesion epoxy potting compounds were reported.

The anomalous behavior of these novel epoxy systems exhibiting low crosslink density, high adhesion combined with low moisture and low thermal stress but a high Tg may be attributable to several structural characteristics of the novel epoxy resins and hardeners. Use of monomeric epoxy resin rather than an oligomeric epoxy component results in a low viscosity resin and allows more silica filler in the composition, hence a higher Tg for the higher silica epoxy cured system. Further biphenyl and bis naphthyl are "skewed" molecules exhibiting low viscosity. Either material with an appropriate hardener accept increased silica filler and lead to cured systems with low thermal expansion coefficents (TEC) of less than 10 ppm/K. Another feature of the study was the use of Xylok type hardeners which according to the autors control crosslink density in such a manner that structures with low crosslink density are obtained. Thermal stress was reduced by the introduction of a siloxane phase which separated upon cure of the highly filled epoxy/hardener system. Thus many desireable advantages such as low moisture, low thermal stress, low TEC, high adhesion and high Tg were achieved by the judicious selection of monomeric epoxy resins and Xylok type hardeners. These structural features may be applicable to other epoxy related areas where cured epoxy systems are the required resin system such as electrical laminates, epoxy molding compounds, and the use of epoxies in Advanced Composite components such as load bearing structures, RTM, and others.

Epoxy novolak resins with different siloxane modifications leading to low stress potting materials have been investigated with respect to the mobility of different structural elements as well as morphology [57]. Using various solid state NMR techniques, relaxation experiments allowed probing the mobility in a range of motional frequencies. Authors were able to correlate low stress modification with microphase separation.

Positron annihilation has been shown to be a method to characterize free volume characteristics and water absorption behavior of different epoxy novolak resins cured by novolak [58].

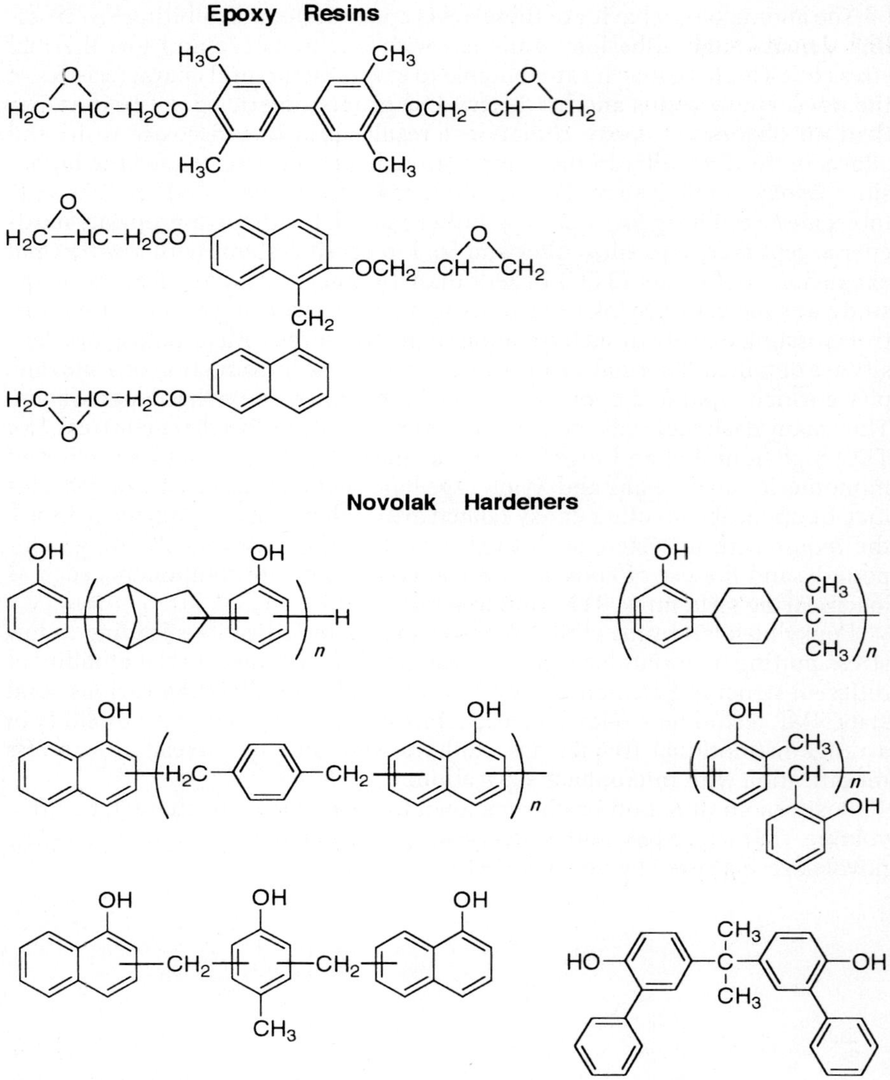

Fig. 2.17. Epoxy resins and novolak hardeners for IC packages

2.3.1.2
Allyl, Benzyl Group

The allyl or benzyl group can be introduced into the phenolic resin by reacting novolak with the appropriate halide and base (Eq. 2.22):

$$\text{Novolak-OH} \xrightarrow[\text{Base}]{CH_2=CHCH_2Cl} \text{Novolak-OCH}_2CH=CH_2 \qquad (2.22)$$

The allyl ether can be thermally rearranged (Fries) into an allyl substituent and/or chromene ring system (Eq. 2.23):

$$\text{Ar-OCH}_2CH=CH_2 \xrightarrow{\Delta} \text{Ar(OH)(CH}_2CH=CH_2) + \text{chromene} \qquad (2.23)$$

The presence of an allyl ring compound allows the introduction of siloxane functionality into phenolic systems (Eq. 2.24):

$$\text{Allyl-phenol} \xrightarrow[\text{Chloroplatinic acid}]{\text{Siloxane}} \text{CH}_2CH_2\text{Siloxane-phenol} \qquad (2.24)$$

Use of vinyl benzyl halide leads to the introduction of a styryl group into the phenolic resin [59] (Eq. 2.25):

$$\text{Novolak-OH} + \text{CH}_2Cl\text{-C}_6H_4\text{-CH=CH}_2 \xrightarrow{\text{Base}} \text{Novolak-OCH}_2\text{-C}_6H_4\text{-CH=CH}_2 \qquad (2.25)$$

The vinyl benzyl ether system can be cured under free radical or Lewis acid conditions. The latter conditions lead to polyether type material (6).

Interpenetrating (IPN) structures (see later) of phenolics/polystyrene are possible.

2.3.1.3
Cyanate Ester

The development of cyanate esters occurred in a manner analogous to epoxy resins. Both difunctional and multifunctional cyanate esters were commercialized based on bisphenols and phenolic novolaks. Monomeric cyanate esters based on phenol or diphenolic compounds were reported by Grigat of Bayer in 1969 [60]. High yields of bisphenol A dicyanate ester are obtained by using cyanogen chloride and base (Eq. 2.26):

$$\text{HO-C}_6\text{H}_4\text{-C(CH}_3)_2\text{-C}_6\text{H}_4\text{-OH} \xrightarrow[\text{Base}]{\text{ClCN}} \text{NCO-C}_6\text{H}_4\text{-C(CH}_3)_2\text{-C}_6\text{H}_4\text{-OCN}$$

$$\xrightarrow{\Delta} \text{BisAO-}\underset{\underset{\text{OBisA}}{|}}{\text{triazine ring}}\text{-OBisA} \quad (2.26)$$

The cyanate ester trimerizes into a cynaurate ring structure at elevated temperatures catalyzed by metal salts yielding a highly crosslinked resin with $T_g > 250\,°C$.

Early application areas that showed promise for the dicyanate ester were electrical laminates and frictional components for aircraft brakes. After a series of setbacks, Bayer in 1978 licensed the cyanate ester technology to Mitsubishi Gas Chemical and Celanese. Mitsubishi has commercialized a family of resins known as BT resins which consist of bis A dicyanate and bismaleimide of 4,4′-methylene dianiline. BT resins are used in high T_g electrical laminates. Eventually the Celanese cyanate ester activities were acquired by Ciba which has expanded the range of cyanate ester products [61, 62].

Dicyanate esters are particularly attractive resin systems for electronic applications due to the unusual combination of properties such as low values for dielectric constant (D_k) and dissipation factor (D_f) with an accompanying high T_g [63]. Only Teflon and polyethylene possess lower D_k and D_f values than cyanate esters but are low modulus materials. Low values of D_k and D_f are highly desirable allowing increased signal speed and circuit density while lowering power requirements and amount of heat generated relative to epoxy or BMI resins. These benefits are extremely important for circuits operating at microwave frequency. Further D_k and D_f allows increased transmission of microwaves through radome walls, antenna housings, and stealth aircraft composites. Dicyanate esters are established thermosetting resins for insulating high speed, high density electronic circuitry, matrix resins for aircraft composites, geostationary broadcast satellites, radomes and antennas, versatile adhesives, as well as passive waveguides or active electro-optic components for processing light signals in fiber optic communications.

High performance thermoplastic resins such as PES, PSF, PPO, PEI, and PI with T_gs ≥ 170 °C are soluble in dicyanate esters but phase separate when concentrations are above 15 % into co-continuous morphologies during cure [64]. The resulting toughened materials exhibit effective impact damage resistance without sacrificing high T_g or elevated temperature modulus.

In 1991 Allied Signal commercialized a multifunctional cyanate ester based on phenolic novolak [65]. The product is known as Primaset PT (phenolic triazine) and undergoes crosslinking at elevated temperatures in a manner analogous to the dicyanate esters (Eq. 2.27):

$$\left(\!\!\begin{array}{c}OH\\ \\ \text{—}\!\!\bigcirc\!\!\text{—CH}_2\!\!\end{array}\!\!\right)_{\!n} \xrightarrow[\text{Base}]{\text{ClCN}} \left(\!\!\begin{array}{c}OCN\\ \\ \text{—}\!\!\bigcirc\!\!\text{—CH}_2\!\!\end{array}\!\!\right)_{\!n} \xrightarrow{\Delta} \left(\!\!\begin{array}{c}\text{triazine}\\ \\ \text{—}\!\!\bigcirc\!\!\text{—CH}_2\!\!\end{array}\!\!\right)_{\!n} \quad (2.27)$$

PT family of resins consist of a viscous liquid, semi-solid and a powdered material. These liquid and solid forms provide broad processing capabilities of PT resin by hot melt (prepreg/adhesives), filament winding, RTM or powder coating/compression molding [66]. PT resin requires lengthy cure and post cure at elevated temperatures for maximum T_g (350 – 400 °C). It exhibits good thermal stability with little change in mechanical properties after aging 1000 h at 288 °C or 200 h at 315 °C. Char and OSU values are quite attractive and responsible for evaluation in aircraft interior applications.

In 1995 Lonza acquired Primaset PT family of resins from Allied Signal. It was an obvious acquisition for Lonza since Lonza conducted the cyanation reaction for Allied Signal under toll arrangement. PT resins are used in aircraft interior duct components, plastic ball grid arrays, encapsulation of electronic components, super abrasives, and friction elements [67].

2.3.2
Ring Formation with Phenolic

2.3.2.1
Benzoxazine

A reaction that lay dormant for many years was reexamined by Ishida and coworkers at Case Western. In a series of papers [68 – 76], Ishida and coworkers have shown that with diphenolic compounds, primary amine and aqueous formaldehyde, difunctional benzoxazine, and other oligomers are obtained [68, 69] (Eq. 2.28):

$$\text{HO-}\bigcirc\text{-X-}\bigcirc\text{-OH} + RNH_2 \xrightarrow{CH_2O} \text{R-N}\bigcirc\text{O-X-}\bigcirc\text{O-N-R} \quad (2.28)$$

X = C(CH$_3$)$_2$, C=O
R = CH$_3$, Phenyl

The use of aniline favors a higher yield of monomeric product than methylamine. Benzoxazine undergoes facile ring opening with active hydrogen containing materials (unreacted starting material, oligomeric phenolic materials, etc.). Using a simplified reaction scheme of a monofunctional benzoxazine, ring opening mechanism occurs by breaking of a C–O bond of the oxazine ring [70] (Eq. 2.29):

$$\quad (2.29)$$

With difunctional benzoxazines highly crosslinked networks result. It has been shown that the ring opening reaction is autocatalytic and can be catalyzed by acids, preferably carboxylic acids and not inorganic acids which are too severe [71, 72]. Polymerization occurs with near zero shrinkage or slight expansion and is quite uncommon for ring opening polymerization [73]. The uniqueness of expansion of benzoxazine materials on polymerization is attributed to molecular packing via hydrogen bond formation [74]. A comparison of thermal properties of benzoxazines with other high performance thermosetting resins is shown in Table 2.9 [75].

Both polybenzoxazine and phenolic properties are favorable based on T_g and char yield as compared to expensive BMI.

A high T_g (350 °C) dibenzoxazine is reported for product from 4,4'-dihydroxybenzophenone and aniline combined with CH$_2$O [76]. Fiber reinforced composites of its polymer with carbon fiber resulted in mechanical properties similar to PMR-15/CF composite. PMR-15 is a high performance thermosetting aerospace material [125].

Table 2.9. Comparison of benzoxazines with thermosetting resins [75]

Resin Family	T_g (°C)	T_d [a] (°C)	Char Yield (%)
Polybenzoxazines	150–260	250–400	45–65
Phenolics	~175	–	40–50[b]
Epoxies	150–261	–	≤40
Bismaleimides	250–300	450–500	50–70

[a] Temperatures when 5% weight loss during TGA in N$_2$, 20 °C/min heat rate.
[b] 55–70% (see Pyrolysis of Phenolics, Sect. 6.4.1, p. 417).

Recently Jang and Seo [77] evaluated the effect of reactive rubbers in dibenzoxazine ($X=C(CH_3)_2$; $R=$phenyl) (Eq. 2.28). Amine terminated (ATBN) and carboxyl terminated butadiene acrylonitrile (CTBN) rubbers were examined. ATBN exhibited better distribution of rubber particles in the matrix phase than CTBN. Higher fracture toughness and smaller rubber particle size was observed in dibenzoxazine modified with ATBN.

Investigators at Hitachi Chemical have developed an in situ method of preparing copolymers of benzoxazine and novolak by conducting the reaction with novolak and a deficient amount of aniline and CH_2O [78] (Eq. 2.30):

$$\text{(structure)} \quad (2.30)$$

These resulting materials have been examined in many different applications such as electrical laminates, friction, gaskets, fiber reinforced composites, and molded products.

Benzoxazine materials possessing high crosslink density were prepared by the Hitachi group [79] using phenylene diamine, bisphenol A, aniline, and CH_2O (Eq. 2.31):

$$\text{(structure)} \quad (2.31)$$

A T_g of 164 °C was reported for the cured product.

2.3.3
Alkylation

2.3.3.1
Friedel-Crafts

Alkylation with various alkylating groups, (dicyclopentadiene [80]), (divinyl benzene/ethyl styrene [81]) are examples of Friedel-Crafts reaction (Eq. 2.32):

$$(\text{Novolak-CH}_2)_n + \text{"Alkylating group"} \xrightarrow{\text{Acid}} (\text{Novolak-CH}_2)_n \text{ Alkylating group appendage} \quad (2.32)$$

2.3.3.2
Hydroxymethylation

Novolaks can be transformed into resoles by reacting novolaks with CH_2O and base catalyst [82] (Eq. 2.33):

$$(\text{Novolak-CH}_2)_n + CH_2O \xrightarrow{\text{Base}} (\text{Novolak(CH}_2\text{OH)-CH}_2)_n \quad (2.33)$$

2.3.4
Miscellaneous

2.3.4.1
Polyimide

A polyimide structure is proposed when novolak is reacted with melamine, pyromellitic anhydride, and hexa [83] (Eq. 2.34):

$$(\text{Novolak-CH}_2)_n + \text{melamine} + \text{pyromellitic anhydride} + \text{HEXA} \quad (2.34)$$

(2.34 cont.)

2.3.5
Natural Products

Natural products containing phenolic groups within the structure (lignin, tannin, and cashew nut shell liquid) are reported to undergo similar reactions as were shown for novolaks. Epoxy group is introduced into lignin [84], hydroxymethylation occurs with lignin [85] and tannin [86]; alkylation or co-reaction of novolak with lignin [87] is reported. Chemistry of cashew was reviewed recently [88].

2.4
Mechanism

Mechanistic considerations related to formation of resoles and novolaks were described in earlier sections when the preparation of resole and novolak resins was discussed. This particular section will consider mechanistic pathways involved in the cure of resole and novolak. Methods involving non-hexa cure, level of cure and post cure conditions will also be discussed.

2.4.1
Cure

Due to their mode of preparation, different curing conditions are necessary to crosslink resole and novolak resins. Resoles are readily cured by acid, base, or thermal conditions. There are some other special resole curing systems such as carboxylic acid esters, anhydrides, amides, and carbonates which are reported to accelerate the cure of resole resins [89] (see Foundry chapter). The conditions under which these selective systems operate are not completely understood.

Novolaks require a source of formaldehyde which is usually hexamethylene tetramine (hexa). Other novolak curing methods besides hexa (see Sect. 6.1.2.3.4) consist of solid resole, bismethylol cresol, bisoxazolines, and bisbenzoxazines (see Sect. 2.4.1.3.).

2.4.1.1
Resole

As the cure of resole commences, MW advancement occurs leading to a gel state (B-staged intermediate) which is no longer soluble in precursor medium (water or alcohol). While the solvent is volatilized during heating, the flexible phenolic intermediate increases in modulus and becomes rigid and crosslinked. Powdered novolak combined with hexa undergoes a similar "melt phenomenon" on heating to a flexible phase and ultimately to a rigid, high modulus, crosslinked system. Complete cure of phenolic resole leads to a high modulus, high crosslink density, moderately high T_g (~150 °C) material exhibiting excellent moisture and heat resistance. These features coupled with attractive economics are the motivating factors for the use of phenolics in many applications. The largest market segment for resole resins is wood adhesives where greater than 60 % of the total volume of phenolics (North America) is used as wood binder. Hence many cure studies related to resoles as wood adhesives have been reported. These resoles are mainly NaOH catalyzed resin system with pH > 9, low viscosity, 40–50 % solids, with low residual phenol and formaldehyde, and molar ratio of F/P of ≥ 2. These resins are primarily used in plywood and OSB panels (see Wood Composite chapter).

2.4.1.1.1
Thermal/Viscoelastic Analyses

Many different analytical techniques have been examined to monitor the cure of resoles. The dynamic curing process can be examined using thermal or viscoelastic methods to follow the curing behavior of resoles. The use of DSC, TMA and/or DMA and DEA as thermal/viscoelastic methods has been effective in establishing cure parameters such as time/temperature conditions for optimum cure.

DSC has been evaluated [90–93] as a method for assessing the total heat that is evolved during thermal cure of resole and relating it to level of cure (see ISO 11 409, Chapter 7). Efforts by Myers et al. [91] and Riedl et al. [92] indicate that the use of DSC for resole cure provides sometimes only qualitative information. Since it is difficult to reproduce resoles with similar level of advancement and water content (especially as wood adhesives), only qualitative comparisons between resole resins can be obtained when considering catalyst content, moisture content, and pre-conditioning prior to DSC analyses.

TMA has been utilized as an analytical probe for thermoset resin cure. Pizzi et al. [89] used TMA as a means of determining crosslink density of different thermosetting wood adhesives such as PF, MF, MUF, RF, and Tannin F by considering the average number of degrees of freedom of polymer segments between crosslinking nodes during hardening of networks. Schmidt [94] used TMA for the determination of T_g of post cured wood adhesive resins.

A modified version of dielectric analysis (DEA) has been reported as a convenient method to monitor the in situ cure of thermosetting resins. It is known as microdielectric spectroscopy and has been used to monitor the cure of

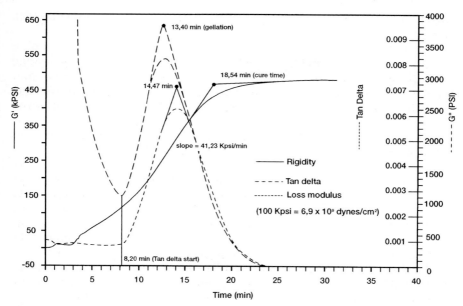

Fig. 2.18. DMA of resole resin

phenolic resins [95] and modified resorcinol formaldehyde resins [96]. It, like DMA, monitors rheological transformations during resin cure by measuring permittivity and dielectric loss.

A dynamic method that has been effective in evaluating the cure characteristics of powdered novolak/hexa resins and is gaining in importance in the evaluation of resole resins is DMA. Several investigators [91, 93, 97–100] have evaluated DMA for cure of resoles as wood adhesives. By monitoring storage modulus, resin rigidity or high modulus occurs when maximum cure is obtained. Under isothermal conditions, the area under the tan δ curve is a measure of residual cure of resin that remains. Kim et al. [98] have shown that one can obtain values for cure time based on storage modulus curve and gelation times from tan δ curve. Isothermal DMA showing storage modulus curve and tan δ curve and how cure time and gelation time are obtained is shown in Fig. 2.18 [98]. These DMA-related values assisted in optimizing the synthesis and cure parameters of resole resins for plywood and OSB wood composites.

Christiansen and coworkers [93, 100] combined both DSC and DMA techniques to identify chemical cure and mechanical cure of wood adhesives. Authors used DSC to follow chemical cure by measuring the exothermic nature or heat of polymerization of resin by DSC while mechanical cure was obtained from stiffness measurements by DMA. These studies were conducted in conjunction with "Steam Injection Pressing" (SIP, see Wood Composites chapter). The continuing study identified chemical cure (DSC) occurring at a slower rate than mechanical cure (DMA) with both cure rates increasing with increasing relative humidity and temperature. However resins with high moisture content retarded cure.

Gardner and coworkers [101] have also used both DSC and DMA to monitor the cure of wood adhesives with fillers or extenders. Fillers such as clay and pecan shell flour and wheat flour extenders in resole were compared with control resole. Little or no effect on the cure properties of the resin was observed. The mechanism of the curing process is essentially unchanged by the additives.

2.4.1.1.2
NMR Analyses

These dynamic processes have aided investigators in developing optimum cure conditions of resoles as wood adhesives. Yet there is an abiding interest to determine the structural characteristics of the crosslinked resin product. Due to complete intractability upon crosslinking, structural identity of the crosslinked resin has relied on solid state NMR [102–110]. Early solid state NMR by Maciel et al. [102] implied the direct involvement of the hydroxyl group of phenol, condensation of methylene bridge with hydroxymethyl group, and crosslinking of methylene bridge with F which is liberated during cure. Many of these proposed components are based on NMR spectra peaks with minimal structural confirmation.

Recently Kim et al. [110] reexamined the structural characteristics of cured resole with respect to resin mole ratio and oxidative side reactions using solid state NMR. They claimed that an F/P ratio of 2.1 is optimum as a commercial wood adhesive resin providing high cure speed with low F emission. A wood adhesive resole with a 2.1 mole ratio of F/P was thoroughly cured and studied by solid state ^{13}C CP/MAS NMR spectroscopy. They reasoned that a fully cured resin would exhibit a methylene to phenol ratio of F/P = 3/2, or 1.5. A value of 1.5 is the theoretical ratio with actual values of 1.35 to 1.46 being observed by NMR. The remaining 0.6 equivalents of F was determined as hydroxymethyl and methylene ether groups, oxymethylene species, and oxidation products of F, with only a very small amount of F being emitted during resin cure. Evidence for carboxaldehydes (A) and quinone structures (B) was presented. Small amounts of methylene ether (~4 mol%) were proposed as hemiformal (C) rather than phenoxy methylene ether as suggested by early studies of Maciel.

Thus the structural characteristics of cured resole resin are consistent with many early (pre-solid state NMR) proposed structures and to a minimal extent to those structures based on phenolic hydroxyl group reaction, methylene bridge reactions with F or hydroxymethyl.

One of the main features of NMR analyses for resoles is resin functionality and the ability of the resole to perform satisfactorily in the end use application by controlled resin advancement. This is particularly important in phenolic prepreg manufacture. Recent NMR studies of resoles identify a trend to examine many high performance phenolic resin intermediates that are subsequently transformed into composites. ^{13}C NMR analyses of Borden resin SC 1008, a popular resin used in many high performance phenolic/carbon fiber composites qualified in many military and aerospace applications, and prepregs based on SC 1008 (Fiberite, Tempe, AZ; products: MX 4926, FM 5939LDC, FM 5055) have provided detailed structural analyses of resin/prepreg and resin advancement as it is transformed into prepreg [111]. These studies provided an understanding of the structure characteristics of the resin and the carbon fiber prepreg system used in rocket nozzles and the NASA Space Shuttle nozzle.

Functional group content within the resole resin can be examined by both solution and solid state ^{13}C NMR. Specific groups or functionalities, such as methylol to methylene, can be spectrally resolved both in solution and solid state. Characteristic chemical shifts for methylol are 62–66 ppm and 35–40 ppm for methylene. Furthermore transformation can be monitored during resin cure. Besides ^{13}C, ^{1}H wide line solid state NMR has been used to measure rate of sample curing "in situ" by monitoring the sample T_2 (spin-spin relaxation time) as function of heating time. These T_2 values are correlated with ^{13}C NMR of methylol disappearance and methylene appearance and provide excellent agreement. Studies (T_2) were conducted at very rapid speed with "on-line" NMR spectrometer for monitoring liquid resole resins with carbon black filler [112]. Several resoles with varying viscosities, from 300 to 60,000 cps, were examined. These studies were conducted by Hercules to monitor the resin system used in aerospace industry to produce rocket motor components with strength and rigidity by filament winding.

Qureshi and MacDonald at Georgia Pacific [113] used ^{13}C solid state NMR to develop optimum post cure conditions of glass phenolic laminates that result in minimum weight loss and excellent hot strength at elevated temperatures (260 °C). By monitoring methylol and methylene functionalities, ^{13}C NMR rapid screening of small glass laminates provided a quick technique to arrive at optimum post cure conditions rather than labor intensive individual laminate prepregging-cure-testing procedures.

Another NMR technique that has been effective in describing resole cure and extent of crosslinking is proton spin-lattice rotating frame relaxation time ($T_{1\rho H}$). Schmidt [94] used $T_{1\rho H}$ relaxation behavior to demonstrate level of crosslinking in wood adhesive resoles. He was able to distinguish crosslink density between resins cured with differing relative humidity. Lower crosslink density of resin cured in the presence of high humidity led to a lower $T_{1\rho H}$ value whereas higher crosslink density exhibited higher $T_{1\rho H}$ value. These relaxation behaviors also provided a correlation with T_g values obtained by TMA.

Curran and coworkers [114] used $T_{1\rho H}$ response to monitor uniformity of cure of carbon-carbon composites based on phenol/phenylphenol formaldehyde polymer used as matrix material in C-C composites for aircraft brakes. $T_{1\rho H}$ response showed wide variation within the same part with relaxation behavior in the center of the part differing from outside surface. The interior of the part exhibited a multiple decay due to phase separation and inhomogeneous distribution of free phenylphenol in the solid matrix. By an unspecified treatment of the part, the relaxation behavior $T_{1\rho H}$ for all locations of the molded part was identical and consistent with homogeneous distribution of phenylphenol throughout the matrix.

Besides structural characterization and functionality, the versatility of NMR lends itself to the use of different relaxation techniques (T_2, T_1, $T_{1\rho}$) to monitor cure, crosslink density, and uniformity of cured resole systems.

2.4.1.2
Novolak

Similar to the difficulties encountered in understanding resole cure, a complete understanding of the mechanism of hexa cured novolak has been lacking due to crosslinking and intractability of the product. In 1987 Maciel and Hatfield [115] identified 15 possible intermediates involved in hexa cured novolak. This initial publication presented solid state ^{13}C and ^{15}N CS/MAS spectra using ^{13}C labelled and ^{15}N labelled hexa to study the curing reaction. More recently Solomon and coworkers [35, 36, 116–123] expanded Maciels's proposed mechanistic scheme by undertaking a major study devoted to the hexa/novolak curing mechanism by synthesizing many phenolic oligomers with known configuration [35, 36, 116, 117], reaction studies of hexa with model compounds [117, 118], fate of hexa/novolak intermediates generated from hexa/model compound studies [120, 121], and finally applying these studies to analyze the novolak/hexa cure reaction by ^{13}C and ^{15}N NMR [119].

The rationale of Solomon's initial strategy was to propose suitable novolak models (monomers, dimers) and conduct specific hexa reactions with each of these models (monomers, dimers) and eventually with the fully characterized novolak. As it will be shown subsequently, these monomer, dimer reactions resulted in a consistent pattern for "initial point of attack" by hexa.

For suitable novolak "monomer" units, it was proposed that the following types of monomer units are present in an uncured novolak:

1. Internal phenolic rings are viewed as *ortho, ortho* (2,6-xylenol) or *ortho, para* (2, 4-xylenol):

2. Terminal phenolic end groups as either *o*-cresol or *p*-cresol:

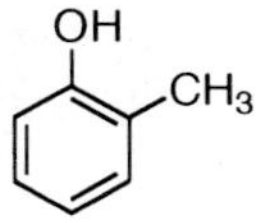

o-Cresol 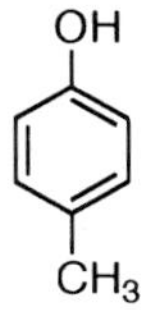   *p*-Cresol

3. Presence of some free phenol:

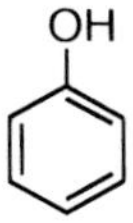

Thus *o*- and *p*-cresols and 2,4- and 2,6-xylenols were considered novolak model compounds. Individual reactions of hexa with the above-mentioned xylenols, cresols, and phenol showed that when an *ortho* site is unoccupied a benzoxazine structure is formed. Benzoxazines formed with 2,4-xylenol, *o*- and *p*-cresol and phenol whereas 2,6-xylenol led to benzyl amines. The monomeric compound studies suggested the "first point of attack" by hexa to novolak with open *ortho* sites favors benzoxazine while open *para* site favors benzyl amine. The study was extended to all three bisphenol F dimers. Similarly benzoxazine structures were observed for all three dimers. When two dissimilar *ortho* positions are present as in *o,p'*-bisphenol F, the preferred attack occurred at the *ortho* "b" position.

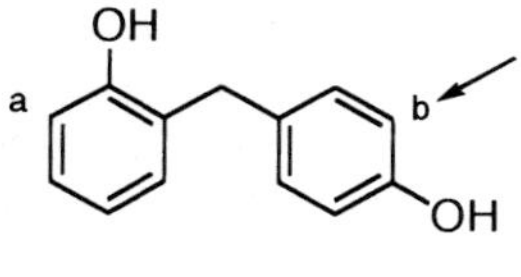

Solomon established the intermediacy of benzoxazine by attack of hexa at vacant *ortho* position. But if *ortho* positions are occupied, benzyl amine is formed by reaction of hexa at the open *para* position.

The curing reaction of novolak resin with hexa was conducted with a conventional novolak with novolak/hexa weight ratios of 80/20, 88/12, and 94/6. The novolak resin with 0.15% free phenol possessed eight phenolic repeat units and comprised 25% *o,o'*; 53% *o,p'*; and 22% *p,p'* methylene bridges. Unsubstituted reactive sites were mainly *ortho* or 88:12 *ortho*:*para*. The curing reaction is proposed to occur in two stages (Fig. 2.19):

1. Formation of initial intermediates such as benzoxazines and benzyl amines, (relative amount of each intermediate is dependent on vacant *ortho* and *para* sites)

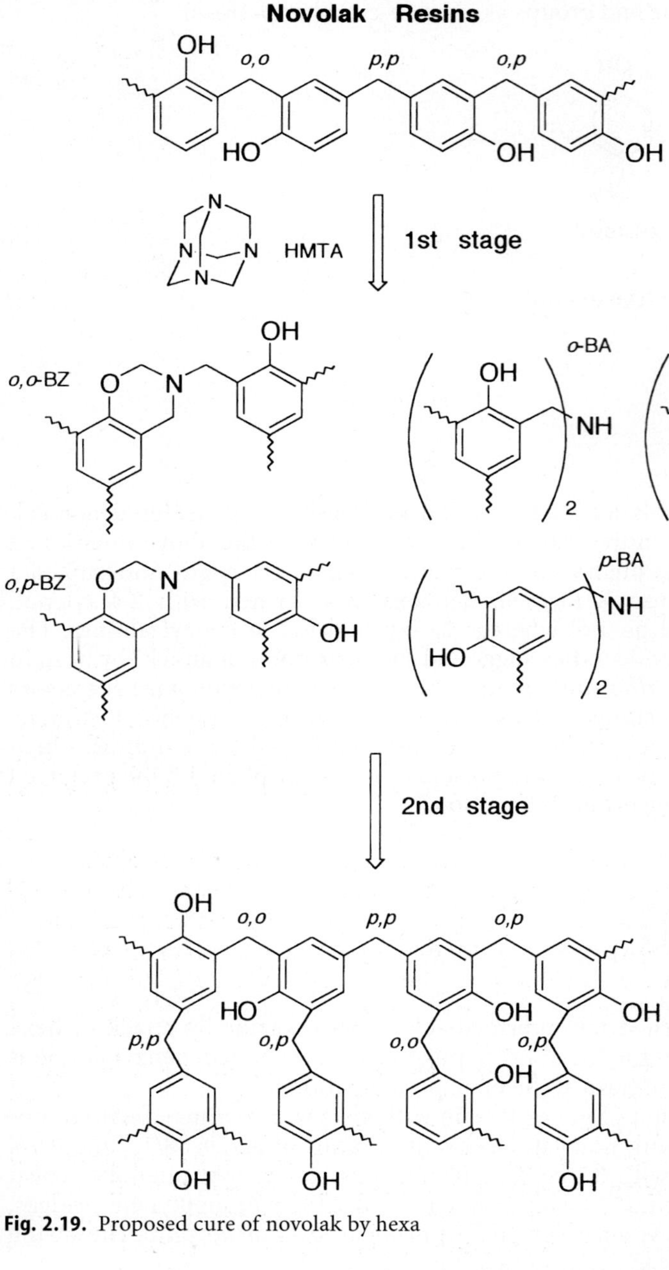

68 2 Phenolic Resins: Chemistry, Reactions, Mechanism

Fig. 2.19. Proposed cure of novolak by hexa

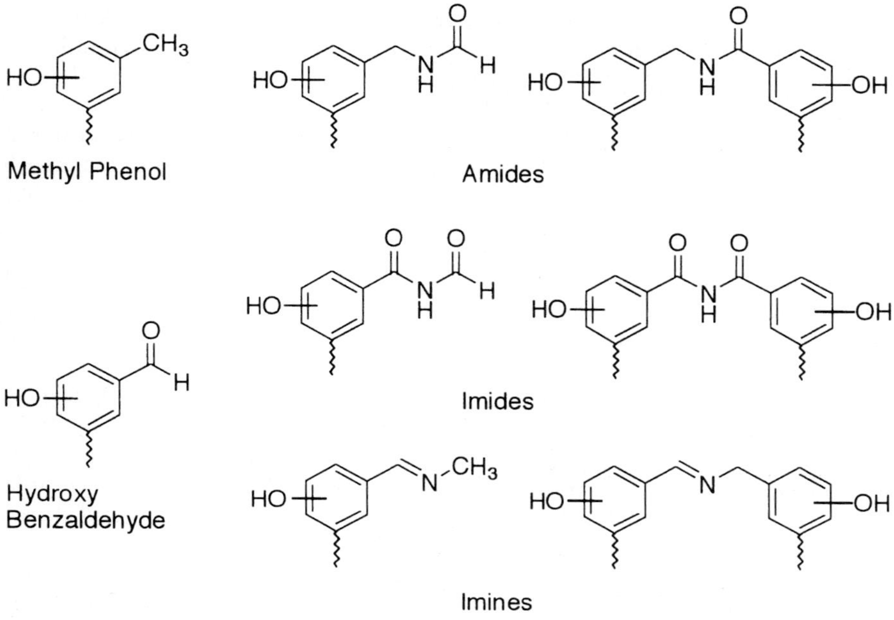

2.4 Mechanism

Methyl Phenol

Amides

Hydroxy Benzaldehyde

Imides

Imines

Fig. 2.19 (continued)

2. Decomposition, oxidation, and/or further reactions of these initial intermediates into methylene bridges between phenolic rings together with various amines, amides/imides, imines, methyl phenol, benzaldehyde, and other trace materials

Besides the generation of benzoxazines and benzyl amines intermediates, triazine, diamine-type structures and ether-type structures are also formed during the initial curing stage. The combined use of ^{13}C and ^{15}N high resolution solution and solid state NMR methods provided a comprehensive and dynamic observation of the entire curing reaction from starting materials, intermediates, even structures emanating from high temperature, post cure conditions.

The novolak to hexa weight ratios of 80/20, 88/12 and 94/6 corresponded to 1:1, 2:1, and 4.4:1 mole ratios of reactive sites in the novolaks to hexa methylenes. Generally each methylene from hexa should join two vacant sites (*ortho* or *para*) in facilitating crosslinking, assuming methylenes are only responsible for forming crosslinks. Three methylenes from hexa combine with two sites to form benzoxazine, while other intermediates such as benzylamines, oxidation/decompositions products such as amides/imides/imines combine a resin site with one hexa methylene. These intermediates (benzoxazine and benzylamine) can react with free resin sites before or after decomposition. Thus when the amount of methylene is low (94/6), hexa provides only one methylene bridge to 4.4 sites. After the initial intermediate is formed, many reactive sites remain and react with intermediates to form more methylene bridges for crosslinking. Disappearance of most initial intermediates and the formation of a large number of methylene bridges occurs at lower temperature in 94/6 composition than in those systems with higher hexa content, e.g., 88/12 or 80/20.

In the 80/20 system, hexa provides one methylene to each site resulting in all sites being completely occupied. Hence there are no free sites to react with intermediates. Thus more nitrogen-containing intermediates remain after curing to 205 °C for 80/20 composition as compared to 88/12 or 94/6.

The pH of the Novolak/hexa composition is also important. Novolaks are moderately acidic while hexa is basic; the use of low hexa content (94/6) would favor a lower pH system and higher reactivity of the system to form initially cured intermediates. A lower pH value is also beneficial to decomposition and further reactions of initial intermediates.

Isomeric structure of methylene bridges is noteworthy. *para, para*-Methylene always increases faster than *o,p* or *o,o* as cure temperature increases. This suggests that *para*-bridged intermediate is less stable and easily decomposes or undergoes further reactions to form other methylene linkages. Favorable stability of *ortho* bridged intermediates are attributable to structural intramolecular hydrogen bond interactions. All *ortho* intermediates form six-member ring structures through intramolecular hydrogen bond and stabilize the intermediates (Fig. 2.20). *para*-Bridged intermediates are not able to form six-membered rings through intramolecular hydrogen bond formation. Unpublished oligomer model studies support *para*-linked curing intermediates as less stable and decompose and/or undergo further reactions at relatively lower temperatures.

A summary of the effects of low and high hexa cured novolaks is given in Table 2.10.

The authors state that the chemical structure of the final cured novolak is controlled by the amount of initial hexa introduced and ratio of *ortho/para* sites of the starting resin. To obtain a crosslinked network with more stable nitrogen-containing structures, a relatively large amount of hexa combined with novolak containing a high amount of *ortho* sites (usually conventional resins are statistically 25:50:25 of $p,p':o,p:o,o$ and o/p active site is >80%) should be used. Alternately with low hexa and resin with a high amount of vacant *para* sites (e.g., high *ortho* novolak), a crosslinked resin network with low nitrogen content should be generated.

Table 2.10. Role of hexa content on cured novolak

Novolak/ hexa Ratio	Reaction Temp.	Hexa Amount	pH	Ratio CH_2 to sites	Reaction Rate	Active Sites Remaining	Fate of Intermediate	Nitrogen Retained Structures	X-link Density
94/6	Low	Low	Lower	4.4:1	Faster	Many	Further reaction	Low	Low
80/20	High	High	Low	1:1	Slower	None	Remains at 205 °C	High	High

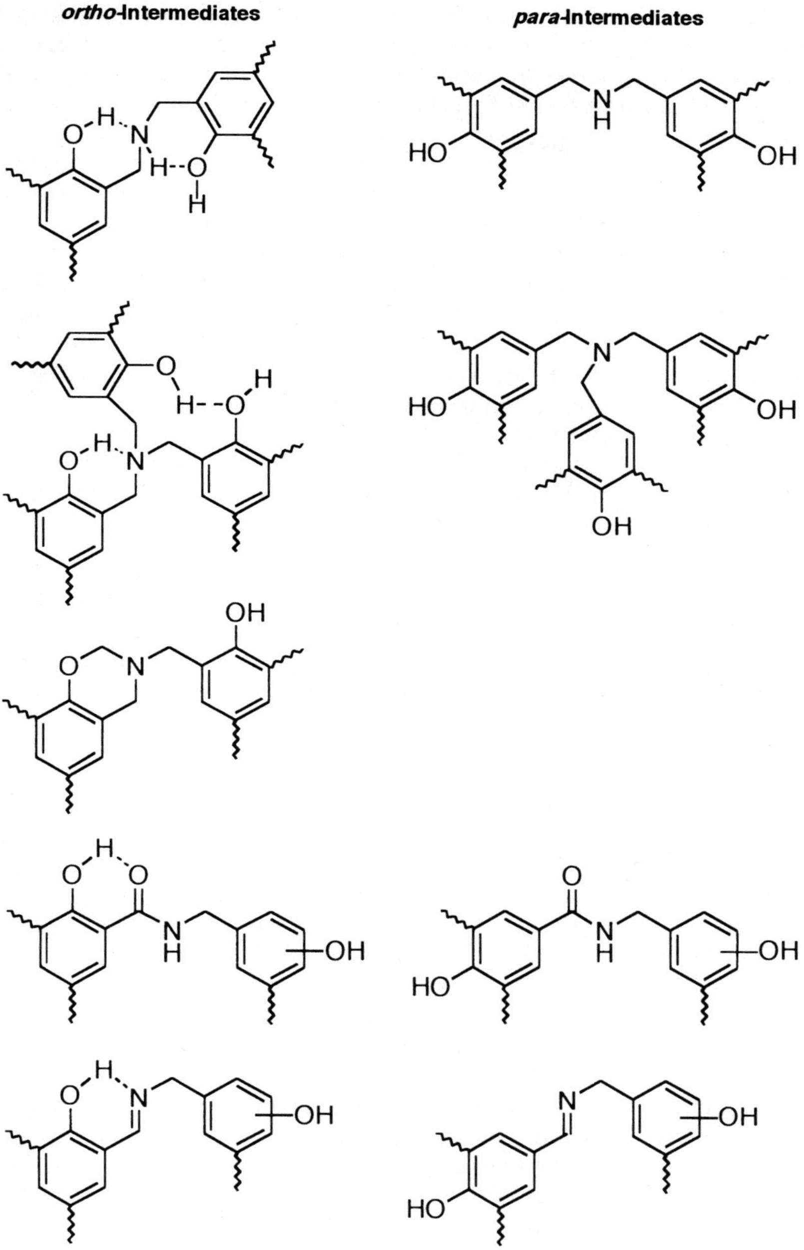

Fig. 2.20. Comparison of *ortho* and *para* intermediates

These mechanistic studies by Solomon may provide further incentive in developing better or improved bond strength in those applications where permanent bonds are required with hexa/novolak such as molding materials, advanced composites, textile felts, and in temporary bonds such as friction and abrasives.

2.4.1.3
Non-Hexa Cure

It was mentioned earlier that non-hexa curing methods consist of the use of hydroxymethyl derivatives of phenol, bisoxazolines, bisbenzoxazines, and solid resole.

Bismethylol cresol was identified as a co-reactant in the preparation of various cresol novolaks as photoresist materials (see Photoresist section). Sergeev and coworkers [124] report the use of a variety of hydroxymethyl derivatives of phenols as curing agents for novolaks. Besides bismethylol cresol (**11**) other bismethylol compounds based on cresol trimer (**12**), cresol novolak (**13**) and Bisphenol A tetramethylol (**14**) were evaluated:

A comparison of curing behavior of hydroxymethyl derivatives and hexa with novolak is shown in Table 2.11.

Using amounts which were comparable in reactive group content as hexa control (Table 2.11), all hydroxymethyl materials were satisfactory curing agents for novolak with low extractibles as hexa control. Impact strengths and ultimate bonding strengths of the hydroxymethyl curing agents were higher than the corresponding values for hexa cured novolak. Bismethylol cresol was superior to hexa and all others in impact and ultimate bond strength of novolak cure product.

Studies initially reported by Culbertson of Ashland [125, 125a] indicated that bisoxazolines (A), known as 1,3 PBOX, co-react with novolak using triphenyl phosphine catalyst (Fig. 2.21).

2.4 Mechanism

Table 2.11. Comparison of hydroxymethyl derivatives of phenol and hexa in cure of novolaks

Curing Agent	% Wgt[a]	Cure Time[b]	% Extractibles[c]	Impact strength (kJ/m^2)	Ultimate bending strength (MPa)
11	36	6.0	2.8	6.2	108.0
12	87	7.5	2.6	2.6	41.5
13	100	6.5	2.9	5.0	64.5
14	38	7.0	2.6	3.6	69.3
Hexa	10	5.5	2.7	1.8	38.9

[a] Comparable amount of reactive groups as hexa.
[b] 180 °C/min.
[c] Acetone extraction/1 h.

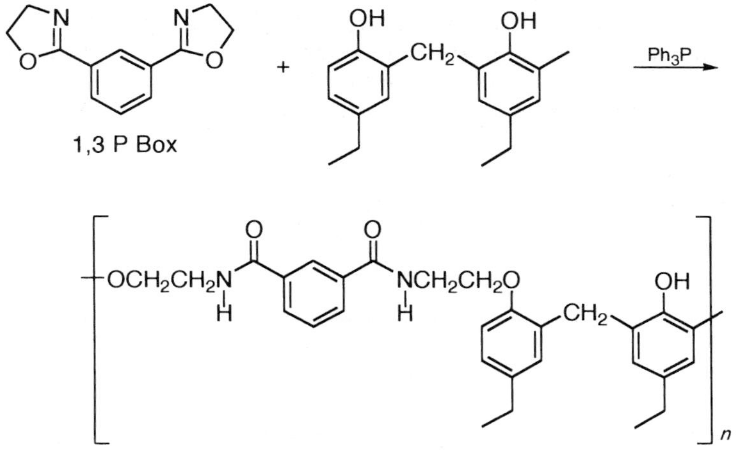

Fig. 2.21. Reaction of 1,3-PBOX with novolak

Bisoxazoline is prepared by reaction of ethanolamine with 1,3-dicyanobenzene. The ring opening reaction of the bisoxazoline is facilitated by the phenolic hydroxyl group with no volatiles being evolved. These fiber reinforced polymers based on 40/60 1,3 PBOX and novolak are claimed to exhibit low smoke and low heat release (OSU value) and suitable for aircraft interiors.

Recently Devinney and Kampa [126] described resin transfer molding and compression molding of 20 cm. thick glass reinforced composites. The resulting polymeric structure, polyester amide resin is named PEAR by Ashland. Recently the technology was sold to Pear Development Co. of Toronto, Canada. PEAR resin has some favorable properties such as high neat resin modulus >4.8 GPa, excellent toughness, excellent adhesion to glass and carbon fibers and metals, excellent long term thermooxidative stability (>10,000 h aging in air at 177 °C), low shrinkage, and favorable F/S/T properties. PEAR compares favorably with polyimides.

Bisbenzoxazines co-react with novolaks (see Reaction section). A study [127] comparing the curing behavior between bisbenzoxazine and hexa with novolak showed that at low cure temperatures the use of bisbenzoxazines in the cure of novolak does not emit ammonia, and a lower weight loss of the cured product results.

Solid resole is also a curing agent for novolaks. Special formulations based on solid resole/novolak systems are used in molding compounds for electrical applications and in textile felts for reduced odor in felt.

2.4.1.4
Level of Cure

Previously is was stated that the area under the tan δ curve (obtained during isothermal DMA) is a measure of residual cure of resin that remains. It implies that both temperature and cure time were insufficient for maximum cure of the phenolic resin. But in most instances cure conditions of resole or novolak/hexa are adequate since the resulting cured phenolic component exhibits satisfactory mechanical properties.

There are, however, some occasions, particularly for molding materials, when higher or maximum cure is desired. Landi [128] has described the importance of developing a sufficiently high T_g and its significance to phenolic molding material technology. T_g relates to mechanical strength, modulus,

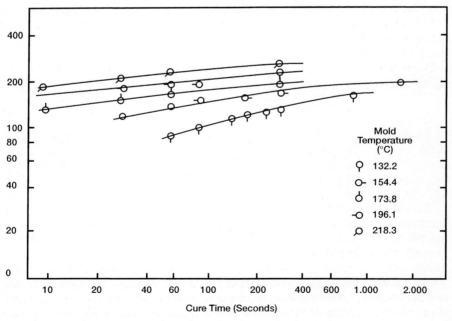

Fig. 2.22. T_g of molding compound (RX611) vs time, temperature

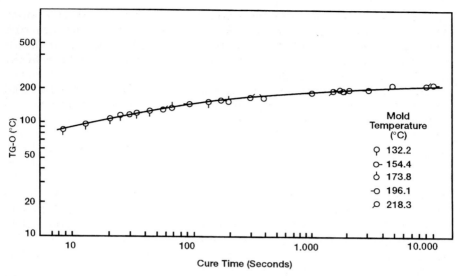

Fig. 2.23. T_g/time/temperature master curve of RX611

creep, and other viscoelastic properties of cured phenolic resin. Volatile by-products such as ammonia (hexa cured novolak) and water (resole) are emitted during cure of molding materials. These volatiles result in internal pressure which expands a hot molded part causing swelling and/or blistering. Resistance of volatile expansion can be achieved by curing the part at a sufficiently high temperature resulting in high modulus/high T_g.

The determination of sufficient degree of cure for molded materials for optimum properties and minimal distortion/blistering was developed by optimizing post cure times through modeling T_g. Five molding temperatures from 132 °C to 218 °C and molding times from 10 to 1800 s were examined (Fig. 2.22) using resin RX 611, Rogers Company, Rogers, CT. All the data were combined on a single "master curve" (Fig. 2.23) using time-temperature superposition. T_g-time curves can be generated at other temperatures. A mathematical expression that relates to the experimental master curve was developed and is successful in predicting T_g. A user friendly form of proprietary Rogers Corp. algorithm has been effective in determining more economical post curing cycles in many commercial applications.

2.4.2
Toughness/Alloys/IPN

Brittle characteristics of phenolic resin have been a constant problem in seeking an easy solution, resulting in mediocre improvements. Several methods have been used and relate to the specific phenolic intermediate being toughened. A method that has been applied mainly to novolak is the direct introduction of a rubber modifier. Either a rubber emulsion or latex is added to crude novolak and then dehydrated (removing water and unreacted phenol) so that the rubber component is uniformly dispersed in the novolak phase. These rubber modified novolaks are used in molding materials and friction linings by curing with hexa. They exhibit marginal toughness. Similarly ABS and MBS (methacrylate butadiene styrene) particles have been dispersed into novolaks [129]. With 20% ABS or MBS, phenolic molding materials exhibit increased charpy impact.

Another method which is primarily applicable to resoles consists of a co-reaction of resole methylol with the double bond/acid end group of nitrile butadiene rubber or resole methylol reacting with polyol/isocyanate (PUR rubbery phase). The flexible rubbery phase is introduced into the resole phase by proposed co-reaction. Studies conducted by Böttcher and Pilato [130] resulted in the introduction of an elastomeric phase into the phenolic continuous phase by modification of the phenolic microstructure. Upon curing, the elastomeric modified phenolic composition led to a resin whose property profile approached that of an epoxy resin. The Bakelite developmental resin known as "PLB" exhibited similar flexural strength, interlaminar shear strength and peel strength as shown by flame retarded epoxy, and co-cured epoxy-phenolic resin systems (see High Peformance and Advanced Composites chapter). The F/S/T characteristics of PLB were comparable to phenolic values.

Friedrich and Folkers of Ameron [131] described an interpenetrating network system (IPN) based on phenolic resin combined with oligomeric silanol or siloxane leading to a toughened phenolic resin system. Filament wound pipe based on this novel silicone modified phenolic resin showed better impact and burst strength than phenolic control. Cure conditions of Ameron siloxane modified phenolic resins and composites were discussed recently [132]. A somewhat similar phenolic siloxane resin was reported by BP [133].

A technique that has been successful in improving the toughness of epoxy resins has been the introduction of thermoplastic, ductile phase into an epoxy resin [125a]. By utilizing selective high performance, high T_g, engineering thermoplastics such as PES, PEI and others, co-reaction or uniformly dispersed systems of thermoplastic/epoxy compositions were obtained. A two-phase co-continuous composition of the thermoplastic, ductile phase dispersed into the continuous epoxy matrix is proposed. An important factor that must be considered in the design of a suitable high performance thermoplastic resin for use in phenolics is its F/S/T characteristics in addition to providing an optimum balance of toughness, solvent resistance, heat resistance, and processability. Polyaryl ethers are high T_g (~200 °C), high modulus/ strength polymeric materials with attractive F/S/T properties. Polyaryl ethers with and without hydroxyl end group were evaluated in hexa cured novolak [134].

In Table 2.12 the effect of co-reacting phenolic terminated polyaryl ether resin B resulted in better toughness values and ductility factor values as compared to polyaryl ether resin A which lacked hydroxyl end group functionality. The authors claimed a semi-IPN resin structure occurred in various stages of cure as viewed by TEMs of fractured specimens. They reasoned that the

Table 2.12. Properties of polyaryl ether added to phenolic resin

% Thermo-plastic Resin	Modulus GPa (E)	Yield Strength MPa (σY)	Fracture Strength Mpa·m$^{1/2}$ (K_{IC})	Fracture Toughness KJ/m^2 (G_{IC})	Ductility Factor µm ($K_{IC}/\sigma Y)^2$
0	6.44	343	0.52	0.07	2.2
15 Resin A	5.60	278	0.97	0.35	12.2
15 Resin B	5.75	276	0.99	0.28	14.3
20 Resin A	5.58	270	0.96	0.31	12.6
20 Resin B	5.58	258	1.23	0.34	22.6

thermoplastic resin becomes immobilized by reaction during early stages of cure and is then locked into a "ribbon-like" or co-continuous morphology. Analyses of fracture toughness of these thermoplastic toughened phenolic resins as short fiber glass reinforced composites is reported not to be straightforward [135]. Comparison of user-industry type tests (notched Izod and Instrumental Falling Weight Impact (IFWI)) vs fracture mechanics type-tests indicate notched Izod test results are erratic when used to evaluate short fiber glass reinforced composites based on thermoplastic resin toughened phenolics. Toughness monitored by energy absorbed by IFWI is more meaningful although sensitivity in the use of total energy absorbed to fracture is observed as low. Authors claim that both initiation and propagation toughness measurements via fracture mechanics analyses are required to measure accurately composite toughness. Slow testing speed measurements are easily obtained but at high testing speeds (impact type speeds) some mechanical dampening of specimen is required to minimize dynamic complications. Measurement of propagation toughness can be achieved but the "R" curves (plots of toughness vs crack length) are too steep, suggesting multiple fracture mechanisms, and complicate toughness interpretation. Thus in some cases phenolic resin toughness is maintained in FRP systems based on "PLB" and silicon modified resole resins while novolak/hexa systems toughened with thermoplastics are anomalous. Non-FRP toughened hexa cured novolaks are indeed tough (Table 2.12) whereas short fiber glass reinforced toughened materials are difficult to measure by a variety of impact or fracture toughness tests. Obviously more testing is necessary to measure these short fiber reinforced composites based on toughened phenolics.

2.4.3
Hazardous Occurrences

Phenol formaldehyde resin preparations are very exothermic and circumstances such as loss of cooling, loss of agitation, and mischarge of catalyst and/or reactants can lead to uncontrollable self-heating conditions. Kumpinsky [136] has examined resole type runaway reactions and separately pH effects on phenolic resin runaway reactions. Using "reactive system screening tool" (RSST) many small scale experiments were conducted to explore many of the variables that contribute to runaway reactions. He found that "potential runaway" reactions are very mild at a pH region between 2 and 7 regardless of molar ratio of P to F. However the extent of the runaway reaction is highly sensitive at a pH below 2, becoming quick and energetic at low pH. At high pH region of runaway reaction, he conducted several experiments at pHs of 7.54, 8.07, and 9.40 and noted that the addition reaction (F added to P is dominant up to 90–95 °C) was followed by condensation reaction above 100 °C. The author recommends that commercial resin compositions be tested for worst case scenario, e.g., at the low pH limit of the operating procedure for acidic catalysis and at the high pH limit for alkaline catalysis with RSST conditions.

2.5
Summary/Trends

The chemistry of phenolic resins is summarized by considering new synthetic techniques for the preparation of monomeric/dimeric components for resoles and novolaks followed by oligomerization leading to resole resins and novolak resins. New mechanistic studies of resoles identify different degrees of reactivity of various methylolated phenols and substitution patterns of the resulting oligomers. Similarly substitution pattern, branching, and residual phenol content is observed by monitoring novolak preparation via ^{13}C NMR. Different types of novolaks such as random, high *ortho*, and several modified novolaks are described along with enzymatic preparation of novolaks without formaldehyde.

Reactions directed to the phenolic hydroxyl of novolak resin by appending epoxy, allyl, benzyl, cyanate, or ring formation to benzoxazine result in many potentially large volume products for many market areas. The newly functionalized materials such as cyanate ester and benzoxazine offer promise in many high temperature demanding applications (electronics, primary structural components for aircraft, aerospace/military) which are not attainable by phenolic resins since no by-products (water, etc.) are formed.

New mechanistic evidence for resole cure and hexa/novolak cure is discussed and consists of recent studies showing mechanistic pathways and structural identity of most components present in fully cured phenolic resins. The level of hexa and the ratio of *ortho* to *para* active unoccupied sites in novolak dictate cured product structure. Non-hexa cure of novolaks and level of cure/post cure conditions provide additional options in novolak cure and improved mechanical properties.

New toughened phenolics using co-reaction conditions of elastomeric or silicon containing materials with resole or co-cure of selected thermoplastics in novolak/hexa composition exhibit improved toughness in the final product.

Recent studies by Kumpinsky provide guidelines for avoiding exothermic runaway reactions under acidic or basic conditions through the use of "reactive system screening tool" in conjunction with small scale experiments.

References

1. Knop A, Pilato LA (1985) Phenolic resins. Springer, Berlin Heidelberg New York
2. Knop A, Böhmer V, Pilato LA (1989) Phenol-formaldehyde polymers, chap 35, vol 5. In: Sir Allen G, Bevington JC (eds) Comprehensive polymer science. Pergamon, Oxford
3. Gardziella A, Haub H-G (1988) Phenolic resins (PF). In: Becker G, Braun D (eds) Handbook of plastics, vol 10, Thermoset (German). Carl Hanser, München Wien, pp 12–40
4. Adolphs P, Haub H-G (1988) Formaldehyde-phenol-amino resins. In: Becker G, Braun D (eds) Handbook of plastics, vol 10: Thermoset (German). Carl Hanser, München Wien, pp 86–89
5. Adolphs P, Giebeler E, Stäglich P, Houben-Weyl (1987) Phenolic resins of Methoden der Organischen Chemie, vol E 20, pt 3, 4 Aufl (German). Georg Thieme, Stuttgart, pp 1794–1810
6. Gardziella A, Müller R (1987) Phenolharze. Kunststoffe 77(10):1049
7. Gardziella A, Adolphs P (1989) Phenolharze. Kunststoffe 79(10):938
8. Gardziella A, Adolphs P (1992) Phenolharze. Kunststoffe 5
9. Gardziella A, Müller R (1992) Phenolharze PF. Kunststoffe 5
10. Shipp DA, Solomon DH (1997) Polymer 38(16):4229
11. Komiyama M (1989) J Chem Soc Perkin Trans 1:2031
12. Komiyama M (1993) Prog Polym Sci 18:871
13. C.A. (1991) 114:142,859; (1991) 115:158,711; (1992) 117:130,922, 212,131
14. Higuchi M, Nohno S, Tohmura S (1998) J Wood Sci 44:198
15. Higuchi M, Nohno S, Morita M, Tohmura S (1998) Forest Products Annual Meeting, June 21–24, 1998, Merida, Yucatan, Mexico, Session 10–3
16. Grenier-Loustalot M-F, Larroque S, Grenier P, Leca JP, Bedel D (1994) Polymer 35(14):3046
17. Grenier-Loustalot M-F, Larroque S, Grande D, Grenier P, Bedel D (1996) Polymer 37(8):1363
18. Grenier-Loustalot M-F, Larroque S, Bedel D (1996) Polymer 37(6):939
19. Grenier-Loustalot M-F, Larroque S, Grenier P, Bedel D (1996) Polymer 37(16):955
20. Astarlon-Aierbe G, Echevarria JM, Egiburu JL, Ormatexea M, Mondragon I (1998) Polymer 39(14):3147
21. C.A. (1992) 117:131,714; (1994) 121:255,395
22. U.S. Pat 5,395,915 (March 7, 1995)
23. C.A. (1997) 126:31,157f
24. Epoxy resin brochure (1996) Bakelite AG, Duisburg, Germany
25. C.A. (1996) 124:57,869
26. Chem Week (1991) p 25 (August 14)
27. C.A. (1997) 127:110,083c
28. U.S. Pat 5,679,305 (October 21, 1997)
29. C.A. (1997) 126:186886
30. Casiraghi G, Cornia M, Ricci G, Balduzzi G, Casnati G, Andreeti GD (1981) Makromol Chem 182:2151, 2973; (1983) 184:1363
31. Katritzky AR, Zhang Z, Lang H, Lan X (1994) J Org Chem 59:7209
32. Pethrick RA, Thomson B (1986) Brit Polym J 18(6):380
33. Ottenbourgs BT, Adriaensens PJ, Reekmans BJ, Carleer RA, Vandergande DJ, Gelan JM (1995) Ind Eng Chem Res 34:1364
34. Podzimek S, Hrock L (1993) J Appl Poly Sci 47:2005
35. Dargaville TR, Guerzoni FN, Looney MG, Shipp DA, Solomon DH, Zhang X (1997) J Poly Sci Part A Poly Chem 35:1399
36. de Bruyn PJ, Foo LM, Lim ASC, Looney MG, Solomon DH (1997) Tetrahedron 53(40):13,915
37. Yamagishi T-A, Nomoto M, Ito S, Ishida S-I, Nakamoto Y (1994) Polymer Bulletin 32:501
38. Yamagishi T-A, Nomoto M, Yamashita S, Yamazaki T, Nakamoto Y, Ishida S-I (1998) Macromol Chem Phy 199:423

39. Laszlo-Hedwig Z, Szesgtay M, Tüdos F (1996) Angew Makro Chemie 241:57
40. Yamagishi T-A, Enoki M, Inui M, Furukawa H, Ishida S-I (1993) J Poly Sci Pt A, Poly Chem 31:675
41. Yamagishi T-A, Tani K, Shirano K, Ishida S-I, Nakamoto Y (1996) J Poly Sci Pt A, Poly Chem 34:687
42. Zampini A, Fisher RL, Wickman JB (1999) Proc SPIE 1086:85
43. Zampini A, Turci P, Cernigliaro GJ, Sandford HF, Swanson GJ, Meister CC, Sinta R (1990) Proc SPIE 1262:501
44. Jeffries A, Honda K, Blakeney AJ, Tadros S (1994) U.S. Pat 5,302,688
45. C.A. (1993) 118:170262i
46. C.A. (1994) 121:84,806v, 15,8410k
47. C.A. (1993) 119:251,323c
48. C.A. (1992) 117:172,736v
49. C.A. (1994) 120:32201d
50. U.S. Patent 4,647,952 (1987); 4,900,761 (1990); 5,153,298 (1992); 5,110,740 (1992); 5,178,762 (1993)
51. Uyama H, Kurioka H, Sugihara J, Komatsu I, Kobayashi S (1997) J Poly Sci Pt A Poly Chem 35:1453
52. Ayyagari M, Akkara JA, Kaplan DL (1996) Polym Mat Sci Eng 72:4 ACS Spring Meeting, New Orleans, LA
53. Bisphenol A (1991) In: Phenol derivatives: Ullman's encyclopedia of industrial chemistry, vol A19. VCH, Weinheim, p 348
54. Gannon J (1994) Epoxy resins. In: Kroschwitz JI (ed) Kirk-Othmer encyclopedia of chemical technology, vol 9, 4th edn. Wiley, New York, p 730
55. Asai S-I, Saruta A, Tobita M (1996) J Appl Poly Sci 51:1946
56. Nagai A, Egucvhi S, Ishii T, Numata S, Ogata M, Nishi K (1994) Polym Mat Sci Eng 70:55, ACS Spring Meeting, San Diego, CA
57. Domke W-D, Halmheu F, Schneider S (1994) J Appl Poly Sci 54:83
58. Suzuki T, Oki Y, Numajiri M, Miura T, Kondo K, Shiomi Y, Ito Y (1996) Polymer 37(14): 3025
59. C.A. 122:107693t (1995); 1076f94u; 215,690y
60. U.S. Pat 3,448,079 (1969)
61. C & EN September 12, 1994, p 30
62. Fang T, Shimp D (1995) Prog Polym Sci 20:61
63. Shimp D (1994) Polym Mat Sci Eng 70:561, ACS Spring Meeting
64. Shimp D (1994) Polym Mat Sci Eng 70:623, ACS Spring Meeting
65. U.S. Pat 4,831,086 (1989)
66. Allied Signal, Morristown, New Jersey (1994) PT Literature
67. Das S, Al group Lonza (1997) Communication, December 5, 1997
68. Ning X, Ishida H (1994) J Poly Sci Pt A Poly Chem 32:1121
69. Ning X, Ishida H (1994) J Poly Sci Pt B Poly Phy 32:921
70. Dunkers JP, Zarate A, Ishida H (1996) J Phy Chem 100:13,514
71. Ishida H, Rodriguez Y (1995) J Appl Poly Sci 58:175
72. Ishida H, Rodriguez Y (1995) Polymer 36(16):3151
73. Ishida H, Allen DJ (1996) J Poly J Sci Pt B Poly Phy 34:1019
74. Shen SB, Ishida H (1996) Poly Comp 17(6):710
75. Shen SB, Ishida H (1996) J Appl Poly Sci 61:1595
76. Ishida H, Low HY (1997) Macromol 30:1099
77. Jang J, Seo D (1998) J Appl Poly Sci 67:1
78. C.A. 126:212,579r (1997); 22,6314n (1997); 127:206,384c
79. C.A. 126:277,927x (1997)
80. C.A. 120:9720p (1994); 32,140h
81. C.A. 119:204,757n (1993); 122:57,824p (1995)
82. C.A. 127:33,973p (1997); 33,974g; 33975r; 33976s; 33977t; 33979v; 33980p; 33981q; 33982r; 33984t; 33985u

83. C.A. 119:251,388 (1993)
 84. Wang J, St. John Manley R, Feldman D (1992) Prog Polym Sci 17:611
 85. Peng W, Riedl B, Barry AO (1993) J Appl Poly Sci 48:1757
 86. Pizzi A (1994) Advanced wood adhesives technology, Chap 5. Dekker, New York
 87. U.S. Pat 5,300,593 (1994); 5,373070 (1994); 5,382,608 (1995)
 88. Gedam PH, Sampathkumaran PS (1986) Prog Org Coatings 14:115
 89. Pizzi A, Garcia R, Wang S (1997) J Appl Poly Sci 66:255
 90. Kay R, Westwood A (1975) Eur Polym J 11:25
 91. Myers GE, Christiansen AW, Geimer RL, Follensbee RA, Koutsky JA (1991) J Appl Poly Sci 43:237
 92. Wang X-M, Riedl B, Christiansen AW, Geimer RL (1994) Polymer 35:5685
 93. Geimer RL, Christiansen AW (1996) Forest Prod J 46(11/12):67
 94. Schmidt RG (1997) Forest Products Society 1997 Annual Meeting, Vancouver, June 22–26, British Columbia
 95. Rials TG (1992) ACS Symposium Series 489. In: Glaser WG, Hatakeyama (eds) Viscoelasticity of biomaterials. Amer Chem Soc D.C., pp 282–294
 96. Rials TG (1991) Wood adhesives 1990. Forest Prod Res Soc, pp 91–96
 97. So S, Rudin A (1990) J Appl Poly Sci 40:2135
 98. Kim MG, Nieh WL-S, Meacham RM (1991) Ind Eng Chem Res 30:798
 99. Peng W (1994) Poly Preprints, ACS Fall Meeting, p 490
 100. Lorenz LF, Christiansen AW (1995) Ind Eng Chem Res 34:4520
 101. Waage SK, Gardner DJ, Elder TJ (1991) J Appl Poly Sci 42:273
 102. Maciel GE, Chuang I-S, Gollub L (1984) Macromol 17:1081
 103. So S, Rudin A (1985) J Appl Poly Sci Lett 23:403
 104. Kelusky EC, Fyfe CA, McKinnon MS (1986) Macromol 19:329
 105. Sinha BR, Blum FD (1989) J Appl Poly Sci 38:163
 106. Kim MG, Amos LW, Barnes EE (1990) Ind Eng Chem Res 29:2032
 107. So S, Rudin A (1990) J Appl Poly Sci 41:205
 108. Chuand I-S, Maciel GE (1991) Macromol 24:1025
 109. Grenier-Loustalot M-F, Larroque S, Grenier P (1996) Polymer 37(4):639
 110. Kim MG, Wu Y, Amos LW (1997) J Appl Poly Sci Pt A Poly Chem 35:3275
 111. Fisher TH, Chao P, Upton CG, Day AJ (1995) Magn Reson Chem 33:717
 112. Neiss TG, Vanderheiden EJ (1994) Macromol Symp 86:117
 113. Qureshi S, MacDonald R (1994) International SAMPE Tech Conf 26:599
 114. Curran SA, Walker TB, Brambilla R (1997) Poly Preprints 38(1):856 ACS, San Fransico, CA
 115. Hatfield GR, Maciel GE (1987) Macromol 20:608
 116. de Bruyn PJ, Lim ASC, Looney MG, Solomon DH (1994) Tetra Lett 35(26):4627
 117. Looney MG, Solomon DH (1995) Aust J Chem 48:323
 118. Dargaville TR, de Bruyn PJ, Lim ASC, Looney MG, Potter AC, Solomon DH, Zhang X (1997) J Appl Poly Sci Poly Chem 35:1389
 119. Zhang X, Looney MG, Solomon DH, Whittaker AK (1997) Polymer 38(23):5835
 120. Zhang X, Potter AC, Solomon DH (1998) Polymer 39(2):399
 121. Zhang X, Solomon DH (1998) Polymer 39(2):405
 122. Zhang X, Potter AC, Solomon DH (1998) Polymer 39(10):1957
 123. Zhang X, Potter AC, Solomon DH (1998) Polymer 39(10):1967
 124. Sergeev VA et al. (1995) Poly Sci Ser B 37(5/6):273
 125. Cuthbertson BM, Tilsa O, Devinney ML, Tufts TA (1989) SAMPE 34:2483
 125a. Pilato LA, Michno MJ (1994) Advanced composite materials. Springer, Berlin Heidelberg New York
 126. Devinney ML, Kampa JJ (1997) SAMPE 42:24
 127. C.A. 113:133,439y (1990)
 128. Landi VR (1993) RETEC, March 8–10, 1993, Research Triangle, NC
 129. U.S. Pat 5,216,077 (1993)
 130. Böttcher A, Pilato LA (1997) Int SAMPE Tech Conf 28:1353

131. Folkers JL, Friedrich RS (1997) Proc International Composites Expo '97, January 27–9, Nashville, TN, Session 22A
132. Tregub A, Inglehart L, Pham C, Friedrich R (1997) International SAMPE Tech Conf 28:787
133. C.A. 127:66,360q (1997)
134. McGrail PT, Street AC (1992) Makromol Chem Macromol Symp 64:75
135. Lowe AC, Moore DR, Rutter PM (1995) Impact and dynamic fracture of polymers and composites. ESIS, vol 19, Mech Eng Publications, London, pp 383–400
136. Kumpinsky (1994) Ind Eng Chem Res 33:285; Ind Eng Chem Res 34:3096

CHAPTER 3
Production of Phenolic Resins

Commercial production of phenolic resins is presently carried out in batches of about 1–50 tonnes; the reactors used for this purpose can have capacities of about 1–60 cubic meters. From Baekeland's tiny cooker unit from 1910 (Fig. 3.1) to large-scale, computer controlled production facilities of today (Figs. 3.2 and 3.3), a long arduous route of development has been achieved through extensive diligence and technical competence. Since appropriate, safe handling and control of the heat evolved by the exothermic reaction represent the main problems encountered in phenolic resin synthesis when the batch

Fig. 3.1. Bakeland's original reactor, 1910

Fig. 3.2. Phenolic resin production reactor, 1998 (Photo: Bakelite AG, Iserlohn)

Fig. 3.3. Computer controlled production of phenolic resin, 1998 (Photo: Bakelite AG, Iserlohn)

size is increased, the utilization and removal of this heat must be considered as the key issue of all phenolic resin technology. The water contained in the 30–55% formaldehyde solutions used mainly as reactants together with the phenolic raw materials represents a heat sink for the liberated heat of condensation that is utilized together with other energy to sustain the reaction.

As shown in the flow diagram in Fig. 3.4, the core of a resin production line is thus a sealed reactor that is capable of operating under vacuum, includes an agitator and heating jacket, and is further equipped with a reflux condenser, chiller, and receiver. The materials currently used for these vessels are, ideally, various grades of alloyed steel that prevent discoloration of the resins that can occur by even traces of iron. The reaction may be carried out in a single step or as a multistage process; the water may be decanted if appropriate, but is generally removed by distillation (for example during production of novolaks), or may in part remain in the resin (aqueous resoles). At present, reduction of the free phenol level by special distillation procedure (particularly in the case of novolaks) and reduction of the free formaldehyde level in aqueous resoles also represent important process considerations.

The size of reactor to be used also depends on the reactivity of the reactants and that of the finished resin at elevated temperatures. Thus, solid resoles are produced in small batches that can be quickly discharged into special chilling equipment (such as slab chillers). Liquid aqueous resoles and phenolic resin solutions are generally produced in medium-size batches; high volume utilization of the resin reactor is possible, the amount depending on the resin content of the product.

Decanting and washing processes may be necessary to achieve particularly high final product purity levels in production of special-purpose products such as coatings and adhesives resins, for example alkylphenolics and etherified resins.

Today, standard phenolic novolaks and aqueous resoles (for example wood adhesives or fiberbonding resins) can be produced in large batches, although this requires detailed control of all individual operations involved in the overall synthesis by means of computer programs. Molten phenolic novolak is discharged onto chilled conveyer belts, possibly following interim storage in heated containers. The resins are then crushed, and if appropriate for further processing, mixed with curing agents and miscellaneous additives before being milled in powder resin grinding equipment to yield highly dispersed powder resins. Based on the above description it follows that a modern large-scale plant for production of phenolic resins can be divided into the following segments: novolak plant, plant for aqueous resoles, resin solution plant, plant for solid resoles and modified resins, and possibly a powder resin plant.

Novolaks (molar ratio of phenol to formaldehyde about 1:0.7–1:0.9) were previously manufactured using the one-stage process (small batch sizes), i.e., the reaction components and catalyst were charged to the reactor, and the reaction initiated. As the reactor volume increased, it became necessary to control the temperature by means of gradual addition of formaldehyde.

Dilute hydrochloric acid was frequently used as a catalyst for production of novolaks, but this led to an increased tendency to corrosion of the reactor

86   3 Production of Phenolic Resins

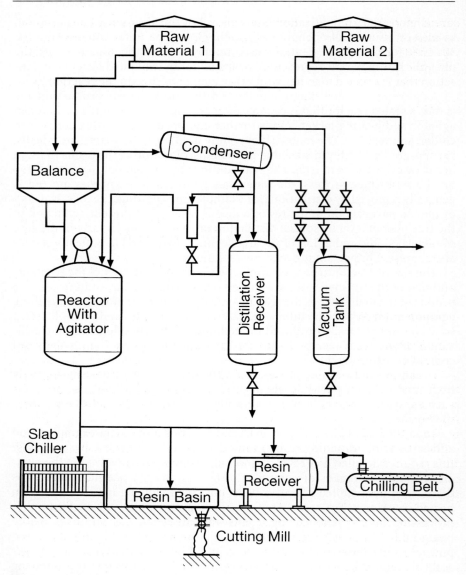

Fig. 3.4. Flow diagram of a resin production line

components. Sulfuric acid is less corrosive, but requires careful neutralization. A similar statement applies to phosphoric acid. Ortho-condensed novolaks are produced using acidic salts of divalent metals, for example zinc acetate. Various organic acids are also used as catalysts.

A modern novolak synthesis includes the following general process stages:

- Raw material charging and heating phase
- Formaldehyde addition
- Post-condensation
- Normal pressure distillation
- Vacuum distillation
- Discharge of the reactor

In-line analyses such as the index of refraction and melt viscosity of the resin can indicate when the individual stages of the reaction have been attained. If a resin is to be brought into solution, this is best achieved following the distillation stage, possibly in a separate mixing tank.

The following batch represents an example of a novolak synthesis.

3.1
Production of Novolak

Phenol (1410 g, 15 mol) is heated to 100 °C in a three-neck flask equipped with a stirrer, dropping funnel, and reflux condenser. Oxalic acid (14 g) is charged to the phenol melt. A 50% solution of formaldehyde (720 g, 12 mol formaldehyde) is added dropwise with stirring at 100 °C. After the dropwise addition is complete, the reaction mixture is refluxed until the level of free formaldehyde is below 0.1%. The post-condensation period is about 1.5 h. The water is then removed by distillation and the reaction mixture finally vacuum distilled at temperatures up to about 180 °C until it reaches a melt viscosity of about 1000 mPa s at 175 °C. The resin melt is poured out on a metal sheet to cool. The resin exhibits a melting range of 90–100 °C (ISO 3146, A) and a "B" (gel) time of 1.5–2.5 min at 150 °C with 10% hexa (ISO 8987).

Production of phenolic novolaks using a continuous process [1] is only of interest when the grades of resin produced in a plant are rarely changed. Figure 3.5 shows a simplified flow diagram of a continuous novolak synthesis (Euteco process). In this method, an aqueous resin is produced in several columns connected in series, the mixture separated in a storage tank, and the resin phase dewatered and freed of phenol by distillation in one of the two alternately used parallel reactors. The novolaks produced in this manner are very uniform in quality.

Powdered resins (mostly with hepa) produced in powder resin plants are required for various phenolic novolak applications. As shown in the flow diagram (Fig. 3.6), such a milling plant is made up of metering equipment, a precrusher, premixer, mill, and a finishing mixer. Special precautions involving antistatic equipment, flue gas treatment, and other measures must be carefully observed for reasons of workplace safety and environmental protection.

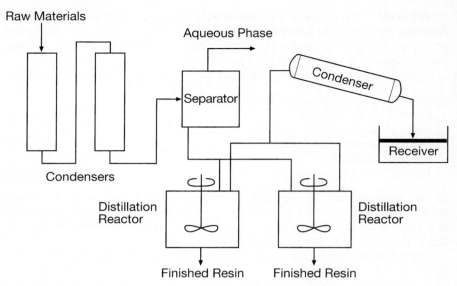

Fig. 3.5. Simplified diagram of a continuous novolak production line

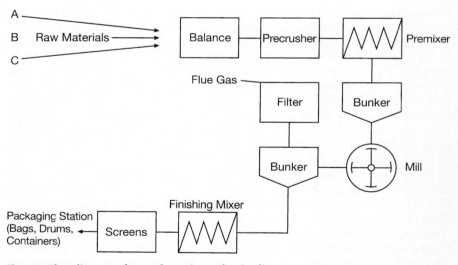

Fig. 3.6. Flow diagram of a powder resin production line

Resoles are classified into three types with their own individual resin formulations and production processes:
1. Solid resoles discharged in the form of molten resin
2. Resoles that are brought into solution in the final stage of the synthesis
3. Aqueous resoles obtained without further processing.

Critical temperature control during the synthesis and a very exact determination of the end points of the individual stages of the process-condensation, post condensation, distillation, etc. – are required in all cases. In the case of aqueous resoles, the temperature at which the product is placed in a container is also important, since such resins undergo temperature-dependent postcuring and generally exhibit only limited storage life. Depending on the type of resin, the molar ratio of phenol to formaldehyde varies greatly, and can range from 1:1.1 to 1:3. Catalysts include alkali metal hydroxides, sodium carbonate, alkaline earth hydroxides, ammonia, various amines, and for production of ortho-condensed resoles [2, 3] specific heavy metal salts such as zinc acetate. Alkali metal hydroxides are generally left in the resin, whereas use of calcium or barium hydroxide offers the possibility of precipitating and separating the metals as their sulfates. The following resole batch provides an example for production of a liquid phenolic resole.

3.2
Production of Resole

Phenol (940 g, 10 mol), 37% formalin (1216 g, 15 mol) and 50% sodium hydroxide solution (18.8 g) are charged to a 4-l three-neck flask equipped with a stirrer, reflux condenser, and contact thermometer. The refractive index of the solution is about 1.4410 at 25 °C. The mixture is heated to 60 °C and the temperature allowed to rise to 100 °C over a period of about 30 min; if necessary, the temperature is adjusted by cooling with water. The reaction mixture is then allowed to react for a further 15-min period (refractive index 1.4800 at 25 °C). Following this reaction period, the resin is vacuum distilled until the viscosity is about 2000 mPa·s at 20 °C. The approximate gel time is 65 min at 100 °C and 9 min at 130 °C (ISO 9396, A); the nonvolatiles level (ISO 8618) is approximately 78% by weight.

3.3
Continuous Processes

Continuous processes for production of aqueous phenolic resoles are also described in the patent literature [4–10] and are suitable for manufacture of novolaks (Figure 3.5). This continuous production is generally initiated in multiple reactors of various design connected in series. Continuous resole synthesis is economically and technically appropriate in situations where large quantities of products exhibiting highly uniform quality are required, for example in the case of binders for chipboard or mineral wool.

References

1. Societata Italiana Resine (1972) US–PS 3,687,896
2. Ashland Oil (1970) DE–PS 2,011,365
3. Ashland Oil (1973) DE–PS 2,339,200
4. St. Regis Paper (1951) US–PS 2,688,606
5. Allied Chemical & Dye (1953) US–PS 2,658,054
6. Nischne-Tagilskij sawod Plastmass (USSR) (1964) DE–AS 1,595,035
7. BASF (1967) DE–AS 1,720,306
8. Butler Manufacturing (1972) US–PS 3,657,188
9. Texaco (1977) DE–OS 2,538,100
10. VEB Sprela-Werke (1977) DD–PS 134,354

CHAPTER 4
Structure (Methods of Analysis)

4.1
Introduction

Structure property relationships provide a description of the many resin characteristics which guide the resin designer in the selection of phenolic resins for the intended application. Resin characteristics such as hydrodynamic volume or size, functionality and/or molecular configuration, solution properties describing the structural attributes of the phenolic resin in various solvents, mechanical property guidelines and fire behavior of phenolics especially fire/smoke/toxicity (FST) criteria are described.

Structural characteristics of phenolic resins differ for resole and novolak resins and relate to method of preparation, monomer/oligomer functionality, level of condensation, and final form of the resin (liquid or solid). Novolaks are solids while resoles can be liquid materials in aqueous or solvent media or solid resin.

Compositional structure and methods of determination are discussed as well as solution properties of resins. The importance of mechanical properties and fire behavior of phenolic resins is discussed since phenolic materials are utilized in a myriad of applications.

4.2
Structural Characteristics

4.2.1
Hydrodynamic Volume Characterization

Many chromatographic techniques have been used to analyze phenolic materials in relation to the hydrodynamic volume (size) of the solvated species. Different methods of chromatography and types of separation/resolution of phenolic resins are shown in Table 4.1.

4 Structure (Methods of Analysis)

Table 4.1. Chromatographic methods

	Gas(GC[a])	HPLC[b]	SEC[c]	Gel(GPC[d])	SCF[e]	Paper/thin layer [1]
Phenol	•	•	•	•	•	•
Dimer, Trimer	•	•	•	••	•	
Low M_w oligomer	•	•	•	••	•	
Med M_w oligomer			•	•	•••	
High M_w			•	•		

[a] Gas Chromatography.
[b] High Performance Liquid Chromatography [2].
[c] Size Exclusion Chromatography [3].
[d] Gel Permeation Chromatography.
[e] Supercritical Fluid Chromatography [4].

4.2.1.1 Chromatography

Chromatographic methods are applicable to the analyses of both resole and novolak resins (see ISO 11 401). HPLC has been effective in establishing and identifying many components contained in resins. Grenier-Loustalot [2] has prepared many known polymethylolated phenols, dimers and identified the phenolic materials through ^{13}C and ^{1}H NMR, FTIR and UV methods, unambiguously identifying most components in liquid resoles in a series of publications related to resole mechanism (see chapter on Chemistry, Reactions, Mechanisms).

Gel Permeation Chromatography (GPC) is the method of choice for examining MW distribution of phenolic resins. Usually MW determination is made using monodisperse polystyrene as calibration standard and provides a qualitative measure of phenolic resin MW. The reported MW values for resoles and novolaks are approximate and relate to polystyrene values. The large number of isomers due to methylene linkage of phenol rings (*o,o′*, *o,p′*, and *p,p′*) for either resoles or novolak as well as the perishability of resoles presents a challenge to anyone desiring to develop calibration standards for phenolic resins. Yet Solomon and coworkers [5] have developed a novolak composite calibration standard (see Chemistry, Reactions, Mechanisms chapter) by developing three calibration curves, each curve representing methylene linkage of *o,o′*, *o,p′*, and *p,p′* type oligomers. Improved MW values of commercial novolaks were obtained through the use of the newly developed curves.

4.2.2 MALDI

A new technique for MW determination and end group analysis is MALDI-MS-matrix assisted laser desorption-ionization mass spectrometry. The MALDI-MS soft ionization technique developed by Karas and Hillenkamp [6] has been used successfully to determine the mass of large biomolecules and synthetic resins. Pasch et al. [7] and Hay and Mandal [8] applied MALDI to examine

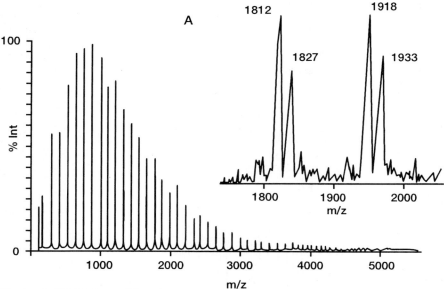

Fig. 4.1. MALDI-MS, novolak resin

phenolic resins. A MALDI spectrum (Fig. 4.1) from Pasch's studies show a spectrum for a novolak resin. Peak to peak mass is 106 g/mol and equivalent to the mass of the phenolic resin chain repeat unit. Peaks represent M + Na$^+$ molecular ions which occur by Na$^+$ attachment to the end group. Na$^+$ emerges from the substrate. According to Pasch the masses of the oligomers is M + Na$^+$ equal to 223 + 106 m with m representing the degree of polymerization, and 223 g/mol is the mass of the end group with the attached Na$^+$ cation. The insert portion of the Fig. 4.1 spectrum shows an accompanying peak of lower intensity and is due to the mass difference between Na$^+$ and K$^+$ or 15–16 g/mol. The peak is the mass plus K$^+$ molecular ions. MALDI-MS of resoles is more complicated as is shown in Fig. 4.2. In a similar manner molecular ions based on M+Na$^+$ are obtained. Structures for compounds **1–8** are shown in Fig. 4.2. MALDI-MS identified hemiformal structures **3** and **4** with **4** differing from **3** by an additional formaldehyde unit. Compounds that are structurally different but contain the same mass cannot be distinguished by MALDI-MS (**5a** and **5b**, others). Pasch suggests that the resole may be regarded as a copolymer of average structure, $P_A F_B$ where A represents the number of phenol rings and B the number of attached formaldehyde units:

$$A = m + 2$$
$$B = (x + 1)_n + (y + 1) + z \quad x, y : 0 \text{ or } 1$$

94   4 Structure (Methods of Analysis)

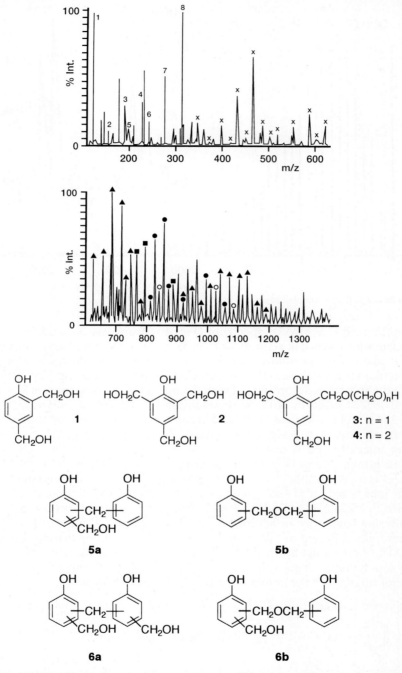

Fig. 4.2. MALDI-MS, resole resin (structures 1–8a, 8b below), ×, ■, ▲, ○, ● = Dimers, Trimers, etc.

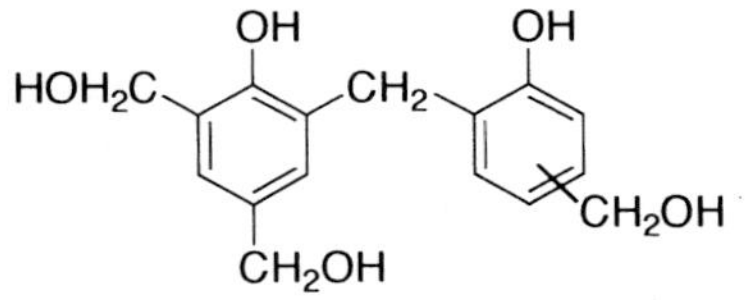

Fig. 4.2 (continued)

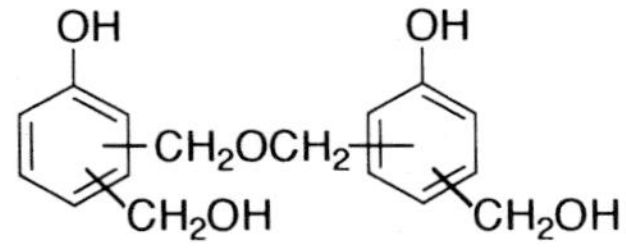

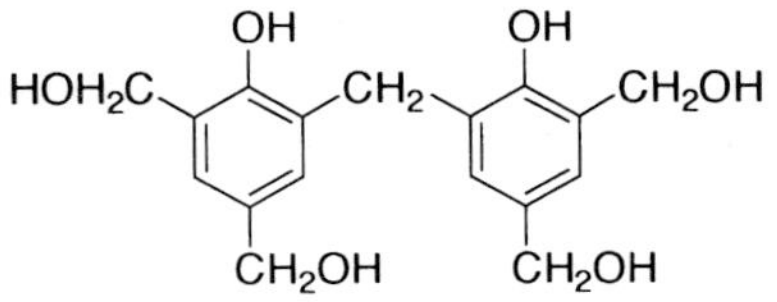

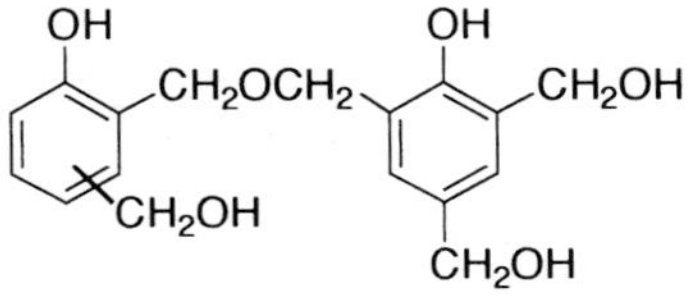

Thus for structure **9a** or **9b**, A = 2 and B = 6 leading to

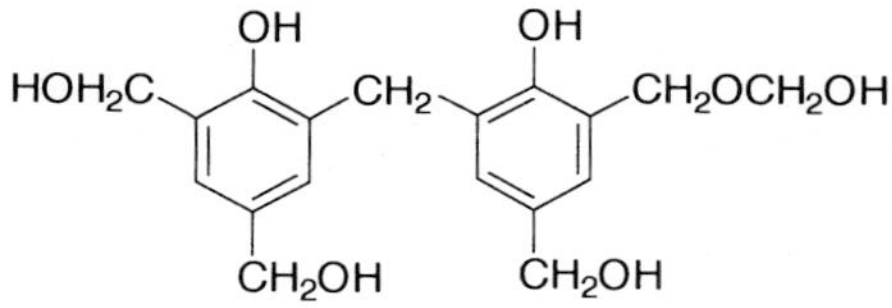

Other structures represented by various symbols in the spectrum follow the same scheme whereby different peaks (M/Z) are tabulated in the publication.

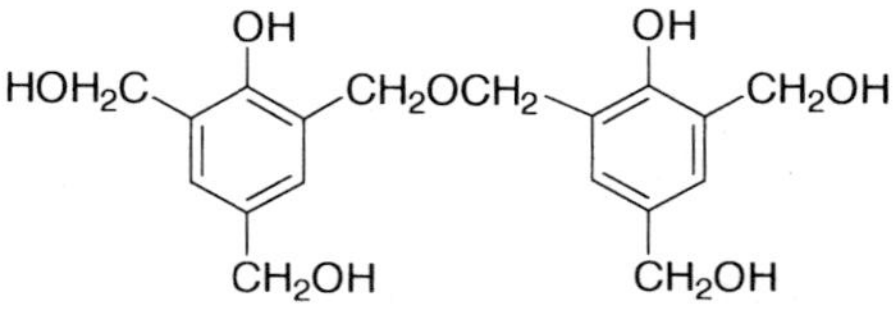

Hay and Mandal [8] prepared novolaks under conditions which resulted in the oligomer being end-capped with m-xylene. The sulfuric acid catalyzed reaction of p-t-butyl phenol with formaldehyde was conducted in m-xylene to facilitate the removal of water as a m-xylene azeotrope. MALDI spectrum (Fig. 4.3) exhibited the presence of two series of oligomers when n varied from 2 to 14. One series – 643, 805, 967, etc. – is attributed to 4-t-butyl phenol novolak with lithium cation while the other series (beginning with 599 and followed by smaller peaks preceding 805, 967, etc.) are those novolak oligomers which are end-capped with xylene and lithium cation. The following equation describes the end-capping reaction:

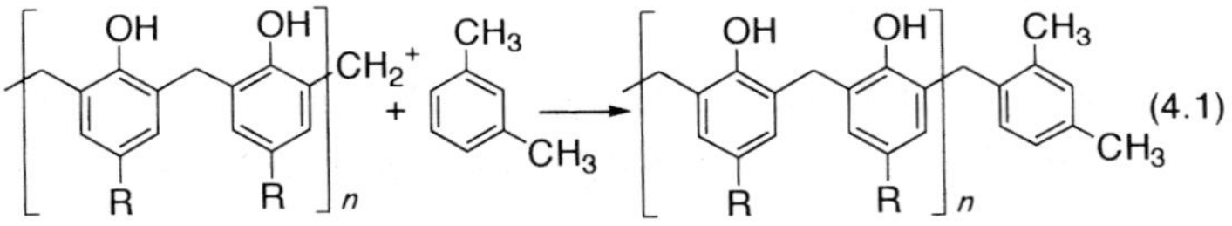

(4.1)

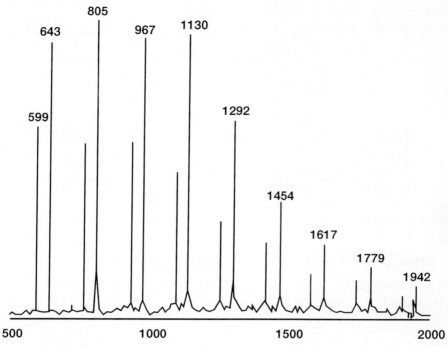

Fig. 4.3. MALDI-TOF 4-*t*-butyl phenol novolak/*m*-xlene end capped

When the novolak preparation was conducted in chlorobenzene with ion exchange resin, only 4-*t*-butylphenol novolak was obtained with no trace of any side products (end capped groups) present in the MALDI spectrum.

4.2.3
Compositional Structure

4.2.3.1
Oligomer, Uncured Resin

Resin compositional structure and functionality is determined by methods involving NMR, FTIR, and UV. NMR is the most desirable technique encompassing such nuclei as ^{1}H, ^{13}C, ^{15}N and ^{31}P. Different types of structural/functional features of soluble resins can be analyzed by NMR (Table 4.2).

4.2.3.2
Cured Resin

Solid state NMR has been widely used to determine mechanistic pathways and ultimate structure of wholly crosslinked resole and novolak resins (see chapter on Chemistry, Reactions, Mechanism) with ^{1}H, ^{13}C, and ^{15}N being used [10].

Table 4.2. NMR characterization of soluble phenolic resins [2, 9]

	^1H	^{13}C	^{13}N	^{31}P
o,p-methylene	•	•		
Branching		•		
End group		•		•
Functional group (CH$_2$OH)	•	•		
Hexa/amine resin	•	•	•	
M$_n$		•		
Degree of polymerization		•		

Further different NMR relaxation techniques (T_2, T_1, $T_{1\rho}$) have been effective in monitoring cure, crosslink density and uniformity of cured resole composites [11–13].

4.3 Solution Properties

Dilute solution properties of phenolic resins depend on solvent type, solvent interaction, and molecular configuration of the resin. The Mark Houwink Sakarada relationship, $[\eta] = Km^\alpha$ resulting from a log log plot of intrinsic viscosity and molecular weight, provides values for Huggins constant, K, as intercept and α as slope. Numerical values of α describe whether the polymer is linear, branched, or coiled while K describes a linear or associated resin:

$\alpha > 0.5$ – Linear polymer
$\alpha < 0.5$ – Branched, coiled polymer
K ~ 0.3–0.4 – Linear polymer
K > 0.5 – Molecular associated polymer

4.3.1 Novolaks

In Table 4.3, MHS data for several novolaks are presented. Values for α vary from 0.2 to 0.5 and indicate branched, coiled polymer configuration for all the novolaks. Huggins constant (K) values are quite variable from low values of 0.02, 0.08, to above 1.0. Low values would indicate polymer linearity; however the value of 0.08 for high *ortho* novolak may be anomalous and relates to a later discussion of all *ortho* novolak.

4.3.2 Resoles

Determination of MHS data for resoles is more challenging due to variable solubility and instability of resole resins. MW determination by GPC or other methods have led to high or fluctuating values attributable to apparent mole-

Table 4.3. MHS data of novolaks

$[\eta]$	Novolak	Solvent	TEMP °C	M_n	Author
$0.019 M_n^{0.47}$	Random	Acetone	25	1000–8050	Tobiason et al. [14]
$0.631 M_n^{0.28}$	Random	Acetone	30	370–28000	Kamide and Miyakawa [15]
$0.80 M_n^{0.28}$	Random	THF	25	540–4700	Ishida et al. [16]
$1.075 M_{vd}^{0.20}$	Random	Acetone	20		Ishida et al. [17]
$0.0813 M_n^{0.5}$	High *ortho*	Acetone	30	690–2600	Kamide and Miyakawa [15]
$0.406 M_w^{0.35}$	Resorcinol	Methanol	30	720–4110	Kim et al. [18]

Table 4.4. MHS data of acetylated resoles

Resole	$[\eta]$	Solvent	Temp. (°C)	Mn	Author
Particleboard	$1.196M_n^{0.23}$	CHCl$_3$	25	1650–5600	Kim and Amos [19]
	$1.735M_n^{0.12}$	Benzene	25	1650–5600	
Plywood	$0.77M_w^{0.21}$	THF	25	2200–114,000	Kim et al. [20]
	$1.16M_w^{0.15}$	Ethyl Acetate	25	2200–114,000	
NaOH Catalyst (non-acetylated)	$0.099M_d^{0.18}$	THF	25	800–111,000	Larent and Gallot [21]

cular association when solution medium is changed from strongly alkaline to a less alkaline aqueous solution or a wholly organic solvent [3]. This behavior is specific to wood adhesive resoles which are strongly alkaline aqueous solutions and deteriorate on storage. Hence Kim and coworkers [19, 20] relied on acetylation as a means of improving resole stability during solution studies (Table 4.4).

Acetylated resoles used in manufacture of particle board and plywood panels yielded reasonably similar exponent values. Higher exponent values of 0.21 and 0.23 were obtained in polar solvents (CHCl$_3$, THF) and indicated a highly branched, coiled polymer. Similarly Huggins constant, K, values ranged from 0.77 to 1.7 and were indicative of a highly associated polymer structure. An early study of a non-acetylated, NaOH catalyzed resin reported by Laurent and Gallot [21] exhibited a similar α value of 0.18 but an unusually low Huggins constant (K) of 0.1 suggesting a linear polymer.

4.3.3
Amine Catalyzed Resoles

Resoles catalyzed by primary amines have also been examined by Ishida and coworkers [16]. The reaction of phenol and formaldehyde with either ethylene diamine (ED) or hexamethylene diamine (HD) results in resoles that incorporate the amine within the resole microstructure:

$$P + F \xrightarrow[\text{or HD}]{ED} \left(\text{(OH-Ph)}-CH_2\right)_m\left(\text{(OH-Ph)}-CH_2NH(CH_2)_xCH_2-NH-CH_2\right)_n \quad (4.2)$$

$$x = 1 \text{ or } 5$$

Data contained in Table 4.5 show MHS values for resoles catalyzed with ED and HD that are non-acetylated and acetylated.

These amine catalyzed resoles are also compared with an ammonia catalyzed resole. Non-acetylated amine catalyzed resoles as well as ammonia catalyzed resole exhibit similar α and K values indicating branched, coiled, and associated polymers while acetylated products have α values approaching 0.5 and greater suggesting linear polymers. Similarly K values below 0.3 suggest linear resin structures. Random novolak (entry 5, Table 4.5) followed a similar pattern as the non-acetylated and acetylated amine resoles.

The effect of substituents present in the amine or phenol was also examined [16]. N,N'-dimethyl hexamethylene diamine (MHD) catalyzed the phenol/formaldehyde reaction yielding a resin identified as P-MHD while p-cresol/formaldehyde was catalyzed by either HD of MHD yielding PC-HD and PC-MHD designated resins (Table 4.6).

Data for α and K indicate that resins are coiled, highly associated, and unaffected by the presence of substituents in either the amine or phenol.

Table 4.5. Amine catalyzed resoles[a]

Ethylene diamine (ED) and Hexamethylene diamine (HD)

Resin type	$[\eta]$ Non-acetylated	$[\eta]$ acetylated
HD	$0.55 \, M_n^{0.32}$	$0.19 \, M_n^{0.43}$
ED	$0.45 \, M_n^{0.34}$	$0.14 \, M_n^{0.44}$
Ammonia resole	$0.73 \, M_n^{0.27}$	$0.04 \, M_n^{0.57}$
Random novolak	$0.80 \, M_n^{0.28}$	$0.09 \, M_n^{0.48}$

[a] THF, 25 °C.

Table 4.6. Effect of substituents on amine catalyzed resoles

Resin type	$[\eta]$
HD	$0.40 \, M_n^{0.34}$
PC–HD	$0.39 \, M_n^{0.32}$
P–MHD	$0.57 \, M_n^{0.32}$
PC–MHD	$0.50 \, M_n^{0.29}$

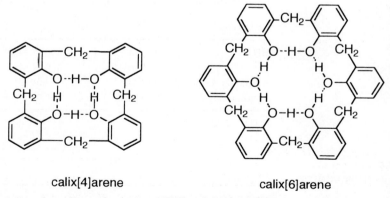

Fig. 4.4. Pseudo cyclic conformations of "all *ortho*" linear novolaks

4.3.4
"All ortho" Novolaks

The Yamagishi/Ishida group extended solution property studies to "all *ortho*" novolak system by preparing wholly *ortho* oligomers from dimer to heptamer of phenol formaldehyde resin [22]. Strong hydrogen bonds were formed when these oligomers were examined in chloroform. Both phenol and 4-*t*-butyl phenol oligomers exhibited hydrogen bond phenomenon. Evidence for dimer and trimer emerging as bimolecular conformations such as tetramer and hexamer (Fig. 4.4) via intermolecular hydrogen bond and analogous to calix[4]arene and calix[6]arene was presented. Similarly intramolecular hydrogen bonds within the linear tetramer and hexamer resulted in pseudo cyclic conformations in chloroform. These strong hydrogen bond observations of "all *ortho*" novolak may have contributed to the low K value observed by Kamide (Table 4.3).

4.4
Complexation

4.4.1
„All ortho" Novolaks

Sone et al. [23] have found that linear all *ortho* novolak oligomers (dimer to hexamer) form inclusion compounds or complexes with selective solvents such as benzene, chloroethanes, cyclohexane, toluene, and others. The Yamagishi/Ishida group [24] appended an ethyl acetate group to the "all *ortho*" oligomers of *p-t*-butyl phenol novolaks:

$$\text{[structure with OH + BrCH}_2\text{COOEt} \rightarrow \text{OCH}_2\text{COOEt structure} \equiv \text{OEs structure]} \tag{4.3}$$

4.4.2
Ethyl Acetate Appendage

Cation extraction properties of these materials were compared with the corresponding ethyl acetate group appended to *t*-butyl calixarene system. These linear ethyl acetate oligomers exhibited an affinity and selectivity toward alkali metal cations but were transient complexes and not permanent as the corresponding calixarene OEs (see later). As an example the authors showed how linear pentamer (5BP-Es) was able to complex sodium cation by forming a "cyclic conformation" similar to calix[4]arene (Fig. 4.5).

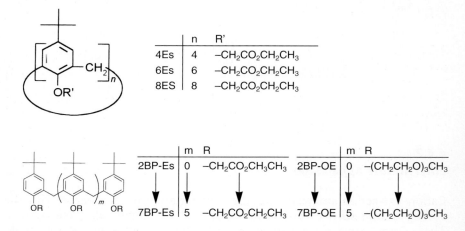

Fig. 4.5. Complexation of sodium ion by linear pentamer (5BP-Es)

4.4.3
Trioxyethylene Ether Appendage

Expanding these structure property relationships of linear "all *ortho*" novolak oligomers to a series of ethyl acetate and trioxyethylene ether derivatives of "all *ortho*" linear oligomers of 4-*t*-butyl phenol (n = 1 – 7) [25], cation extraction properties were determined and compared with the corresponding calixarene (m = 4, 6, 8) materials:

	n	R'
4Es	4	$-CH_2CO_2CH_2CH_3$
6Es	6	$-CH_2CO_2CH_2CH_3$
8ES	8	$-CH_2CO_2CH_2CH_3$

	m	R		m	R
2BP-Es	0	$-CH_2CO_2CH_3CH_3$	2BP-OE	0	$-(CH_2CH_2O)_3CH_3$
7BP-Es	5	$-CH_2CO_2CH_2CH_3$	7BP-OE	5	$-(CH_2CH_2O)_3CH_3$

Extraction results of BP-Es or BP-OE for alkaline earth metal cations showed that the affinity of BP-Es was greater for alkaline earth cations as compared to alkali cations. However the heptamer oligomer (7BP-Es) exhibited a preferable affinity for Sr^{+2} and Ba^{+2}. Authors claim that selectivity of BP-Es oligomers may

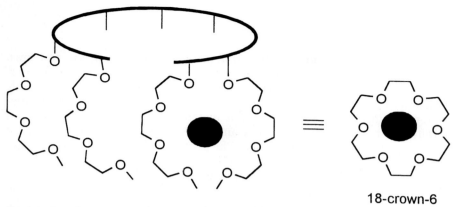

Fig. 4.6. Complex structure of n BP-OE with alkali metal cation

be dependent on ionic size as well as ionic charge density. The linear heptamer (7BP-Es) displayed a higher extraction ability for larger alkaline earth cations (Sr^{+2}, Ba^{+2}) and similar to calix[8]arene. (8Es).

Both the heptamer and calix[8]arene are flexible and the higher affinity of 7BP-Es for alkaline earth cations is probably due to its greater conformational flexibility.

The extraction profile of the trioxyethylene ether derivatives (nBP-OE) was completely different to the nBP-Es series. The nBP-OE preferred the larger alkali cations (K^+, Rb^+, Cs^+) rather than Li^+ or Na^+. Furthermore affinity was higher for even numbered BP-OE than odd numbered BP-OE. Extraction is facilitated by a complex structure related to a cylic crown ether structure (Fig. 4.6) rather than cavity formation.

4.5
Mechanical Properties

The determination of mechanical properties is of fundamental importance for all resin systems. Although other properties such as flame resistance, thermal stability, and chemical resistance are important for most applications, all polymers regardless of use must exhibit a specified range of mechanical properties suitable for the end use application. Tensile, compressive, flexural strength (and corresponding moduli), impact resistance and T_g or HDT are most important. Related properties provide a measure of how much stress a resin sample will withstand before failing and further provide guidance in the design of the fabricated part. In most instances phenolic resole or novolak resin is combined with various substrates such as wood (to plywood, particleboard), sand (foundry products), mineral fillers, wood flour, cellulose or glass (molding materials), alumina (abrasives), mineral or glass (insulation), paper (laminates), woven glass or carbon fiber (FRP), and the mechanical properties of these resulting cured compositions are determined.

4.6
Fire Properties

Most thermoplastic and thermosetting polymers are flammable and burn as long as a source of flame is present but cease burning when the flame is removed. Few polymers are inherently non-flammable and these include phenolic, PMR-15, polyphenylene sulfide, PEEK, polyarylethers, and polysulfone. Some newer non-flammable materials within the phenolic resin family are novolak cyanate esters and polybenzoxazines (see chapter on Chemistry, Reactions, Mechanisms). The prime focus of phenolic resins and their fire properties relates to fiber reinforced composites and their corresponding FST behavior. Schematically the dissipation of heat and flame is shown in Fig. 4.7 for a fiber reinforced phenolic composite. Burning/charring occurs in several stages. The flame front or external heat source increases the composite temperature leading to the development of a glassy, dense carbon char with some gaseous materials being emitted. The char region becomes a heat shield by insulating the composite interior. FRP systems based on phenolic/carbon fiber are the prime materials for rocket propulsion and for shuttle reentry surfaces and other aerospace applications exposed to extreme temperatures.

Since synthetic polymers are used increasingly as the matrix resin for FRP products in construction and transportation, non-flammable polymeric materials are desired. Besides flammability the polymeric materials should emit little or no smoke and/or toxic gases during combustion. Fire safety is a

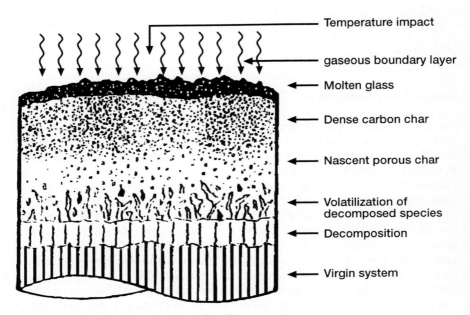

Fig. 4.7. Ablation of a fiber reinforced phenolic composite

priority especially in facilities where persons congregate or modes of transportation such as aircraft, rail, bus, or marine conveyances are used and limit passenger egress. A familiar acronym has evolved within the last decade being identified with flame, smoke, toxicity. It is FST and applies to fire safe or resistant composites (see High Performance and Advanced Composites chapter for fire testing of composites; see Insulation Chapter for fire testing of foams).

Fire safety and testing have evolved due to the rapid expansion of civil aviation as the dominate means of long distance travel. From the rudimentary Bunsen burner testing protocol to the current extensive fire testing procedure, the combined efforts of material suppliers, aircraft manufacturers, and the regulatory agencies have resulted in the development of many fire resistant composites that have been incorporated within the aircraft interior, many of which are based on a phenolic resin matrix.

Fire tests specific to composite for transportation and construction are Tunnel Test (ASTM E-84), OSU Total Heat Release and Peak Heat Release, Surface Flammability (ASTM E-162), Smoke Density (ASTM E-662), Cone Calorimeter (ASTM E-1345), and Oxygen Index (ASTM-D-2863).

4.7
Summary/Trends

Structural aspects of phenolic resins encompass hydrodynamic volume determinations via various chromatographic techniques. Molecular weight determinations obtained by GPC are approximate and due to the use of polystyrene calibration methods. A new calibration method based on a novolak calibration standard of three curves has yielded improved M_w values for commercial novolaks. A similar calibration method for resole remains elusive due to the difficulty of preparing pure resole isomeric configurations with a corresponding resole limited stability. A new method for M_w determinations of phenolic resins is known as MALDI-MS and provides masses of novolak and resoles oligomers and end groups other than the phenolic repeat unit.

Compositional and functional characteristics of uncured and cured phenolic resins rely mainly on ^1H and ^{13}C NMR in solution or solid state. Different NMR relaxation techniques (T_2, T_1, $T_{1\rho}$) are used to monitor cure, crosslink density, and uniformity of cured resole composite. Dilute solution properties of novolaks and resoles in conjunction with the Mark Houwink Sakarada relationship indicate, in most cases, phenolic resins (novolaks, resoles, amine catalyzed resoles) are branched, coiled, and molecularly associated polymers. Attempts to apply MHS conditions to "all *ortho*" novolak oligomers resulted in the observation that these materials in $CHCl_3$ form bimolecular conformations (dimer, trimer) or pseudo cyclic conformations (tetramer, hexamer) analogous to calixarenes. By appending either ethyl acetate group or trioxyethylene ether group to the "all *ortho*" novolaks selective cation extraction of alkali and alkaline earth cations was observed. The appended ethyl acetate "all *ortho*" novolak closely resembled calixarene in alkali cation extraction while a proposed cyclic crown ether structure formed by two adjacent trioxethylene ether groups in the "all *ortho*" novolak facilitates extraction of alkaline earth cations rather than cavity-calixarene type material.

The importance of mechanical and fire properties in the final phenolic composition was emphasized. Fire properties or FST characteristics are critical in those FRP applications which are directed to construction and transportation industries.

References

1. Haub H-G (1997) Bakelite AG (private communication)
2. Grenier-Loustalot MF, Larroque S, Grenier P, Leca JP, Bedel D (1994) Polymer 35(14): 3046; Grenier-Loustalot MF, Larroque S, Bedel D (1996) Polymer 37(8):1363; Grenier-Loustalot MF, Larroque S, Bedel D (1996) Polymer 37(6):939; Grenier-Loustalot MF, Larroque S, Grenier P, Bedel D (1996) Polymer 37(6):955
3. Sellers T Jr, Prewitt ML (1990) J Chromatogr 513:271
4. Allen RD, Chen RJR, Gallagher-Wetmore PM (1995) SPIE 2438:250, 261
5. Dargaville TR, Guerzoni FN, Looney MG, Shipp DA, Solomon DH, Zhang X (1997) J Poly Sci Part A Poly Chem 35:1399
6. Karas M, Hillenkamp F (1988) Anal Chem 60:2299
7. Pasch H, Rode K, Ghahary R, Braun D (1996) Die Angew Makrom Chemie 241:95
8. Mandal H, Hay AS (1997) Polymer 38(26):6267
9. Otxenbourgs BT, Andriaensens PJ, Reckmans BJ, Caleer RA, Vandergunde DJ, Gelan JM (1995) Ind Eng Chem Res 34:1364
10. Zhang X, Looney MG, Solomon DH, Whittaker AK (1997) Polymer 38(23):5835
11. Neiss TG, Vanderheiden EJ (1994) Macromol Symp 86:117
12. Schmidt RG (1997) Forest Products Society 1997, Annual Meeting, June 22–26, Vancouver, British Columbia
13. Curran SA, Walker TB, Brambilla R (1997) Poly Preprints 38(1):856, ACS San Francisco, CA
14. Tobiason FL, Chandler C, Schwarz FE (1972) Macromol 5:321
15. Kamide K, Miyakawa Y (1978) Makromol Chem 179:359
16. Sue H, Nakamoto Y, Ishida S-I (1989) Polyn Bull 21:97
17. Ishida S-I, Nakagawa M, Suda H, Kaneko K (1971) Konbushi Kagaku 28:250
18. Kim MG, Amos LW, Barnes EE (1993) J Poly Chem Part A Poly Chem 31:1871
19. Kim MG, Amos LW (1991) Ind Eng Chem Res 30:1151
20. Kim MG, Nieh WL, Sellers T Jr, Wilson WW, Mays JW (1992) Ind Eng Chem Res 31:973
21. Laurent P, Gallot Z (1982) J Chromatogr 236:212
22. Yamagishi T, Enoki M, Inui M, Furukawa H, Nakamoto Y, Ishida S-I (1993) J Poly Sci Part A Poly Chem 31:675
23. Sone T et al. (1989) Bull Chem Soc Jpn 62:1111; (1991) Bull Chem Soc Jpn 64:576; (1993) Bull Chem Soc Jpn 66:828
24. Yamagishi T, Tani K, Ishida S-I, Nakamoto Y (1994) Poly Bull 33:281
25. Yamigishi T, Tani K, Shirano K, Ishida S-I, Nakamoto Y (1996) J Poly Sci Part A Poly Chem 34:687

Part B
Applications of Phenolic Resins

CHAPTER 5

Thermosets: Overview, Definitions, and Comparisons

The history of thermosets [1–6] and their application began around 125 years ago (Table 5.1) with phenolic resins, and is thus, historically speaking, closely linked with the name "Baekeland".

Leo H. Baekeland first discovered a way to produce high-polymer substances from oligomeric phenolic resins and fillers/reinforcing agents. This concept, which may be termed the "Baekeland principle" (Fig. 5.1), has been subjected to continued technological development over the years. It has provided a very broad scope of application and has remained (in some cases in modified form) the basic principle of thermoset production and processing up to the present. Monomers are used to synthesize defined oligomers ("resins") that are transformed (and shaped) to yield highly crosslinked, insoluble polymers in a second step that occurs:

- With or without application of heat;
- With or without a catalyst;
- With or without fillers and reinforcing agents;
- With or without pressure.

These highly crosslinked products can accommodate up to 80% fillers and reinforcing agents, a fact that affords them significant advantages over

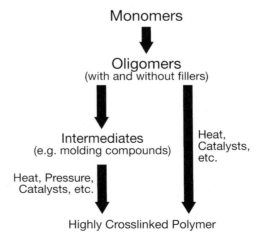

Fig. 5.1. The "Baekeland principle" of thermosetting systems

Table 5.1. Chronological history of thermosets (selection)

1872	Adolph von Baeyer: first concrete experiments to produce phenol-formaldehyde condensates
1902	First synthetic resin developed: "Laccain", phenol-formaldehyde resin, substitute for shellac; invented by C. H. Meyer, Zwickau (Louis Blumer Co.)
1907	L. H. Baekeland's "heat and pressure" patent on curing of phenolic resins: the first thermosetting plastics. Further patents: the "base" patent, "varnish" patent and the "grinding wheel" patent
1910	Bakelite Company founded in Berlin. Fabrication of thermosetting molding compounds and industrial (Bakelite) resins
1910	L. Berend (Dr. Kurt Albert Co. in Wiesbaden) develops the first oil-soluble synthetic paint resin "Albertol" (rosin acid-modified phenolic resin)
1922	First patent on production of organic solvent soluble urea-formaldehyde resins (BASF)
1928	Invention of oil-reactive phenolic resins by H. Hönel (RCI, Detroit at that time, later Vianova)
1934	Marketing of furfuryl alcohol for production of furan resins begins
1934	Development of epoxy resins by Pierre Castan (de Trey Frères Co., Switzerland)
1935	Development of melamine-formaldehyde resins (CIBA, Casella, Henkel)
1936	C. Ellis discovers cure of unsaturated polyesters in presence of polystyrene
1937	Plasticized and etherified phenolic resins for use in paints (A. Greth und K. Hultzsch, Dr. Kurt Albert Co. in Wiesbaden)
1937	Polyaddition products based on diisocyanate and polyols (Bayer AG and others)
1948	First patent on production of water-soluble thermosetting resins by H. Hönel (Vianova Co., Graz)
1960	Introduction of furan resins as binders in the foundry industry (Quaker Oats, USA)
1962	Market introduction of weather-resistant, phenolic resin bonded chipboard
1964	Development of thermosetting molding compounds for injection molding
1970–1975	Furan and phenolic resins for production of heat shields for the aerospace industry (USA)
1982	Phenolic resin bonded FRP components in automotive and aeronautic engineering (e.g., in Airbus A 300)
1990	Use of epoxy-phenolic resin combinations for composites in the transportation industry
1992	Increased use of UP FRP in power generation by wind energy (rotor blades)
1993	Use of tannin-modified phenolic resins for production of wooden materials
1994	Introduction of solventless manufacture of epoxy resin electrical laminates (NEMA FR4)
1995	Introduction of phenolic resins for production of FRP by pultrusion

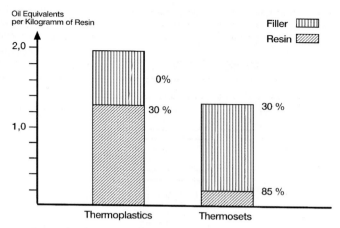

Fig. 5.2. Energy demands of thermosets and thermoplastics

thermoplastics with respect to the raw material demand (for example the consumption of petroleum, a non-renewable raw material component used to produce phenol and styrene), and the capability for (internal and external) modification. The compounding level or an amount of filler in the case of thermoplastics is generally only up to 30% (Fig. 5.2).

The Baekeland symbol "B∞" ("Resins for an infinity of applications"), which he granted to his licensees, displayed an insight bordering on clairvoyance. As we know today, the message of this symbol is generally applicable to thermosetting materials and thermosetting base resins, since the Baekeland phenolic resins only represented the beginning of the thermoset group of materials [7–9].

However, with his "pressure and heat" patent [10], Baekeland developed the first practical entry to the basic technology in the age of plastics that followed.

Today, in contrast to the situation in Baekeland's time, a wide variety of raw materials is available for plastics manufacture (Table 5.2), and they serve as a basis for the production of a broad range of thermosetting plastics [11–17]. In the cured state, thermosets – in contrast to the linear, convoluted thermoplastic polymers and loosely bonded elastomers – exhibit a high degree of crosslinking. Thermosets may be classified into at least ten different varieties (Fig.

Table 5.2. Thermosetting resins and their raw material basis (selection)

Resins	Raw Materials
Phenolic resins (PF)	Phenol, formaldehyde
Melamine resins (MF)	Melamine, formaldehyde
Urea resins (UF)	Urea, formaldehyde
Furan resins (FF)	Furfuryl alcohol, formaldehyde
Epoxy resins (EP)	Bisphenol A and F, epichlorohydrin
Unsaturated polyester resins (UP)	Unsaturated and saturated dicarboxylic acids and polyols

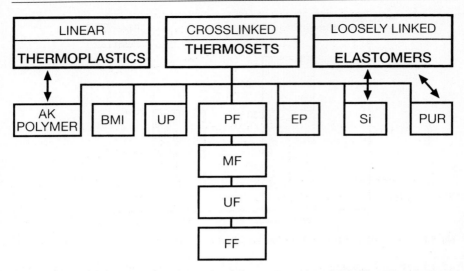

Fig. 5.3. Survey of thermosetting resins (AK = acrylate, BMI = bis-maleimide, UP = unsaturated polyester, PF = phenolic, EP = epoxy, Si = silicone, PUR = polyurethane, MF = melamine, UF = urea resin, FF = furan resin)

5.3). The resin matrix in these may represent a phenolic, urea, melamine, unsaturated polyester, or epoxy resin. Moreover, the "thermosetting" or crosslinking polyurethanes, silicone resins, and acrylate systems are also considered members of this group. However, the latter three classes may only be partially classified as thermosets. BMI = bismaleimide in Fig. 5.3 represents an example of related "miscellaneous" products that only occupy a small segment of the market.

In the area of thermosets, a certain variety of terms and definitions are prevalent both domestically and internationally, and can confuse not only laymen but experts as well. It is desirable to arrive at a general definition by first considering crosslinked polymers or "thermosets" (Fig. 5.4) and defining how they arise as follows: *"Thermosets are fully synthetic, generally highly crosslinked polymers that arise (1) by further condensation of an oligomeric resin, (2) by initiated self-polymerization or copolymerization of an oligomeric unsaturated resin or (3) by polyaddition of functional oligomers and/or monomers."* Definition (1) describes how products such as phenolic, urea, and melamine resins as well as furfuryl alcohol polymers arise, (2) considers UP resins and some of the methacrylate resins, whereas (3) defines epoxy and thermosetting polyurethane systems.

This is a simplified consideration, but many combinations also exist. The three principles of structural development may be regarded as descriptions of "thermosetting", "crosslinking", or "reactive" systems that in turn lead to the final, crosslinked polymeric state and may be directly or indirectly assigned to the Baekeland principle mentioned above.

The above definition is significant when **comparisons of resins** are considered (cf. below). Prepregs or thermosetting molding compounds may be

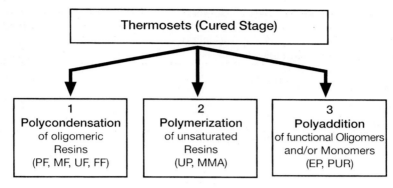

Fig. 5.4. Origin of thermosets (final cured stage)

produced from the oligomers and represent precursors to the final polymeric stage.

In addition, thermoset binders are used in the production of many industrial products which are, in some cases, vital to the national economy and general welfare of the nation (Table 5.3). Although they are invisible in some applications, it is hard to imagine being without them in practice since they are absolutely necessary and cannot be replaced by other materials such as thermoplastics, metals, wood or glass for reasons of quality, cost and performance.

Due to the high degree of crosslinking and the use of fillers and reinforcing agents, thermosets display their "profile" (Table 5.4) in all applications, a "profile" that can feature particularly desirable combinations of properties:

Table 5.3. Thermosetting binders for various industrial applications (PF = phenolic resin; UF = urea resin; MF = melamine resin; UP = unsaturated polyester; EP = epoxy resin; PUR = polyurethane resin; MMA = methyl methacrylate; FF = furan resin)

Applications	Binders
Fiber composites	UP, EP, PF, FF
Refractories	PF, FF
Molding compounds	PF, UF, UP, MF, MF/PF, MF/UP, EP
Foundry binders	PF, UF, EP, FF
Wood materials	PF, UF, MF
Friction linings	PF, MF, EP
Acid-resistant cements and compounds	PF, EP, FF, PUR
Foams	MF, PF, UF, PUR
Laminate moldings, electrical laminates	MF, PF, EP
Abrasives	PF, UF
Grinding wheels	PF, FF
Textile auxiliaries	UF, MF
Thermal and acoustic insulation	PF
Coatings	EP, UP, MMA, PUR, MF, PF

Table 5.4. Thermoset "profile" (typical properties)

Good electrical and thermal insulating capabilities
High strength, stiffness and surface hardness
High dimensional stability, low coefficient of thermal expansion
Favorable thermomechanical properties, nearly no cold flow
High resistance to aggressive media, no stress corrosion cracking
High thermal resistance
Favorable behavior when exposed to fire without halogenated flame retardants
Safety reserves at peak temperatures

a high strength level, excellent thermal behavior, long-term thermal and mechanical stability, and good fire and flame resistance, combined with good electrical and thermal insulating capabilities and – the "top industrial parameter" – a generally acceptable price level for applications in large-scale production. Which parameter is of the greatest significance varies, and relates to different criteria that the design engineer places on properties, process, and economics.

Even today, after they have undergone nearly 90 years of development, the technical development potential of thermosets is far from exhausted. Thermosets (phenolics, epoxies, urea resins, melamine, furan, and unsaturated polyester resins as well as thermosetting polyurethane, acrylate, and silicone resins) make up around 18–20% of the world production of plastics (approx. 100–120 million tonnes). The approximate distribution of phenolic, amino, epoxy, and unsaturated polyester resins in western Europe, based on the production figures (100%), is (Fig. 5.5) 49% urea resins, 22% phenolics, 14% unsaturated polyester resins, 8% epoxies, 6% melamine resins, and 1% furan polymers (based on furfuryl alcohol). These figures represent around 65% of the thermoset volume; the remainder is mainly composed of thermosetting polyurethanes, acrylates and silicone resins. The industrial application should also be kept in mind as thermosetting resins are compared (Table 5.5).

Fig. 5.5. Distribution in %:
UF, PF, UP, EP, MF, FF
(FF = furan resin)

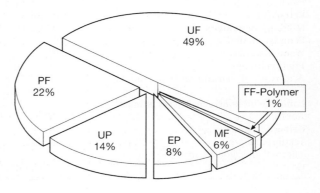

Table 5.5. Fields of application for thermosets (selection)

Fields of use	Synthetic Resins						
	UF	PF	UP	EP	MF	PUR	MMA
Automotive construction and accessories		x	x		x	x	
Aircraft construction		x	x	x			
Wood materials processing	x	x				(x)[a]	
Building industry (above ground, subterranean, streets)		x	x		x	x	
Mechanical engineering and equipment construction	x	x	x	x	x		
Abrasives industry	x	x		x			
Foundry industry, refractories, steel	x	x		x	(x)[a]	(x)[a]	(x)[a]
Paint and adhesives industry	x	x	x	x	x	x	x
Electrical and lighting industry	x	x	x	x	x	x	

[a] (x) = only usable in some cases.

The various areas include (as headings and examples) the particularly important applications in transportation systems, electronic components [18, 19] and energy generation.

An attempt will be made below to compare phenolic resins with other thermosetting resins, specifically urea, melamine, unsaturated polyester and epoxy resins, on the basis of a number of aspects in their application technology. As the Germans say, the comparison "limps" (is inept), which in this case means that all the resin systems considered here can generally be adapted for an application using modifiers or fillers/reinforcing agents. Comparisons can really only be made under essentially identical conditions. Sometimes technical advantages can even simply vanish when confronted with economic reality, i.e., certain advantages are sometimes not utilized for reasons of cost.

The technical assessment of applications thus naturally includes a consideration of the costs of a product group (Fig. 5.6). Nonetheless, a price index comparison must be regarded objectively and can only serve as an indication. Variations in the prices of raw materials render such price index comparisons highly dependent on the specific raw materials, and vulnerable to the prevalent price and raw material considerations. However, the price index comparison shows that phenolic and urea resins exhibit a relatively low price level. If the variety of possible areas in which phenolic resins find application is also considered (Tables 5.3 and 5.5), an alternate resin matrix must exhibit very significant advantages to be considered as a replacement. Although urea resins are less expensive, their strengths lie in the areas of textiles, wood materials,

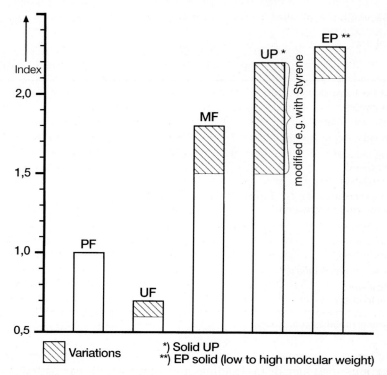

Fig. 5.6. Price index comparison of PF = 1 with UF, MF, UP, EP (prepared in January 1998)

and (earlier) insulating foam as well. However, their disadvantage lies in the possible liberation of formaldehyde (in foam and wood materials). Thus, UF insulating foam is no longer used to any extent, and it was necessary to take far-reaching measures (including some at the expense of reactivity) to reduce formaldehyde liberation in wood materials [10, 20, 21]. The advantages of urea resins are based on their light color (allow dyeing) and relative good textile compatibility.

Melamine resins are superior quality urea resins that, in contrast to phenolics, are widely used in the area of coatings (for example in automotive coatings) and in manufacture of (decorative) paper. Their outstanding features include light fastness, light color, and good scratch resistance.

A comparison of amino (melamine and urea) resins and phenolics with unsaturated polyester and epoxy resins must consider the fact that the latter cure without liberation of by-products, as shown in Fig. 5.4, a circumstance that offers significant advantages that can positively affect the curing and particularly (in the case of epoxy resins) shrinkage behavior. In contrast to phenolics, both systems (epoxies and UP resins) can be used as casting resins, whereas this is only possible within narrow limits in the case of phenolic resins. Epoxy resins represent outstanding electrical casting resins [22, 23], a circumstance that particularly emphasizes their high electrical quality compared to that of other systems.

An indirect comparison of the properties of PF, MF, UP, and EP resins is made in the chapter on molding compounds (containing mineral and/or organic fillers). These four resin systems are compared with respect to their use as molding compounds. The essence of this comparison is that phenolic resin molding compounds exhibit good electrical and mechanical parameters and can be used in many areas. They can be easily modified, exhibit excellent thermal resistance, and (in combination with appropriate fillers) feature outstanding resistance to flame and glowing heat. Urea resins are less expensive, can be easily pigmented, and are light fast. Melamine resins are similarly easy to color and light fast, and their electrical parameters exhibit (as do those of UP resins) particularly high tracking resistance. A negative point is that when urea and melamine resins are used as a matrix in molding compounds, they exhibit considerably higher processing shrinkage and post-shrinkage results than a phenolic resin matrix. On the other hand, special melamine resin molding compounds are the only members of the indicated product classes (molding compounds) that are approved for use in food packaging.

Due to the fact that they cure by way of self-polymerization, monomer-free UP molding compounds feature relatively low processing shrinkage and post-shrinkage. However, this does not apply to styrene-modified (BMC) molding compounds, which can only be processed without distortion and shrinkage ("low profile") in the presence of added thermoplastic polymers. Depending on the resin composition, UP molding compounds are also relatively resistant to weathering. Epoxy resin molding compounds exhibit particularly great superiority with respect to their extremely low processing shrinkage and post-shrinkage, high electrical quality (except for the tracking resistance), and high strength and toughness. On the other hand, their price level is also significantly higher than that of other systems.

Phenolics are superior to all other resin systems with respect to their particularly good thermal behavior (for example when used in brake and clutch linings and grinding wheels), their high level of flame resistance, low smoke density, and their use as a carbon donor in pyrolytic applications. They provide high carbon yields in various applications such as refractories and carbon and graphite materials. All phenolic resin applications are coupled with an excellent cost/performance relationship.

The good flame-related properties of phenolics, including their low smoke density, is utilized in composites intended for aircraft construction. In determination of the smoke density [24] in an NBS smoke density chamber as specified in ASTM E 662, the attenuation of a beam of light by the smoke collected in the chamber is measured and the specific optical density D_s calculated from this. This results in D_s figures of 300–600 for epoxy resin-based composites, 50–100 for polyester resins containing flame retardants, and less than 20 for phenolic resins (see later in High Performance and Advanced Composites). Due to these properties, phenolic resin FRP components are increasingly used in the transportation industry [25, 26]. These presently include combinations of epoxy and phenolic resin prepregs that are simultaneously compression molded and are used to combine the outstanding fire-related behavior of the phenolic resins with the high level of mechanical properties offered by epoxies.

A completely new class of phenolics has rendered it possible to reach such high levels even without combination with epoxy resins using a phenolic resin binder alone (Figs. 5.7 and 5.8). This new product class exhibits the strength

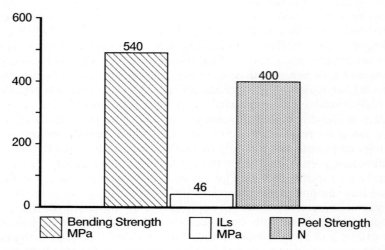

Fig. 5.7. Mechanical parameters of a flame-retardant epoxy resin (ILS = interlaminar shear strength)

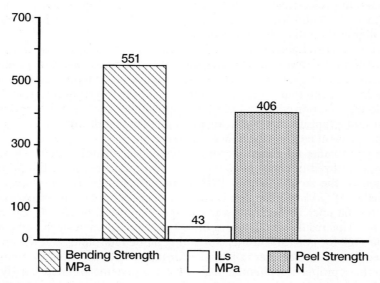

Fig. 5.8. Mechanical parameters of the PLB phenolic resins (PLB = internal term of Bakelite AG, Iserlohn, Germany)

Table 5.6. Carbon content and carbon yield of different thermosetting binders

	Carbon-content (%) (theoretical)	Carbon-yield (%) (after pyrolysis)
PF	<80	55–70
FF	<75	50–60
EP	<75	20–30
UP	<60	15–25

inherent to epoxy resins, and the fire-related behavior of phenolics. Phenolic resins have thus acquired a completely new significance as binders for fiber reinforced composites used in the aircraft and transportation industries.

Pyrolysis and formation of bonding carbon from phenolics (PF), furan resins (FF), unsaturated polyesters (UP), and epoxy resins (EP) achieves carbon in varying yields. Up to 70% is obtained in the case of the phenolics (a high yield based on the cured resin matrix), whereas among the other three systems only the furan resins afford a relatively high carbon yield (Table 5.6). For this reason, and due to their capability for technologically versatile application, phenolic resins are favored as binders for refractories and graphite materials [27].

Such a comparison of resin systems can be further pursued and many additional advantages of phenolic resins recognized; this similarly applies to the chemical resistance as determined in industrial chemically resistant coatings [28]. Table 5.7 illustrates a comparison of coating systems [29, 30] based on

Table 5.7. Properties of solvent-free thermosetting coatings, excerpt from [30]

	Epoxies	Poly-urethanes	Phenolics	Furan resins	Unsaturated polyesters	Vinylester resins
Type of curing reaction	Polyaddition	Polyaddition	Polycondensation	Polycondensation	Polymerization	Polymerization
Strengths	Good all-round properties; very good resistance to alkaline solutions and solvents; extremely high adhesive strength levels	Elasticity; optical appearance; relatively good acid resistance	Very good resistance to hot water, solvents, and acids	Very good resistance to solvents and alkaline solutions	Good resistance to acids (including oxidizing acids)	Good resistance to acids (including oxidizing acids) even at elevated temperatures; hot water resistance
Weaknesses	Attack by organic acids and concentrated inorganic acids	Attack by solvents and alkaline solutions	Attack by alkaline solutions	Shrinkage; attack by oxidizing agents	Shrinkage; attack by alkaline solutions	Shrinkage

epoxies, polyurethane resins, phenolics, furan resins, unsaturated polyesters, and vinyl ester resins. This indicates that phenolic resins exhibit good resistance to solvents, hot water, and non-oxidizing acids. Thus, phenolic resin systems are mainly employed for tiling adhesives and grouts used for acid-resistant tile flooring or in (acid-resistant) coatings for industrial laminates.

References

1. Schönthaler W (1991) Duroplaste: Zukunft von Anfang an. Brochüre of the "Technische Vereinigung", Würzburg, November 1991
2. Schwenk E (1982) 80 Jahre Kunstharze; fast vergessene Erfinder. Company brochure, Hoechst AG, Frankfurt a.M.
3. State Museum of Koblenz (1993) Bakelit, ein Werkstoff mit Zukunft. Publication of the State Museum of Koblenz
4. Hansen A (1993) Von der chemischen Grundlagenforschung zum synthetischen Werkstoff and Duroplastische Werkstoffe heute. In: Bakelit, ein Werkstoff mit Zukunft. Publication of the State Museum of Koblenz
5. Schröter S, Eckert A (1993) Phenolharze und Formmassen als erster Duroplast - Anwendungen heute. In: Bakelit, ein Werkstoff mit Zukunft. Publication of the State Museum of Koblenz
6. Green H (1993) Bakelit-Kunststoffe, Technologie der chemischen Großproduktion. In: Bakelit, ein Werkstoff mit Zukunft. Publication of the State Museum of Koblenz
7. Becker W, Braun D (1988) Duroplaste. In: Handbuch der Kunststoffe, vol 10. Carl Hanser, Munich
8. Brandau E (1993) Duroplastwerkstoffe, Technologie, Prüfung, Anwendung. Verlag Chemie, Weinheim
9. Ehrenstein GW, Bittmann E (1997) Duroplaste: Aushärtung-Prüfung-Eigenschaften. Carl Hanser, Munich
10. Baekeland LH (1908) DRP 23:3803
11. Gardziella A (1996) Duroplaste. Kunststoffe 86(10):1566-1578
12. Gardziella A (1996) Duroplaste - Statusreport zum Stand der Technik. Address at the Kunststoffe '96 Congress, April 1996 in Würzburg (Organizers: Süddeutsches Kunststoffzentrum in Würzburg, Carl Hanser in Munich and KI Verlagsgesellschaft in Bad Homburg)
13. Gardziella A, Schirber H (1992) Anwendungen von duroplastischen Werkstoffen im Automobilbau. Congress of the SKZ, September 23-24, Würzburg
14. Gardziella A, Baumgärtel K (1995) Duroplastische Werkstoffe aktuell. Congress of the Süddeutsche Kunststoffzentrum (SKZ), April 26-27, Würzburg
15. Gardziella A, Baumgärtel K (1997) Duroplastische Werkstoffe in der Elektronik- und Elektroindustrie. Congress of the SKZ, 4-5 June, Würzburg
16. Gardziella A (1996) Duroplaste - Statusreport zum Stand der Technik, Kunststoff-Kautschuk-Produkte, Yearbook 1996/97. Hoppenstedt, Darmstadt, pp 224-227
17. Gardziella A (1997) Prädestiniert. Plastverarbeiter 48:92-95
18. Böttcher A (1995) Zähelastisch und flammfest. Kunststoffe 8:1142-1144
19. Böttcher A, Pilato LA (1997) Phenolic resins for FRP systems. SAMPE J 33(3):35-39
20. Wittmann O (1997) Einsatzgebiete für Aminoharze. In: Gardziella A, Baumgärtel K (1997) Duroplastische Werkstoffe in der Elektronik- und Elektroindustrie. Congress of the SKZ, 4-5 June, Würzburg, pp X1-27
21. Roffael E (1997) Duroplastgebundene, umweltfreundliche Holzwerkstoffe. In: Gardziella A, Baumgärtel K (1997) Duroplastische Werkstoffe in der Elektronik- und Elektroindustrie. Congress of the SKZ, 4-5 June, Würzburg, pp XI1-25
22. Scheuer C (1997) Die neue Generation von Epoxidharzsystemen für Mittel- und Hochspannungsanwendungen. In: Gardziella A (1996) Duroplaste - Statusreport zum Stand der Technik, Kunststoff-Kautschuk-Produkte, Yearbook 1996/97. Hoppenstedt, Darmstadt, from Ref. 15, pp F1-14

23. Rogler W, Swiatkowski G (1996) Gießharze für Hochspannungsanwendungen. In: Gardziella A (1996) Duroplaste – Statusreport zum Stand der Technik, Kunststoff-Kautschuk-Produkte, Yearbook 1996/97. Hoppenstedt, Darmstadt, from Ref. 15 pp B1–24
24. Troitzsch J (1983) Brandverhalten von Kunststoffen. Carl Hanser, Munich, pp 471–473
25. Gardziella A (1985) Duroplastische Werkstöffe aktuell. Kunststoffe 10:1736–1741
26. Böttcher A (1997) Phenolharze für Faserverbundwerkstoffe. In: Gardziella A, Baumgärtel K (1997) Duroplastische Werkstoffe in der Elektronik- und Elektroindustrie. Congress of the SKZ, 4–5 June, Würzburg, pp II 1–10
27. Gardziella A, Solozabal R, Suren J, in der Wiesche V (1991) Synthetic resins as carbon forming agents for various refractories (carbon yields, analytical methods, structures and emissions). UNITECR Congress 1991, Aachen, Reprints, pp 260–264
28. Fries A (1995) Chemikalienbeständige Beschichtungen auf Basis duroplastischer Harze. In: Gardziella A, Baumgärtel K (1995) Duroplastische Werkstoffe aktuell. Congress of the Süddeutsche Kunststoffzentrum (SKZ), April 26–27, Würzburg, pp XII 1–17
29. Bühler HE (1991) Moderne Säureschutz-Systeme. Vulkan, Essen
30. Rauhut K (1991) Säurefeste Kitte. In: Bühler HE (1991) Moderne Säureschutz-Systeme. Vulkan, Essen, pp 34–45

CHAPTER 6

Economic Significance, Survey of Applications, and Six Bonding Functions

The following remarks will review the uses of phenolic resins and their distribution throughout the various areas of application, making reference to the six "bonding functions" that phenolic resins mainly assume in applications. The organization of the application-related chapters of this book is oriented toward these bonding functions. Tables 6.1 and 6.2 summarize the main types of resins – resoles and novolaks – that are commercially available and are used for various applications (solid and liquid resoles, solid and solution novolaks).

According to the chemical definition in ISO 10082, phenolic resins [1–3] are the condensation products of phenols and aldehydes, specifically formaldehyde, and are converted into high molecular mass polymers in a secondary (curing) reaction. The question as to a definition for the term *phenolic resins* or the historical product bakelite that arose from these can also yield other responses. Karl Ruisinger [4] writes that "bakelite is the stuff of which collector's dreams are made." In his article in Magazin Sammeln (Collecting Magazine) he showed numerous photos of utensils and toys made of bakelite (Fig. 6.1). The author ends his article with the nostalgic observation that "Objects made of bakelite disappeared from the scene at the end of the fifties." This is not entirely correct! Even today, the German Bakelite Company,

Table 6.1. Survey of resoles

	Examples of applications
1. Solid resoles	Molding compounds
2. Resole solutions	Interior protective varnishes, laminates
3. Aqueous resoles	Abrasives, refractories (e.g. MgO-C-bricks)

Table 6.2. Survey of novolaks

	Examples of applications
1. Solid resins (crushed)	Tire and molding compound production
2. Novolak solutions	Refractories and impregnating applications
3. Aqueous novolak dispersions	Coatings
4. Powder resins with HMTA	Brake linings, grinding wheels

Fig. 6.1. Toy horses of "bakelit" (source: collection of P. Flier)

founded in 1910, fabricates articles that maintain the positive features of bakelit: insulators and handles for irons as a final memory of a material that industry has stopped using and driven it into the arms of collectors".

Such words could cause one to burst into tears – but the tears would soon dry, since phenolic resins, the oldest fully synthetic resin products, form the basis for a very large number of materials of outstanding economical and technical importance, and have experienced continued favorable market acceptance in the 1980s and 1990s [5, 6].

The question as to a definition of what phenolic resins are can even be asked a third time. If one were to believe some articles in the press or the words of certain opinion makers [7], phenolic resins – because of their phenol and formaldehyde raw materials – contribute to human hazards due to their toxicity. Unfortunately, some branches of industry accept such occasional unqualified and biased remarks at face value, and initiate campaigns against the use of thermoset products with the intent of banning them from specific areas of application. Such attempts can be met by noting the fact that the level of monomeric residues in the cured, high molecular mass phenolic plastics is extremely low, and apart from that, phenol and formaldehyde – the starting materials of phenolic resins – have for years been among the products subjected to the most intensive toxicological studies. Furthermore, modern phenolic resins, when properly processed, are toxicologically trouble free if certain technical requirements are observed in processing. Practice has shown this to be true in the countless areas [8] in which this group of products finds applications (Table 6.3).

Table 6.3. Percentage distribution of phenolic resin production among important uses and percentages of resins used in various materials (based on various German and USA statistics)

Application	Percentage consumption of production		Percentage binder in material	Main types of resin
	USA	Germany		
Wood materials	55	27	approx. 10	Aqueous resoles

Insulating materials
  Inorganic fibers
  Organic fibers
	17.5	19	approx. 2 to 3
			approx. 30
				Aqueous alkaline resoles
				Novolak powder resins

Molding compounds	6	12	approx. 40	Solid novolaks or resoles
Laminates	6	10	approx. 30–50	Resoles (aqueous and solution)
Paints, adhesives	2.5	7	approx. 50	Resole and novolak solutions
Foundry, refractories	4.5	10	approx. 2	Solid novolaks, aqueous resoles, novolak/HMTA powder resins
Abrasives	1.5	3	approx. 12	Unmodified and modified phenolic novolak/HMTA powder resins, aqueous resoles
Friction linings	2	4	approx. 10	Unmodified and modified phenolic novolak/HMTA powder resins, aqueous resoles
Micellaneous	5	8	5–50	All types of resins

Phenolic resins exhibit a very wide range of uses. As a basic composition, it is necessary to expand application areas listed in Table 6.3 because in some cases they encompass very different fields, and are defined with difficulty. Applications as varied as decorative laminate moldings, textile and paper-based laminates, electrical laminates, and special-purpose laminates for aircraft construction are grouped under the heading of "laminates". Inorganic materials such as mineral wool that are used for acoustical and thermal insulation in the building trade, organic insulating materials and moldings for installation in motor vehicles, such as textile mats, and insulating foam are all considered as "insulating materials". Similarly varied situations also exist in the other listed areas.

Table 6.3 reveals significant differences in the main applications when the markets in Europe and the United States are compared. On the US-American market, about 72–75% of the phenolic resin volume is used in the areas of wood materials and insulating products, which mainly involves the building industry. In Germany, a situation that also largely applies to the remainder of western Europe, only 45–47% of the volume goes into the building sector. On the other hand, the volume of phenolic resin used for molding compounds (in Germany 15%) is higher than the percentage in the USA (6%).

As varied as the applications of phenolic resins are, the levels of binder in the materials produced from them are equally diverse. These levels range from barely 2% in foundry molding sands to more than 50% in coatings and laminates. Based on the consumption figures in Table 6.3 and the binder levels in the materials, it may be estimated that a quantity of nearly 3.5 million tonnes of materials can be produced using phenolic resin binders at an annual rate of consumption of about 250,000–260,000 tonnes of phenolics in Germany (about 1.4–1.6 million tonnes in the USA). This volume of phenolic resin bonded materials is of the same order of magnitude as that of the mass-produced thermoplastics.

The "miscellaneous applications" are also relatively varied. These include chemical-resistant building components and acid-resistant cements, fiber-reinforced plastics (composites), fishing rod prepregs, floral foam, brush and lampbase cements, adhesives, industrial filters, rubber additives, carbon and graphite materials, auxiliaries for petroleum production, binders for carbonless copy paper, and phenolic resin fibers. Modifications relating to raw material availability and environmental questions have also been carried out in these areas of application in recent years. On the whole, the application-related versatility of phenolic resins as a component of important and highly varied materials is outstanding.

This versatility arises from circumstances including the fact that phenolic resins as binders are encountered in six different main functions with various application-related attributes (Table 6.4). These functions are:

1. *Permanent bonding,* which ensures dimensional stability and resistance to external influences and mechanical stress even after the phenolic-bonded materials are put to use.
2. *Temporary bonding,* which provides the materials with the highest possible resistance to wear when these are subjected to abrasive or destructive use, coupled with high thermal resistance over the longest possible period of time and where appropriate lengthy resistance to sudden heat development or mechanical effects.

Table 6.4. Six phenolic resin bonding functions with examples of applications and relevant properties of these

Bonding function	Example of application	Required properties (examples)
Permanent	Molding compound	Dimensional stability
Temporary	Abrasive	Good grinding rate
Intermediate	Refractory	Thermosetting
Complementary	Adhesive additive	Good adhesion
Carbon forming	Carbon material	High carbon yield
Chemically reactive	PUR foundry binder	Cures by amine gassing

3. *Intermediate bonding*, which affords the greatest possible accuracy in shaping and ensures that dimensional relationships exactly correlate with those of the final products during later curing and carbonization processes.
4. *Complementary bonding*, which in combinations significantly improves the range of properties of other thermosets, elastomers, or thermoplastics, for example the strength, wear and chemical resistance, and adhesion.
5. *Carbon-forming bonding*, which allows production of polymeric carbon by carbonization – if appropriate following prior shaping – to afford a carbon yield of around 50–70%, and opens paths to production of ultramodern carbon-based materials and refractories.
6. *Chemically reactive bonding*, which permits phenolic resins to be transformed into compounds of other important synthetic classes, allows production of other functional products from phenolics, or enables them to be used in non-binder functions, for example as dye developers for carbonless copying paper.

In some cases, for example in binders for foundry cores, all functions can even be related to a single application. Thus, the binders should provide the cores with good permanent dimensional stability as long as they are in storage. Intermediate handling of the core should be good, and the intermediate dimensions transferred to the casting in an inverse but dimensionally accurate form when the core is poured off. When pouring the molten metal, the cured binder system should withstand temperatures as high as 1500 °C for a short time (temporarily). In the uncured state, the binder system should be capable of being adapted with modifiers (complementary bonding); it should form a considerable amount of carbon under pyrolytic conditions, and should exhibit a certain degree of chemical reactivity. The main function in this case is temporary bonding.

On the other hand, the different functions of phenolic resins naturally result in development of various special-purpose resins, groups of resin, and resin systems that may then be very specifically used to achieve an explicit application function. Examples of such applications will be mentioned in the tables of the following application-related chapters, and in some cases discussed. Although they do not represent simple "standard" resins, phenolics that can be used in multiple bonding functions are of particular interest. Examples of these are heterogeneously modified phenolic powder resins that can be used to produce special flame-resistant textile mats used in automotive construction, and in slightly modified form to fabricate particularly heat-resistant brake linings for motor racing. Specific low-melting novolaks can also be used for various application/binding functions when processed by the warm mixing method.

6.1
Permanent Bonding

The products in which phenolic resins afford permanent, i.e. continuous bonding in a narrow sense include moldings and sheet goods made with molding compounds; molded laminates, wood materials, inorganic and organic insulating materials, fiber composites, chemically resistant components, and various cements. Permanent bonding means that the resin bond produces a stable range of properties over a lengthy period of time. This refers to properties such as the long-term resistance of moldings to thermal effects or environmental exposure. The principal of permanent bonding is that the original properties of products are retained unchanged, or change only slightly, over an extended period of time and under a wide variety of environmental conditions.

These properties include the resistance to moisture and aggressive chemicals. Phenolic resin bonded wood materials such as compreg exhibit good resistance to moisture and environmental effects, and retain this range of properties even after lengthy exposure. Acoustic and thermal insulating materials should retain a constant level of acoustical and/or thermal properties over long periods of time. Phenolic resin bonded textile mats that are used as insulating materials and in automotive construction exert a load bearing and/or noise reduction function in the passenger compartment, the trunk, or under the hood should ideally be even more permanent than the vehicle itself, or at least last throughout its lifetime without any loss in function.

Particularly long-term service is expected in the case of the properties phenolic or epoxy resins provide to electrical laminates, the base material for printed circuits. Long-term electrical insulating properties and dimensional stability are of great importance for trouble-free operation of electronic equipment. Glass fiber-reinforced interior components for aircraft construction made with specially modified phenolic resins must feature long-term material reliability in their area of application.

6.1.1
Wood Composites

6.1.1.1
Introduction

Phenolic resins are one of the major adhesives in binding wood of all sizes and shapes into a wide array of products such as panels, molded products, lumber, and timber products. Besides phenolic resins, other adhesive resins consist of F reacted with urea and/or melamine leading to UF, MF, and combinations thereof. Methylene diisocyanate (MDI) is also used as a wood adhesive. Pizzi has published two recent books which describe wood adhesives in greater detail [1, 2].

New developments of raw materials from regenerable materials as either adhesives (lignin or tannin) or as panel components to be bonded (straw, jute, bamboo, bagasse, etc.) are discussed. Panels or wood composite products consist of plywood, particleboard (PB), waferboard, medium density fiberboard (MDF), and oriented strand board (OSB). Glulam, laminated veneer lumber (LVL), and recently engineered products such as laminated strand lumber (LSL), timberstrand, and parallel strand lumber (PSL), or parallam, constitute engineered lumber products. These wood composite products represent a volume in excess of 50.6 million m³ in North America for 1997.

6.1.1.2
Wood Structure/Surface

The composition of wood is reasonably similar in most trees with cellulose ranging from 40–45%, hemicellulose 20–30%, and lignin 20–30%. Wood is a fibrous material with a porous network structure. The cellulosic component of wood has a high degree of compatibility with many resins. Some resins penetrate into the wood "pore" structure to facilitate a mechanical bond as well as bond directly with cellulose.

A fundamental understanding of wood surface chemistry is necessary to perceive changes in wood sources (early harvested young vs mature trees) as well as different wood species. Surface characterization of these and other woods can assist in the design of new and improved adhesive systems. Chemical functional group characterization and/or types of chemical components can provide a better understanding of the exposed surface of the wood. The relative ease or lack of wettability of their surface is important and may relate to hydrophilic or hydrophobic behavior as the wood is treated with the adhesive. Functional groups located on the surface contribute to chemical interactions that occur between the wood substrate and the adhesive. Low molecular weight extractibles diffuse to the surface and affect wettability. Extractibles are many and include starches, sugars, tannins, fatty acids, alcohols, sterols, proteins, terpenes, resinous materials as well as inorganic compounds and minerals. These latter materials are absorbed in the wood system from soil. Contact angle measurements [3, 4] and XPS [3] indicate that the wood surface behaves as typical polymer surfaces when exposed to environmental changes and suggest that extractibles influence wood surface. Depending on the type and remaining extractibles, molecular orientation and reorientation facilitated by the extractibles may occur depending on environmental conditions to which the wood is exposed. Furthermore, differences in polar characteristics [solubility parameter, δ] of aqueous adhesive resins vs less polar MDI will influence wettability as well as depth of penetration of the adhesive into the wood surface.

Surface effects of wood and wood composites have been examined by X-ray microtomography. Studies conducted by Shaler and coworkers [5] using the Brookhaven National Synchrotron Light Source and X2B beamline facility (developed by Exxon Research & Engineering) revealed the 3-D structure of wood specimens viewed from any angle; they also carried out sectioning of the material to examine internal features. Internal features such as intraflake cracks in OSB, 3-D fiber orientations in MDF, and interphase region of pultruded glass phenolic composite bonded to solid wood are observed and provide a unique "picture" of these wood composites.

6.1.1.3
Wood Sources

Another issue which is critical and relates to wood/adhesive interactions is wood source. The wood source for plywood can be either softwood or hardwood with the former in general purpose industrial use while the latter is used in those areas where appearance is important. Mature softwood (Douglas fir, southern pine) is generally the wood source for plywood. Softwood undergoes periodic shortages of logs due to depletion and governmental restrictions in the Pacific Northwest (North America). Environmental protection of North American public lands continues to reduce the annual harvest of softwood timber. Scarcity of wood source plus only about 75% log utilization of useful veneer for plywood prompted the development of a new wood composite product in the early 1960s.

A wood panel product based on wafers (later strands) known as waferboard was produced from fast growing small diameter trees such as aspen, young southern pine, and other species. Virtually the entire tree could be transformed into wafers or strands. Eventually this technology led to the development of oriented strands and a composite panel known as oriented strand board (OSB). OSB significantly displaced waferboard and with suitable strand alignment competes successfully in plywood markets. Construction of North American style residential housing consumes large quantities of plywood for uses such as sheathing, flooring, and roofing. From the 1980s to the present, 69 OSB mills are operating in North America and Europe with more planned even in the Far East and South America. Schematic process for OSB is shown in Fig. 6.2.

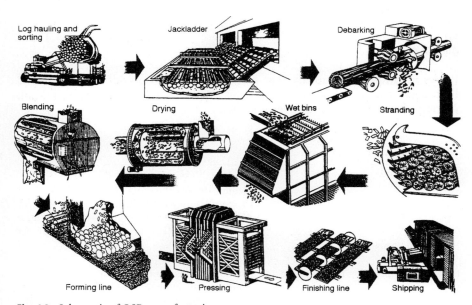

Fig. 6.2. Schematic of OSB manufacturing process

There is also a further shift in raw material source, and it relates to nonwood materials such as straw and/or bast fibers (flax, hemp, kenaf, ramie, jute). Strawboard (wheat or rice) particleboard plants with wheat straw as raw material source are operating in Wahpeton ND, Le Center MN, and Elie, Manitoba, Canada with many more in the planning stages. Thus different wood sources as well as raw materials derived from annual agricultural products are factors which identify with current and newly developed phenolic resins for panel manufacture. Turnkey operations are available (Compak, see later) to transform agriwaste into panels.

6.1.1.4
Wood Composite Materials

6.1.1.4.1
Wood Panel and Engineered Lumber Products

Wood composites involve quite an array of products ranging from panels, molded products to large lumber products. Maloney [6] proposed the following classification of these materials (Table 6.5). The listing provides an identity of familiar products or panels as well as newly evolving engineering lumber materials such as LVL, Parallam®, and Timberstrand®. Table 6.6 summarizes panel and lumber products, sales volume, wood sources, and adhesive types. Physical properties of many of these wood composite materials are listed in Tables 6.7 and 6.8. Table 6.8 lists allowable design stresses for code approved products of engineered lumber compositions.

6.1.1.4.2
Adhesives

It was mentioned previously that several types of wood adhesives are used in the fabrication of wood composites. These are UF, MF, PF, and MDI. A comparison of these adhesives is shown in Table 6.9.

Table 6.5. Wood composite materials

I	Panels	Plywood
		Medium density fiberboard (MDF)
		Oriented strand board (OSB)
		Waferboard
		Particleboard (PB)
II	Molded products	Automobile panels
		Door skins
III	Lumber/timber engineered products	Laminated veneer lumber (LVL)
		Parallel strand lumber (PSL) Parallam®
		Laminated strand lumber (LSL) Timber Strand®

Table 6.6. Wood panel and lumber products, 1997 North America [7]

		Sales vol (million m³)	Wood sources	Adhesives
Panels	Plywood	20.2	Hardwood, softwood	PF
	OSB	13.0	Aspen, pine	PF, MDI
	PB	10.3	Wood waste	UF[a]
	MDF	6.0	Wood fibers	UF[b]
Engineered lumber Products	LVL	1.12	Veneer	PF, PRF
	Parallam®	Small	1.25 × 94 cm strands	PF, MDI
	Timber Strand®	Small	30.5 cm strands	MDI

[a] Small amounts of phenolic, for construction use.
[b] Some PF, lignin phenolic and MDI.

Phenolic Resins

Phenolic resins, primarily resoles, are the preferred adhesive for exterior wood products. Hence for plywood, particle board (water resistant), MDF, OSB, and WB, different types of powdered and liquid resoles are utilized. Powdered resins originate from spray dried resoles. Many different factors are considered in the design of a suitable phenolic resin for each of these wood composites. These factors consist of resin characteristics and process conditions in the manufacture of the wood product.

Resin Characteristics:
- Molar ratio of F/P and type/content of catalyst
- Viscosity of liquid resole
- Low free F and P in resin
- Shelf life/storage stability at room temperature
- Water dilutability

Process Conditions:
- Wood moisture content
- Additives, extenders
- Method of introduction of adhesive (liquid or solid)
- Press temperature and residence time
- Pressing cycle: s/mm board thickness (6–12 s/mm)

Panels

Plywood

An excellent review of plywood resins and additives was published by Sellers [8]. Plywood consists of at least three layers of wood, mostly veneers, with the application of adhesives, pressure, and heat. Plywood is identified as either interior or exterior grade. Exterior plywood retains its glue bond when repeatedly moistened and dried or subjected to vigorous weather conditions for constant exposure.

Table 6.7. Mechanical properties of composite wood materials

Type of board Properties		Plywood		Fiber board		Particle Board	
		Veneer Plywood	Core Plywood	HDF	MDF	Furniture V 20	Construction V 100
Thickness	mm	12–15	16–19	4	6–19	19	19
Specific gravity (SG)	Kg/m³	550–700	450–650	900	680–750	620	700
Flexural strength ∥	N/mm²	60–100	10–15	40–60	10–40	19	22
Flexural strength ⊥	N/mm²	20–40	–	43	18–35	18	20
Mod. of elast. (MOE) ∥	N/mm²·10³	7.0–12.0	1.0–1.5	–	2.0–2.2	3.0	4.5
Mod. of elast. ⊥	N/mm²·10³	1.5–3.0	–	–	2.0–2.1	2.8	3.3
Internal bond (IB)	N/mm²	–	–	0.45	0.18–0.70	0.38	0.45
Internal bond, wet	N/mm²	–	–	0.10	0.02–0.20	–	0.20
Int. bond, dry again	N/mm²	–	–	0.35	0.15–0.65	–	0.32
Thickness swelling, 2 h	%	–	–	12	3–12	4	8
Thicknes swelling, 24 h	%	–	–	18	6–17	11	14
Thickness swelling, 120 h	%	–	–	22	22–80	16	20
Water absorption, 24 h	%	–	–	45	15–35	–	–
Water absorption 240 h	%	–	–	70	–	–	–
Swelling (lengthwise)	%	0.1–02	0.2–0.3	0.2–04	0.2–0.3	0.2	0.3
Shrinkage	%	0.2	0.2	0.3	0.3	0.2	0.2

Table 6.8. Specifications for engineered lumber products[a]

	LVL	Parallam	Timber strand
Design stress			
Flexure (mPa)	18	20	10–14
Shear modulus of elast.(MPa)	821	863	518
Modulus of elasticity (GPa)	13.1	13.8	8.3
Compression $\perp$ to grain:			
‖ to glueline (MPa)	5	4.5[b]	4.3[b]
Compression ‖ to grain (MPa)	16	20	10
Horizontal shear $\perp$			
Glueline (MPa)	2	2[c]	2[c]

[a] Allowable design stresses for code approved products, Trus Joist McMillan brochures.
[b] Parallel to wide face of strands.
[c] $\perp$ wide face of strands.

Table 6.9. Comparison of wood adhesives

	Process conditions		Panel properties			
Type	Cure catalyst	Press speed	Moisture content[a]	Water resistant	Free CH_2O	Relative cost
UF	Yes	fast	Effect	No	~0.1 ppm(E1)	1
MF	Yes	Medium to fast	Effect	Yes	Low	3
PF	No	Medium	Effect	Yes	Low	2
MDI	No	Fast	No effect	Yes	None	4–5

[a] Wood.

In North America typical plywood specifications are product standards established by the industry with assistance from NIST, U.S. Department of Commerce. Plywood is covered by U.S. Product Standard, PS1–95 [9]. These specifications include requirements for species, grade, thickness of veneer and panels, glue bonds, moisture content, etc. Standardized definitions of terminology, suitability, performance test specifications, minimum test results, and code explanations stamped on each panel are also provided. German classification according to DIN 68705 is contained in Table 6.10. Typical application uses of plywood include containers, marine-grade materials, concrete forms, exterior sidings, kitchen and bathroom cabinets, door skins, interior parts in transportation vehicles, and foundry patterns.

Resins, Additives, and Formulations. Sodium hydroxide catalyzed resoles are used for weather resistant plywood. Resin viscosity varies between 700 and 4000 mPa·s and relates to either a dry or wet gluing process. Typical resin properties are shown in Table 6.11 [10]. Free phenol and free formaldehyde are low (0.1%), and the resins are low in air emissions. Kim [11] et al. have

Table 6.10. Plywood classification (DIN 68705) according to humidity resistance and adhesive

Class according to DIN 68705	Adhesive	Property	Pre-treatment of samples
IF 20	Urea resins	Resistant to low humidity, usual to closed rooms	24 h in H_2O at $20 \pm 2\,°C$
IW67	Urea and urea/melamine resins	Resistant to higher humidity, resistant to water up to 67 °C for a short time	3 h in H_2O at 67 ± 0.5 and 2 h in H_2O at $20 \pm 3\,°C$
A 100	Urea/melamine resins	Outdoors limited resistance	6 h in H_2O at 100 °C
	Melamine resins	resistant to cold and hot water	2 h in H_2O at $20 \pm 5\,°C$
AW 100	Phenol resins Phenol/resorcinol resins Resorcinol resins	Resistant to all climatic influences	4 h in H_2O at 100 °C, 16–20 h in air at $60 \pm 2\,°C$ and 4 h in H_2O at 100 °C, 2–3 h in H_2O at $20 \pm 5\,°C$

Table 6.11. Resole resins for plywood manufacture [10]

	Predrying	Without predrying
Solids	43–45	45–47
Viscosity @ 20 °C (mPa · s)	3500–4500	550–750
Alkaline content (as NaOH)%	2.7–3.0	7.5–8.0
Gel Time @ 100 °C min	23–29	30–40
Water dilutability	Indefinite	Indefinite

shown that plywood resins have MWs up to 114,000 daltons (determined by light scattering). Mark Houwink Sakarada (Structure chapter 4) values indicate a compact or coiled molecular structure due to molecular association and branching of the polymer chain. Evaluation of plywood resin by DMA (Chemistry, Reaction Mechanisms Chapter 2) by Kim et al. [12] involves the cure behavior of plywood resin on glass cloth and provides cure parameters in the manufacture of plywood. Since resin exhibits a high MW, a high temperature of about 150 °C is required to develop high rigidity and full cure of the resin.

To adjust wetting and avoid excessive penetration for uniform joint thickness, diluents and/or fillers are introduced into the formulation. Rye or wheat flour is used in combination with UF and MF resins while $CaCO_3$ or non-swelling fillers such as coconut shell flour are used for exterior grade plywood formulations. An excessive amount of fillers should be avoided since they lead to a reduction in plywood properties. Adhesive formulations are listed in Table 6.12.

Production of Plywood. Either softwood or hardwood can be used in plywood manufacture. Type of wood determines final use of plywood. Veneers are care-

Table 6.12. Adhesive formulations for the production of weather-resistant plywood [10]

Formulation and processing		Dry process		Wet process	
Phenol resin	Pbw	100	100	100	100
CaCO$_3$	Pbw	–	10	10	10
Coconut flour (300 Mesh)	Pbw	5–10	8	8	8
Water	Pbw	10–20	–	–	–
Hardener (paraformaldehyde)	Pbw	–	–	2	2
Accelerator (resorcinol resin)	Pbw	–	–	–	6
Adhesive formulation service life		1 d	1 d	1 d	4 h
Press temperature	°C	130–150	130–140	120–125	100–110

fully dried prior to coating with adhesive formulation. Moisture content should be 2–5% for dry and wet process. The formulated adhesive (Table 6.12) is placed on both sides of the veneer which is then inserted between two plies of uncoated veneer in the panel assembly. Roll spreaders apply the adhesive with spread rates of 160–200 g/m². A rate of 130–140 g/m² is sufficient for low porosity veneers. A thin adhesive or "glue" line reduces costs but may contribute to possible glue defects. After coating, the veneers are dried in a veneer dryer to a 8–12% RH (predrying) with temperature not exceeding 70–75 °C. Veneers are assembled and pressed in a multi-daylight press. Depending on the adhesive formulation, press temperature can be 100–140 °C with pressures of 0.8–1.5 N/mm² for softwood and 1.5–2.5 N/mm² for hardwood. The use of accelerators allows a lower cure temperature. Resorcinol resins (see later) will facilitate cure at room temperature. After pressing, the rough plywood is trimmed to size and sanded to the desired thickness. Testing of exterior type plywood consists of vacuum-pressure tests, boiling tests, scarf and finger joint tests, and heat durability tests [8]. Mechanical properties of plywood are shown in Table 6.7.

Processing changes are occurring in plywood gluing systems. Spreader curtain coater and spray methods are being converted in some cases to foam extruders. New plywood formulations of phenolic foam glue mixture result in less water on the glue line so that high moisture veneers can be used [13]. Gluing with higher moisture content veneer and foaming the glue results in reduced adhesive use and lower production costs. Thus 10% moisture content veneer instead of 3–4% moisture allows higher veneer dryer productivity, lower glue spreads, and better pre-pressing. Gluing veneer with high moisture content (up to 15%) results in a wood product which is very similar to the equilibrium moisture value in actual use and thus reduces warp and dimensional changes.

Co-Reaction of PF with UF. In North America virtually all exterior grades of plywood, particleboard, hardboard, and OSB are manufactured with phenolic resins. A wide range of formulations of both liquid and solid resins is available that meet product and property performance characteristics of the final panel. Still there are other desirable improvements which are requested by wood panel manufacturers. Most prevalent requests are low cost resin and faster cure. It is these two features that have been the motivating factors for an ongoing study conducted by Professor Tomita and coworkers [14–20]. Tomita has examined the mechanism of the co-reaction of urea formaldehyde resin with phenolic resins. The resulting hybrid resin system could be an attractive wood adhesive due to beneficial attributes of each resin segment:

U/F	P/F
Low cost	Moisture resistance
Fast cure	Low F emission
	Heat resistance

Since 1985 Tomita and co-workers have been actively involved in developing a resin system as UFP to meet these objectives. Using torsional braid analysis (TBA) to compare the relative amount of rigidity or cure (Gt/Go) vs temperature, Tomita was able to distinguish the cure differences of UF, PF, blend of UF and PF, and UFP (Fig. 6.3). In Fig. 6.3 curves for PF resin cure (curve 1), UF cure (curve 4), blend of UF and PF (curve 3) and co-condensed UFP (curve 2) are shown. A temperature greater than 100 °C is required to cure the PF resin whereas the use of a UF catalyst (NH_4Cl) facilitates the cure of the UF resin below 80 °C. The blend of UF and PF exhibited separate segments of UF and PF undergoing cure (curve 3). The cure pattern of the co-condensed resin UFP (curve 2) demonstrated that both UF and phenolic segments undergo cure with increasing temperature with the cure composition exhibiting high heat

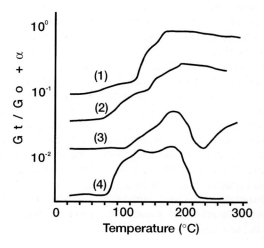

Fig. 6.3. Torsion braid analysis of curing reaction of various resins. *1)* Commercial resole; *2)* Co-condensed resin UFP; *3)* Blend of UF resin and resole wih NH_4Cl; *4)* Commercial UF resin with NH_4Cl

resistance due to the phenolic segment. To achieve this co-cure reaction required an understanding of the different condensation reactions (UF, PF, and co-condensation) with model compounds. The composition of the co-condensed resin is F/U/P: 3.5/1/1. In evaluating the co-condensed composition in plywood, shear tests at 160 °C were required to obtain properties of 18.2 kg f/cm^2 with 85% wood failure. But 72 h boil test results were inferior to plywood with normal resole. More work is required on co-condensed resin microstructure to reduce water sensitivity.

Color Improvement. For color improvement and reduction in cure temperature or lower press cycle, melamine is introduced into the phenolic composition leading to a PMF [21, 22]. This method is used in Japan and to some extent in Europe. Both phenol and melamine are comparably priced in the Far East. Lighter plywood and fast cure is obtained but some urea or UF resin must be added to reduce F emissions.

Plywood is faced with severe competition from OSB (see later) and thin MDF. Thin MDF (see MDF section) is a competitor to hardwood plywood. With high quality overlays or prints, MDF appears the same as hardwood plywood and is capturing a portion of the hardwood plywood market.

Oriented Strand Board (OSB)/Waferboard
Most of the early adhesive development studies for waferboard were applicable to OSB since both wood wafers and wood strands were treated with powdered or droplets of phenolic resin. As the OSB market became more focused in capturing some of the plywood market, a greater effort was directed to OSB resin development and less to WB. Prior to discussing OSB resins, it is important to distinguish the differences between gluing plywood and OSB/WB. The plywood panel is fabricated with a known amount of gluelines. Each of these gluelines contains a uniform coating of liquid phenolic resin. During hot pressing, all the wood at the surface of the glueline comes in contact with the adhesive. Thus processing conditions during application of the adhesive, moisture content of the veneer, and drying of the adhesive layer all influence glueline strength. If failure occurs through the adhesive rather than the wood, a dark color will be observed throughout the failure surface – the dark color is due to the phenolic resin cured appearance. If failure occurs through the wood, the failure surface is light or the color of the wood surface. On the other hand, OSB/WB phenolic resin is applied as discrete particles (powder) or droplets (liquid). Some flow of the phenolic resin may occur during hot pressing, but it does not result in a continuous coating as in plywood. Obviously resin content in plywood is greater than in OSB/WB. Further panel property differences exist between plywood and OSB. Deflection under load, creep and linear expansion in a humid environment, and thickness swell are all greater for OSB as compared to plywood. However OSB is less expensive than plywood. A comparison of properties of plywood and OSB are shown in Table 6.13.

Table 6.13. Mechanical properties of U.S. composite wood materials used in sheathing (ASTM test methods)[a]

	Plywood	OSB
Linear hydroscopic expansion (1%) (30–90% RH)	0.15	0.15
Linear thermal expansion (cm/cm/°C)	6.1×10^{-6}	6.1×10^{-6}
Flexure		
MOR (Mpa)	21–48	21–28
MOE (GPa)	7–13	5–8.3
Tensile Strength (MPa)	10–28	7–10
Compressive strength (MPa)	21–35	10–17
Shear through thickness (edgewise shear)		
Shear strength (MPa)	4–8	7–10
Shear modulus (GPa)	0.5–0.8	1.2–2.0
Shear in plane of plies (rolling shear)		
Shear strength (MPa)	1.7–2.1	1.4–2-1
Shear modulus (GPa)	0.14–0.21	0.14–0.34

[a] Source: APA, Tacoma WA.

Core and Surface Resins. During manufacture the OSB panel experiences different temperature conditions from the core to the surface during hot pressing. Hence two different resin systems were developed for OSB: core resin and surface resin. The core resin is for the middle layers of the panel where a slow temperature rise occurs and the final temperature is lower than the face (surface) layers on the outside of the panel. Most OSB manufactured in North America utilizes liquid and powdered (spray dried resoles) and some MDI. The MDI is primarily used in the core whereas phenolic resin can be used on the face/surface panel. Major disadvantage for using MDI as surface resin is that it adheres strongly to press plates. Approximately 2–2.5% powdered phenolic resin is employed with wood strands containing moisture content of 6% whereas with faster reacting liquid phenolic resins, higher moisture content of 8–9% can be tolerated but a higher amount of liquid resin (3.5–4%) is necessary.

A liquid phenolic resin for OSB (outer layer) is shown in Table 6.14.

Efforts of Steiner and Ellis [23–27] have been effective in providing an understanding of cure and flow parameters for powdered phenolic resins as they are used as face resins, core resins, and resins for use throughout the panel. Flow properties determined by TMA, fusion diameter values, and stroke cure time were compared with GPC and NMR data. Both laboratory and

Table 6.14. OSB resin (outer layer) [10]

Solids (%)	43.5–45.5
Viscosity @ 20 °C (mPa · s)	200–300
Alkaline content (as NaOH) %	2.3–2.8
Gel time at 100 °C, min	14–19
Water dilutability	Indefinite

commercially available resins were evaluated in the studies. NMR analyses indicated that the commercial resins possessed a molar ratio of 2:1 of F to P and pH values of approximately 9.0. Flow was related to MW whereby reduced flow was attributable to increased resin MW. Flow data indicated that the face resins exhibited the highest flow; resin for throughout the panel showed intermediate or reduced flow while core resin had the least flow. Further examination of the role of MW and MWD was conducted by Ellis [25] who spray dried low and high MW resins (Table 6.15, Fig. 6.4) and subsequently prepared mixtures of both resins for evaluation as OSB/WB adhesives. A bimodal resin composition containing 20% low MW resin and 80% high MW resin provided optimum internal bond (IB) strength and low thickness swelling. The high MW resin portion facilitates rapid bond formation and gap-filling properties while the low MW resin is necessary for wood substrate penetration. Studies (Ellis and Steiner [24]) of optimum particle size of spray dried phenolic resins

Table 6.15. Characteristics of liquid resins

	Resin	
	Low	High
Resin solids (%)	38.8	38.0
pH	10.1	10.1
Free formaldehyde (%)	0.30	0.29
Specific gravity	1.16	1.15
Viscosity at 25 °C (mPa · s)	34	4990

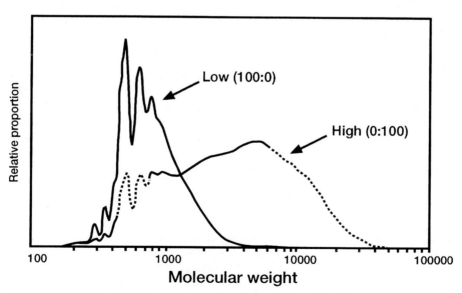

Fig. 6.4. MWD of low (100%) and high (100%) MW resins (acetylated)

and spray drying variables (inlet temperature, feed rates, resin solids content, etc.) provided valuable information on spray drying methodology and parameters leading to useful powdered resoles.

Resin Distribution by Imaging. Both Ellis [27] and Kamke et al. [28] report a novel imaging method to measure the distribution of resin/wax on wood surface. By employing a fluorescence microscope, video camera, frame grabber board, and PC/image processing analysis software, Ellis succeeded in determining wood failure values for internal bond specimens and obtained a correlation between wood failure and internal bond. Kamke observed depth of resin penetration of both liquid and powdered phenolic resins as well as MDI. His observations showed that powdered resin remains on the surface of the wood flake during the blending process with resin flow occurring during heat up; liquid resin wets the surface of the wood flake on contact but also partly penetrates into the porous structure of the flake; MDI with lower viscosity and "strong attraction" to wood exhibits the smallest volume on wood surface. Kamke's studies encompassed OSB strands from several panel manufacturers and showed MDI to be drawn very tightly into the longitudinal cell elements of the wood. Powdered phenolic resin was observed as clusters of fine particles on the wood flake surface while liquid phenolic resin was distributed between the flake surface as well as some penetration within the cell of the wood. Kamke claims the method has merit as a QC procedure for efficiency of resin spraying and flake blending processes.

The continuing interest in developing adhesives to bond wood with high moisture content (12–15%) is recognized as providing improved economics in panel manufacture as well as improved panel quality. Andersen and Troughton [29] identify some of the benefits of using higher MC in wood: improved dryer capacity, reduced flake breakage, improved wood appearance/yield, and better panel dimensional stability. It is also plausible that the development of fast curing, moisture tolerant phenolic adhesive may also be of value to the newly evolving steam injection pressing method (see later). The authors relied on some earlier observations [30] whereby plywood resin was transformed into dispersion resin and combined with an alkaline soluble phenolic continuous phase. The resulting system was effective in bonding veneer with MC of 10–12% and suppressed the problems associated with excessive adhesive migration ("wash out") from the glueline and resin cure retardation. Thus the strategy involved the transformation of a plywood resin with an OSB resin into the moisture tolerant adhesive. The rationale for combining the OSB resin to the plywood dispersion resin was to facilitate the precipitation of the OSB resin onto the plywood resin dispersion. The OSB resin is a relatively low MW resin and becomes an oily phase on normal precipitation. When it is precipitated in the presence of dispersion plywood resin, a granular composition is obtained. It is believed the high MW plywood resin serves as a site for the OSB resin to precipitate. OSB core and surface resins were examined with dispersion plywood resin resulting in intermediate flows when compared to thermal flows of commercial OSB surface and core resins. Using wood with MC of 12–15% these newly developed adhesives were superior in comparison to resin controls, resulting in excellent bonds exceeding 80–90% after 4–5 min pressing at 204 °C with a press load of 1.1 MPa. These results were also applicable to waferboard with high MC.

MDI. The use of MDI is beneficial when used as a core resin in OSB. MDI rapidly cures even with high MC in wood, providing a strong, moisture resistant bond with no formaldehyde emitted. It cannot be used as an OSB surface resin because MDI strongly adheres to the hot platens. The amount of MDI used as wood adhesives in the manufacture of wood composites is about 8–10% of the total wood adhesives volume or approximately about 100 million pounds/yr.

Particleboard
More than 95% of the total amount of adhesives used to bond wood waste materials into particleboard in North America is urea formaldehyde (UF). They are low in cost, easy to use, and are fast curing but F emissions must be controlled. Approximately 10.3 million m^3 of PB was manufactured in the US (1997). Most of the PB is used for indoor applications such as furniture and shelving. In Europe some PB is utilized as construction panels for outdoor use and requires water resistant bond by employing phenolic resin. A typical phenolic resin for PB is shown in Table 6.16:

Classification of PB. The use of UF for PB has led to a classification of PB as E1, E2, and E3. These relate to the amount of formaldehyde emitted using the Perforator test. Value of E1 is ≤10 mg F/100 g material and equates to 0.1 ppm F value using 1 m^3 chamber. Recently Wolf [31] reported the development of UF resins with low formaldehyde release similar to natural wood. The amount of formaldehyde released by natural wood is dependent on the wood species, humidity, and drying conditions. The value varies from 0.2–0.6 mg/100 g or 0.002–0.006 ppm (1 m^3 chamber). An E-zero product (EO) which is below the current limit of 0.1 ppm has been developed and utilizes either MUF or UF + MUF. EO emits 0.01–0.015 ppm (1 m^3 chamber) and ≤1.0 mg/100 g (perforator).

Most PB meets E1 classification and is due to improvements in the manufacture of UF resins (reduction of molar ratio of F:U) and the use of formaldehyde scavengers.

Table 6.16. Particleboard resins [10]

	Core layer	Outer layer
Solids (%)	47–49	43–45
Viscosity @ 20 °C (mPa · s)	700–800	200–300
Alkaline contact (as NaOH)%	8–9	2.5–3.0
Gel time @ 100 °C, min	30–37	15–20
Water dilutability	Indefinite	Indefinite

Emissions Studies. Studies related to VOC emissions during PB panel manufacture indicate that lower emissions are observed when lower press temperature and longer press times are employed in panel production. The VOCs of PB with UF binder were identified as F, methanol, and ammonia using TGA/FTIR [32, 33].

In North America, round robin testing of press emissions emanating from UF and PF resins is being conducted by large wood adhesives manufacturers (Neste, GP, and Borden) using similar laboratory caul plates, sampling devices, and analytical methods. Data is shared between resin producers and panel manufacturers [32].

Medium Density Fiberboard (MDF)
MDF is the most important fiberboard product with an SG of 0.6–0.8 g/cm^3. A recent review by Chapman [34] mentions different plant species for MDF panels. In 1997 the volume of MDF in the US was 6.0 million m^3, 6.5 million m^3 in Europe with continued growth worldwide. It is used in furniture and some similar PB areas. A comparison of properties of PB and MDF is shown in Table 6.17. In the Far East it is also used in the wood working industry. Thin MDF competes with hardwood plywood. UF is the binder of choice but some MF resins are used when improved moisture resistance and clear panel color is desired. The convenience of using MF is due to the compatibility of MF with UF and requires no equipment modification in the plant employing UF. In Europe, Japan, and Australia blends of UF with low MW MF (containing high melamine content) are used for MDF [35]. In Asia, Pacific region, MDF made with melamine modified adhesives leads to panels with moisture resistance comparable to plywood.

Recently Riedl and Park [36] reported that a 40/60 low to high MW phenolic resole was effective in preparing MDF with unusually high internal bond (IB) strength. Riedl's resin studies paralleled early work of Steiner and Ellis [25] for waferboard.

Hardboard
Hardboard or high density fiberboard (SG 0.90–1.20 g/cm^3) is used in the furniture industry, for doors, automobile interiors, and house trailers. With

Table 6.17. Mechanical properties of U.S. composite wood materials (MDF and PB)[a]

	MDF	PB (flooring)
Thickness, mm	21	–
Density (kg/m^3)	640–800	–
MOR (MPa)	24	11–20
MOE (GPa)	2.4	1.8–3.1
Internal bond (MPa)	0.6	0.4–0.55
Screw-holding face (N)	1,400	–
Edge (N)	1,050	–
Linear expansion (% max. average)		0.3–0.35

[a] Source: National Particleboard Association.

Table 6.18. Fiberboard resin (dry process) [10]

Solids (%)	28–30
Viscosity @ 20 °C (mPa · s)	150–200
Alkaline content (as NaOH)%	0.9–1.1
Gel time @ 100 °C, min	25–35
Water dilutability	~1:1

phenolic resin as the adhesive, improved mechanical properties and reduction in water absorption/swelling characteristics are obtained. Both wet and dry processes can be employed to manufacture hardboard. The dry process requires more resin for hardboard production than the wet process (<1% added resin). Properties of liquid resin are contained in Table 6.18.

6.1.1.4.3
Engineered Lumber Products

Laminated Veneer Lumber (LVL)

A total of 1.12 million m^3 of LVL was sold in 1997. LVL consists of billets made by parallel lamination of veneers into thicknesses common to solid-sawn lumber (Fig. 6.5). Veneers are 2.5–3.2 mm thick, hot pressed with phenolic resin to lengths from 2.4 to 18 m, with thicknesses of 38–45 mm and widths of 69–127 cm. Radio frequency cure, when used, reduces cure time. LVL is slowly emerging as a growth product due to scarcity of structural wood. Phenolic resin is the predominant adhesive with some MDI. Most LVL is used in wood I-joists, residential beams, and headers. An excellent summary of LVL markets was published by Guss [37].

Recently Stahl [38] described continuous production of 122 cm-wide LVL using a Dieffenbacher designed continuous LVL production line. The line consists of several veneer sheet feeders, moisture detection, veneer sheet grading, precision alignment of veneer sheets, a resin curtain coater, a continuous LVL layup station, a microwave pre-heating system, a continuous zone heated press, a billet crosscut saw, and a billet stacker. Besides widths of 122 cm, billet thickness can vary from 2 to 10 cm. Billets exit the press continuously at speeds ranging from 0.76 to 3.7 m/min. As of 1998 new LVL continuous lines are operating in the United States and Canada.

Parallel Strand Lumber (Parallam®)

Parallel strand lumber or Parallam® [39] is a remarkable development that evolved from MacMillan Bloedel laboratories and was commercialized through a joint venture of Trus Joist and MacMillan Bloedel in 1986 (Fig. 6.6). The first plant producing Parallam was in Vancouver, B.C., followed by a second plant in Colbert, GA. Wood strands in varying lengths (Douglas fir, Vancouver; Southern Pine, Georgia: strength properties are similar for both species) are transformed continuously by a combined microwave/pultrusion

Fig. 6.5. LVL process schematic

Fig. 6.6. Parallam®, Truss Joist/MacMillan

process into lengths of 15 m or longer. The unique Parallam® process involves microwave cure of the phenolic resin adhesive/wood mixture combined with high temperature pultrusion and leads to billets with dimensions of 30 × 50 cm and variable length.

Parallam® Features

- Strong product – free of knots, grain deviations, and weak wood
- Very consistent – 10 % coefficient of deviation as compared to 25 – 50 % for solid wood
- Very durable – performs like solid wood, little effect of water, remains strong when wet with minimal swelling

Properties of Parallam are listed in Table 6.8. Parallam was used in Vancouver's World Fair, Expo '86, in every foreign visitors' pavilion or a total of 35 buildings. It was the infrastructure of the 7.3 m high wall system where strength and stiffness were critical for performance. Current market activity is directed to beams, headers, and posts allowing large spans, high loads, more rigid, dimensionally stable structures for generally more attractive and functional buildings. These features have resulted in code approval acceptances by all major applicable code regulatory bodies in North America.

6.1.1.5
New Process and Equipment Developments in Panel Manufacture

6.1.1.5.1
Steam Injection Pressing (SIP)

By critically examining the cure characteristics of resin bonded wood flakes, it is possible to improve or increase the rate of panel production. Press time, temperature uniformity within the panel during cure, and wood moisture content all relate to resin characteristics. A new process development applicable to PB, MDF, and OSB has involved the introduction of steam injection pressing (SIP) to reduce the time lag necessary for the interior of the board (core) to reach a uniform resin curing temperature and hence facilitate rapid cure for thick panels. Both UF and MDI adhesives respond favorably to the SIP technique but phenolic resins are erratic. Phenomena known as "wash out/wash in" and dilution/over penetration causing premature resin cure (short press time) and poor quality panels are reported. Geimer and co-workers [40, 41] have examined phenolic resins in flakes under conventional and SIP conditions. Various analytical methods (DMA, DSC, lap shear tests) were evaluated in an attempt to distinguish different types of cure (mechanical vs chemical) and board bond strength under dynamic conditions. The authors sought to distinguish resin cure from bonding strength of phenolic bonded flakes under dynamic conditions whereby mats with 7 – 12 % moisture content could be compared between conventional pressing and SIP with 10 – 20 s steam duration. Resin cure was examined by DMA while lap shear tests provided board bond strengths. Data showed that resin curing and bond strength did not proceed simultaneously. In conventional pressing higher moisture content (12%) favored resin curing but slightly retarded lap shear bonding when compared to 7% moisture content. With SIP, results were inconsistent. Authors indicated that high moisture (present in wood flakes and from steam injection) and temperature affect both wood T_g and phenolic resin to a level at which corresponding strengths of these board components are insufficient to overcome vapor pressure and internal stresses. More work is necessary to establish the value of SIP with phenolic resins.

6.1.1.5.2
Wider Presses

Wider forming lines and presses resulting in OSB panels with dimensions of 3.7 × 7.3 m using multiple daylight presses with up to 14 openings are becoming operational.

The larger width accommodates the North American and Japanese markets (120 cm for North America and 91.5 cm for Japan) and provides flexibility in cutting either width from master panels. Additional advantages include reduced trim waste and lower line speeds to achieve the same production volume as either 2.4 or 2.7 m wide lines.

6.1.1.5.3
Increased Length – OSB Flakes

Presently the disc flaking machine is the standard of the industry for OSB flakes. The ring strander, introduced in 1993, is being used to produce longer strands (> 15 cm.) for other engineered wood products beyond OSB.

6.1.1.5.4
Uniform Mats

Improvement in classification/uniformity of mats for PB and OSB is resulting in improved forming accuracy, good surface quality, and homogeneous cores with an accompanying removal of undesirable material.

6.1.1.6
Natural Resins

6.1.1.6.1
Introduction

There have been numerous attempts to develop and transform renewable biomass resources into wood adhesives. The most desirable components derivable from wood are those which contain phenolic-like structures such as lignin (I) and tannin (II) – see below (see also Sect. 8.6).

A review of the large number of publications using lignin and/or tannin as wood adhesives is beyond the scope of this book. Excellent reviews of lignin by

Glasser and Sarkinen [42], Seller [43], and Pizzi [1] are available. Chapter 5 of Pizzi's book [1] contains a review of tannin-based wood adhesives.

There are, however, some recent developments that provide an overview of the current status of these naturally occurring materials.

6.1.1.6.2
Lignin

Lignin (I)

Lignin, a polyphenolic component present in wood (about 20%), is a by-product of the paper pulping process. It is estimated that over 50 million tonnes of lignin are produced worldwide with the majority burned as fuel. Small amounts are used as dispersants, oilfield drilling muds, concrete additives, and adhesive extenders. Lignin sulfonates, by-products of the sulfite pulping process, are used as extenders for wood adhesives.

Since lignin must be removed from wood for paper, there have been many attempts to delignify wood into low MW soluble components. Delignification [44] of wood in non-aqueous media, also known as organosol pulping, has been a subject of considerate interest since the concept was first proposed in the early 1900s. Mild acid hydrolysis of lignin cellulose bonds under solution conditions (alcohol, dioxane, phenol, acetic acid) is organosol pulping and leads to lignin as a polyphenol characterized by high purity and low MW, being used as a partial replacement of phenolic resin in plywood, OSB, and waferboard. Analyses of lignin hydroxyl group (phenolic and aliphatic OHs) is quite complex. Six different soluble lignins were recently subjected to an international round robin analyses of hydroxyl groups by ^{1}H, ^{13}C, ^{31}P NMR, FTIR, and wet chemical methods [45, 46]. The agreement between all methods is only approximate. The most rapid method to determine hydroxyl ratio (phenolic/aliphatic) uses FTIR or ^{1}H NMR of acetylated samples.

Advantages and disadvantages of organosol pulping processes have been reviewed and proposed as complementary systems to pulping rather than separate pulping operations [47].

Alcell Technologies of Repap (Valley Forge, PA) have developed a process for soluble lignin in a demonstration plant located at Miramichi, N.B., Canada, with a capacity of 120 tonnes/day. Up to 20% of powdered phenolic face resins for waferboard can be replaced by Alcell lignin; 15% of resin solids plywood adhesives can be replaced based on successful manufacturing trials [48]. Similarly the use of Alcell lignin (up to 20%) in powdered or liquid phenolic resin for waferboard or PB leads to similar properties of the panels with an accompanying reduction in CH_2O emissions released from the press vent [49]. Authors claim that lignin is acting as an F scavenger. Thus pure lignin can be directly substituted for a portion of the phenolic resin used as wood adhesive. When lignin is methylolated (in situ reaction of lignin with CH_2O) the resulting product is reported to replace up to 35% of phenolic resin solids in MDF [43].

6.1.1.6.3
Tannin

Tannin (II)

Depending on wood source, there are a variety of tannins which have been evaluated as wood adhesives. The most common commercial tannin extracts are mimosa bark tannin, quebracho wood tannin, pine bark tannin, and pecan nut pith tannin. Various extraction conditions are employed to isolate these materials in low to medium yield. Tannins are phenolic in structure but are more reactive toward CH_2O than the similar reaction of phenol with CH_2O. While resoles are readily isolable from the reaction of phenol with CH_2O, "tannin resoles" are not stable and continue to condense/oligomerize due to tannin reactivity. Thus the use of tannins as wood adhesives involves the addition of a CH_2O source prior to its use as an adhesive. Formaldehyde sources are paraform, hexa and U/F concentrates. Without the CH_2O source, tannin is unreactive with a long shelf life in liquid and powder forms. Depending on open positions, 6 and 8, and ratio of pyrogallol to catechol (ring B) which is obtained by ^{13}C NMR, comparative gel times at 94 °C of tannin extracts with 40% CH_2O leads to the ranking: pecan nut > pine > mimosa > quebracho. According to Pizzi [1] other factors such as degree of polymerization and branching/rearrangement may also contribute to these differences in gel times. By adjusting the pH to 9–10 slower gelling tannins (mimosa, quebracho) can approach gel speeds of pine and pecan nut tannin extracts.

6.1.1.7
Non-Wood Materials

Agricultural waste materials or "agriwaste" are becoming a viable raw material in the manufacture of PB and MDF panels. Several agriwaste products have been transformed into panels; these include wheat straw, rice straw, and bagasse. These waste materials are extremely attractive for those countries which lack adequate wood supply such as Pakistan and India. Other countries including the United States and Canada have plants or plan to build plants to produce panels from agriwaste. The agriwaste panel business concept is based on regional availability of annual crop waste at a low price ($ 24–40/tonne) and low freight. Compak (UK) provides turnkey plant operations in transforming agricultural waste into commercial panels using MDI adhesive. These are relatively small plants with annual capacity of 12,000 m^3/year as compared to a volume in excess of 100,000 m^3/year for a dedicated panel plant based on wood materials. Mott [50] of the Biocomposites Centre (University of Wales, UK) has reported a variety of adhesives (U/F, MUF, PF, and MDI) that are satisfactory adhesives for agriwaste panels.

6.1.1.8
Structural Wood Gluing

6.1.1.8.1
Introduction

The attachment of wood components via RT curing adhesives rather than metal fasteners (nails or screws) is dominated by resorcinol formaldehyde (RF) resins. The resulting bond has high structural strength exceeding the strength of wood, is resistant to most conditions that wood is exposed to, and maintains original load bearing strength during exposure to all types of climatic changes. Depending on whether the bonded wood component is for indoor or outdoor use, cold setting adhesives are based on urea, resorcinol, or phenol/resorcinol combinations. UF resins are satisfactory for interior components.

RF adhesives were important during the 1940s for airplane construction, laminated wood propellers, and wooden hulled ships [51]. As a follow-up to these wartime efforts, the use of RF adhesives gained prominence in construction of laminated wooden beams, arches, and other structural wooden elements. These glue-laminated timber beams (glulams) are widely used for large span structures such as arches, domes, and bridges. Features are light weight, economical production of tapered and curved components, excellent energy absorption characteristics, and attractive appearance.

6.1.1.8.2
Resorcinol Adhesives

RF resins are costly because of the raw material cost of resorcinol. There are only three producers of resorcinol: Indspec in the United States, and Mitsui Petrochemical and Sumitomo in Japan. Resorcinol is highly reactive in the presence of formaldehyde whereby a 0.5–0.7 mol/l ratio of formaldehyde to resorcinol is combined under mildly basic conditions for a water soluble product. ^{13}C NMR analyses by Werstler [52] and Kim et al. [53] provide structural assignments of resorcinol formaldehyde resins to show that the 4-position is the favored position.

Formation of dimers, trimers, etc. occurs via 4,4′-; 2,4′-methylene units with a small amount of 2,2′- methylene linkage as the MW increases.

Indspec 2170 resin (used as wood adhesive, tire cord adhesive, etc.) contains 17.6% free resorcinol, and a ratio of 17.5:7.4:1.0 for mono-, di- and trimer species [52].

The majority of RF wood adhesives are modified with phenol to reduce the cost. Phenol and excess F are pre-reacted under alkaline conditions to form a low MW resole with subsequent addition of resorcinol to react with the methylol groups.

Recent work by Christjansen et al. [54] identifies *p*-methylol phenol as reacting much faster with resorcinol as compared to *o*-methylol phenol. Further improvements in reactivity of methylol phenols with resorcinol occur with NaOH or zinc acetate catalysts.

Depending on use, RPF resins can vary from 20 to 50% resorcinol. Recent studies by Pizzi and Scopelitis [55] indicate that resorcinol content can be reduced to about 10% in an RPF resin by introducing small amounts of urea as a branching unit during resin preparation. Good adhesive performance is maintained as is typical of RPF cold setting adhesives.

The convenience of room temperature cure of RF or RPF adhesives requires the introduction of a CH_2O source. Paraform with a filler such as wood flour or nutshell flour facilitate RF or RPF resin cure under neutral or alkaline conditions. The resulting bond possesses high strength, creep resistance, high humidity/water resistance, and excellent long term durability. Depending on the wood component and the urgency of assembly, superior performance coupled to convenience of bonding large parts at ambient temperatures justify the use of these costly adhesives.

6.1.1.8.3
New Areas

FRP Glulam Adhesive

A recent use of RF resin as an adhesive to bond fiber reinforced plastic (FRP) to glulam beams appears quite promising. The resulting bonded FRP glulam system serves to improve the relative stiffness and strength of the glulam [56, 57]. The FRP glulam system utilizes an FRP pultruded flat profile of unidirectional glass, carbon fiber, or aramid in a thermosetting resin (polyester, vinyl ester, epoxy, or phenolic) or thermoplastic resin (see High Performance and Advanced Composites chapter 6.1.5) joined to the glulam. Integrating the pultruded component to the glulam results in a wood/composite beam that can be customized for tension, compression or shear strength. Figure 6.7 shows the use of FRP glulam system for a bridge (Clallam Bay Bridge, Clallam River, Sekiu, WA). Aramid provides best tension reinforcement while CF optimizes compression reinforcement.

Fig. 6.7. Use of FRP glulam on Clallam Bridge, Sekiu, WA

Coupling Agent

There are many wood bonded systems which are not sufficiently durable for structural exterior applications or do not develop sufficient adhesion. Although epoxy adhesives develop dry shear strengths that exceed wood strength, epoxy bonds fail in delamination once exposed to severe stresses of water soaking followed by drying. Besides epoxy adhesive wet/dry weaknesses, common wood adhesives do not adhere to wood treated with chromated copper arsenate (CCA) preservative sufficiently to meet rigorous industrial standards for resistance to delamination.

The development of a chemical or physical bond in the interfacial region between the adhesive and the adherent could provide the necessary adhesion between these two dissimilar regions. Coupling agents are known to promote fiber-matrix adhesion in FRP composites. Silanes are recognizable coupling agents used in glass/matrix resin systems. Vick and coworkers [58–60] have observed that hydroxymethylated resorcinol (HMR) is a suitable coupling agent to couple epoxy and other thermosetting adhesives to wood. HMR is prepared as a 5% aqueous solution by reacting 1.5 moles of F to R under mild base conditions. Partial analysis suggests hydroxymethyl resorcinol along with dimer, trimer are the components contained in HMR. One half percent of dodecyl sodium sulfate is added to the resulting aqueous resin to aid in wetting of the wood surface. HMR treatment of the wood surface and RT evaporation of water leads to a primed surface which exhibits increased delamination resistance and shear strength of various epoxy resin systems to softwood and hardwood species as well as wood to FRP, such as FRP glulam above [61]. CCA treated southern pine lumber primed with HMR facilitated excellent adhesion with epoxy, RPF, and isocyanates.

"HMR" + Dimer, Trimer

6.1.1.9
Summary/Future Directions

Phenolic resin continues to be the preferred adhesive for bonding wood components into exterior grade products. The role of wood surface, extractibles, remaining volatiles within the wood, and wood functional group orientation after drying are all important wood characteristics to be considered in bonding wood segments into panel products. Changing wood source and the impact of the change has resulted in the evolution of new wood products, waferboard, and OSB, the later looming as a serious contender to plywood. Either liquid or solid resoles are effective adhesives for OSB. Lumber products, namely LVL and Parallam®, are expected to be large volume wood products due to favorable product properties and scarcity of similar lumber products from harvested trees.

New process developments which are expected to improve overall OSB line speed and economics involve the use of steam injection pressing and wider presses. Better quality and improved panel properties are expected from new developments in OSB flake preparation and uniform mats.

Regenerable modified adhesives based on lignin and tannin are briefly described. Depending on modification of these materials, up to 40% can replace phenol/phenolic resin in phenolic resin adhesive formulation for different panels.

Annually renewable agricultural waste is being processed into panels that may compete with wood-based PB and MDF. Global activity in establishing regional plants that rely on low cost "agriwaste" with low freight cost has resulted in panel manufacture based on wheat straw, rice straw, bagasse, and others in many countries worldwide.

Resorcinol resins continue to be the adhesive of choice for room temperature cure of wood products. New areas include the use of RF or RPF as adhesive to bond FRP to glulam and the use of hydroxymethylated resorcinol as wood coupling agent.

6.1.2
Insulation

6.1.2.1
Inorganic Fibers

6.1.2.1.1
Introduction

Inorganic and organic materials can be transformed into attractive thermal and sound insulation products. Thermal and acoustical insulation continues to be an important market segment for phenolic resins as either binders for inorganic/organic fibers or as resin transformed into a foam structure. Phenolic resins are used as the binder for fiber glass and/or mineral wool thermal insulation and for reclaimed natural and organic fibers used as thermal/acoustical insulation in automobiles. The conversion of phenolic resin into foam encompasses a manifold of foam applications from insulation, floral, orthopedic to mining-tunnel use. The thermal insulation (λ value) characteristics of phenolic foam are quite favorable as compared to other polymeric foams. Thermal insulation materials, especially polymeric foams, compete on a cost performance basis when many factors are considered in the selection of materials. These include density/strength, thickness, thermal insulation rating (usually expressed as "R" value or thermal resistivity) and service temperature. In Table 6.19, R values and thicknesses for several materials are listed [1].

Table 6.19. "R" values for various insulation materials: "R"-value[a], thickness in cm

Insulation material	λ	K-Factor[b]	7	5.3	3.5	1.8	Maximum Service Temp °C
Phenolic	0.017	1.73	12.2	9.1	6.1	3.0	150
PUR/PIR[c]	0.026	2.60	18.3	13.7	9.1	4.6	125
Extruded polystyrene	0.029	2.88	20.3	15.2	10.2	5.1	75
Expanded polystyrene	0.035	3.46	24.4	18.3	12.2	6.1	75
Glass fiber	0.040	4.03	28.4	21.3	14.2	7.1	450
Mineral fiber	0.050	5.04	35.6	26.7	17.8	8.9	450
Perlite	0.056	5.62	39.6	29.7	19.8	9.9	–
Vermiculite	0.069	6.91	48.8	36.6	24.4	12.2	–

[a] $m^2 \cdot K/W$.
[b] $(W \cdot cm)/h \, m^2 \cdot K$.
[c] Polyurethane/polyisocyanurate.

The "R" values protocol was developed in North America and is related to the thickness of insulation material (in inches). It is the reciprocal of the K factor (thermal conductivity) whose units are $BTU \cdot in/h.ft^2 \, °F$. R values of 10, 20, 30, and 40 were tabulated with corresponding foam thicknesses. These R values were applicable to different geographical regions of North America. Recommended R values for ceilings/walls/floors for Chicago are listed as R 33/19/22 respectively while for New York values are R 30/19/19. Thus polyurethane/polyisocyanurate (PUR/PIR) foam thickness for Chicago would be 5.4/3.4/4 (inches) and for New York 5.4/3.4/3.4 (inches). These values and thicknesses were guidelines for the home construction industry to use sufficient foam thickness or inorganic materials for satisfactory home and warehouse insulation in ceilings/walls/floors. By examining the R value table, North American builders could determine the thickness of insulation for ceiling/wall/floor that would be suitable and could specify the type of material (foam or inorganic material).

R values listed in Table 6.19 relate to the metric system with arbitrary R values of 7, 5.3, 3.5, and 1.8 which equate to 40, 30, 20, 10 R values (in U.S. units) and indicate insulation thicknesses in cm. Thus phenolic foam is the most effective thermal insulation material while vermiculite is the least with glass fiber/mineral wool intermediate.

Market share and companies involved with inorganic fibers for building insulation in North America and Europe are shown in Fig. 6.8. Volume of inorganic fibers continues to be greater than polymeric foam materials (Table 6.20).

6.1.2.1.2
Types of Inorganic Fibers

Inorganic fiber-based products that are transformed into thermal insulation consist of either glass or mineral wool fibers. Different continuous processes

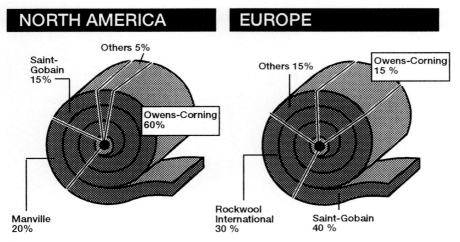

Fig. 6.8. Market share for North America and European companies engaged in inorganic fibers for insulation (1993)

Table 6.20. Volume of insulation materials in Europe (1994)

Material	Volume (million m³)
Mineral wool	55.1
Polystyrene	21.3
Polyurethane	3.9
Miscellaneous	3.0

[2] are involved whereby the molten inorganic composition flowing from a melt furnace is divided into streams and attenuated into fibers. The fibers are collected in a felted irregular manner to form a mat. To produce most thermal insulating products, the fibers must be bonded together in an integral structure. Thus, thermal insulation batts, acoustical tiles, and similar structures comprising glass, rock wool, or other mineral fibers are held together by a resinous binder of 5–20 wt% to provide strength and resiliency to the structure and preserve insulating, acoustical, and dimensional properties.

The most commonly employed binder is phenolic resin which is water soluble, easily blended with other components (urea, silane, etc.), and diluted to low concentrations which are readily sprayed onto the fiber. The optimum amount of binder for most thermal insulation products should be sufficient to "lock" intersecting fibers into a mass by bonding the fibers as they cross or overlay. Binder compositions should exhibit good flow characteristics so that the binder solution can be applied in low volume and flow into the fiber intersections. Glass fiber insulation materials comprising very small diameter filaments (3–6 µm) are bonded with phenolic resins to reduce the brashness of glass fiber and provide flexibility to the fibers as they regain their original thickness after unfolding the compressed glass fiber material from its packag-

Fig. 6.9. Glass wool insulation products, Pfleiderer, Delitzsch, Germany

ing container. Brashness is associated with dust and breakage of filaments during handling and shipping. Cured glass fiber/binder compositions are normally very bulky and voluminous. Batts and rolls used as insulation in buildings have densities ranging from 8 to 16 kg/m^3 and require binder content of 3–7 wt%. Since it is expensive to ship bulky materials in an uncompressed state, batts and rolls are bundled and compressed in packages 8–25% of manufactured thickness. Once unfolded the material should recover to 100% of its original volume. Insulation materials not achieving these recovery values would have difficulty meeting advertised thermal resistivity (R) values. Methods are available to fiber glass manufacturers that determine the amount of binder present by loss on ignition test (LOI) which should be greater than 4%. With LOI determinations combined with photomicrographs, one can determine binder junction quantity (μ^2) and relate to recovery of the material on unfolding.

Different types of glass fiber compositions are shown in Fig. 6.9.

6.1.2.1.3
Resins for Inorganic Fibers

Highly water soluble resole resins are prepared by reacting phenol with an excess of formaldehyde at temperatures below 70 °C. The F/P ratio is usually 3.5–4.0:1. Generally alkaline earth hydroxides (calcium or barium) are used as catalysts. High quality resins are normally ash free with catalyst precipitated as carbonate or sulfate and removed by filtration.

Excess F, as much as 7%, is available to react with urea to a urea extended resole. Not only does urea react with F but it may [3, 4] undergo a co-conden-

Table 6.21. Phenolic resins binders for inorganic fibers [5]

	PF resin (modified with urea) for mineral wool	PF resin (unmodified) for glass fiber
Refractive index	1.4610	1.4650
Solids content (%)	48.0	49.0
Viscosity (mPa·s)	<25	<25
Water dilutability	∞	∞
pH	8.8	8.8
Free phenol	<0.5	<0.3
Free formaldehyde	0.4	7.0
Density g/cm^3	1.175	1.190
B-time (130°) min	9	5

sation reaction with methylol phenols (Wood Composites chapter 6.1.1). Urea extended PF binders are more cost effective than pure PF resins but exhibit some loss in properties as urea content increases. If added urea is too high, product performance, particularly recovery from compression/rigidity, and especially after storage under humid conditions, will be decreased.

Properties of phenolic resin binders for glass fiber and mineral wool are contained in Table 6.21 [5].

6.1.2.1.4
Resin Formulation

The resin is applied as a 10–15% aqueous solution. High water dilutability is attributable to the presence of mainly polymethyolated phenols with the prevailing species being trimethylol phenol and very little dimer, trimer oligomer components.

Composition for treatment of fiber glass or mineral wool is:

- 100 pbw phenolic resole
- 7 pbw 20% ammonia solution
- 0.02 pbw amino silane (γ-aminopropyl triethoxysilane)
- 0–800 pbw water

Ammonia reacts with free F and transforms it into hexamethylene tetramine as well as pH adjustment to a low alkaline value. Amino silane acts as a coupling agent to improve the moisture resistance and increase mechanical strength. Urea is added to the resin at the end of resin manufacture or immediately prior to use by the glass fiber producer. Lignin or lignin sulfonate salts are additives to improve PF/urea blend and reduce cost.

Properties of mineral fiber mats are shown in Table 6.22.

Table 6.22. Physical properties of mineral fiber mats

Property	Test method	Unit	Test value
Density	DIN 18165	Kg/m³	20–80
Behavior in fire	DIN 4102	–	Fire-proof A2
Thermal conductivity	DIN 4108	W/mK	0.040
Spec. thermal capacity C_p	–	KJ/kg °C	0.84
Noise reduction coefficient	DIN 52212		
125 Hz			0.17
500 Hz			0.76
1,000 Hz			0.86
4,000 Hz			1.00

6.1.2.2
Phenolic Foam

6.1.2.2.1
Introduction

The three basic mechanisms of heat transmission that are involved in thermal insulation are radiation, conduction, and convection. Thermal conductivity (λ in W/mK) or heat flow of various components is shown in Fig. 6.10. Thermal insulation materials attempt to minimize and control these mechanisms. Conduction and convection are minimized by incorporating a large number of

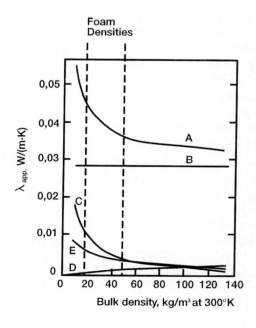

Fig. 6.10. Thermal conductivity components vs density at 300 K: *A*, total conductivity; *B* air conduction; *C* radiation, *D* convection; *E* conduction of solid polymer

small, low density, low thermal conductivity elements (foam closed cells with low λ blowing agent). Radiation is reduced through the use of low emittance – high reflectance barrier film. Thus low density closed cell foams with low λ blowing agents ($\lambda \leq 0.02$) exhibit λ values of 0.018–0.035, have significantly lower than the total conductivity of solid polymer, and are attractive materials for insulation.

Phenolic foam is recognized as a versatile foam composition which can be utilized in a variety of diverse market areas such as thermal insulation, fresh flower support, mine/tunnel use, or orthopedic foot impressions. These applications are dependent upon the cellular structure of the phenolic foam whereby a high closed cell content is necessary for thermal insulation while the completely open cell is critical for the floral application. Orthopedic foot impression type foam is mainly dependent on the friable nature of phenolic foam and ease of replication of a foot imprint. Mine or tunnel application requires an extremely rapid foam generation with corresponding fire resistant properties.

The attractive characteristics of phenolic foam for insulation were the motivating factors for its commercialization by Koppers in 1983. It was known as Exeltherm Xtra and reported by us in a previous publication [6]. Although possessing favorable fire/smoke/low toxicity (F/S/T) properties, high upper temperature usability at 150 °C, and an excellent K factor, phenolic foam was withdrawn from the market by Schuller in 1992. Schuller purchased Koppers's foam technology in the late 1980s but subsequently terminated the foam business activity due to corrosion problems emanating from roof decking components. Yet R & D efforts by many organizations continue in developing phenolic foam for thermal insulation by alleviating some of the past problems (see later).

The components required to prepare phenolic foam are a liquid resole resin, surfactant, a blowing agent, and an acid catalyst. Depending on the end use application, surfactant and wetting agent (floral foam) may or may not be similar in the foam formulation.

6.1.2.2.2
Resins

Phenolic resins are predominantly liquid resole resins. There are several literature references that report the use of novolak resins with a chemical blowing agent. To date none of these disclosures has resulted in a commercial foam product [7].

Aqueous resoles consist of phenol and formaldehyde in molar ratios of 1:1.5–2.5. The reaction is conducted to yield resin with low amounts of residual formaldehyde and phenol. Low residual monomer content is beneficial during handling as well as to minimize monomer emission during the foaming process. Sodium hydroxide or alkaline earth hydroxides are used as catalysts. Condensation is conducted within a temperature range of 60–90 °C. The desired resin solids content and viscosity are adjusted by distillation. The properties of a typical resole for foam preparation consists of about 80%

solids, with volatile components being water, formaldehyde, and phenol. A moderate viscosity of 3000–8000 in mPa · s is desirable with a residual water content of 5–10%.

Many variables relate to the final foaming behavior of the resulting resole. Within the prescribed solids, viscosity, and water content of the resin, a reactivity test is usually conducted with the resin. Many different resin reactivity tests have evolved for the evaluation of resole resins for foaming. These tests are proprietary and are mutually developed between resin manufacturer and foam producer. For example, one particular reactivity test consists of a known amount of resin at 20 °C catalyzed with 5 ml. of phenol sulfonic acid and the time to reach 50 °C by acid catalyzed reaction gives the number of seconds of resin reactivity. This quantity in seconds provides the basis for determining whether the resin will foam favorably in the customer's foam operation. If the resin reactivity is satisfactory, one can formulate the necessary components within the resin and conduct foaming.

Foam density is regulated by the type of surfactant, amount of blowing agent, catalyst content, temperature, and reactivity of the resin. Reactivity is influenced by storage time of the resin. The storage life of high solids resoles (~80% solids) is limited, and generally 2–4 months storage at 5–10 °C is optimum. Resins reactivity decreases with increasing storage time. Similarly resin viscosity increases considerably; maximum viscosity should be below 10,000 mPa · s.

6.1.2.2.3
Catalysts

Moderately strong organic acids (sulfonic acids) are effective in catalyzing the resole resin into foam. Typical sulfonic acids such as p-toluene, phenol, cumene, xylene, and methane sulfonic acids provide optimum curing/foaming rates. Inorganic acids such as H_2SO_4 and HCl should be avoided due to severe corrosion problems. An exception is the use of small amounts of phosphoric acid which is non-corrosive and provides improved flame resistance of the foam. Lesser common acids consist of oligomeric sulfonic acids [8] and sulfonated novolaks [6]. These latter acids allegedly exhibit minimal foam corrosion by incorporation into the foam network structure by co-reaction. Acid free systems can be obtained by using resorcinol novolak [6]. A newer catalyst system based on resorcinol, diethylene glycol, and xylene/toluene sulfonic acid mixture was recently reported [9].

6.1.2.2.4
Blowing Agents

The role of the blowing agent is to reduce the exothermic heat of reaction that occurs during acid catalysis of the resole resin. Heat of vaporization of the volatile blowing agent moderates the exothermic reaction and results in a uniform expansion of the growing phenolic resin network. Liquids with a low boiling point and corresponding low heat of vaporization are desirable blowing agents. Liquids which are non-solvents for the resole resin are highly desirable. Slightly soluble blowing agents result in a coarse, friable cell structure. Newer environmentally acceptable HCFC solvents exhibit poor foaming performance in phenolic foam as compared to banned CFCs. Similarly perfluoroalkanes used alone are poor. However a blend of HCFC and perfluoroalkane is suitable [10]. Suitable blowing agents are:

Hydrocarbons B.pt. (°C)
- Isopentane   28
- n-Pentane   36
- Cyclopentane   49
- Isohexane   60
- n-Hexane   69
- Petroleum ether   30–60

Ethers
- Diisopropyl ether   68

Fluorocarbons
- HCFC-141b (CCl_2FCH_3)   32
- HCFC-22 (Cl_2FCH)   −40.8
- HCFC-142b (ClF_2CCH_3)   −9.8

The liquid blowing agent must volatilize as soon as the viscosity or the phenolic "growing network" increases. Foam expansion occurs as the exothermic crosslinking reaction continues with either the blowing agent entrapped within the cellular structure (closed cell) or the cell is ruptured resulting in a partly open/closed cell foam or a fully open cell foam. In addition to volatilization of the blowing agent, water and monomers are volatilized during the foaming operation. Flammability of a non-halogen blowing agent is an issue and requires the necessary safety precautions. Safe use of pentane as blowing agent in PUR is reported [11].

The use of CO_2 as an "in situ" blowing agent has been reported [12]. By using small amounts of isocyanates in an acid catalyzed resole foam formulation, CO_2 is generated by the reaction of water with isocyanate and leads to phenolic foam with densities of 16–320 kg/m^3.

6.1.2.2.5
Surfactants

Surfactant as the name implies must act as a surface active agent by lowering the surface tension of the resin formulation and by providing an interface between the highly polar phenolic resin and the non-polar blowing agent. It must be miscible or dispersible in the resin, non-hydrolyzable, and resistant to the acid catalyst. The surfactant prevents the developing foam from collapsing or rupturing. Proper selection and/or design of the surfactant generally yields a phenolic foam with a uniform, fine, cellular structure. Nonionic and silicone-containing surfactants are commonly used [6]. The amount added may vary from 1 to 5 %. Depending on the amount added, too little surfactant fails to stabilize sufficiently the foam, resulting in very coarse cell structure or foam collapse. When too much surfactant is used the foam may collapse or yield a plasticized foam. Newer reports mention a gelling surfactant which accepts water formed during curing/crosslinking reaction [13].

6.1.2.2.6
Wetting Agents

Wetting agents are necessary only in the production of phenolic floral foam. Besides the usual components in the resin formulation (resole, surfactant, blowing agent, and catalyst), a wetting agent is introduced prior to catalysis. Suitable wetting agents are detergent-like compounds such as alkyl ether sulfonate salts, alkyl benzene sulfonate salts, and fatty alcohol ether sulfonate salts.

6.1.2.2.7
Fillers

For high density foams (>50 kg/m^3) fillers such as $Al(OH)_3$, ammonium polyphosphate, graphite, calcium silicate, and calcium carbonate are used [14]. Moderately viscous resin solutions are obtained when fillers are introduced.

6.1.2.2.8
Foam Stabilizers

Adjuvents or viscosity modifiers are identified and claimed to improve efficiency of blowing agent, reduce brittleness, and maintain high closed cell structure. Butyrolactone [15], ethylene or propylene glycol, dipropylene glycol, polar solvents (DMSO, NMP), different morpholines [16], and perfluoro-N-methyl morpholine [17] are mentioned.

6.1.2.2.9
Foam Applications

Thermal Insulation

Low Density Foam (<50 kg/m^3)
There are many desirable properties of phenolic foam that warrant its acceptance as a material for thermal insulation:

- F/S/T properties
- High service temperature of 150 °C
- Lowest λ value for organic polymeric foam
- Satisfactory strength at economical density
- High closed cell (under special conditions)
- Reduced friability

yet recurring problems associated with corrosion and closed cell are discussed (below) and must be resolved for successful reintroduction of phenolic foam for insulation.

Corrosion. The main deficiency that let to the withdrawal of phenolic foam as thermal insulation was corrosion. It is not clear whether the quality or quantity of sulfonic acid caused the corrosion problem. Use of oligomeric sulfonic acids which would co-react and become part of the foam network offers potential as well as the recent use of combined resorcinol/glycol/sulfonic acid system [19].

Closed Cell. The low λ value is contingent on high closed cell. Attaining a high amount of closed cell requires either conducting foam expansion under pressure or conducting the foam expansion at temperatures below 100 °C followed by post cure. A pressure of about 14 Pa is recommended [18] to obtain high closed cell during foam expansion. This pressure minimizes any cell rupture that can occur and maintains volatile emissions, water, and blowing agent within the closed cell. Thus batch and continuous foam operations require some opposing pressure (continuous) or closed mold (batch) to manufacture foam with high closed cell.

Newer technology developed by Owens Corning [19] describes the foaming operation being conducted below 100 °C until foam is cured followed by post cure conditions. High closed cell (>90%) is obtained with moderately constant λ value of 0.0188 and increasing to 0.0195 after 28 days at 100 °C.

Other methods for closed cell include the use of "foam stabilizers" which are polar solvents such as butyrolactone, DMSO, NMP, and various morpholine compounds, and facilitate closed cell by plasticizing the phenolic resin network, allowing egress of water without cell rupture. Long term λ values have not been reported. It is uncertain whether these foam stabilizers compromise the favorable F/S/T characteristics of phenolic foam.

Medium to High Density Foam
Highly filled foam systems ($\leq$ 300 phr filler) can be formulated and extruded through a slot die using a twin screw extruder. The extrudate from the slot die is foamed on a continuous double belt press at a speed of 2 m/min. Foam density is 92 kg/m^3 [20]. Although no applications are mentioned in the publication, high density, fire resistant foam product would be attractive in wall paneling, flooring materials, or interiors of mass transportation vehicles.

Other highly filled foam systems are extruded onto steel plate or gypsum board and then laminated with another plate or aluminum treated paper. These fire resistant laminates with high density phenolic foam core are reported to be used for interior/exterior walls, partitions, and building materials in general [21].

Floral Foam

By foaming resole into a completely open cell material with crisp friability and high water absorption, a unique phenolic foam product enjoys favorable acceptance as a fresh flower support for floral arrangement. The product must saturate water within 1–2 min by floating the dry brick whose dimensions are 23 × 11 × 8 cm in water. The dry brick weighs 40 g (20 kg/m^3 density) and absorbs 1900 g, almost 2 l of water or about 50 times its weight in water! The

Table 6.23. Physical properties of phenolic resole resin for floral foam [5]

Dry solids	77.5–79.0	%
Viscosity, 20 °C	2300–3000	MPa s
Specific gravity, 20 °C	1.28–1.32	g/cm^3
Reactivity, T_{max}	104–107	°C
Content of free phenol	5.4–6.0	%
Content of free formaldehyde	0.5–1.0	%

proper balance of components such as resole, surfactant, wetting agent, blowing agent, and catalyst leads to floral foam. The crisp friability of the wet foam allows easy penetration of flower stems into the foam, even soft stemmed flowers such as tulips, anemones, and other fragile, spring flowers.

By modification of the foam composition, floral foam can be used as growing medium for propagation of bedding plants such as poinsettias, chrysanthemums, and other greenhouse plants.

Properties of phenolic resole resin for floral foam are listed in Table 6.23 [5].

Orthopedic Foam

Another attractive and unique application for phenolic foam is its use for cast impression or replication of a foot imprint. Phenolic foam for the orthopedic application must be very crisp, friable, and moderately low density, 8–12 kg/m^3. The dimensions of the orthopedic foam for the foot imprint are 30 × 15 × 5 cm. By assisting the patient and pushing the entire foot into the foam, a uniformly replicated footprint is obtained that is later duplicated in the finished orthotic device. Orthopedic foam is very effective in providing an imprint which is later duplicated by a plaster of Paris casting. The plaster of Paris foot casting facilitates the fabrication of custom made shoes for those individuals who require comfort during long hours while on their feet.

Properties of phenolic resole resin for orthopedic foam are listed in Table 6.24.

Mine/Tunnel Foam

This is a two component, room temperature stored composition which is available to fill voids or openings in mines/tunnels. The two components must instantly react and foam on mixing and cure rapidly in the mine opening or void. Usually an inorganic carbonate is dispersed in the resin along with

Table 6.24. Physical properties of phenolic resole resin for orthopedic foam [5]

Dry solids	78–81	%
Viscosity, 20 °C	3500–4100	MPa s
Specific gravity, 20 °C	1.23–1.24	G/cm^3
B-time, 130 °C	12	Min
Content of free phenol	13.0–15.0	%
Alkali content	1	%

a surfactant and some filler. This represents the resin component while the other component is the acid catalyst. Upon mixing the two components, the "expanding" mixture is rapidly thrust into the mine opening as CO_2 is liberated from the carbonate. New efforts in this area have identified the use of thixotropic agents to maintain resin uniformity [22].

Hybrid Phenolic Foams

Many hybrid phenolic foams have been described in the literature and include a variety of components that co-react with the phenolic segment to yield modified phenolic foam systems. The rationale behind the hybrid approach is the combination of favorable features of the phenolic foam segment (FST, high heat and chemical resistance) with another component which would improve phenolic foam limitations such as reduced brittleness, friability, and emissions.

Polyurethane
Using high ortho liquid resoles (see chapter on Chemistry, Reactions, Mechanisms) as the polyol, and subsequently reacting it with diisocyanates yields the corresponding polyurethanes [23]. In some instances, chloro phosphate esters are added to provide flame retardant foam.

Urea Formaldehyde
A co-reaction between resole and UF resin is claimed to yield the corresponding PF/UF foam [24].

Furfuryl Alcohol
Furfuryl alcohol co-reacts with resole [25] or resole/novolak mixture [26] and acid catalyst to yield a PF/FA foam.

6.1.2.2.10
Foaming Equipment

Foaming conditions and equipment required for the various phenolic foams are summarized in Table 6.25.

Table 6.25. Foam conditions/equipment

Foam type	Density (kg/m^3)	Batch	Continuous
Insulation low density	32–50	Closed mold	Double belt with pressure
Insulation medium to high density	>50	–	Extrusion/double belt
Floral	24–32	Open/closed mold	Belt with free rise
Orthopedic	8–12	Open mold	Same as floral
Mine/tunnel	8–32	No mold	–

Table 6.26. Properties of various phenolic foams

Foam Type	Density (kg/m³)	% Closed Cell	Friable (% wgt loss)[a]	Comp. Strength (kg/cm²)	Water Absorption
Insulation[b]	32–50	>90	<30	2.8–7	Very low
Floral	24–32	None	High	0.7	Very high
Orthopedic	8–12	Open/closed	Very high	0.2–0.3	Low
Mine/Tunnel	Irregular	Open/closed	Irregular	Irregular	Low

[a] ASTM C421.
[b] $\lambda = 0.018$ W/m K; K Factor 1.73 (W · cm)/hm² · K.

Batch

In many cases discontinuous or batch conditions are employed to produce phenolic foam in most of the above applications. For floral foam and orthopedic foam satisfactory foam is obtained with low investment in equipment, molds (open or closed molds can be used, Table 6.25), and other components. For insulation foam, technology to date indicates that a closed mold is necessary for high closed cell content in batch operation.

Continuous

For insulation foam, double belt continuous line with pressure for low density closed cell foams (32–50 kg/m³) is recommended. For medium to high density and highly filled foams (>50 kg/m³), a twin screw extruder into a heated slot die and continuous foaming on a double belt press are recommended.

Floral foam and orthopedic foam can be foamed continuously without pressure, i.e., free rise allowing fully open cell (floral) or partial open cell for orthopedic applications.

6.1.2.2.11
Properties of Various Phenolic Foam Products

An overview of the properties of various phenolic foams is contained in Table 6.26.

6.1.2.2.12
Foam Testing

Fire Conditions

The propensity of phenolic resins to undergo char formation and maintain structural integrity at elevated temperatures is unique to phenolic resins (see High Performance and Advanced Composites chapter) and is a major factor in the renewed interest in phenolic resins.

Table 6.27. Ignition and opposed flow flame spread data for rigid foams (LIFT, NIST)

Foam	(q_{ig}) [a] (kW/m^2)	(t_{ig}) [b] (°C)	$K_{\rho c}$ [c] $(kW/m^2K)^2_s$	Φ [d] (kW^2/m^3)	(q_s, min) [e] (kW/m^2)	t_s, min [f] (°C)
PUR	21.0	445	0.037	8.8	7.7	176
PIR	30.0	445	0.021	28.0	10.8	201
Phenolic	30.0	524	0.11	0.15	28.0	509

[a] (q_{ig}) minimum heat flux for ignition.
[b] (t_{ig}) temperature of ignition.
[c] $K_{\rho c}$ thermal inertia of material (product of thermal conductivity, density and heat capacity).
[d] Φ flame heating parameter.
[e] (q_s), min minimum heat flux for flame spread.
[f] (t_s), min minimum surface temperature for flame spread.

Cleary and Quintiere [27] conducted an intensive characterization of foam plastics as a means of developing an alternate testing protocol to Steiner Tunnel Test (ASTM E-84). Although the E-84 is a widely accepted surface burning test for the building industry, there are many problems associated with the test such as lack of correlation to actual fire conditions, variability for repeat testing, and thickness limitation. Moreover the test requires large quantities of material for full size specimen of 7.3 × 0.5 m.

The authors determined that the Cone Calorimeter (ASTM E-1354) and "Lateral Ignition and Flame Spread Test" apparatus (LIFT) were satisfactory to completely characterize foam flammability. Stevens et al. [28] similarly observed that the Cone Calorimeter (E-1354) provided good correlation in various small scale fire tests such as oxygen index (ASTM D-2863), NBS Smoke Chamber (ASTM E-662), and modified ASTM D-3806 (2 ft tunnel test) instead of the Steiner Tunnel Test (E-84) to characterize novel polymeric systems completely. Recently Grand and co-workers [29, 30] have favorably compared the use of the Cone Calorimeter vs the OSU heat release calorimeter and favor its use in investigating fire resistant components for aircraft interior materials.

Testing by Cleary and Quintiere using the LIFT apparatus was conducted on a bench scale level with optimum size samples (15.5 × 15.5 cm, with 50 mm thickness). Testing allowed evaluation of foam flammability and smoke obscuration. Other important fire characteristics that are determined are ignitability, opposed flow flame spread rate, rate of heat release, and light extinction from smoke over a range of external heat fluxes that can occur in a fire. In Table 6.27 Class I polyurethane (PUR), Class I polyisocyanurate (PIR), and Class I phenolic foam are compared relative to ignition and opposed flow flame spread. As expected, phenolic foam exhibited highest temperature of ignition, highest heat flux for flame spread, and highest surface temperature for flame spread than PUR and PIR.

Comparison of similar materials via the Cone Calorimeter (Table 6.28) were inconsistent. Heat effects of PIR and phenolic foam were variable; only ignition time (t_{ig}) and smoke obscuration data (σ_m, ave) were favorable for phenolic when compared to PIR or PUR.

Table 6.28. Cone Calorimeter test data of rigid foams (heat flux = 50 kw/m^2)

Foam	(t_{ig})[a] (s)	PK HRR[b] (kW/m^2)	THR[c] (MJ/m^2)	H_c[d] (kJ/g)	σ_m, ave[e] (m^2/kg)
PUR	4	147	13.9	10.0	403
PIR	3	79	4.7	9.1	264
Phenolic	9	111	36.0	14.2	72

[a] (t_{ig}) ignition time.
[b] PK HRR Peak heat release.
[c] T HR total heat release.
[d] H_c heat of combustion.
[e] σ_m, ave specific extinction (obscuration).

Thermal Conductivity

With changes in blowing agents from CFC to HCFC and possibly other aliphatic types as well as new types of foam being commercialized, long term thermal performance (constant or variable λ) of foam products is an important characteristic which will determine the type and thickness of foam to consider for satisfactory thermal insulation.

Bomberg [31] has developed accelerated test procedures to predict long term thermal performance of various foams and can compare them with field measurements. A thin layer scaling (slicing) method is used for unfaced and homogeneous foam. The scaling technique provides reasonably good agreement for K values between predicted and field values (within 6%) over a 4 year period. For inhomogeneous foam (faced foam or skin/core layers are significantly different), an extrapolative model known as "distributed parameter continuum" (DIPAC) is used [32]. The method requires measurement of changes in thermal resistance of thin surface and core layers. Various foams, including phenolic foam, were monitored under field conditions for 2.5 years, samples being periodically removed from the field (different roofs) and measured. The DIPAC model agreed favorably with field measurements.

Foam structural parameters such as cell size and cell shape also relate to thermal conductivity and are discussed by Booth and Grimes [33] with respect to a study of extruded polystyrene foams and thermoset foams (polyurethane modified isocyanurate and phenolic foams).

6.1.2.2.13
Aerogels

Early aerogels were prepared from oxides of silicon, aluminum, and zirconia using the sol-gel technique. Aerogels of silica [34] exhibited a density as low as 0.01 g/m^3. For a given volume of this aerogel, more than 99% is air and less than 1% is silica. High porosity and low density characteristics of aerogels are responsible for the extremely low thermal conductivity ($\lambda \sim 0.007$ W/mK). Efforts by Pekala and others at Lawrence Livermore National Laboratories,

Livermore, CA have been successful in developing carbon aerogels based on resin systems of resorcinol/formaldehyde or novolak/furfural by sol-gel polymerization followed by supercritical drying from CO_2 and subsequent pyrolysis in an inert atmosphere. Materials can be thin films, powders, or microspheres [35]. Carbon aerogels exhibit an open cell structure with an ultra fine pore size (<100 µm), high surface area (400–1100 m^2/g), and a solid matrix composed of interconnected particles, fibers, or platelets with characteristic dimensions of 10 nm.

The many potential applications for these unique materials include thermal insulation, chromatography packing material, water filtration, ion exchange, batteries, and capacitors. The ability to tailor the structure and properties of porous carbon suggests its use as electrodes in energy storage devices such as batteries and double layer capacitors.

Recent thermal conductivity measurements of R/F aerogels conducted at the Bavarian Center for Applied Energy Research [36] are reported as $5-8 \times 10^{-3}$ W/mK in vacuo and $11-13 \times 10^{-3}$ in air at room temperature. The measured conductivities may well be the lowest thermal conductivities ever measured for any solid material in air.

Aerojet (Sacramento, CA, USA) has produced aerogels with dimensions up to 30×30 cm by 2.5 cm with densities of $0.1-0.2$ g/cm^3 [37].

Although there have been several reports of improved processing to accelerate aerogel drying [38], carbon aerogels remain expensive.

6.1.2.2.14
Summary/Trends

Thermal and sound insulation materials derived from inorganic or organic fibers continue to be bonded by phenolic resins. Higher temperature binders are desirable to increase the upper temperature use of inorganic fiber systems. A comparison of λ values and R values identifies phenolic foam as the most efficient thermal insulation material among organic foams whereas products based on glass fiber or mineral wool are intermediate in thermal performance.

Phenolic foam is shown to be a versatile material. By judiciously formulating various components with the appropriate phenolic resole resin, a wide range of foam products can be produced to service such application areas as insulation, floral, orthopedic, and mine/tunnel. New fire testing of foam products for the building industry is achievable with the Cone Calorimeter and possibly the newly developed NIST apparatus, LIFT. Satisfactory fire testing data are being generated with small samples using both of these pieces of apparatus.

Determination of accelerated long term λ values by either the scaling or the DIPAC method provides λ values within a few hours and results in satisfactory agreement with values obtained from field samples during the lifetime of the foam. Cell size and shape of foam are additional structural parameters that influence λ value of foam. A new interacting foam structure, aerogel, which is based on phenolic chemistry possesses the lowest thermal conductivities for solid materials and shows promise as a high technology material for challenging foam applications.

6.1.2.3
Phenolic Resin-Bonded Textile Felts (DIN 61210)

6.1.2.3.1
Introduction (Textile Recycling, Raw Materials, Applications)

For many decades, the textile industry has devoted considerable effort to recycling textile scraps (trimmings and fibers) in an economically and environmentally acceptable manner [1]. This is accomplished either by recycling textile scraps into the production process, or by using them to manufacture new products such as textile fiber-reinforced plastics. Many textile materials may be reprocessed using such methods. Long before glass and carbon fibers were used to reinforce plastics on an industrial scale, textile fiber-reinforced molding compounds were employed in automotive and mechanical engineering for purposes such as manufacture of gears for printing presses and textile machines [2].

Phenolic resin-bonded textile fiber felts containing 25–40% of a characteristic binder have been familiar for more than 35 years, and also represent fiber-reinforced thermosets. They are used as thermal and acoustical insulating materials for numerous applications in the automotive market area, and as padding and insulating materials for household appliances [3, 4]. The fibers used in such applications are recycled products from the textile industry, and in the case of cotton represent renewable raw materials. These materials compete with thermoplastic or PUR foams in some cases, and the phenolic resin-based binders with thermoplastic melt binders or melt fibers. Phenolic resin-bonded textile fiber felts are set above the other products by their outstanding range of insulating and acoustical properties, low weight, and the ability to be combined with many other materials. They are economical, and when properly fabricated, meet current automotive industry demands for lack of odor and no-fogging properties. All parameters relating to optimization of fiber and binder quality as well as process conditions are presently observed in production of low or zero-emission textile fiber felts. The fibrous material and powdered resin are generally combined at proportions that vary according to the specific demands on the finished article, for example over a range of 75:25–60:40 expressed as weight percentages.

These materials, which have been primarily used in the automotive industry for many years, represent recycled products containing lower levels of synthetics than alternate materials. Numerous development and optimization technical programs in the binder, fiber, felt, and automotive industries have demonstrated that the ecological and economical objectives determined for these materials can be quite satisfactorily achieved [5].

6.1.2.3.2
Definitions

Phenolic resin-bonded textile felts may be considered fiber-reinforced plastics with a high (60–75%) level of fibers. According to Eisele [6, 7], such materials may also be referred to as three-phase composites (fibrous material, cured phenolic resin and air) that are produced with bulk densities of 50–1000 kg/m³ and thicknesses of 5–30 mm. DIN 61210 would term this type of material a binder-containing matting with cotton as the fiber of choice and phenolic resins as binders.

6.1.2.3.3
Composition of Textile Felts

For economical reasons, fibrous material derived from treatment of textile scraps and trimmings in a shredding process are almost exclusively used. Cotton is particularly well suited for production of textile felts, since its structure (Fig. 6.11) affords a high level of airborne noise absorption in finished products of low bulk density, or high bending strength in the case of those with high bulk density. The melt binders must exhibit excellent wetting and impregnating abilities as well as adhesion [8]. The textile scraps may contain a certain fraction of polyester fibers, since the clothing industry generally uses

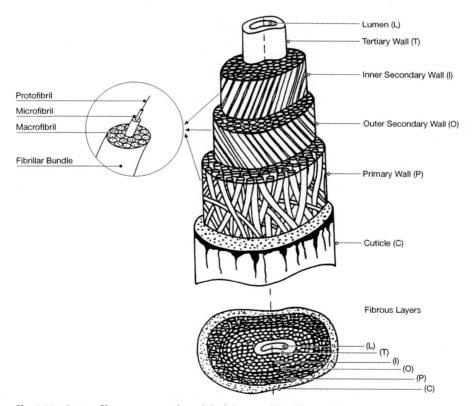

Fig. 6.11. Cotton fiber – structural model of discrete fiber [from 14]

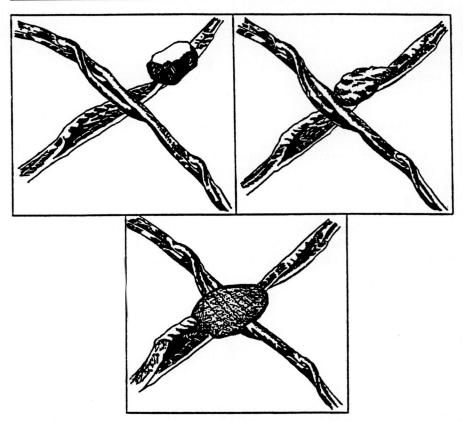

Fig. 6.12. Formation of point bond between two fibers by phenolic resin (from [1])

fiber blends. The possibility that wool, polyamide, and acrylic fibers are present as components in shredded bulk fiber from textiles also cannot be overlooked, which are less satisfactory for mat production than polyester fibers. The binders used are powdered phenolic resins that melt and cure by application of heat, leading essentially to punctiform linking of the fibers by resin at their intersections (Fig. 6.12). In addition, the phenolic resins impregnate and coat the fibers, thus providing a supportive and reinforcing function.

6.1.2.3.4
Types of Resins for Textile Felts

Phenolic resins for textile felt production are finely milled, powdered phenolic (generally novolak-hexa-based) resins that exhibit a specific particle size range and relate to the processing properties of free flow and metering ability. Powdered phenolic resins to which around 9–10% hexamethylene tetramine had been added (Table 6.29) were initially satisfactory. These resins, for

Table 6.29. Various powder resins used for manufacture of phenolic resin-bonded textile felts

Properties	A Standard[a], conventional	B Reduced level of free phenol	C Reduced level of hexa	D Catalytically accelerated cure	E Free of hexa	F Lignin- modified[b]
Melting point in °C, ISO 3146	85±8	90±10	90±10	90±10	105±5	95±5
Flow distance in mm, ISO 8619	40–50	30–40	40–30	30–40	45–60	25–35
Gel "B" time in s, ISO 8974	100–150	100–150	60–100	30–60	50–100	120–180
Percentage free phenol, ISO 8974	1–4	<0.2	<0.2	<0.2	<0.9	<0.1
Percentage free hexa Bakelite FPH 6.16	9–11	8–10	5–7	5–7	–	4–5

[a] Conventional resin with elevated phenol level.
[b] Resin made by condensation with lignin.

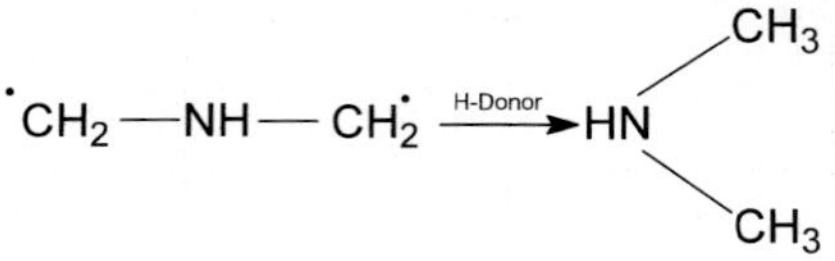

Fig. 6.13. Secondary reaction during curing of novolak with hexamethylene tetramine

$$\overset{\bullet}{CH_2} - NH - \overset{\bullet}{CH_2} \xrightarrow{\text{H-Donor}} HN \begin{matrix} CH_3 \\ \\ CH_3 \end{matrix}$$

example Product A in Table 6.29, resulted in good productivity in the manufacture of sheet goods, and could also be used in production of loadbearing moldings. However, these products contained high levels of hexa as well as 1 % or more free phenol in the base resin, and developed an undesirable odor caused by substances such as dimethylamine, which may be produced in a secondary reaction during curing (Fig. 6.13). These nuisance odors were increased by the odors of free ammonia and volatile phenolic components, and depended upon the storage conditions of the moldings. In the case of cotton-containing matting, the volatile components are absorbed by the fibrils and can be partially released (for example under the influence of a humid environment) to cause a "fishy" odor. This tendency is increased and enhanced when the core of low-density or variable-density material is not completely cured or even still contains completely unreacted hexamethylene tetramine. The free phenol levels of the phenolic novolaks – the base resins of the powdered binder – were reduced to below 0.2 % (Resin B in Table 6.29). A further reduction of the hexa level to values below 6 % represented another significant step to counteract this trend. However, such binders tended to afford reduced stiffness of the cured material if the curing time and temperature were not modified (cf. gel "B" time of Resin C in Table 6.29). Further development work thus involved introduction of various accelerators to reduce the curing time and increase productivity (Resin D in Table 6.29). Figure 6.14 illustrates the accelerated curing properties of such resins as a function of the temperature on the basis of a DSC conversion curve. Such a resin also allows work to be performed at low temperatures, and despite this affords good through curing and simultaneously permits milder thermal treatment of the surfaces of the moldings or sheets.

The odor potential may be reduced to nearly zero [9] by appropriate co-curing of the novolaks with methylol group-containing resins (Resin E in Table 6.29) and/or addition of a certain fraction of epoxy resin. However, such resin systems and particularly straight epoxy resins lead to an increase in the cost of the felts. New low-emission systems (Resin F in Table 6.29) have been developed using lignin-modified resins as the binder [10].

The tendency of the resins to cause sticking during felt production and processing may be reduced by addition of releasing parting agents. Selected grades of high melting waxes are appropriate for this purpose. Dusting problems during processing are reduced to a minimum by uniform addition of various antidusting agents [11]. Flame-resistant resin systems containing various levels and combinations of fire retardants are used for felts required as insulating materials in automotive engine compartments or household appliances, or as insulation for air conditioners. Synergistic combinations such as those of nitrogen, phosphorus, and boron compounds, if appropriately combined with aluminum hydroxide, have been found suitable [12].

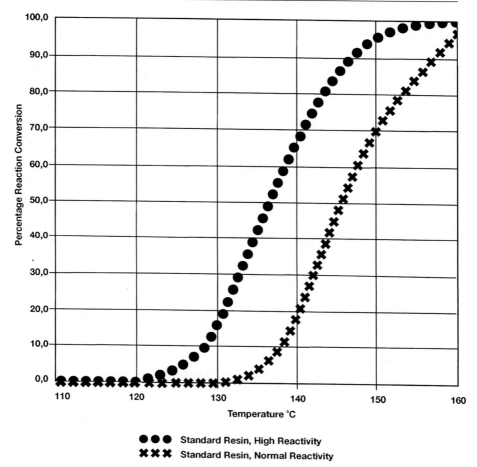

●●● Standard Resin, High Reactivity
✖✖✖ Standard Resin, Normal Reactivity

Fig. 6.14. DSC conversion curves of accelerated and unaccelerated phenolic powder resin (in each case with 6% hexamethylene tetramine)

Binders other than phenolic resins are also occasionally considered for textile felts [6]. However, the range of properties of the final product can be considerably altered by use of binders such as thermoplastics and elastomers in the form of powders, fibers, or emulsions. The advantages offered by the basic properties achieved with phenolic resins are generally not attainable by these other binders. Moreover, the fabrication techniques must be completely or partially altered, and costs can markedly increase.

When using thermoplastic or elastomeric binders, undesirable changes may occur in the area of temperature-dependent strength levels and in the range of acoustical properties. Phenolic resins represent an attractive material component exhibiting greater versatility and suitability than other binders, particularly for applications in automotive engineering.

6.1.2.3.5
Manufacturing Processes

The main method employed for production of fiber felts is aerodynamic matting (cf. Fig. 6.15 for process diagram). Figure 6.16 shows the schematic of a felt production line, and Fig. 6.17 depicts the raw materials and finished product.

The material, which has been spread out and finely divided by the shredding process, is first deposited as loose, fluffy batting on a conveyer belt (generally following a mixing operation designed to homogenize the bulk fiber), and transported to the powdering station. The powder resin is dispersed over the fibrous material using proportioning rollers, vibratory chutes, or similar metering equipment. The binder must be nebulized in a stream of air in a subsequent closed system to ensure that it is intensively and uniformly distributed throughout the fibrous material. The binder-containing dispersed fiber is then collected to the width of the fiber batting by aspiration in the gap formed between two punched screen rollers, and deposited as a felt.

The felt can also be produced using the mechanical carding method, or by a combination of the two (mechanical and aerodynamic) processes. In the carding method, the fibers are combed out to extremely fine webs that are then alternately deposited lengthwise and crosswise over each other until the desired weight per unit area is obtained [1]. The carding process is rarely used for automotive felts.

The uncured felts are further processed to manufacture either flatware (by complete through curing) or by way of precured flatware to yield moldings. In production of flatware, the fiber felt is held between two screen belts and passed into a hot air tunnel in which the resin is completely cured by air heated to a temperature of around 160–200 °C. The web must be cooled at the exit of the curing tunnel. The product is trimmed and rolled (Fig. 6.18), or immediately cut into sheets.

For production of moldings, the material is only precured. This is achieved by briefly melting the resin in the curing tunnel at lower temperatures and high throughput speed to fix it on the felts. The raw mat obtained in this manner is later shaped to yield the finished article by application of pressure in heated compression molds (Fig. 6.19); the pressure, temperature, and pressing time depend on the thermal resistance of the fibrous material used, the curing properties of the phenolic resin (whether it is accelerated or unaccelerated), and on the required bulk density (weight and dimensions) of the finished article. The pressing temperatures normally range from 160–200 °C at pressing times between 1 and 5 min. Following the shaping operation, the molding can be subjected to coating or cladding processes. The curing process provides the molding with the required stability, stiffness, and permanence of shape that must be ensured even under varying conditions of climate and at elevated temperatures.

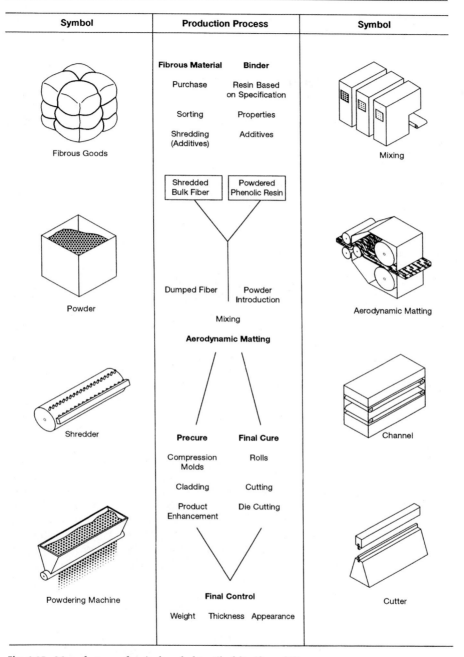

Fig. 6.15. Manufacture of resin-bonded textile felts (from [1])

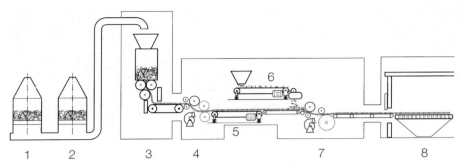

Fig. 6.16. Schematic of a production line for phenolic resin-bonded textile felts (source: V. Schirp GmbH & Co. KG, Hildesheim). *1* Fiber Opener I, *2* Fiber Opener II, *3* Fiber Feeder, *4* Aerodynamical Web Formation I (Fiber), *5* Weighing Belt, *6* Phenolic-Resin Feeder with Weighing Belt, *7* Aerodynamical Web Formation II (Fiber + Phenolic-Resin), *8* Sieebelt Oven

Fig. 6.17. Powder resin, shredded fiber, sheet goods – raw materials and final textile felt product (source: J. Borgers GmbH & Co. KG, Bocholt)

Fig. 6.18. Textile felt fabrication – semifinished goods outlet (source: J. Borgers GmbH & Co. KG, Bocholt)

Fig. 6.19. Stripping of textile felt compression molding (source: J. Borgers GmbH & Co. KG, Bocholt)

6.1.2.3.6
Properties and Applications

Textile felts are produced with bulk densities of 50 – 1000 kg/m³ and weights per unit area of 500 – 3000 g/m² (Table 6.30). Textile felts possess high sound absorption and acoustical insulating capabilities, and their efficacy in this area generally exceeds by far that of other insulating materials such as polyurethane-based foams (Fig. 6.20), crosslinked polyethylene, modified polystyrene, or PVC. Other outstanding typical properties of phenolic resin-bonded textile felts are their exceptional resistance to splintering under all conditions of climate and particularly in the cold, their high bending and surface strength levels, their good aging and thermal resistance, their recycling capabilities, and their ability to establish an effective heat/moisture balance. The versatility of textile felts is particularly evident in their ability to be combined with many cladding and decorative materials, with heavy-duty films to form laminates designed to dampen vibration, and even with glass cloth, metal films, embedded metal frames, bolt anchors, and other components to yield multilayer composites. The benefits of this wide and versatile range of properties, are identified primarily in

Table 6.30. Bulk densities and properties/applications of various types of felts

Flatware at bulk densities of 50 – 150 kg/m³ with sound absorption and padding functions is generally used below surfaces

Moldings as functional components at bulk densities of 150-kg/m³ are self–supporting, absorb sound and are generally applied in visible areas

Rigid compression moldings (RCM) as carrier components at bulk densities of 300-kg /m³ serve as substrates for decorative materials and are thus not visible

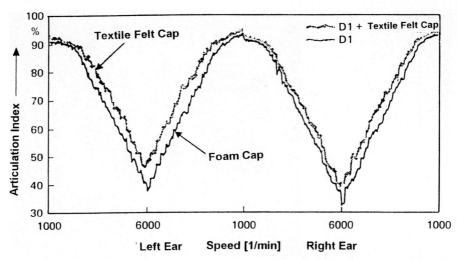

Fig. 6.20. Analysis of noise level (articulation index) in an automotive interior using a roof liner made of phenolic resin-bonded textile felts (Triflex) or glass fiber-reinforced polyurethane foam (from [4])

the automotive industry (Fig. 6.21). Figure 6.21 shows an automobile and illustrates 20 different possible applications, Fig. 6.22 depicts the paneling of the driver's cabin in a truck, and Fig. 6.23 the hatrack in a car is illustrated.

To an ever increasing extent, textile felts are being used in other fields of application, for example to suppress the noise caused by household appliances such as washing machines, dishwashers, and vacuum cleaners, and to insulate loudspeaker enclosures and air conditioners. Flame-resistant resin systems that can contain up to 30% fire retardants are preferred for use in most of the latter applications.

6.1.2.3.7
Property and Quality Testing

Property and quality testing of textile felts, particularly those intended for use in automotive engineering, is very extensive [13–15]. Table 6.31 surveys the mechanical properties of textile mat flatware exhibiting various specific weights. Further parameters used to assess the suitability for use are the aging and thermal resistance, flammability, physiological padding properties such as the heat transfer resistance, heat/moisture balance, the ability to be clad, and the results of acoustical tests.

The outstanding absorption results of textile felts, determined according to DIN 52215 or DIN EN 29053, are related to the structure and morphology of cotton, whose basic properties are retained in the resin-bonded textile material.

The tests for moldings are similar to those performed for flatware, and determine such parameters as the mechanical properties, surface character-

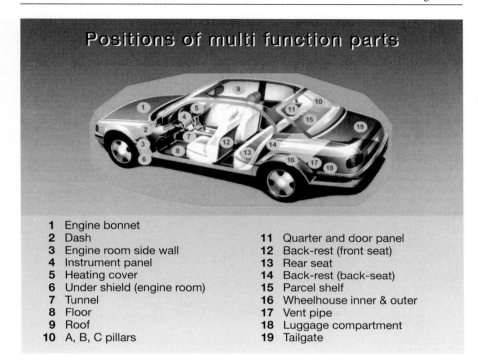

Fig. 6.21. Model of automobile and outlines of various applications for textile felt flatware and moldings (from [14])

1. Engine bonnet
2. Dash
3. Engine room side wall
4. Instrument panel
5. Heating cover
6. Under shield (engine room)
7. Tunnel
8. Floor
9. Roof
10. A, B, C pillars
11. Quarter and door panel
12. Back-rest (front seat)
13. Rear seat
14. Back-rest (back-seat)
15. Parcel shelf
16. Wheelhouse inner & outer
17. Vent pipe
18. Luggage compartment
19. Tailgate

Fig. 6.22. Roof and wall panels of textile felt moldings for a truck interior (source: J. Borgers GmbH & Co. KG, Bocholt)

Fig. 6.23. Hatrack and trunk lining made of phenolic resin-bonded textile felts (source: J. Borgers GmbH & Co. KG, Bocholt)

istics, water absorption, thickness increase due to swelling, surface smoothness, and heat-sealing properties.

Beside tests for mechanical and acoustical properties, the very subjective odor test was employed as a measure for emissions for many years [5]. Taken at first sight, this test method appears logical, since the buyer of an automobile also possesses no other assessment criteria. However, such a subjective method is completely inadequate for identification of emission causes (odors, nuisances). A wide knowledge of materials and analytical methods was necessary to assess objectively and properly the emission characteristics of construction components [6].

Emission tests are performed in a 1 m^3-glass chamber held at a defined temperature and humidity (Fig. 6.24), which permits the determination of the long-term emissions of substances such as formaldehyde, phenols, phenolic oligomers, and ammonia. The exhaust air in the test chamber may be uniformly sampled to determine the pollutants present using a system of gas washing bottles. The tests on textile felts may be performed at various temperatures and relative humidities. Under certain circumstances, this method can simulate the conditions present in the interior of an automobile quite well, but is very complicated and time-consuming. This lengthy controlled-climate chamber test can be used in specific cases where it is vital for technical reasons, for example, in preliminary testing of new models. For fast in-plant

Table 6.31. Mechanical properties of 12 different standard grades of textile felts (source: J. Borgers GmbH & Co. KG, Bocholt)

Parameter methods and units	1	2	3	4	5	6	7	8	9	10	11	12
Weight in g/m^3 DIN 53854	450	450	600	600	750	800	1.100	1.100	1.300	1.300	1.600	1.600
Thickness in mm DIN 53855	6.5	7	8	9	11	13	16	18	22	25	28	29
Bulk density in kg/m^3 DIN 52350	69	64	75	67	68	62	69	61	59	52	57	55
Compression stress value, N/m^2 DIN 53577	350	320	400	340	340	300	360	300	300	240	250	240
Tensile strength, N/5 cm DIN 53857												
Lengthwise	75	70	83	77	68	65	85	60	58	50	55	56
Crosswise	65	67	79	70	60	55	80	55	57	46	53	49
Cleavage resistance, N/5 cm DIN 53 357												
Lengthwise	7	6	9	8	7	6	8	6	5	4	5	5
Crosswise	7	5	9	6	6	4	6	5	5	3	4	5
Tear propagation resistance N/5 cm DIN 6335												
Lengthwise	22	18	25	23	22	17	20	17	16	13	15	15
Crosswise	20	17	23	23	19	15	20	16	14	10	14	15

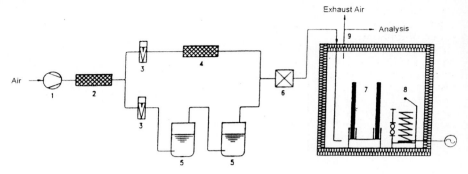

Fig. 6.24. One cubic meter test chamber for determination of emissions liberated by construction components. *1* Pump, *2* C-filter, *3* Flow meter, *4* Silica filter, *5* Absorption bottles, *6* Mass regulator, *7* Test material, *8* Thermostat with ventilation, *9* Exhaust air pipe

testing, however, the VDA 275 bottle method (Fig. 6.25) developed by the German automotive industry affords results that largely correlate with those obtained in the controlled-climate chamber. In any case, the bottle method permits relative analyses and differential assessments of construction components from various origins. It is suitable for production tests and spot checks.

6.1.2.3.8
Recycling

Considerable future growth potential may be predicted for phenolic resin-bonded textile fiber felts, assuming the question of emissions and odors in automotive interiors is satisfactorily addressed, not only by development of new resin systems, but also by observance of certain specifications (Table 6.32)

Table 6.32. Specification of powder resin used for manufacture of textile felts[a]

Chemical characterization: Delivery form: Parameters	Industrial novolak-hexa-based binder Finely milled powder	
1. Hexa level	DIN ISO 8988	4–6%
2. Free formaldehyde	DIN ISO 9397	Not detectable
3. Free phenol	DIN ISO 8974	<0.1–0.2%
4[b]. Melting range	DIN ISO 3146A	75–100 °C
5[b]. Flow distance at 125 °C	DIN ISO 8619	25–85 mm
6[b]. B transformation time at 150 °C (plate)	DIN ISO 8987	15″–120″
7[b]. Screen analysis, >0.09 mm fraction	DIN ISO 8620	1–15%
8. Storage life at 15–25 °C in dry areas		>Six months

[a] Source: German automotive industry.
[b] The specific parameters 4–7 are matched to the technical requirements. The physiology and toxicology of the modifiers must be scrutinized in the case of modified binder systems such as those containing curing accelerators. The possibility of additional emissions – including odors – must be excluded.

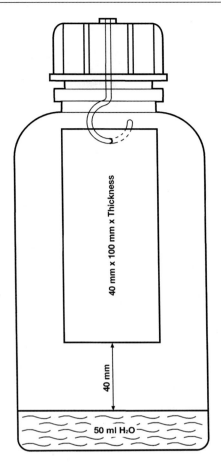

Fig. 6.25. Determination of phenol and formaldehyde emissions from phenolic resin-bonded moldings (textile felts). A 40 × 100 mm specimen is suspended over distilled water. The bottle is kept in a laboratory oven for 60 min, after which the solution is analyzed for formaldehyde and phenol

established by the automotive industry. This also includes adaptation and review of curing and processing technologies to match the curing characteristics of the resin systems [16]. However, the question of recyclability is also of great importance, particularly since claims that thermosetting materials cannot be recycled are repeatedly encountered.

Products exhibiting low bulk densities may be reprocessed by shredding, and then reused for applications outside the passenger compartment, for example, in engine compartment insulation. High-density products can be milled using a special process (Fig. 6.26) and used as fillers in various applications (Table 6.33).

In cases where separation of the felts from decorative materials and similar components is difficult, the material should be subjected to energy recycling [6] and be used, for example, as a supplementary fuel in household waste incineration, since the energy content of the products corresponds to that of wood (fuel equivalent about 26 MJ/kg).

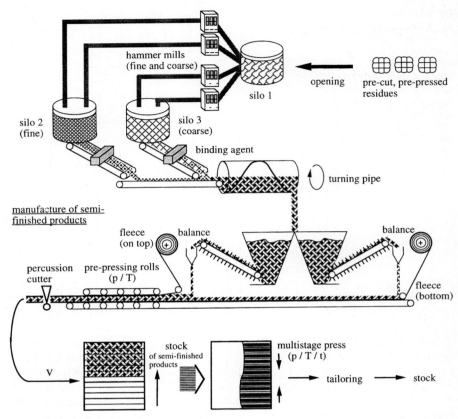

Fig. 6.26. Recycling methods for textile felts: grinding, bulk and compression processes (from [16])

Table 6.33. Possible recycling applications of milled textile felts (from Ref. [16])

Transport Pallets
Transport Packaging for Returnable Systems
Workboards for Household Appliances
Spare Tire Cavity Covers
Chipboard Substitute for Furniture Industry
Floor Leveling Boards in Building Trade
Partitions in Horse Stalls

6.1.3
Phenolic Molding Compounds

6.1.3.1
Introduction: Economic Aspects

Thermosetting phenolic molding compounds emerged at the onset of the age of plastics. They were a direct technological and application-related consequence of Baekeland [1] discoveries (1907–1910). They have retained their prominence as an important synthetic material up to the present day, and have furthermore met and adapted to many new technical and economical demands over the ensuing years [2–9].

The originally relatively modest group of plastics has meanwhile been transformed into a wide range of thermosetting, thermoplastic, and elastomeric materials (estimated annual world production of at least 100 million tonnes) of which around 20% may be considered thermosets. These include phenolic resins and phenolic molding compounds [10–15].

Of the phenolic resins currently manufactured in Western Europe, about 13–15% are used in the production of molding compounds. Thus, phenolic molding compounds, together with synthetic resin bonded wood materials and inorganic/organic thermal and acoustic insulation, represent the three largest areas of application for phenolic resins in Western Europe.

In the area of thermoset molding compounds [16, 17], the phenolic resin-based molding compounds that long led the field have been joined by additional products (Table 6.34), for example, products based on urea, unsaturated polyester, melamine, melamine/phenolic and epoxy resins, and the overall range of thermoset molding compounds. Based on their market fraction in Western Europe, phenolic molding compounds presently are slightly behind the generally light-colored urea resin molding compounds.

Table 6.35 shows the production of phenolic molding compounds from 1988 to 1996 (with a projection for 2002) by regions. According to these data, Western Europe and the USA exhibit marked production decreases, whereas the production levels are rising in Southeast Asia. The figures for Asian production do not include the People's Republic of China [18].

Table 6.34. Percentage breakdown of various thermoset molding compounds in western Europe

Resin basis	Market fraction in percent		
	1988	1991	1994
Phenolic	46	43	42
Urea resin	40	44	44
Melamine, melamine/phenolic	9	7	8
Unsaturated polyester	6	6	6
Totals	100	100	100

Table 6.35. Production of phenolic molding compounds according to regions, 1988–1996 (prognosis for 2002)

Region	Production (in thousands of tonnes)				
	1988	1991	1994	1996	2002
Asia	195	199	213	220	250
North America	90	71	67	60	60
Western Europe	72	67	64	60	60
Totals	357	337	344	340	370

Table 6.36. Percentage breakdown of phenolic and urea resin molding compound consumption in Western Europe (status of 1996)

Application	Percentage	
	PF-MC[a]	UF-MC[a]
Electrical engineering	38	15
Household/electrical appliances	36	44
Automobile	15	0
Sanitation	5	33
Closures	2	4
Others	4	4
Totals	100	100

[a] MC = molding compounds.

The main fields of application for phenolic and urea resin molding compounds, the most important thermoset molding compounds, continue to be in electrical engineering as well as in the household/electrical home appliance areas [19–23]. Urea resin compounds dominate in the latter area. Phenolic molding compounds have highlights in both sectors, and are also strong in the automotive area (Table 6.36). In the last few years, glass fiber-reinforced phenolic molding compounds with high levels of thermal and mechanical properties combined with outstanding dimensional stability have opened a new market in the electric and automotive areas for molding compounds that withstand high mechanical, thermal and chemical stress (Fig. 6.27).

6.1.3.2
Composition of Phenolic Molding Compound (Resins, Fillers, and Reinforcing Agents)

Due to their thermosetting binder, thermoset molding compounds (and in this particular case phenolic molding compounds) generally fully cure during the shaping process, i.e., during compression, transfer, or injection molding (Fig. 6.28) to yield high molecular mass, highly crosslinked materials contain-

Fig. 6.27. Applications of a glass fiber-reinforced molding compound (e.g. impeller and pulley)

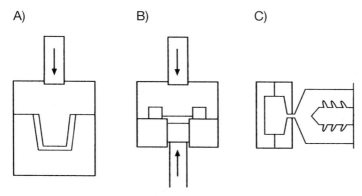

Fig. 6.28A–C. Outline of: A compression; B transfer; C injection molding of phenolic molding compounds

Fig. 6.29. Differences between thermosetting and thermoplastic molding compounds (MC = molding compound)

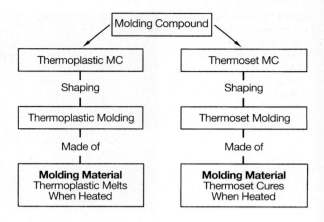

ing fillers and reinforcing agents. The products are insoluble, nearly impervious to swelling, and do not melt. The latter property differentiates them from thermoplastics (Fig. 6.29), which can be remelted as often as desired. This results in two important differences between thermosets, for example, phenolic molding compounds, and thermoplastics such as polyethylene. Cured thermoset molding compounds exhibit far superior thermal behavior, since they simply do not melt, but may decompose and carbonize at elevated temperatures. On the other hand – a contemporary aspect – thermoplastics such as polyethylene can be more easily recycled due to the fact that they can be remelted, whereas thermosets can only be recycled by a "roundabout" way, for example, by "particulate" recycling (to be discussed in Sect. 8.5).

Molding compounds (Table 6.37) contain about 25–40% binder, depending on the type of resin. The type of binder and the binder level have a

Table 6.37. Compositions of molding compounds based on various binders

	Raw materials[a]
Binder (25–40%)	Phenolics, urea resins, melamine resins, unsaturated polyester and epoxies
Catalysts (0.5–1%)	Peroxides for UP molding compounds
Curing agents (2–6%)	Hexamethylene tetramine for phenolic novolaks
Accelerators (0.10–5%)	Various, depending on resin system
Fillers (15–50%)	Wood flour, rock flour, flake mica
Reinforcing agents (10–50%)	Glass fibers, synthetic fiber, cotton and cellulose fiber, textile flakes
Modifiers (3–6%)	Rubber, synthetic resins, graphite, etc.
Additives (1–3%)	Lubricants, pigments and similar

[a] Selection.

Table 6.38. Typical specifications for resins used to produce phenolic molding compounds

Type of resin	Novolak[a]	Resole[a]
Melting range, °C (ISO 3146)	75–85	55–65
Flow distance, mm (ISO 8619)	55–75[b]	50–90
Viscosity at 175 °C, mPa · s (ISO 9371)	250–350	–
Free phenol,% (ISO 8619)	0.5–0.9	3–4
B-Time, min/sec (ISO 8987)	1′30″ ± 30″[b]	2′30″ ± 30″
Water in %, Karl Fischer (ISO 760)	max. 1	max. 2

[a] Bakelite AG, Iserlohn, Germany.
[b] With 10% HMTA as a curing agent.

major impact on the properties of the moldings, and their effect on molding compound production technology (to which they are matched) is equally important. Catalysts, curing agents, and accelerators are used to cure the binders. Thus, the phenolic novolaks generally used (Table 6.38) require hexamethylene tetramine (HMTA) as a curing agent to achieve adequate crosslinking of the product. The curing reaction can be optimized using *accelerating agents* as "external" or "internal" promoters. In Fig. 6.30, the rate of cure (percentage conversion as the temperature rises) of the unaccelerated novolak-HMTA system (A) is compared to those of the accelerated system (B) and a standard resole system (C), which cures more slowly, using

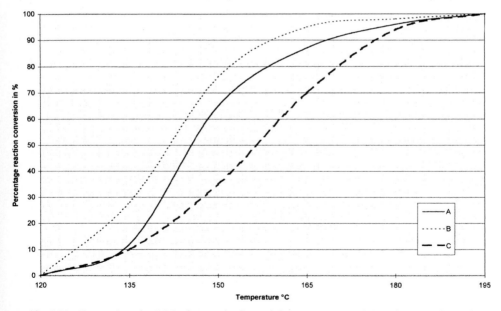

Fig. 6.30. Comparison by DSC of a standard novolak-HMTA system (A) with an accelerated system (B) and a resole (C), percentages of conversion

DSC (differential scanning calorimetry). Similarly selective accelerating agents are of major importance in increasing the productivity of epoxy molding compounds, for example, ones based on phenolic novolaks and epoxidized novolaks (see section 6.1.3.11).

Fillers and reinforcing agents are frequently used in combination [24–27]. An important organic filler for phenolic molding compounds is fibrous wood flour; when combined with materials such as powdered mineral fillers, this permits modification of the hardness and surface appearance. Softwood (such as spruce) flour or alternatively hardwood (such as beech) flour were earlier used in Germany as fillers and reinforcing agents. Since the dust from hardwood flour has been classified as a carcinogen in Germany, only softwood flour is utilized. Depending on the intended use, the mechanical strength of phenolic molding compounds is increased by use of synthetic organic fibers, cotton fiber, cellulose fiber, or textile flakes. Glass fiber meanwhile performs an important role as a reinforcing agent. Asbestos fibers – which have also been classified as a carcinogen and are thus no longer available for use – have been replaced with glass fiber and other mineral-based fibers.

Fibrous reinforcing agents are basically used to increase the notched impact strength. This is frequently representative for examples of applications in which stresses arise in compression molded articles, particularly at points where a transition to a lower wall thickness takes place. Compression molding compounds exclusively containing wood or cellulose flour as fillers are generally unsuitable for such applications. In such cases, fibrous fillers can be used as a supplement. A variety of standardized and non-standardized molding compounds have been developed to increase the impact resistance. Addition of various levels of fibers, for example textile-based fiber, make it possible to produce a continuous range of molding compounds from those including only wood flour to ones containing mainly textile fiber. Textile flakes can be used as additives in molding compounds including those from which large, robust machine or equipment housings are manufactured.

Special molding compounds containing inorganic materials such as siliceous or carbonaceous rock flour in addition to an organic filler can be used in applications where the molding is exposed to permanent thermal stress – although the glow heat resistance and thermal deformation resistance of a compression molding compound containing exclusively mineral fillers are not required – and where there may be a desire to avoid excessively rapid wear of the mold. Secondary shrinkage due to lengthy thermal stress is a significant factor in some applications. Appropriate combination of organic and mineral fillers can lead to selective reduction of post shrinkage. A similar situation applies to reduction of water absorption.

Among compression molding compounds with straight mineral fillers, phenolic compounds containing mica fillers (Table 6.39) are of particular importance due to their insulating power. Mixtures of mica and other (powdered) mineral fillers are generally used to improve the processing properties. Due to the interactions of their fibrous, powdered, and lamellar structures, combinations of various inorganic fillers offer certain advantages under conditions of long-term thermal stress.

Table 6.39. Suggested formulations for type 13 (mica filler) and 31 (wood flour filler) molding compounds (DIN 7708)

Type 13 (DIN 7708)		Type 31 (DIN 7708)
34 %	Novolak + HMTA	47 %
2 %	Accelerator	2 %
2 %	Parting agent	2 %
–	Wood flour	30 %
20 %	Inorganic filler	17 %
40 %	Mica	–
2 %	Color pigment	2 %

Chopped glass fiber has long been of great importance as a filler and reinforcing agent for polyester resin and epoxy compression molding compounds. Glass fiber-reinforced phenolic compression molding compounds have achieved great prominence since the use of asbestos fibers ceased. Short glass fiber is generally used, since destruction of longer fibers may be expected in customary manufacturing and processing operations.

Aside from the fillers and reinforcing agents mentioned above, certain *modifiers* are also in use. Rubber is used to provide elasticity to formulations for special applications. There are several ways to incorporate rubber into a molding compound, whether by way of the resin (by incorporation of lattices) or by direct addition to the premix in the powder, pellets, or flake delivery form followed by customary processing by the melt flow process on mastication extruders. The range of application of such molding compounds is mainly in situations where the molded article is exposed to shock and impact stress, for example in automotive industry.

Addition of graphite represents a further possibility to provide a significant effect on the final properties of a phenolic compound-based molding. As a modifier, graphite provides two properties that can also be of use in production of phenolic molding compounds: (1) graphite is an electrical semiconductor and (2) it exhibits a certain slip and lubricant effect.

The latter property is utilized in production of molding compounds for self-lubricating articles, for example, bearing shells. The required levels of filler and reinforcing agent depend on the expected stresses. Rock flour or metal powder are adequate for simple, small slip rings, whereas textile fibers or textile flakes are used in larger bearing shells or seals.

Various processes for surface metallization of plastics exist. Phenolic molding compounds containing added graphite are also capable of being electroplated, and can be used to manufacture certain consumer articles [28].

To summarize, *fillers and reinforcing agents* for phenolic molding compounds can be organic and/or inorganic in nature, and are used in the form of powder or materials with lamellar or fibrous structures. *Organic reinforcing agents* such as wood flour or cellulose fibers supply a positive effect on the processing properties of the molding compound since they reduce peak friction levels and reduce mold wear. *Inorganic reinforcing agents* such as glass fibers, if appropriately combined with powdered minerals, increase the mechanical strength and thermal resistance of the molding compound at the same time as they reduce the water absorption.

Molding compounds generally also contain 0.5–2% lubricants. Lubricants facilitate the shaping operation and particularly easy ejection of the compression molded articles. Fatty acids such as stearic acid, stearates such as magnesium, calcium, or zinc stearate, and the esters and amides of fatty acids are used as lubricants. Selected waxes, such as montan wax, are also well suited for this application.

Lubricants are differentiated according to whether they are *"internal"* or *"external"* in nature. However, an exact assignment is impossible since many of the indicated materials can be used both as internal, (i.e., soluble) and external, (i.e., insoluble) active ingredients depending on their solubility in the resin and the addition level. Internal lubricants improve the flow of the molding compound, whereas external lubricants act as a parting agent in the mold and thus facilitate ejection of the molding.

Both soluble dyes and pigments may be used as *colorants*. Various organic dyes are used as soluble colorants for phenolic molding compounds. Pigments that can be used include carbon black, finely milled coal, iron oxide pigments, earth pigments, and organic pigments.

6.1.3.3
Phenolic Molding Compound Standardization (Current Status)

Phenolic molding compounds are standardized in DIN 7708 (Table 6.40), and the standard is continuously revised to reflect changing requirements, new

Table 6.40. Types of phenolic molding compounds as defined in DIN 7708, part 2, draft of 1993

Molding compound type	Type of filler
12	Mineral fibers
13	Mica
31	Wood flour
31.5	Wood flour
31.9	Wood flour
51	Cellulose and/or other organic fillers[a]
52	Cellulose
71	Cotton fibers with or without addition of other organic fillers[b]
74	Cotton cloth flakes with or without addition of other organic fillers[b]
83	Short cotton fibers and/or wood flour[a]
84	Cotton cloth flakes and/or cellulose[a]
85	Wood flour and/or cellulose[a]

[a] The term "and/or" means that both fillers as a mixture, or the individual fillers themselves can be used.
[b] The term "with/or without" means that the compound can contain additives other than the first-mentioned filler.

types of molding compounds, and the use of modern units [29]. Standardization into types, for example, Types 31.5 and 31.9, is oriented toward the reinforcing agents and fillers as well as the minimum level of physical properties that can be achieved with the indicated type. Type 31 represents a phenolic molding compound mainly containing wood flour as a filler. The supplementary ".5" indicates that this compound achieves specific electrical data, and the supplementary ".9" that the compound is free of ammonia (after compression molding). Types 12 and 13 contain mineral fillers (mineral fibers, flake mica) that increase the thermal deformation resistance and improve the electrical properties. The earlier DIN 7708 Type 12 contained asbestos as a filler.

Types 51, 71, 74, 83, 84, and 85 are made with cellulose fibers, cotton fiber, and textile flakes as reinforcing agents, in some cases (Types 83 and 85) combined with wood flour, and offer increased notched impact strength.

ISO 800 (Table 6.41) uses a simplified classification scheme, differentiating according to resole and novolak molding compounds, and subdividing these groups according to the types of filler and reinforcing agent employed. Thus, PF 1 E indicates a resole molding compound with a mica filler, offering premium electrical properties. To permit worldwide processing in data bases such as "Campus" [30, 31], uniform type classifications which incorporate salient features of the national standards [32] have been agreed under the auspices of the International Standardization Organization (ISO, with headquarters in Geneva). This enables improved comparisons of thermoset with thermoplastic molding compounds.

The DIN type classifications originally also included information on the minimum resin content and specified a standard color shade. ISO 14526, that will soon represent the main international standard, specifies ranges for the filler content. Table 6.42 surveys the phenolic molding compounds described in ISO 14526-3. This table shows that DIN 7708 Type 31 corresponds to Type PF WD molding compound (wood flour containing phenolic molding compound). The content of wood flour can range between 30% and 40%, and the corresponding level of (MD) mineral fillers between 20% and 10%. ISO 14526 includes a new type of molding compound (Type PF GF, glass fiber-extended phenolic molding compounds) that was not listed in DIN 7708.

The required minimum levels of (mechanical, physical, etc.) properties for the individual types of compounds represent an important criterion for all

Table 6.41. ISO 800 (1977) molding compounds (example: PF 2A designates a novolak molding compound with a wood flour filler for general use)

Resin	Fillers and Reinforcing Agents	Property, Application
PF 1 Single-stage (resole) resin	A Wood flour	1 General purpose
PF 2 Two-stage (novolak) resin	C Mineral fillers and reinforcing agents	2 Thermally resistant
	D Organic fibers	3 Impact resistant
	E Mica	4 For electrical applications

Table 6.42. Phenolic molding compounds as defined in ISO 14,526-3 compared to different national standards

ISO/FDIS 14,526-3									
Type/Grade	WD30 / MD20 through WD40 / MD10	WD30 / WD20,E through WD40 / MD10,E	E WD30 / MD20,A through WD40 / MD10,A	LF20 / MD25 through LF30 / MD15	SC20 / LF15 through SC30 / LF05	SS40 through SS50	PF40 through PF60	LF20 / MD15 through LF40 / MD05	GF30 / MD20 through GF40 / MD10
ISO 800; BS771; P1	PF 2A1	PF 2A2	PF 1A1	PF 2D2	PF 2D3	PF 2D4	PF 2C3	–	–
DIN 7708; P2; 1975	31	31.5	31.9	51	84	74	13	83	–
JIS K 6915	PM-GG	PM-GE	PM-EG R	PM-ME	PM-MI	–	–	–	PF6507 [c]
NF T 53-010	PF 2A1	PF 2A2	PF 1A1	PF 2D2	PF 2D3	PF 2D4	PF 2C3	–	–

Nature[a]	Form[b]		Percentage (%)		spec. properties	
C Carbon	C	Chips, cuttings	5	<7.5	A	Ammonia free
D Alumina trihydrate	D	Dust/Powder	10	7.5<12.5	E	Electrical properties
G Glass	F	Fibre	15	12.5<17.5	FR	Flame resistance
K Calcium carbonate	G	Ground	20	17.5<22.5	N	Nutrition (food contact)
L Cellulose	S	Scale, Flake	25	22.5<27.5	M	Mechanical properties
M Mineral			30	27.5<32.5	R	Containing recycling material
P Mica			35	32.5<37.5	T	Thermal resistance
S Synthetic, Organic			40	37.5<42.5		
W Wood			45	42.5<47.5		
			50	47.5<52.5		
			55	52.5<57.5		
			60	57.5<62.5		

[a] = first letter.
[b] = second letter.
[c] = glass fiber reinforced (Bakelite AG).

Table 6.43. Selection of minimum requirements for DIN type 13 and 31 molding compounds including new nomenclature of ISO 14526

DIN 7708; P3; 1975	PF 13	PF 31
ISO 14526	PF40–PF 60	IWD30/MD20 WD40/MDl0
Filler	40–60% Mica	30–40% wood powder 10–20% mineral powder
Tensile strength (ISO 527-1/2)	$\geq 30^a/40^b$ MPa	$\geq 40^a/50^b$ MPa
Flexural strength (ISO 178)	$\geq 50^a/60^b$ MPa	$\geq 70^a/80^b$ MPa
Charpy impact strength (ISO 179–1/2 eU)	$\geq 2.5^a/3.5^b$ kJ/m^2	$\geq 4.5^a/5.0^b$ kJ/m^2
Temperature of deflection under load (ISO 75-2 HDTA-1,8 Mpa) (ISO 75-2 HDTE-8 MPa)	$\geq 170\,°C$ $\geq 130\,°C$	$\geq 160\,°C$ $\geq 115\,°C$
Surface resistivity (IEC 60093)	$10^{11}\,\Omega$	$\geq 10^9\,\Omega$
Water absorption (ISO 62) 24 h/23 °C	30 mg	≤ 100 mg

^{a} Compression molding
^{b} Injection molding

standards, both the original DIN 7708 and the future international ISO 14526 standard. This is shown in Table 6.43, which shows a selection of data for DIN 7708 Types 31 and 13 that are defined in ISO 14526 and lists a number of (selected) test parameters in the ISO standard.

The minimum values specified for phenolic molding compounds in ISO, DIN or other national standards can only provide an indication of a specific range of properties; in the case of special applications, this range must be adapted to the specific use (assuming this is possible with a specific type-classed compound), or use must be made of special-purpose compounds [25].

6.1.3.4
Production of Molding Compounds

Phenolics were the first completely synthetic resins used in production of molding compounds. Originally – in the initial decades following Baekeland's discovery – a *wet conditioning process* involving the use of liquid phenolic resoles (Fig. 6.31) was used. The resultant technology afforded "cold-compression molding compounds" – doughy, occasionally plastic compounds that following mixing were shaped by compression in unheated molds and then dried and cured by heating. The resultant moldings, for example, terminal strips, did not exhibit particularly good surface quality.

The *melt flow process* (Fig. 6.31) afforded a significant improvement. This process has been practiced since around 1925, and has experienced constant improvement over the course of time. The process was originally carried out in heated masticators and calendar mills; continuous mastication extruders are mainly used at present. Milled solid resins – novolaks with the required

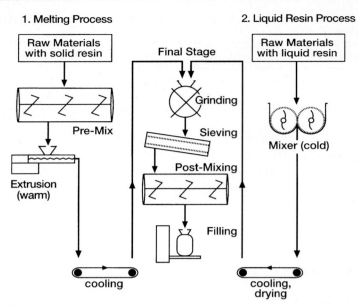

Fig. 6.31. Melting process (1) and liquid resin process (2) for manufacture of molding compounds (scheme)

amount of HMTA or solid, fusible resoles – are used as binders. In this process, a premix is prepared with the milled resin and other components. The mix is then continuously or discontinuously homogenized on a heated melting/mastication unit such as a masticator or mastication extruder. In this process, the resin melts and impregnates the fillers and reinforcing agents. The resin system undergoes further condensation due to the heat developed in this operation and as a function of the mastication or homogenization time to reach a desired rate of flow oriented toward the subsequent processing technology and the molding to be produced. Today, the melt flow process is still occasionally carried out on heated roller mills – generally equipped with automatic stripping plate systems – for continuous production of phenolic compression molding compounds.

The *"turbomix" process* (Fig. 6.32), in which the resin impregnates the fillers and reinforcing agents in a high-speed mixer, represents a variation of the melt flow method. When this method is applied to production of phenolic compression molding compounds, the solid resins melt due to the heat of friction (thus making this method a "melt flow process"), leading to impregnation of the other components and formation of pellets. In principle, this method is used for production of UP molding compounds, but for technical reasons has not been adopted for manufacture of free-flowing, low-dust phenolic molding compounds.

Fig. 6.32. Outline of the turbomix process

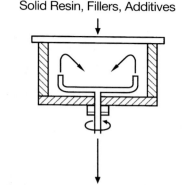

6.1.3.5
Processing of Molding Compounds

After technology to produce pellets from the phenolic resin/additive mixture had been developed, the resultant "compression molding compounds" were processed by the *compression molding method* [33–37]. Later the *transfer molding* method was developed. This latter process offered far fewer technical and economical advantages than those seen in injection molding of thermoplastics. Concentrated efforts and development work by the thermoset manufacturers and processors in the 1960s led to the use of thermoset molding compounds for *injection molding* as well. This represented an outstanding process engineering development success [38–40].

Generally all types of molding compounds – including amino, UP, and epoxy molding compounds – can be processed by the compression, transfer, and injection molding methods. The injection molding process also includes two modifications, *injection embossing* and *sprueless injection molding*, that can make processing of thermoset molding compounds even more attractive.

Today, without the outstanding, dedicated, and continued development work leading to the possibility of injection molding, thermosetting molding compounds would probably only lead a marginal existence in the shadow of thermoplastics and elastomers. Such development work has ensured that phenolic molding compounds, products with extraordinarily versatile uses, are available to the final customer at relatively low cost, both from raw material and productivity considerations resulting in an attractive cost-performance relationship.

The compression molding process is the conventional method of processing phenolic molding compounds. In this process, the compound is transformed into the molding (compression molded article) under heat (heated mold) and pressure (regulated at the press) in a compression mold (Fig. 6.33). The pressure forces the heated, melted compound into the relief of the mold. The exposure to heat causes essentially complete curing of the resin to the

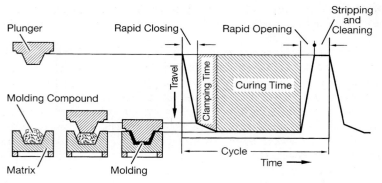

Fig. 6.33. Three-stage pressing operation for processing of phenolic molding compounds

"resite" state. The minimum temperature required for curing and the pressure required for shaping depend on the history of the molding compound (resin content, percentage of HMTA based on resin fraction, precure, level of filler). The shaping pressure can be influenced by additional "precuring", i.e., preheating of the compound. This furthermore affects the curing time, molding quality, and productivity.

Within certain limits, the *curing time* can be reduced by increasing the mold temperature, or influenced by internal or external acceleration of the resin system. On the one hand, the resin can cure more rapidly due to its composition (by „internal" acceleration) or due to the addition of an external accelerator to the system.

The crosslink density of the system also increases during curing. However, this increase itself limits further crosslinking. This situation is demonstrated by methods such as determination of certain parameters, for example, the thermal deformation resistance, and repetition of the determination after a lengthy period of heat treatment. A rise in the magnitude of the result is generally observed. A similar circumstance is observed in practice during manufacture of hot compression molded brake linings – the lining compounds in principal also represent molding compounds – in order to achieve optimum crosslink density (refer to Friction Materials Sect. 6.2.3).

The *compression force* (the other important compression molding parameter) mainly influences the shaping operation. It may be considered as opposing pressure to the vapor pressure produced by emissions such as water and ammonia that form and escape during the curing process. Thus, it prevents formation of bubbles or pores in the molding. The compression force must therefore be maintained until the strength of the molding is adequate to withstand the pressure of gases such as trapped water vapor.

The *cycle time* of the press is the sum of the fill time, clamping time, the curing time (including possible venting operations) and the ejection time (including cleaning). Efforts are made to decrease the time required for the individual operations as much as possible to reduce the total cycle time. This can result in measures such as reduction of the fill time by compound preforming, preheating outside the mold, shortening of the clamping time by use of modern presses, and decreasing the curing time by chemical means.

Transfer compression molding (transfer molding) originated in the thirties, and represents a process engineering hybrid between compression and injection molding [41, 42]. The compound is generally preformed, preheated, and/or premelted (using a melt extruder) and conveyed from the filling reservoir or prechamber through distribution channels and a special gating system into the mold for cure (cf. Fig. 6.28) by a plunger. Transfer molding with preformed and preheated (by microwaves) compression molding compounds is widespread. In this method, the preformed tablets are quickly preheated to 110–120 °C, and thus already exhibit a pasty consistency when they enter the injection cylinder. For optimal processing using the transfer molding method, it is important that the preheating times be matched to the pressing cycle and that the transfer times of the tablets be minimized. Preheating of the tablets reduces injection times and protects the gating.

Transfer molding of phenolic molding compound offers a series of advantages. Thus, the extreme preheating it involves permits shorter curing times than in compression molding. Metal inserts or pins in the mold are not as easily displaced or damaged due to the very effective premelting of the compound. Good mixing of the compound in the gating channels and nozzles leads to very uniform preheating. This is of particular benefit in production of compression moldings with varying wall thicknesses, and represents an advantage that is of special significance in processing amino resins (urea and melamine molding compounds). Due to the good degassing, transfer moldings generally exhibit lower secondary shrinkage levels than compression molded articles made with the same compound. Disadvantages in transfer molding are the slightly higher compound consumption caused by loss in the gating system, and the greater mechanical outlay of equipment.

Due to its great economy [43], *injection molding* (cf. Fig. 6.28) is presently the most prevalent method of processing [44–47]. In this method, all necessary operations can be automatically performed by a single system. The screw of the melting unit meters and homogenizes the molding compound that has been melted by friction and heat transfer, and acts as a plunger in the injection operation. Due to the benefits of high material temperature at injection, only little additional energy (heat transfer) is required to cure the molding. This results in short cycle times.

In practice, processing of phenolic molding compounds on injection molding machines proceeds in the following manner (Fig. 6.34). The powdered or granulated material is drawn in by the rotating screw and transported to the head of the screw. In this process, the screw is displaced backwards against an adjustable back pressure [48–56]. Transfer of heat from the extrusion barrel heater and the developed heat of friction feed enough energy into the material to melt it. Metering is accomplished by limitation of reverse screw movement. To inject the material into the mold, the melting unit advances until the nozzle is pressed against the gating of the mold. Like a piston, the screw then presses the molten material into the mold cavity. The screw should

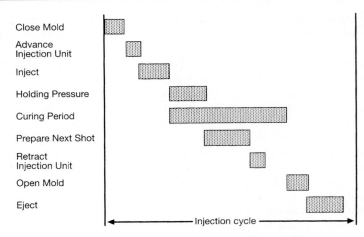

Fig. 6.34. Example of operational sequence in injection molding machines

be locked to prevent rotary movement during this operation. Injection is only carried out with a rotating screw in rare cases.

Near the end of the filling operation, the injection pressure is reduced to the holding pressure; the time of this changeover is critical, and the holding pressure is maintained until the curing process ensures that the molding compound can no longer flow back out of the mold [57, 58]. The screw then starts to rotate again, melting material for the next shot. The injection unit can return to the retracted position at this point. After the next shot has been melted, the injection unit remains on standby until the curing time of the molding has elapsed. After the pre-set curing time has elapsed, the mold opens and the molding is ejected. The signal to begin the next injection operation is then given. The mold closes again. The injection unit advances once again until the nozzle is pressed against the gating of the mold, and the new cycle begins.

Various systems for *sprueless injection* have been developed to avoid the sprue losses that occur in injection molding. In these systems, the material in the gating is kept molten and is used in the next injection operation.

The *injection compression* method is derived from the injection molding process. This process combines the advantages of injection molding with those of compression molding. The compound is injected into a mold that is not fully closed (Fig. 6.35), and the mold closes following the injection operation. The quantity of injected compound is calculated to fill the closed mold.

Injection compression has been used for thermoplastics processing for years, specifically for production of optical lenses and other optical components. Sink marks due to differing wall thicknesses can be prevented with this method. The advantage in phenolic molding compound processing lies mainly in the more favorable filler orientation. The different directions of flow of the molding compound during injection and compression render the moldings less anisotropic, i.e., their properties are less dependent on the orientation.

Fig. 6.35. The injection compression process

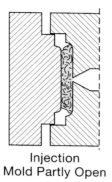
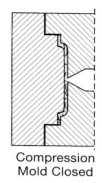

Injection
Mold Partly Open

Compression
Mold Closed

This reduces cracking in the gating zone and distortion. Additional benefits arise from the relatively low required injection pressure and the good venting. The lower injection pressure reduces the wear on the screw, barrel, nozzle, and gating, and increases the true shot volume due to lower backflow losses. The surfaces of injection embossed moldings are frequently better than those of injection moldings. The increased formation of mold marks, approximately corresponding to that seen in compression molding, represents a disadvantage of the injection compression process, since these boost dressing costs. Single molds with interlocking edges are best suited for *injection compression*. Use of multiple molds is complex since the distribution channels would similarly have to be provided with interlocking edges for the embossing operation. The injection molding machines must be appropriately equipped for use in injection compression [59, 60].

6.1.3.6
Test Methods (Application of ISO Standards)

A wide variety of test methods that are standardized on a national or international basis and in some cases are also used for other materials or have been adopted (if necessary with modification) from the testing techniques of other materials are available for quality control [61, 62] and classification of molding compounds. These are used to compare various types of molding compounds both to one another and to thermoplastics, metals or other materials. Tables 6.44 and 6.45 survey the ISO, DIN, ASTM, and Japanese standards. Today, ISO test methods are mainly referenced for both molding compound type classification and for test methods used to determine the physical and chemical parameters, and efforts are made to increase international awareness of these. In Europe, the ISO methods are adopted in European CEN standards where possible, and these are then introduced in Germany as "DIN-EN" standards.

Standardized methods will be briefly described below; the pertinent standards must be consulted for details. The majority of these are ISO standards, although DIN standards are also still used in Germany if these

Table 6.44. Selection of physical tests for thermoset molding compounds as defined in ISO, DIN, ASTM, and JIS

	ISO	DIN	ASTM	JIS 6911–79
Density (g/cm^3)	1183	53479	D-1505	5.28
Processing and secondary shrinkage (%)	2577	53464	D-955	5.7
Tensile strength (N/mm^2)	527	53455	D-638	5.18
Bending strength (N/mm^2)	178	53452	D-790	5.17
Compression strength (N/mm^2)	604	53454	D-695	5.19
Impact resistance (kJ/M^2 or J/M^2)	179/180	53453	D-256	5.20
Ball indentation hardness (N/mm^2)	2039/1–2	53456	D-822	5.16.1 (5.16.2)
Dimensional stability temperature (°C)	75	53461	D-648	5.35.1
Vicat test (°C)	306	53460	D-525	–

Table 6.45. Selection of electrical tests for thermoset molding compounds as defined in ISO, DIN, ASTM, JIS, IEC, and VDE

	ISO	DIN	ASTM	JIS 6911–79
Surface resistance (ohm)	93	53482	D-257	5.13
Specific volume resistance (ohm · cm)	93	53482	D-257	5.13
Dielectric constants (tan δ and ε)	250	53483	D-150	–
Dielectric strength	243–1	53,481	D-149	5.11.3
Tracking resistance	IEC 112	VDE 0303–1	D-3638	IEC 112
Arcing resistance	–	53484	D-495	5.51

are required and have not yet been assigned ISO status or replaced by an ISO (or CEN) standard. This applies analogously to other national standards such as ASTM or BS methods.

UL (Underwriters Laboratory, Northbrook IL, USA) procedures are referenced in the area of flammability testing. In the USA, UL is a government-approved institute for testing of plastics for flammability, thermal resistance, and similar properties. In the area of electrical properties testing, IEC (International Electrotechnical Commission, Geneva, Switzerland) standards are also used, or reference made to VDE (Verband Deutscher Elektrotechniker e. V.) procedures.

Various properties are always determined using the test pieces specified in the pertinent test standard. Test pieces are prepared in standardized compression (ISO 295) and injection (ISO 10724) molds. The values determined with the standard test pieces cannot be simply extrapolated to other forms of moldings, since the manner of attachment and the shape of the molding have a major effect on various properties. However, the results are at least useful for comparative purposes.

6.1.3.6.1
Bulk Density (ISO 60)

The bulk density is determined as specified in ISO 60 to monitor the uniformity of a granulated molding compound and provide a basis for calculation of the true positive mold. The bulk density in g/cm^3 is the quotient of the mass and the volume of a loose heap of molding compound. The test is carried out by pouring the molding compound into a test vessel of specific volume and weighing the material.

6.1.3.6.2
Process Shrinkage (ISO 2577)

Process shrinkage is an important factor in designing molds to achieve the desired dimensional tolerances in production of moldings. It represents the difference between the dimensions of the cold mold and those of the cooled molding. Post-shrinkage is the difference between the dimensions of the cooled molding and those of the same molding following storage at a specific temperature (168 h at 110 °C). Both process shrinkage and post-shrinkage depend on the resin system and pertinent processing conditions as well as the moisture level and flow of the pertinent molding compound.

6.1.3.6.3
Tensile Strength (ISO 527)

As in the case of other materials, such as thermoplastics or metal, measurement of the tensile strength is used to determine information on behavior under tensile stress. The tensile strength (expressed in MPa) is the peak tensile stress, applied uniaxially, that leads to rupture of the test piece. The measurement is performed on a shoulder beam at a constant rate of increase in longitudinal tensile force to the point of rupture (during the process of elongation). Measurement of the tensile strength at elevated temperatures is also of interest to compare behavior with that of other materials, for example, thermoplastics, when these are exposed to heat. Special equipment is used for these determinations.

6.1.3.6.4
Bending Strength (ISO 178)

The bending strength provides information on the strength of a molding when subjected to flexural stress. It is the quotient of the transverse movement of the test piece and its modulus of resistance (expressed in MPa). In this test, the test piece is placed on two supports (Fig. 6.36), a uniformly increasing load applied to its center, and the force required to break the test piece determined.

Fig. 6.36. Bending strength (ISO 178)

6.1.3.6.5
Charpy Impact Resistance (ISO 179/1eU) and Charpy Notched Impact Strength (ISO 179/1eA)

These tests (Charpy impact resistance and Charpy notched impact strength, expressed in kJ/m^2) are used to determine the toughness and notch sensitivity of a molding when exposed to impact stress. The impact resistance is the impact force at break relative to the cross-section of a test piece without a notch, and the notched impact strength the corresponding force applied to a notched test piece. The test is performed using a pendulum ram. The impact force required to break the test piece is determined.

6.1.3.6.6
Heat Distortion Temperature (ISO 75)

The material of a test piece sags when the test piece is heated under a certain load. The heat distortion temperature is the temperature at which the test piece sags to a specific extent. The test is carried out in an oil medium. Although the measurement is performed at different loading stages, the test procedure does not provide information on the maximum application temperature.

6.1.3.6.7
Maximum Application Temperature (IEC 60216, Part 1)

This test method affords limit temperatures for short-term (less than 50 h) and long-term (20,000 h) stress. The maximum application temperature is defined as the temperature at which the molding material can be used without significant reduction in its properties. The maximum application temperature depends on the nature and duration of thermal exposure, the load, and the surrounding medium. The test detects changes in the properties, for example, the change in bending strength after hot storage. This test provides no information on the residual levels of properties or the loss in performance a molding suffers at a specific temperature.

6.1.3.6.8
Flammability Test (UL 94)

This UL method is used to assess the flammability of materials such as phenolic molding compounds by determining the behavior of a test piece when briefly exposed to a flame. The determination establishes whether the test piece burns and, if so, the time required for the flame to go out, i.e., the test piece is briefly exposed to a flame and the length of time it burns after the ignition flame is removed is determined. This test method is also applied to electrical laminates.

6.1.3.6.9
BH Incandescent Rod Flammability Method (IEC 60707)

This test method is used to classify types of molding compounds according to increasing flammability. The compounds are assigned to specific levels, for example, BH 1. The combustion length and the rate at which the flame propagates upon contact of a test piece with an incandescent rod are determined. The significant parameters are development and propagation of a flame.

6.1.3.6.10
Water Absorption (ISO 62)

The possible water absorption under conditions of exposure to cool (23 °C) water is an important parameter for many molding compound applications. It is determined by immersing a defined test piece in water for a storage period of one day and measuring the gain in mass as a percentage and in units of milligrams.

6.1.3.6.11
Specific Surface Resistance (IEC 60093)

Measurement of the surface resistance provides information on the state of insulation at the surface and thus on the insulating power of the molding compound. It is defined as the electrical resistance between two electrodes attached to the surface of the test piece. The measurement is performed following underwater storage of the test piece for a set length of time. It is carried out at a potential of 100 V d.c.

6.1.3.6.12
Specific Volume Resistance (IEC 60 093)

The specific volume resistance provides information on the electrical insulating power of the molding compound. It is defined as the electrical resistance between specific measuring electrodes attached to the top and bottom of a test piece relative to its thickness. The determination is performed by measuring the resistance between the upper and lower surfaces of the test sheet (in units of ohmsxcm) at a potential of 100 V d.c.

6.1.3.6.13
Dielectric Loss Factor and Dielectric Number (IEC 60 250)

The dielectric loss factor (tan δ) of an insulator is the tangent of the loss angle by which the phase displacement between current and potential in a capacitor of $\eta/2$ deviates when the dielectric medium of the capacitor consists exclusively of the insulator. The relative dielectric number (ε) specifies how much the capacitance of a capacitor using the insulator as a dielectric medium exceeds that when the medium is a vacuum. The measurement is performed at a specific frequency with a measuring bridge using a guard ring capacitor, thus determining the dielectric loss of molding compounds (Fig. 6.37).

6.1.3.6.14
Dielectric Strength (IEC 60 243, Part 1)

The dielectric strength provides information on the behavior of the molding compounds when exposed to high voltages. The alternating current potential applied to the electrodes is increased at a constant rate to the point at which the test piece is perforated. The dielectric strength is the quotient of the breakdown potential and the distance between the electrodes located on opposite sides of the test piece. It is dependent on the wall thickness. All results (in kV/mm) are determined using test pieces with a wall thickness of 1 mm.

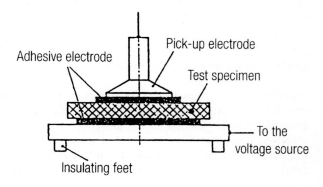

Fig. 6.37. Test system for relative dielectric index and the dielectric loss factor

6.1.3.6.15
Tracking Resistance (CTI and PTI Comparative Tracking Index, IEC 60112)

A knowledge of the tracking resistance makes it possible to assess the surface behavior of insulators when exposed to tracking currents. The results are highly dependent on the resin matrix; phenolic resin, for example, exhibits behavior completely different from that of unsaturated polyester or melamine resins. The tracking resistance is the resistance of the molding compound to formation of electrical creep tracks. A creep track results from local electric current-induced thermal decomposition in the presence of conductive impurities. In this test, drops of electrolyte solutions are deposited between test electrodes charged with a potential of 100, 250, 300, 375, or 500 V, and the number of drops required to produce a current flow at one of the indicated potentials measured. The tracking resistance of an insulator is classified with these test data.

6.1.3.7
Flow Behavior of Phenolic Molding Compounds

Determination of the flow/curing behavior, which can be examined by various methods, is a significant factor in quality control of phenolic molding compounds during production and when they are further processed by the customer [63]. In addition to simple plant methods for direct production control, special discriminating procedures with great rheological capabilities are in use. For example, simple tests are carried out using the DIN 53465 *cup test*, the *disc flow test* or the ISO 7808 *orifice flow test (OFT)*, and more discriminating tests using the DIN 53764 *analytical masticator test*. In the case of many products, supplementary, practice-related special tests also exist. Practical tests on presses or injection molding machines (Figs. 6.38 and 6.39) using the compression or injection molds employed for further processing are particularly important.

In the *cup test*, a standard cup (DIN 53465) is produced by the compression molding process, and the cup clamping time measured in seconds (Fig. 6.40). The compression force and temperature are constants. The determined time and the appearance of the cup give an impression of the degree of flow and curing behavior of the molding compound.

In the *disc flow test*, a specific quantity of molding compound is compressed between two heated plates to yield a disc (Fig. 6.41). The compression force and temperature are also constants in this case. The thickness of the resultant compression molding as measured at four specific points represents a parameter providing information on the flow/curing behavior of the molding compound. The numerical result, known to the processor as "disc flow", is obtained by addition of the four measurements expressed in millimeters and multiplication of the sum by ten.

A compression mold (ISO 7808) consisting of a cylindrical matrix and a punch in the form of specially shaped plunger is used in the *orifice flow test* (Fig. 6.42). Two channels are machined in the punch and permit a certain

210 6 Economic Significance, Survey of Applications, and Six Bonding Functions

Fig. 6.38. Press for processing phenolic molding compounds (source: Bakelite AG, Iserlohn, Germany)

Fig. 6.39. Injection molding machine for processing phenolic molding compounds (source: Bakelite AG, Iserlohn, Germany)

Fig. 6.40. The cup test

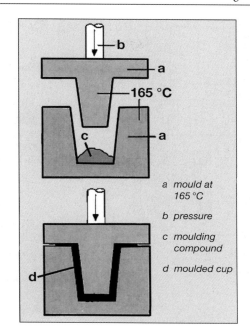

a mould at 165 °C
b pressure
c moulding compound
d moulded cup

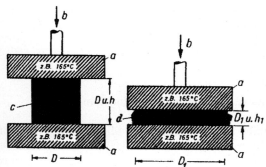

Plate Method
a) Plates heated to elevated temperature, e. g. 165 °C
b) Pressure
c) Moulding compound with periphery D and h before test
d) Moulding compound with periphery D_1 and h_1 after test

Fig. 6.41. The disc flow test

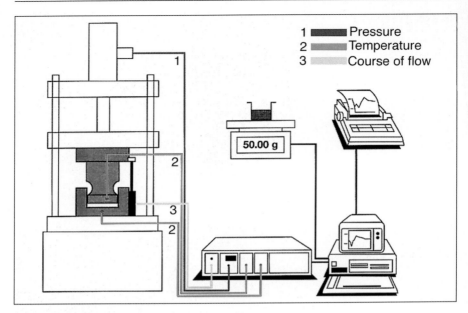

Fig. 6.42. The orifice flow test

quantity of molten molding compound (depending on its curing rate and viscosity) to exude during the closing operation. This fraction, expressed as a percentage of the total sample size, represents the OFT result. The pressure, mold temperature, and sample size are constants in this test. The customary compression force stages are 4, 7, 12, and 19 MPa.

6.1.3.8
Application of Phenolic Molding Compounds

Molded articles made from phenolic molding compounds are presently used in all areas of commerce [64]. Electrical engineering, household appliances, and automotive and mechanical engineering represent particularly important fields of application.

The significance of phenolic molding compounds in the area of *electrical engineering* has advanced from plain insulating materials to high-quality construction materials (Fig. 6.43). The most important functional components in electrical engineering are housings of switches from microswitches to heavy-duty switching equipment, coil forms, contact holders, switch shafts or release levers. Moreover, housings for electric meters and power distributors, domestic junction boxes, and many other articles are fabricated from phenolic molding compounds.

An important application in the electrical industry involves the production of commutators using phenolic molding compounds [65–67] that exhibit high levels of thermal and electrical properties. Commutators, high-performance

Fig. 6.43. Phenolic molding compound applications in electrical engineering (bobbins)

Fig. 6.44. Commutators from special phenolic molding compounds

components in *electric motors*, are used in numerous applications, for example in the household and automotive areas. The reliability and efficiency of the electric motor depends on the operation of the commutator. This applies to equipment such as washing machines in the household area, and to components from the starter through the windshield wiper motor or gasoline pump to the ABS (automotive braking system in automobiles). Commutators (Fig. 6.44) consist of insulating bodies that possess a large number of laminar copper sections. It is important during operation that the copper sections be solidly anchored in the commutator body and have no contact with one another. The commutator is mounted on the driveshaft of the motor. The reverse sides of the copper sections are undercut to anchor them more solidly. In particularly heavy-duty commutators, metal reinforcement is used to strengthen anchoring.

The commutator is exposed to two main forms of stress when the electric motor is in operation: (1) those of *centrifugal force* and (2) *thermal stresses* due to the temperature of the motor. In addition, the commutator must withstand certain stresses even during assembly: (1) during compression molding and (2) when the wire terminals of the windings are welded and attached to the copper sections of the commutator, a process during which extremely high temperatures and forces can develop temporarily.

The range of demands made on (mainly phenolic) molding compounds as possible materials for the commutator body may be summarized as: (1) good electrical insulating power, (2) high mechanical strength, and (3) high thermal resistance. Phenolic resins to which thermally stabilizing modifiers are added in some cases (particularly to novolak-type materials, that are easier to work) are mainly used as binders. Epoxy molding compounds are also used in the case of large commutators. Modification of the phenolics with melamine resins achieves improved copper adhesion. This means that the molding compound exhibits good adhesion to the copper surfaces. Proper processing renders the adhesion of metal to resin greater than the cohesive strength of the molding compound.

A basic change has occurred in the filling and reinforcing materials used to manufacture commutator molding compounds, since it was necessary to replace the asbestos fibers that had been used for many decades by other filling and reinforcing materials such as glass fibers. Commutators are differentiated according to whether they are reinforced with rings, or are not reinforced. The latter are most frequently used. In this case, the molding compound alone assumes the support function to enable the commutator to withstand centrifugal forces, a use for which glass fiber-reinforced molding compounds are particularly suited.

The commutator molding compounds must furthermore assure high dimensional stability, i.e., exhibit favorable shrinkage behavior. Dimensional stability is achieved by pertinent formulation or appropriate thermal posttreatment. Such measures pursue two objectives: (1) completion of the curing reaction and (2) release of volatile reaction products. They afford particularly good electrical parameters, dimensional stability by prior compensation of post-shrinkage, a desirable coefficient of thermal expansion, thermoshock resistance, and appropriate mechanical strength.

Figure 6.45 shows a simplified overview of molding compounds for commutators with information on the thermal resistance and the fillers and reinforcing agents (mica and glass fiber). PF 13 is a standardized molding compound containing mica as a filler, whereas the two specialty phenolic molding compounds PF 4109 and PF 4110 (that exhibit particularly high thermal resistance) contain a glass fiber filler.

Phenolic molding compounds similarly provide insulation characteristics in the area of *household appliances*. In this case, however, they are generally used as thermal insulation. Thus, handles and knobs for kitchen pots and pans are made from high-temperature resistant phenolic molding compounds (Fig. 6.46). Grip handles for use in electric ovens demand high levels of dimensional stability and lack of distortion from the plastics processor. Aside from the demand for thermal resistance to temperatures of up to 280 °C, high-temperature applications of this type also place great emphasis on the dishwasher/dishwashing detergent resistance as well as the color uniformity and

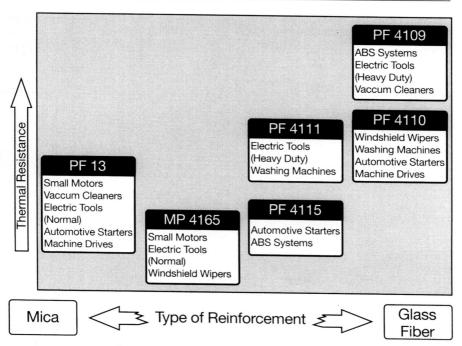

Fig. 6.45. Commutator molding compounds in their typical areas of application (with consideration of fillers and thermal resistance)

surface quality. Only muted colors such as black, brown, red, and green can be achieved using phenolic molding compounds. Brighter shades are produced using melamine or melamine/phenolic molding compounds (also refer to Fig 6.45). These "MP" (melamine/phenolic) molding compounds represent a compromise between color uniformity and enhanced tracking resistance on the one hand, and reduced post shrinkage on the other hand – which is generally greater in melamine than in phenolic molding compounds. The level of melamine resin in MP molding compounds generally exceeds that of the phenolic resin, specifically to achieve the desired improvement in the tracking resistance.

A classical application for thermoset molding compounds in the *automotive area* [68, 69] is in the ignition system, i.e., in spark plugs, spark plug caps, distributor rotors and caps, and ignition coil covers. Type 31.5 phenolic molding compounds are only used to a minor extent for such purposes. This application area is presently the domain of free-flowing polyester molding compounds, and (in the case of particularly high requirements) special epoxy molding compounds. Phenolic molding compounds are used in production of components including ashtrays and insulating flanges (Fig. 6.47).

In the area of braking systems, phenolic molding compounds have long been used to produce the pistons for power brake actuators [70]. In addition,

Table 6.46. Physical properties of standardized phenolic molding compounds as defined in DIN 7708 and ISO 14526 (see also Table 6.42)

Property	Unit	Standard ISO[a]/IEC[b]/UL	PF 13 (PF PS)	PF 31 (PF WD)
Apparent density (molding compounds)	g/cm^3	60[a]	0.56–0.90	0.45–0.75
Post-shrinkage (168 h/110 °C)	%	2577[a]	0–0.1	0.1–0.9
Tensile strength (5 mm/min)	MPa	527[a]	30–50	50–70
Tensile modulus (1 mm/min)	MPa	527[a]	6000–12,000	6000–12,000
Flexural strength (2 mm/min)	MPa	178[a]	50–120	70–120
Charpy impact strength (23 °C)	kJ/m^2	179/1/2 eU[a]	2.5–4.5	4.5–11
Charpy notched impact strength (23 °C)	kJ/m^2	179/1/2 eA[a]	1.5–2.5	1.3–1.8
Temp. of deflection under load HDTA–1.8 MPa	°C	75[a]	170–220	160–200
Max application temp. < 20,000 h	°C	60216/P1[b]	150	140
Surface resistivity	Ω	60093[b]	1E11-1E12	1E9-1E10
Volume resistivity	Ω × cm	60093[b]	1E12-1E13	1E10-1E11
Dissipation factor	100 HZ	60250[b]	0.05–0.1	0.2–0.5
Relative permittivity	100 HZ	60250[b]	3–8	8–13
Proof tracking index test liquid A	PTI	60112[b]	175	125
Flammability UL 94 (thickness tested 1)	Step/mm	UL 94	94 V-1/0.75 (NC)	94 V-1/1.5 (ALL)
Flammability UL 94 (thickness tested 2)	Step/mm	UL 94	94 V-1/1.5 (NC)	94 V-0/3.0 (ALL)
Flammability method BH (Glow bar)	Step ≤	60707[b]	BH1	BH2–10
Water absorption (24 h/23 °C)	mg/%	62[a]	<30/<0.2	<100/<0.75

[a] ISO.
[b] IEC.
[c] spezial electrical values.
[d] ammonia free.

PF 31.5[c] (PF WD, E)	PF 31.9[d] (PF WD, A)	PF 51 (PF LF)	PF 74 (PF SS)	PF 84 (PF SC)
0.50–0.70	0.55–0.75	0.55–0.70	–	0.45–0.65
0.15–1.2	0.15–0.9	0.3–0.8	0.3–0.5	0.2–0.6
50–70	50–70	50–70	30–60	45–65
6000–12,000	6000–12,000	5000–11,000	5600–11,500	5000–11,000
70–120	70–120	70–120	600–100	70–120
4.5–11	4.5–11	4.5–9	7–12	5.5–10.5
1.3–1.8	1.3–1.7	2.5–3.5	7–10	4–7
160–200	160–200	160–200	160–200	160–200
140	140	130	130	130
1E10-1E11	1E9-1E10	1E8-1E9	1E8-1E9	1E8-1E9
1E11-1E12	1E10-1E11	1E9-1E10	1E9-1E10	1E9-1E10
0.005–0.1	0.05–0.35	0.2–0.5	0.5–0.8	0.3–0.6
3–8	8–13	7–12	12–17	8–13
125	125	125	125	175
94 V–0/3.0	94 V–0/3.0	94 V–0/3.2	94 HB/1.6	94 V–0/3.2
94 V–1/1.5	94 V–1/1.5	94 V–1/1.6	94 HB/3.2	94 V–0/1.6
BH2–10	BH2–10	BH2–30	BH2–30	BH1–10
<100/<0.7	<100/<0.7	<150/<1.1	<200/<1.45	<150/<1.1

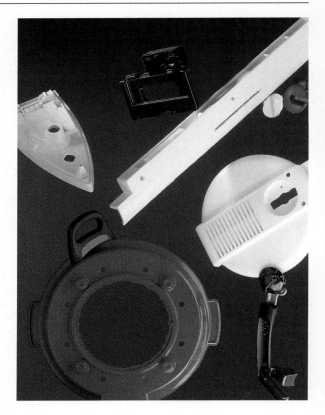

Fig. 6.46. Applications of molding compounds in household appliances (e.g. pan holder *right on the bottom of the picture*)

the hydraulic pistons of disc brakes are fabricated on a large scale (although only in the USA) from special high-temperature resistant phenolic molding compounds. Aside from the requirement that the coefficient of thermal expansion approaches that of steel, the thermal resistance of contact with a brake lining bridging plate that can sometimes reach a temperature of over 400 °C is of prime importance in this application.

The objective of recent developments under the automotive hood is the systematic replacement of metallic materials such as die cast aluminum for ancillary motor components. This includes equipment such as the coolant pump. The impeller and the pump housing are fabricated from glass fiber-reinforced, elastomer-modified phenolic molding compounds; compared to the conventional metal version, the nonmetallic pump made of phenolic molding compound stands out due to its quieter running and longer service life coupled with greater efficiency and lower manufacturing costs. The coolant pump was among the first series-manufactured components to be made using phenolic molding compounds. In this application, the long-term dimensional stability in contact with the water/glycol coolant at a temperature of 140 °C was an important consideration.

Fig. 6.47. Applications of phenolic molding compounds in the automotive industry (ignition, ashtrays, insulating flanges, etc.)

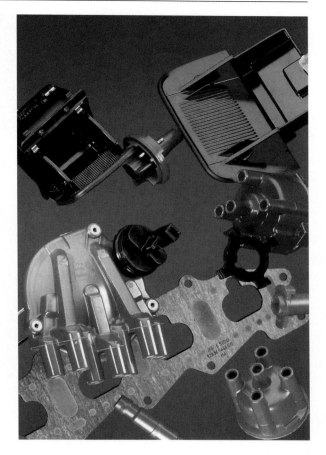

Moreover, the drive and tensioning elements of the belt drive have been produced using phenolic molding compounds. These include the tensioning pulleys and profile or toothed pulleys for the drives of auxiliary equipment such as the coolant pump, generator/alternator, or air conditioner, and for the toothed belt drive controlling the camshaft.

Thermal insulation also represents one of the areas where phenolic molding compounds are used in automotive applications. Insulating flanges represent a typical example of this application in the engine compartment. They are exposed to great thermal stress and to considerable compressive stress due to the bolted connections. Exact requirements that are best met by use of phenolic molding compounds with inorganic fillers are also encountered in components mounted directly on the engine block, such as the coolant pump, thermostat housing, or inlet tube.

In the field of mechanical engineering, phenolic molding compounds are also very important for machine and equipment construction (Fig. 6.48).

Fig. 6.48. Phenolic molding compounds for machine and equipment construction (control elements)

6.1.3.9
Standardized and Non-Standardized Molding Compounds (as Examples for Applications)

Table 6.46 surveys the (minimum required) properties of the main standardized phenolic molding compounds as specified in DIN 7708 and ISO 14526.

These molding compounds may be associated with the following fillers, properties and areas of application (ISO Standard number in parentheses).

PF 13: (PF PS)
– Phenolic molding compound, inorganic (mica) filler, heat resistant, good electrical properties, UL listed [0.75/V-1 (NC)], and standardized molding compound
– *Application areas*: electric motor slip ring commutators, commutators, adjustment motors, truck starter motors, vacuum cleaners, kitchen appliances, electric tools, electric vehicles, and other users of large commutators, bobbins

PF 31: (PF WD)
– Phenolic molding compound, mainly organic filler, standard molding compound for normal stress, UL listed [1.5/V-1 (ALL), 3.0/V-0 (ALL)], and standardized molding compound

- *Application areas*: moldings of all kinds, from screw caps to large casings, electrical installation material, handle casings (bars), pan handles, operating elements, toaster parts and pistons for power brake actuators, carbon brush holders, lamp casing parts, decorative items such as ashtrays

PF 31.5: (PF WD, E)
- Phenolic molding compound, mainly organic filler, electrically high grade, standardized molding compound
- *Application areas*: parts for electrical engineering and telecommunications, automotive electronics, terminal boards and baseplates, housings, mounting parts, engine ignition systems

PF 31.9: (PF WD, A)
- Phenolic molding compound, mainly organic filler, ammonia free, standardized molding compound
- *Application areas*: parts for electrical engineering and telecommunications, automotive ignition systems, mounting parts, automotive electronics, terminal boards, moldings also exposed to warm and humid climates

PF 51: (PF LF)
- Phenolic molding compound, organic filler, reinforced with cotton fibers, enhanced notched impact strength, standardized molding compound
- *Application areas*: switch gears cross bars, solenoid switch covers, magnetic switches, pulleys, tension/deflection pulleys, covers, handwheels

PF 74: (PF SS)
- Phenolic molding compound, organic filler, reinforced with cotton fiber flakes, high notched impact strength, standardized molding compound
- *Application areas*: molding exposed to elevated mechanical stress levels, power sockets, operating elements, gears, rolls, handwheels

PF 84: (PF SC)
- Phenolic molding compound, organic filler, reinforced with cotton fiber flakes, high notched impact strength, standardized molding compound
- *Application areas*: molding exposed to elevated mechanical stress level (housings, rolls, gears, roll bodies, and bearings), pulleys, operating elements, covers

Table 6.47 surveys a number of special-purpose phenolic molding compounds. The listed compounds only represent a selection. The number of special-purpose molding compounds on the market is relatively large, since such compounds are generally matched to specific applications; thus, the following comments can only be representative in nature.

PF 4111
- Phenolic molding compound, inorganic filler, glass fiber reinforcement, free of ammonia and acetic acid, electrically high grade, high temperature stability, high mechanical strength, copper-adhesive, UL listed [0,75/V-0 (BK)] molding compound
- *Application areas*: commutators, windshield wipers, adjustment motors, washing machine motors, kitchen appliances, electric tools

Table 6.47. Physical properties of special-purpose (non-standardized) phenolic molding compounds

Property	Unit	Standard ISO[a]/IEC[b]/UL	PF 4111	PF 5109
Apparent density (molding compounds)	g/cm^3	60[a]	0.75–0.85	0.50–0.60
Post shrinkage (168 h/110 °C)	%	2577[a]	0–0.25	0.2–0.6
Tensile strength (5 mm/min)	MPa	527[a]	60–80	45–65
Tensile modulus (1 mm/min)^M	MPa	527[a]	10,000–16,000	5500–11,500
Flexural strength (2 mm/min)	MPa	178[a]	120–160	70–110
Charpy impact strength (23 °C)	kJ/m^2	179/1/2 eU[a]	8–12	7–11
Charpy notched impact strength (23 °C)	kJ/m^2	179/1/2 eA[a]	2.5–4	4–7
Temp. of deflection under load HDTA – 1.8 MPa	°C	75[a]–2	200–240	150–190
Max application temp. <20,000 h	°C	60216/P1[b]	170	130
Surface resistivity	Ω	60093[b]	1E11-1E12	1E8-1E9
Volume resistivity	Ω × cm	60093[b]	1E12-1E13	1E9-1E10
Dissipation factor	100 HZ	60250[b]	0.02–0.05	0.3–0.6
Relative permittivity	100 HZ	60250[b]	3–8	8–13
Proof tracking index test liquid A	PTI	60112[b]	225	125
Flammability UL 94 (thickness tested 1)	Step/mm	UL 94	94 V-0/0.75 (BK)	94 V-0/3.2
Flammability UL 94 (thickness tested 2)	Step/mm	UL 94	94 V-0/1.5 (BK)	94 V-1/1.6
Flammability method BH (Glow bar)	Step ≤	60707[b]	BH1	BH2–10
Water absorption (24 h/23 °C)	mg/%	62[a]	<30/<0.2	<150/<1.1

PF 7130	PF 7272	PF 7300	PF 7595	PF 7909
0.60–0.70	0.55–0.65	–	0.85–0.95	0.65–0.75
0.2–0.5	0.25–0.7	0.2–0.4	0.05–0.2	0.4–0.9
45–65	45–65	40–60	45–65	50–70
5000–11,000	6000–12,000	5500–11,500	7000–13,000	7000–13,000
75–105	70–110	70–110	70–100	80–120
5–9	5–9	5–9	3–6	5–9
1.3–1.6	1.3–1.5	5–9	1.0–1.3	1.3–1.6
160–200	155–195	170–210	210–240	155–195
150	140	130	150	130
1E9-1E10	1E9-1E10	1E8-1E9	–	1E9-1E10
1E10-1E11	1E10-1E11	1E9-1E11	–	1E10-1E11
0.1–0.3	0.3–0.6	0.4–0.7	–	0.1–0.3
8–13	9–14	12–17	–	5–10
125	125	125	–	175
94 V – 1/1.6	94 V – 1/1.6	94 V – 1/1.6	94 V – 0/3.2	94 V – 0/0.33 (BK)
94 V – 1/3.2	94 V – 1/3.2	94 V – 1/3.2	94 V – 0/1.6	94 V – 0/0.62 (BK)
BH2 – 30	BH2 – 10	BH2 – 30	BH2 – 10	BH1
<50/<0.25	–	–	<20/<0.15	<55/<0.4

PF 5109
- Phenolic molding compound, organic filler, reinforced with textile flakes, high notched impact strength, low abrasion
- *Application areas*: Cross bars

PF 7130
- Phenolic molding compound, inorganic/organic filler, average heat resistance, powder painting possible, high surface quality
- *Application areas*: iron heat shield

PF 7272
- Phenolic molding compound, mainly organic filler, high surface quality, good dimensional quality
- *Application areas:* fittings

PF 7300
- Phenolic molding compound, organic filler, reinforced with textile flakes, high notched impact strength, good slip properties
- *Application areas:* bearings, guide rails and gears

PF 7595
- Phenolic molding compound, inorganic/organic filler, modified with graphite, good heat conductivity, good slip properties (not suitable for exposure to elevated tension levels)
- *Application areas:* supporting bodies for diamond grinding wheels, gas meter parts, pump parts, sliding/guide elements, self-lubricating bearing parts

PF 7909
- Phenolic molding compound, inorganic/organic filler, ammonia free, self extinguishing, UL listed [0.33/V-0 (BK)] molding compound
- *Application areas:* electrical engineering and telecommunications components, bobbins, relay sockets, electricity meters, automotive ignition systems, mounting parts, automotive electronics, terminal boards, moldings also exposed to warm and humid climates

6.1.3.10
Properties of Phenolic Molding Compounds Compared to Those of Other Materials

Possible properties of phenolic molding compounds (depending on the specific fillers and reinforcing agents, and various combinations) are high strength, stiffness and hardness, low tracking, good toughness, high thermal deformation resistance, low coefficient of thermal expansion, little or no flammability, incandescence resistance, and stress cracking resistance. Furthermore, phenolic molding compounds exhibit a good cost-performance relationship. Phenolic molding compounds are resistant to weak acids, weak alkalis, alcohols, esters, ketones, ethers, chlorinated hydrocarbons, benzene, mineral oil, animal and plant-based oils and fats. Depending on the type of phenolic molding compounds they are not resistant to strong acids, strong alkalis, or boiling water in long-term testing.

The advantages of phenolic molding compounds over metals (Table 6.48) are in areas such as the construction design freedom, i.e., benefits in design of

Table 6.48. Advantages of phenolic molding compounds over metals

1. Construction design freedom (functional integrity, reduction of production costs)
2. Longer mold service lives (for example compared to aluminum die casting, cost savings)
3. Material savings (25–30% due to lower density)
4. Maintenance of dimensional tolerances (precisely, and without followup machining)

the molding [71]. Functional components can be integrated and a reduction achieved in the total fabrication cost of the workpiece. Due to the basic properties of plastics, the molds have longer service life than for example in aluminum casting, thus affording considerable cost reduction. Cost reductions are also achieved due to the weight savings resulting from the lower specific weight, particularly in the case of automotive fabrication. These weight reductions in turn directly lead to decreased fuel/energy costs. Dimensional tolerances can be observed more precisely, and without the need for follow-up machining.

A comparison of thermosets [72–74], specifically phenolic molding compounds, with thermoplastics (in other words a comparison of one class of plastics with another) shows the advantages listed in Table 6.49. Moldings made of phenolic molding compounds feature superior dimensional stability under conditions of heat and pressure. They exhibit largely reliable chemical resistance, high surface hardness with low mar sensitivity, good electrical and thermal insulating power, high dimensional stability, favorable flammability properties even in the absence of halogenated or phosphorus-based fire retardants, and sufficient structural integrity under conditions of exposure to high temperatures. From the aspect of process engineering, the facts are also important that they exhibit no sink marks at points where walls of various thicknesses intersect in fabrication of moldings, and that thermoset molding compounds can be processed by transfer molding in a manner similar to that used with thermoplastics.

The high spatial crosslinking density of thermoset molding compounds leads to high compression strength and surface hardness coupled with low elongation at break. Due to this structure, phenolic molding compounds cannot offer the same toughness properties as thermoplastic products. This is logical, since other positive properties such as the thermal resistance or low cold flow would be lost if an attempt were made to increase the elongation at break, for example, by incorporation of modifiers. Figure 6.49 compares stress-strain curves of glass fiber-reinforced polyamide 6 and a PF 6507 glass

Table 6.49. Advantages of thermoset molding compounds over thermoplastics

1. Dimensional stability under conditions of heat and pressure
2. Reliable chemical resistance
3. High surface hardness, low mar sensitivity
4. High insulating power
5. Flame retardency without added halogens
6. No sink marks at differing wall thicknesses

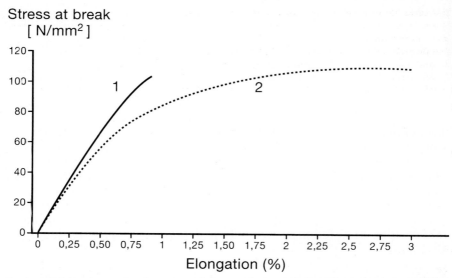

Fig. 6.49. Elongation at break curves of thermoplastics vs phenolic molding compounds (*1* = phenolic molding compound, *2* = thermoplastic molding compound)

fiber-reinforced phenolic molding compound manufactured by Bakelite AG (Iserlohn, Germany). Less than 1% elongation is observed for the phenolic molding compound as compared to high level of elongation for polyamide 6.

The frequently cited main difference in the behavior of phenolic or other thermoset molding compounds and thermoplastics – the thermal behavior – may be best illustrated on the basis of shear modulus curves (Fig. 6.50). In Fig. 6.50, the shear modulus of a phenolic molding compound containing 30% glass fiber (PF GF 30) is compared with those of glass fiber-reinforced polycarbonate (PC GF 30) and polyethylene terephthalate (PET-GF 20). Despite the glass fiber reinforcement, softening occurs in the two thermoplastic plastics at a characteristic temperature, and manifests itself by a severe drop in the shear modulus curve. In contrast, the shear modulus of the thermosetting (phenolic) molding compound remains essentially constant over the entire range of measurement. The glass transition temperature, at which a certain increase in molecular mobility occurs, characterizes the inflection point of the curves. In the case of moldings made of phenolic molding compounds, this mobility is so small that it cannot be detected from a measurement of the shear modulus. The glass transition temperatures of phenolic molding compounds can be far better determined by TMA measurements.

Heat treatment produces an increase in the glass transition temperature. Figure 6.51 shows a curve of the glass transition temperature as a function of the thermal pretreatment using a glass fiber-reinforced molding compound PF 6507 as an example. Depending on the thermal pretreatment, the glass transition temperature ranges between 200 and 350 °C in this case, in other words significantly higher than in thermoplastics (see also chapter on Chem-

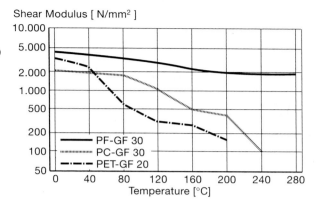

Fig. 6.50. Shear modulus curve (PC = polycarbonate, PET = polyethylene terephtalate, GF = glass fiber)

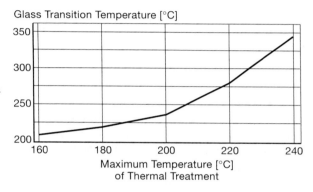

Fig. 6.51. Glass transition temperature dependence on the maximum pre-treatment temperature of phenolic molding compound PF 6507 (thermal treatment: 6 h/150 °C and 2 h/maximum temperature)

istry, Reactions, and Mechanisms). A marked difference between thermoplastics and phenolic based thermoset molding compounds can also be observed in a comparison of the tensile strength as a function of the temperature. Figure 6.52 shows that the glass fiber-reinforced phenolic molding compound still reaches about 50 % of its original strength at 230 °C, determined using test pieces without heat treatment. This level can be further increased by 20 – 30 % (PF 6540 phenolic molding compound with and without heat treatment) using the process of heat treatment described above. A residual strength level of more than around 80 % of that at ambient temperature is thus obtained. In contrast, the tensile strength of the high-tech thermoplastic molding compound PPS-GF 40 undergoes a dramatic fall at 180 °C.

In general, increasing demands for higher thermal stability have been expressed in development of thermoset molding compounds in recent years [75–77]. However, the term "high temperature resistance" can only be illustrated to a reasonably exact extent by a combined consideration of highly diverse thermal stresses. The *"Temperature Index"* (TI) defined in IEC 60216 is frequently used as a basis for the requirement of long-term thermal resistance, and describes the temperature at which moldings still exhibit 50 % of

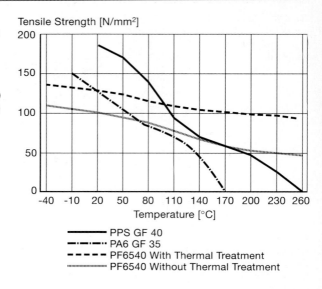

Fig. 6.52. Tensile strength vs temperature; comparison of glass fiber-reinforced thermoplastics (PPS GF 40 and PA6 GF 35) with glass fiber-reinforced phenolic molding compound PF 6540 (heat treated and untreated)

their original ambient temperature strength after 20,000 h storage at 180 °C or 100,000 h at 150 °C. The TI can only be used for comparisons of thermoplastics with thermosets to a limited extent, since the TI provides no information on the residual strength a material exhibits at the indicated temperatures (Table 6.50). For example, a high-performance thermoplastic such as PPS, that displays a temperature index of 220 °C following thermal exposure, exhibits barely 5 % residual strength at 200 °C, whereas a glass fiber reinforced phenolic molding compound (PF GF 25) still provides 52 % residual strength (relative to the original strength level) at this temperature.

As an extension of such material comparisons, Table 6.51 surveys the basic properties of phenolic, unsaturated polyester, epoxy, and melamine/phenolic molding compounds containing organic and inorganic fillers.

Due to their pattern of properties, phenolic molding compounds can be used in a relatively broad range of applications, particularly since they are capable of versatile (organic and inorganic) modification. Their cost-per-

Table 6.50. Residual strength levels (RS) of molding compounds (thermoplastics: PA 66, PETP, PPS; phenolic molding compounds: PF 31 and PF GF 25)

	TI (°C) UL	RS (%) 100 °C	RS (%) 150 °C	RS (%) 200 °C
PA 66	130	51.3	39.7	7.2
PETP	140	38.1	21.1	–
PPS	220	48.3	34.9	4.7
PF 31 (WD30 MD20–WD40 MD 10)	–	83.7	51.2	50.1
PF 6507, 6537 (GF30 MD20–GF40 MD10)	160	71.9	57.8	52

Table 6.51. Basic properties of different molding compounds (resin matrices: PF = phenolic resin; MF/MP = melamine phenolic; UP = unsaturated polyester resin; EP = epoxy) with organic and inorganic fillers

Property	PF organically	PF inorganically	MF/MP organically	UP organically	UP inorganically	EP inorganically
Strength	medium	very good	medium-good	medium	good	good-very good
Modulus of elasticity	medium	medium-high	high	medium	high	high
Surface resistance	good	very good	very good	medium	good	very good
Creep rupture strength	good	very good	good	medium	very slightly	very slightly
Thermal expansion	slightly	very slightly	slightly	slightly	very slightly	very slightly
Reserves by peak temperatures	high	very high	high	medium	very high	very high
Inflammability	medium	very high	high	medium	very high	medium
Electrical isolation properties	good	good	very good	very good	very good	very good
Tracking resistance	medium	medium	very high	very high	very high	medium-very good
Media resistance	good	very good	good	medium	medium	very good
Chromatic spectrum	subdued color	subdued colors	all colors	all colors	all colors	subdued colors

formance relationship is outstanding, and they can be recycled (particulate recycling) relatively well. The disadvantage that phenolic molding compounds yellow when exposed to light and can only be colored to a limited extent (no light colors possible) is not shared by melamine, melamine/phenolic, or UP molding compounds. In addition, melamine (MF) and UP molding compounds exhibit very good tracking resistance. UP molding compounds also feature reduced processing and secondary shrinkage (depending on the type of resin used). These levels can be extremely low in the case of epoxy molding compounds, which offer particularly high-grade electrical properties and very high chemical resistance. Special melamine molding compounds are approved for use in food packaging, whereas this does not apply to phenolic molding compounds for reasons of taste.

6.1.3.11
Phenolic Novolaks and Epoxidized Phenolic Novolaks for Production of Epoxy Molding Compounds

On the subject of phenolic resins/molding compounds and their applications, free-flowing, thermosetting epoxy molding compounds, specifically "electronic molding compounds" used to encapsulate miniaturized electronic components and microcomponents, are produced by combining epoxidized phenolic or cresol novolaks that cure and crosslink in a system with novolaks as curing agents (see chapter on Chemistry, Reactions, and Mechanisms). For use in the area of electronics, it is necessary that the novolaks, the epoxidized novolaks produced from these, and the novolaks used as curing agents exhibit the highest possible purity. This refers to extremely low levels of free phenol in the novolak and low levels of chlorine in the epoxidized novolak. Such highly pure epoxy molding compounds containing special (generally cresol-based) novolaks as a curing agent occupy a secure position as compression molding compounds for encapsulation components. These products afford relatively inexpensive encapsulation of even the smallest components. About 95% of all electronics molding compounds worldwide are formulated on this basis. Table 6.52 shows a proposed formulation for an epoxy molding compound [78, 79]. Epoxy electronic molding compounds based on the combination of an epoxidized phenolic novolak and a novolak curing agent are processed globally by transfer molding. The granulated molding compound is generally first preformed, the tablets premelted in a microwave generator, and finally molded. The molding compounds are generally provided with a relatively high level of fused quartz flour of high chemical purity to reduce coefficient of expansion, increase glass transition temperature, and improve their moisture resistance. The grades of quartz flour used in this application are generally silanized.

Engineering epoxy molding compounds formulated with a similar composition (EP-NOV/NOV) generally contain inorganic fillers, glass fiber, or mica as reinforcing agents, and feature high thermal resistance, good electrical properties even at elevated temperatures, and very low secondary shrinkage [80]. Furthermore, they are suitable for applications where only minor losses in mass may take place at temperatures between 200 and 300 °C. The dimen-

Table 6.52. Composition of epoxy electronics molding compounds [79]

Epoxy resin	Epoxidized novolak, mainly cresol novolak	ca. 20%
Curing agent	Phenolic novolak	ca. 10%
Accelerator	Imidazole derivatives	
Filler	Fused quartz flour	66–75%
Flame retardant.	Brominated epoxy resin, antimony trioxide)	ca. 5%
Adhesion promoter	Epoxy silane	
Parting agent	Synthetic waxes	
Pigment	Carbon black	

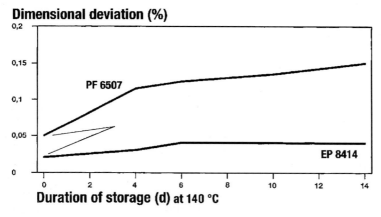

Fig. 6.53. Dimensional deviation after 8 h post stoving at 180 °C and storage at 140 °C (14 days) (EP 8414–PF 6507)

sional changes that take place in hot (140 °C) storage are generally very low compared to those of straight phenolic molding compounds (Fig. 6.53). Such epoxy molding compounds are used for premium products such as special commutators, lamp sockets (as a ceramic substitute), reflectors, electronic components; thermally, chemically and mechanically stressed parts such as magnetic valve actuators, electrotechnical components, coil forms, automotive electrical systems, spark plug caps, and other applications.

6.1.4
Impregnation of Paper and Fabric (Overview)

Impregnation of a carrier material such as paper or fabric based on inorganic fibers generally involves saturating it, whereas in painting or coating, the surface of the substrate is only treated. With this apparent difference between impregnation and coating, specially modified, filled or unfilled resin solutions are used in the impregnation process. Key characteristics are mainly associated with the base resin and its range of properties [1, 2].

Table 6.53. Impregnation applications of phenolic resins

Carrier Material	Resin Type	Application(s)
Glass cloth	Novolak and resole solutions	Cloth inserts for grinding wheels
Paper	Modified and unmodified phenolic resin solutions	Electrical paper laminates, with and without copper cladding
Cotton cloth	Modified and unmodified resole solutions	Cloth laminates for mechanical and electrical engineering
Paper or plastic	Aqueous resoles	Battery separators
Special papers	Modified and unmodified phenolic resin solutions	Linings for automatic transmissions
Honeycomb	Modified phenolic resin solutions	Production of special composites
Glass cloth	Modified phenolic resoles polymer modified phenolic resin solutions	Composites, pultrusion blanks, fishing rod production

The properties of final products produced by impregnation with phenolics or other synthetic resins such as epoxies or melamine resins followed by curing depend on both the type and grade of the carrier material, and on the specific properties of the impregnating resin. Liquid phenolics – resin solutions and aqueous resoles – are used to impregnate paper, textiles, or glass fabric and mats in various areas of application (Table 6.53). Production of "reinforcing glass fabric" (glass fabric impregnated with alcoholic phenolic resin solutions) used as an insert in grinding wheels to increase their stability and burst strength will be discussed in the Abrasives chapter. Impregnated woven or glass filament is also used in production of composites made by pultrusion or shaping of prepregs. Composites in which the matrix is an epoxy resin feature particularly good toughness. Phenolic resin bonded glass fiber composites exhibit particularly good flame retardancy and low smoke density; thus, they are increasingly used in aircraft and automotive applications [3–8] (see Section on High Performance and Advanced Composites 6.1.5). Combination – co-curing of epoxy and phenolic resins – permits useful combination of the two groups of properties. This concept is utilized for applications such as the flooring in construction of aircraft, for example, the Airbus A 340.

In addition to the above impregnation applications, filter papers used to manufacture air, fuel, and oil filters for the automotive industry are impregnated with special phenolic resins, and resin-impregnated paper "separators" used as dividers in lead storage batteries. Other applications include honeycomb and special-purpose impregnated paper linings for automatic transmissions.

6.1.4.1
Molded Laminates (Introduction)

The following discussion will mainly consider production of clad and unclad laminates for the electrical and electronics industries, and molded laminates for industrial use, for example, in mechanical engineering. This includes electrical paper laminates in the form of sheets or rolls (Figs. 6.54 and 55) used for various industrial and insulation applications (sheets or tubes).

As the name suggests, molded laminates – simply expressed – represent materials made up of individual resin-impregnated layers of paper, textile fabric, glass mat, or glass fabric. In an extreme case, wood laminates also belong to this group.

Layers are bonded together by the homogeneously introduced matrix resins (phenolics, melamine resins, epoxies, unsaturated polyesters, and others). This publication understandably concentrates on phenolic resins; a number of comparisons with epoxies develop from these considerations.

Further processing of the thermosetting resin-impregnated carrier materials is generally carried out on a discontinuous basis, i.e., the carrier materials are cut into sections of equal size and converted into homogeneous products by placing them between pressing plates or in molds and applying heat and pressure according to the "Baekeland principle" described in his "heat and pressure patent." Using this procedure, phenolic resin cures and adhesively bonds the individual layers together. The base materials for printed circuits [9] are especially important members of the group formed by industrial laminates. Tubes are also produced from the impregnated carrier materials by the winding or compression molding method. Secondary shaping can afford profiles and rods.

The above comments suggest that the final products (molded laminates, industrial laminates) can be either sheets clad with copper foil for further processing, or tubes and profiles.

Fig. 6.54. Examples of applications for paper- and fabric-based molded laminates (photo: Isola AG, Düren)

Fig. 6.55. Working of a paper-based laminate tube (Photo: Isola AG, Düren)

Most of these materials are used as insulating materials in the electrical industry and in data systems engineering, and must exhibit an outstanding range of mechanical and electrical properties matched to the resin matrix. Many special-purpose unmodified and modified phenolic resins have thus been developed for these application areas. The modifiers that are used mainly affect the flame retardancy, flexibility, or plasticity, allowing the product to function under special processing conditions [10–13]; i.e., they exert a specific effect on the fabrication, processing and end-use application. Among molded laminates, paper-based and fabric-based laminates represent some of the most important products [14]. Their low density, good electrical properties, and outstanding workability are noteworthy.

6.1.4.2
Molded Laminates (Survey of Technologies and Diversification)

Fabrication of molded laminates using thermosetting resins is accomplished by impregnating the fabric or paper carrier materials with phenolic resins or

Fig. 6.56. Horizontal impregnating system for paper-based laminates

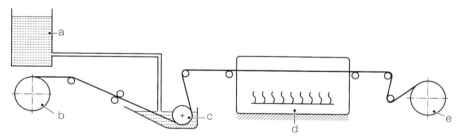

Fig. 6.57. Impregnation of carrier materials (a = impregnating bath; b = carrier web feed; c = impregnation; d = drying; e = coiling of impregnated web) (from [10])

other liquid thermosetting binders such as melamine resins or epoxies (Figs. 6.56 and 6.57), drying the impregnated webs in a festoon dryer, and then cutting them to sections of specific dimensions (Fig. 6.58). Industrial impregnation systems operate horizontally or vertically [15–17]. Horizontal impregnating and drying technology is mainly used for phenolic paper laminates; modern vertical systems equipped with strategically positioned radiant heaters are used for cotton fabric and particularly for high-quality, epoxy-impregnated glass laminates.

The drying sections represent the most important part of the impregnation line. Various drying systems are used, and particular care is devoted to tempe-

Fig. 6.58. Further processing of impregnated carrier materials to afford molded laminates (a = impregnated carrier material; b = cutting to size; c = stage press) (from [10])

rature control since it must be possible to set a temperature or temperature program that best corresponds to the individual curing characteristics of the resin system. The ideal temperature program may be empirically determined.

Continuous double belt presses can also be used to press the impregnated webs of carrier material. In these, the impregnated webs of material are continuously pressed and cured between two steel belts. Such a technique requires rapidly curing resin systems. This process, which is suitable for thin materials exhibiting a homogeneous formulation and quality, is used on a large scale for decorative laminates and to a lesser extent in the area of thin epoxy glass fabric laminates.

After the impregnated material has been cut to size and stacked, it is subjected to hot compression in multi-daylight presses (Fig. 6.58). The layers are converted into a relatively homogeneous material by pressing at a temperature of 150–160 °C. The phenolic resins cure during this process (position c in Fig. 6.58). The layers should not delaminate during working or subsequent use.

The three most important operations in production of molded laminates are thus (1) impregnation (saturation) or coating of the carrier material, (2) drying (evaporation of solvents and water) coupled with precuring of the heat-reactive binder, and (3) final curing in presses [18].

Vulcanized fiber (made using stacks of parchmentized paper as a starting material) can be regarded as a precursor to molded laminates [19]. Today, the main use of vulcanized fiber itself is in production of vulcanized fiber wheels (cf. sections on coated abrasives).

Molded laminates also include the laminated profiles and tubes mentioned above. Tubes can be produced by winding impregnated webs around a mandrel (positions b and c in Fig. 6.59) and compacting these, if appropriate, with secondary shaping. Secondary shaping can afford profiles such as angles or flat bars (positions f and e in Fig. 6.59) (see also Filament Winding in High Performance and Advanced Composites Section 6.1.5).

In the winding process for fabrication of tubes, impregnated or coated carrier materials (paper or fabric) are heated by passing them over a system of rollers, wound around a heated mandrel, and the uncured shape on the mandrel cured by application of additional heat. The winding machines used for this purpose can be constructed using various systems (two-roll or three-roll machines). Another fabrication method is used to manufacture compression shaped tubes and profiles, and has the advantage that the material is subjected to less stress (during stripping). In compression shaping, the web is wrapped

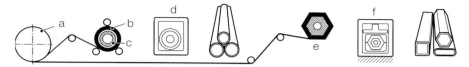

Fig. 6.59. Further processing of impregnated carrier materials to afford profiles and tubes (a = impregnated carrier material; b = coiling of impregnated web; c = mandrel; d = heat curing; e = profile winding system; f = pressing) (from [10])

around the mandrel in either a cold or moderately warm state, and the wrapped mandrel (of any desired cross-section) placed in a mold that corresponds to its profile. Pressing is then carried out under conditions of temperature and pressure appropriate to the molded laminates. The round or angular tubes can also be compression shaped to yield flat bars or profiles.

6.1.4.3
Economic Considerations and Background (Electrical Laminates and Printed Circuit Boards)

Unclad and copper-clad base materials for printed circuit boards (Fig. 6.60) have by far the greatest importance in the areas of fabrication and application of molded laminates, particularly in view of the boom in the electronics and entertainment industry and the corresponding demand for products including computers and television sets.

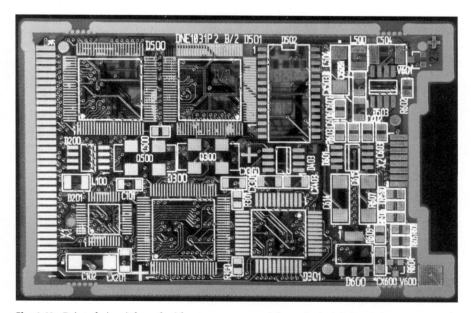

Fig. 6.60. Printed circuit board without components (photo: Isola AG, Düren)

Phenolic resins were the first synthetic impregnating agents used in fabrication of molded laminates. Other binders followed later. Higher demands on quality and development of new resin systems have led to market introduction of epoxy resin-impregnated papers and glass fabric, which exhibit premium electrical properties and increased dimensional stability.

The present situation of the laminate market can be best understood by considering the NEMA (US National Electrical Manufacturer's Association) classes of electrical laminates (Table 6.54) and listing several types of properties and current areas of application. The flame retardant grades of phenolic-bonded laminates FR-1 and FR-2 (the latter exhibiting better electrical properties) have experienced a certain degree of further development to the epoxy-bonded types FR-3, FR-4, FR-5, and higher containing glass fabric as

Table 6.54. NEMA Pub. 1989 classification of electrical laminates by types, properties and applications

NEMA Grade[a] (Resin Basis)	Carrier	Property/Properties	Examples of Applications
FR-1 (PF)[b]	Paper	Flame retardant, not electrically high quality	Automatic toys, pocket calculators
FR-2 (PF)[b]	Paper	Flame retardant, electrically high quality	Standard TV and audio equipment, telephone equipment
FR-3 (EP)	Paper	Flame retardant, high-grade dielectric properties	High-quality TV and audio equipment
FR-4 (EP)	Glass cloth	Flame retardant, electrically very high quality	Multilayer circuit boards, computer systems
FR-5 (EP)	Glass cloth	As FR-4 but higher glass transition point	Electronic equipment exposed to thermal stress
G-10 (EP)	Glass cloth	Not treated for flame retardancy	Electronic equipment exposed to great mechanical stress
G-11 (EP)	Glass cloth	Not treated for flame retardancy, T_g higher than in G-10 strips	As in G-10, but in components exposed to thermal stress such as insulating
CEM-1 (EP)[c]	Core of paper, cover of glass cloth	Flame retardant electrically high quality	FR-3 applications such as TV and audio equipment exposed to mechanical stress
CEM-3 (EP)[c]	Core of glass mat, cover of glass cloth	Flame retardant, electrically high quality	Economical applications, FR4 with lower dimensional stability

[a] FR = Flame retardant; PF = Phenolic resin; EP = Epoxy resin.
[b] FR-1 and -2, straight insulating or copper-clad materials.
[c] Application-related alternatives to FR-3 and FR-4.

Table 6.55. Standard data for copper-clad (PF and EP) paper laminates; actual data according to manufacturer in parentheses

Standard Designation Electrical properties	Units	Phenolic-based paper laminate NEMA-LI 1-1989 XXXPC/FR-2		Phenolic-based paper laminate DIN EN 60249-2-7 PF-CP-Cu		Actual data	Epoxy-based paper laminate NEMA-LI 1-1989 FR-3		Epoxy-based paper laminate DIN EN 60249-2-3 EP-CP-Cu		Actual data
		Pretreatment	Data	Pretreatment	Data		Pretreatment	Data	Pretreatment	Data	
Surface resistance<?1>	Ω	C-96/35/90	10^9	C-96/40/92	$5 \cdot 10^8$	(10^{10})	C-96/35/90	10^9	C-96/40/92	$2 \cdot 10^9$	$(3 \cdot 10^{11})$
	Ω	–	–	C-96/40/92+	10^9	$(6 \cdot 10^{11})$	–	–	C-96/40/92+	$2 \cdot 10^{10}$	$(6 \cdot 10^{11})$
at elevated temperature	Ω	–	–	E-1/100/T-100	$3 \cdot 10^7$	$(2 \cdot 10^8)$	–	–	E-1/100/T-100	10^9	$(5 \cdot 10^9)$
Specific volume resistance	Ωcm	C-96/35/90	10^{10}	C-96/40/92	$5 \cdot 10^9$	$(2 \cdot 10^{12})$	C-96/35/90	10^{10}	C-96/40/92	$8 \cdot 10^{10}$	$(4 \cdot 10^{12})$
	Ωcm	–	–	C-96/40/92+	10^{10}	(10^{13})	–	–	C-96/40/92+	$2 \cdot 10^{11}$	$(2 \cdot 10^{13})$
at elevated temperature	Ωcm	–	–	E-1/100/T-100	$1.5 \cdot 10^9$	$(2 \cdot 10^{12})$	–	–	E-1/100/T-100	10^{10}	$(3 \cdot 10^{10})$
Electrical breakdown voltage parallel to layers	kV	D-48/50	15				D-48/50	30			
Relative dielectric number (ε) at 1 MHz		D-24/23	4.8	C-96/40/92+	5.5	(5.0)	D-24/23	4.8	C-96/40/92+	5	(4.7)
Dielectric loss factor (tan o) at 1 MHz		D-24/23	0.04	C-96/40/92+	0.07	(0.050)	D-24/23	0.04	C-96/40/92+	0.050	(0.031)
Corrosion Edge corrosion	Index	–	–	C-96/40/92		(A/B 1.0)	–	–	C-96/40/92		A/B 1.6 (1.4)
Tracking resistance (CTI-DIN-IEC 112)	Stage	–	–	A		(200)	–	–	A		(300)
Non-electrical properties											
Mechanical properties											
Land removal force	N	–	–	A	50	(100)	–	–	A	60	(130)
Punching property at 20 °C	Index	–	–	A	–	(2.0)	–	–	A	–	(1.9)
Bending longitudinal strength	N/mm²	A	84	A	100		A	140	A	110	
transverse	N/mm²	A	74				A	110	A	110	

Table 6.55 (continued)

Standard Designation Electrical properties	Units	Phenolic-based paper laminate				Actual data	Epoxy-based paper laminate				Actual data
		NEMA-LI 1-1989 XXXPC/FR-2		DIN EN 60249-2-7 PF-CP-Cu			NEMA-LI 1-1989 FR-3		DIN EN 60249-2-3 EP-CP-Cu		
		Pretreatment	Data	Pretreatment	Data		Pretreatment	Data	Pretreatment	Data	
Adhesion power of copper foil (35 μm) after thermal shock	N/mm	A	1.1	A	1	(1.9)	A	1.4	A	1.2	(2.0)
after elevated temperature	N/mm	–	–	E-500/100	1	(1.9)	–	–	E-500/100	1.2	(2.0)
after thermal shock	N/mm	–	–	T100	–	(1.2)	E-1/105	0.9	T-100	0.6	(1.2)
after simulated electroplating treatment	N/mm	–	–	A	0.8	(1.9)	–	–	A	0.8	(2.0)
Thermal properties Solder bath resistance at 260 °C Unetched sample	s	A	10	A	10	(15)	A	5	A	10	(25)
Etched sample	s	A	10				A	5			–
Flammability[a] Mean max. combustion time	s	A	Kl.1		100	–	A	Kl.1	A	–	–
Miscellaneous properties Water absorption	%	–	E/D[b] 0.75	E/D[c]	60	(52)	–	E/D[b] –	E/D[c]	40	(35)
	mg										

[a] UL 94 (vertical) flammability VO, limit temp. As defined in UL 746 is 105 for PF, 110 for EP; for pretreatment refer tio NEMA-LI 1-1989 or DIN 60,249-1.
[b] E/D=E-1/105+des+D-24/23.
[c] E/D=E-24/50+des+D-24/23.

a carrier material. Types G-10 and G-11 are non-flame retardant versions of FR-4 and FR-5, and CEM-1 and CEM-3 are composites that represent alternatives to FR-3 and FR-4 with respect to their physical properties and cost. The list of properties parallels the rising quality of the applications, from products such as electronic automatic toys from east Asia to high-tech computers from Japan, USA, or Europe.

The supplementary Table 6.55 compares the above NEMA-L1 1–1989 laminate Types FR-2 and FR-3 with DIN EN 60249-2 grades PF-CP-Cu and EP-CP-Cu. Nominal, and in some cases, actual data [20] are shown in the table. As expected, the epoxy paper laminates exhibit marked advantages in their electrical data, enabling them to be used in more complex (TV or audio) equipment. Further evolution of molded laminates (matrix and carrier materials) to their present state of development has been associated with optimization of printed circuit boards (Fig. 6.61) and continued development of these to the multilayer circuit board (Fig. 6.62).

The historical development of printed circuit boards (for assembly and configuration with components) dates back to the invention of the transistor by the Nobel laureate Shockley (1910–1963). In the 1940s he had developed an important electronic component, the transistor. Using relatively low potentials and currents, a transistor can assume the digital and analog functions of an electronic tube. It is based on semiconductor technology, mainly that of silicon. This technology permits large numbers of components such as diodes and resonant circuits to be combined and many functions united in a single silicon

Fig. 6.61. Printed circuit board assembly in a telephone (photo: Isola AG, Düren)

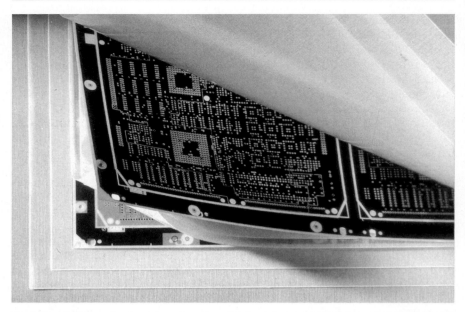

Fig. 6.62. Multilayer laminate-based on glass fabric and epoxy resin (photo: Isola AG, Düren)

chip, the "integrated circuit" or IC. Since such integrated components, together with their associated resistors and capacitors, only require copper connections with very small cross-sections, development concentrated on creating these connections by etching copper foil clad to the surface of a flat insulating material. The printed circuit board thus emerged, and with it the necessity to develop copper-clad molded laminates together with the associated technologies including those required to produce appropriate thin copper foils and bond these to the laminate. The printed circuit board must thus fulfill three functions: it must (1) hold the variety of components, (2) connect these by way of tracks, and (3) insulate connections and components from one another.

The tracks gave rise to the "printed circuit". Printed circuit boards based on copper-clad phenolic paper laminates are generally produced using the screen printing or photoprinting process (the former for high productivity, the latter for high accuracy). The etch-resistant circuit image is first transferred to the board. In the screen printing process used for industrial series production, the printing screen, with open spaces where the tracks should be positioned, is placed on the board and the printing ink forced through the screen. After the ink dries, the board is etched, rinsed, dried, and the printing ink finally removed with solvent or other chemicals. In the photoprinting process, a light-sensitive coating is applied to the copper, the circuit image transferred to the coating by exposing the latter through a negative film, and the coating developed. The printed circuit image is then etched in the board, and the finished circuit cleaned (see Photoresist/Imaging Section 6.3.1.8).

Fig. 6.63. World market for printed circuit boards in 1995 (U.S. $26 billion); source: VDL

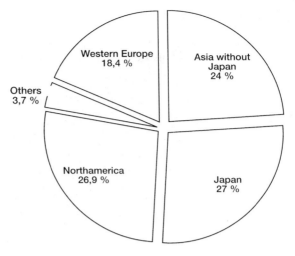

In large-scale production, the component connections are soldered into the board using automatic soldering baths at 260 °C. Due to this circumstance, the bond of the copper to the matrix of the board must temporarily withstand this temperature. The increasing complexity of electronic equipment and associated components rendered printed circuit boards clad on only one side with copper foil inadequate for further development. Thus, it was just a matter of time until circuit boards clad on both sides, and finally multilayer circuit boards were developed. The latter, and increased demands, then led to the use of glass fabric in addition to paper as a carrier, and to special "highly sophisticated" epoxy resins with particularly good adhesion to glass fiber and copper in addition to modified and unmodified phenolics as a matrix.

At present, printed circuit boards generate an estimated annual turnover of more than 26 billion US dollars worldwide, with a main focus in Asia (Fig. 6.63). The consumption of laminates has risen from about 102 million m² in 1984 to more than 200 million m² in 1996. During the same period, the volume of paper laminates (phenolics and epoxies, FR-1 to FR-3) has risen from about 50 million to over 80 million m², and presently leveled at 80–90 million m² (Fig. 6.64). At present, paper-based laminates (FR-1 to FR-3) represent about 40–41 % of the world laminate consumption or 81.4 million m², and phenolic paper laminates make up around 65 million m² of this. This indicates a worldwide resin demand of 65,000 tonnes, 4000–5000 tonnes of this in Europe. A laminate consumption of about 92 million m² results when the CEM-1 paper/glass fiber composites are included in the paper-based laminate total (Fig. 6.65), a large fraction of this in Asia, from where inexpensive toys and consumer articles as well as standard TV sets and audio equipment are exported to Europe and the USA. Of the laminates used worldwide, 55 % are based on epoxies, 3.3 % on polyester, and 1.8 % on polyimide. The paper-based laminate demand in Europe is only 8.7 % of the total (Japan: 38 % and rest of

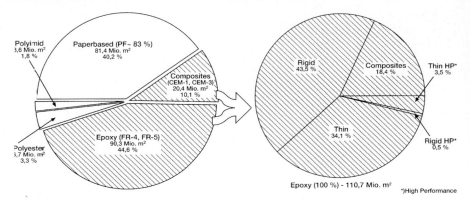

Fig. 6.64. Worldwide consumption of laminates in 1996, **ca.** 157 million square meters (source: bpa)

Fig. 6.65. Worldwide consumption of paper-based laminates (FR-1, FR-2, FR-3, CEM-1) in 1996, 92 million square meters

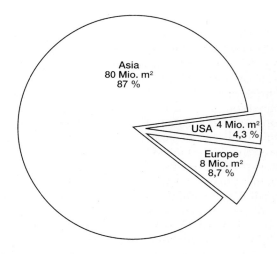

Asia: 49%), and includes about 4.5 million m² of FR-2 (Fig. 6.66). FR-1 (cf. Table 6.54) is mainly used in Asia. The focus in Europe and the USA is on FR-3, FR-4, and similar materials. Although PF (phenolic) laminates are mainly used in less expensive ("consumer") electronics, the demands on the base material and thus on the resins as well have risen considerably over the years.

6.1.4.4
Standardization of Molded Laminates

Various international standards are presently used for quality assurance of molded laminates. In Europe, CEN (European Committee for Standardization)

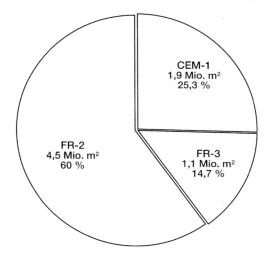

Fig. 6.66. European market for paper-based laminates in 1996, 7.5 million square meters

criteria are becoming established as the principle standards. As in the case of thermosetting molding compounds (cf. section on phenolic molding compounds 6.1.3), both internationalization (in the globalization process of the pertinent industry and suppliers) and a sensible concentration of the various types of materials has taken place during the adaptation process of standardization. Standards that should be observed are, aside from CEN, the international IEC (International Electrotechnical Commission) and ISO (International Organization for Standardization) papers and the US NEMA (National Electrical Manufacturer's Association) standards.

Thus, the European DIN EN 60893 standard "Specification for industrial rigid laminated sheets based on thermosetting resins for electrical purposes," based on the international IEC 893 standard, replaces the DIN 7735 standard that was only valid in Germany. The type classifications (Table 6.56) in DIN EN 60893 specify the resin matrix with an abbreviation such as PF for phenolics (EP, UP, and Si are also possible), and also indicate the carrier material. The serial numbers, such as 201 to 204, indicate changes or enhancement in the levels of properties or application scale, for example, a higher or lower value for the electrical or mechanical properties. Table 6.56 simultaneously lists data from the previous DIN 7735 standard.

In principle, this corresponds to the classifications in NEMA publications (X, XX, or XXX with supplementary abbreviations such as "CP" for cold punching). All standards mentioned here and below also specify minimum required physical properties.

The standard that applies to copper-clad, paper-based laminates is DIN EN 60249 (based on IEC 249), in which phenolic grades (1, 2, 6, 7, and 14) are classified according to electrical and economic criteria as well as their flame resistance. If at all, such differences are only indirectly rendered in the NEMA classifications. Thus, the DIN EN standards may only be compared with the internationally accepted NEMA publications to a limited extent. An

Table 6.56. DIN EN 60893 molded laminates with comparison to DIN 7735 (cotton fabric, paper, glass)

DIN EN 60893 type[a]	Applications	DIN 7735 type
PF CC 201	Mechanical (coarse cloth), electrically poorer than 202	Hgw 2082
PF CC 202	Mechanically and electrically high quality	Hgw 2082.5
PF CC 203	Mechanical (fine cloth), small parts	Hgw 2083
PF CC 204	Mechanical and electrical, small parts as PF CC 203	Hgw 2083.5
PF CP 201	Mechanically high quality, low use at normal humidity	Hp 2061
PF CP 202	High-voltage area, largely penetration resistant	Hp 2061.5
PF CP 203	Mechanically and electrically high quality, also capable of punching	Hp 2061.6
PF CP 204	Electric and electronic uses, also at elevated humidity	Hp 2063
PF CP'205	Like 204, but defined flame retardancy	Hp 2062.9
PF CP 206	Good electrical properties at high humidity, versions with punching capability also possible	Hp 2062.8
F CP 207	Similar to 201, but can be punched even at low temperatures	(Hp 2061)
PF CP 308	Similar to 206, but with defined flame retardancy	(Hp 2063)
PF GC 201	High mechanical strength and good electrical properties at normal humidity, heat resistant	Hgw 2072

[a] Abbreviations: PF = phenolic resin; CC = cotton cloth; CP = cellulose paper, GC = glass cloth.

attempt at such a comparison is made in Table 6.57. Using a practical example (comparison of FR-2 and FR-3 with Cu-PF-CP), Table 6.54 has already shown that such comparisons are applicable in principle, particularly in the case of paper-based laminates. The standard that applies to tubes and profiles is DIN EN 61212. The classifications established there distinguish between the types of products (1–3, rods and tubes), resins (such as phenolic or epoxy), and the carrier material, that in the case of phenolics can represent cellulose paper, cotton, or glass fabric. The numbers of warp and woof threads in the fabric fall into the following rough categories: *fine fabric* has more than 30 threads/cm, *coarse fabric* 18–30 threads/cm, and *very coarse fabric* less than 18 threads/cm.

Modifications in the quality and carrier material reflected in the minimum requirements are found in grades 21–23 and 41–43 respectively.

Table 6.57 Comparison of clad and unclad paper-based laminates in DIN EN 60893-3-4, DIN EN 60249-2 and NEMA Pub LI-1-1989

DIN EN 60893-3-4	DIN EN 60249-2	NEMA Pub. No. LI-1-1989	Properties and end uses of laminates	DIN 7735[a] (for comparison purposes)
PF CP 202	–	XX	Phenolic paper laminate with good mechanical properties and excellent insulation data, particularly suitable for oil-submerged transformers	Hp 2061.5
PF CP 201 / PF CP 207	1/n	X, XP / XPC	Phenolic paper laminate with good mechanical properties, suitable as insulating material for low-voltage applications	Hp 2061
PF CP 203 / PF CP 204	1/n	XX, XXP / XXX	Phenolic paper laminate with low water absorption, resistant to changes in humidity, suitable for telecommunications technology	Hp 2061.6
PF CP 206	–	XXXP XXXPC	Resin-rich phenolic paper laminate, particularly low water absorption, tropics resistant, for high-frequency requirements; can be cold punched and copper clad as a special version	Hp 2062.8
PF CP 205	–6	XXXP	Flame resistant version of this grade, also in PF CP 204 quality as a copper-clad version	Hp 2062.9
PF CP 308	–6	XXXPC	High quality phenolic paper laminate, corresponds to PF CP 206, low dielectric losses, good corrosion resistance and cold punching capability	Hp 2063
–	–6	–	Flame resistant, copper-clad version of PF CP 206 grade, flame resistance meeting UL 94 in the HB version	–
–	–7	FR2	UL 94 stage V1 or V0 version of PF CP 206	–
–	–14	FR 1	Not electrically high quality, but UL 94 stage V1 and V0	–

[a] DIN 7735 has been superseded by DIN EN 60 893 and 60 249.

6.1.4.5
Raw Materials and Impregnation

Unfilled, unsized, practically electrolyte-free soda cellulose paper is used as a carrier material for paper laminates; bleached, highly porous cotton linter, α-cellulose, or mixed papers are used for premium grades to achieve special insulating properties. Table 6.58 lists a selection of commercial impregnating resins. The comments in the "remarks" column note where certain differences in quality may be expected.

Thus, special resins may be used for the cover sheets to achieve optimal electrical properties. For example, resins G and H in Table 6.58, which are prepared as water resistant and flexible by use of special modifiers such as tung oil, are used as cover sheet resins for FR-2 laminates. Products manufactured with such impregnating resins are matched to modern application technologies and offer application-related electrical laminate properties such as cold punching and cutting capabilities, solvent resistance, and dimensional stability. Modification with epoxy resins can further improve the electrical properties of phenolic paper-based laminates as well as their dielectric properties, solvent resistance, and dimensional stability. Such materials also exhibit better tracking resistance than unmodified phenolic laminates.

Less expensive resins (resin K in Table 6.58) are basically adequate for unclad grades of laminate. On the other hand, specially modified resin solutions such as resin G in Table 6.58 have been found best for "potentiometer" grades of laminate with particularly good surface properties. In addition to their use in paper-based laminates, resins C and D are employed as standard resins for fabric laminates. Xylenol-cresol resins (resin E in Table 6.58) are used to impregnate high-voltage resistant laminates.

The resin solution used for impregnation is adjusted to the desired viscosity. In dip impregnation, metering may be accomplished using rollers. Aqueous, low molecular mass phenolic resoles (resin A in Table 6.58) are added to the solution to achieve hydrophobic properties of the laminate. Care must be taken that appropriate means are employed to keep the impregnating bath homogeneous when fillers/solids are added, such as antimony trioxide as a flame retardant. The flame retardancy can also be regulated using melamine resins (resin B in Table 6.58).

A formulation such as that in Table 6.59 is processed in the following manner. The powdered melamine resin is dissolved in a mixture of water and methanol to yield a clear secondary impregnating resin solution that is only slightly warm. This solution is blended into the primary, modified impregnating resin with stirring. The result is a clear to slightly opalescent final mixture. The resin level in the paper is around 56% following impregnation. Pigments such as titanium dioxide and fillers such as aluminum oxide trihydrate and magnesium hydroxide are used to provide additional flame retardancy.

The impregnated carrier material then passes through a vertical or horizontal drying stage. Drying and pre-condensation at temperatures of 100–190 °C produce the required degree of resin flow and desired level of volatile components in the paper web. Depending on how they are dried, the lamina-

6.1 Permanent Bonding 249

Table 6.58. Phenolic resin-based impregnating solutions for production of laminates for various requirements and areas of application

Impregnating resin and additives	Chemical nature	Nonvolatiles (ISO 8618)%	B-Time at 130 °C (ISO 8987) min	Viscosity at 20 °C (ISO 9371) mPa·s	Remarks
A	Aqueous phenolic resole, pre-impregnating resin	71±2	3–5	1250±250	Good impregnating properties for pre-impregnating
B	Melamine resin, powdered additive	98±2	9.5–10.5	–	Flame retardant additive
C	Phenolic-cresol resole in methanol	60±2	6–9	200±100	Suitable for paper and cotton cloth
D	Cresol resole in methanol	60±2	4.5–6	250±100	As C, but with greater high-voltage resistant laminates
E	Xylenol-cresol resole in methanol	51±2	8–10	75±25	For high-voltage resistant laminates
F	Tung oil-modified cresol resole in methanol/toluene	61±2	11–13	300±100	Main impregnating resin for core and cover sheets
G	Tung oil-modified, bromine containing phenolic resole in methanol/toluene	65±2	10–14	500±150	"Potentiometer" grade, i.e. meets high demands on surface quality
H	Tung oil-modified cresol resole in butanol/toluene	60±2	4–5	500±100	Qualitative high grade resin for cover sheets
I	Cresol-modified phenolic resole in methanol	57±2	12–14	60±30	Particularly good impregnating properties for unclad material
J	Phenolic resole in methanol	72±2	3–5	1100±200	Main impregnating resin for high resin levels
K	Phenolic resole in methanol	60±2	10–14	50–70	Economical resin for unclad FR-1 and FR-2

Table 6.59. Formulation of halogen-free FR-2 laminates

160	PBW[a]	Powdered melamine resin solids (resin B in Table 6)
110	PBW	Water
80	PBW	Methanol
235	PBW	Pre–impregnation resin, 70–72% NVM[b] (aqueous resole)
1670	PBW	impregnation resin, 60% NVM[b], e.g. Resin K in Table 6

[a] Parts by weight.
[b] non volatile matter.

tes exhibit a resin content that ranges between 30% and 50% after drying, and can be as high as 60% in the case of particularly high-quality grades of paper laminate.

A two-stage process of impregnation was used earlier in the manufacture of premium laminates. This involved pre-impregnation with 10–25% resin, depending on the requirements and additives, and a main impregnation step with 100–110% resin. However, this method has been displaced by a single-stage process as indicated in the above example. The impregnated carrier material is finally cut to size and fed to a multi-stage press where it is compressed in stacks at pressure of 50–120 bar and programmed temperatures of 100–170 °C to yield laminate sheets.

A special melt impregnation process was introduced [21] to permit solvent-free impregnation in manufacture of FR-4 electrical laminates. This type of impregnation is expected to comply with new emission regulations, particularly in the USA. Highly pure, low-melting novolak systems exhibiting a low level (less than 0.2%) of free phenol are used as curing agents for the epoxy resins in this newly developed process.

6.1.4.6
Fabric Laminates

Fabric laminates are molded laminates produced from resin and fabric either in the form of a semifinished material such as sheets, tubes, solid rods, or flat strips, or as molded articles. In DIN EN 60893 and DIN EN 61212, the various grades are classified according to the type of fabric used – for example cotton or glass fiber – by the delivery form, such as sheets or moldings, and by the stipulated property data. The quality of fabric laminates largely depends on the selection of raw materials, i.e., the types of fabric and the binder. Phenolic resins have been adapted to meet the specific requirements. Fabric laminates manufactured with phenolics feature low density, simple working, high mechanical strength, and good electrical properties, and are established construction materials in many fields of application. Various types of resins are used, such as resins C and D in Table 6.58, and differ with respect to their high-voltage resistance. In principle, fabric laminates are manufactured as described in the case of paper laminates by impregnation and subsequent compression/curing.

6.1.4.7
Applications of Paper- and Fabric-Based Laminates

Industrial fabric laminates are mainly used in areas where high levels of mechanical and thermal stress are encountered, for example, for production of bearing shells. When using fabric laminates, the designer must consider the fact that for structural reasons their properties depend on the direction of stress. In addition to the lower density of fabric laminates relative to that of metal, the relatively low coefficient of friction and the chemical resistance are of particular importance to the mechanical engineer. Additives such as graphite or molybdenum sulfide produce a further reduction in the coefficient of friction of fabric laminates, and improve the "emergency running properties" – in particular in low-maintenance equipment.

Phenolic fabric laminates are also outstanding mechanical engineering construction materials for many additional applications. They can generally be used wherever corrosion resistance, resistance to fats, oils, and many chemicals, coupled with good thermal resistance and mechanical load bearing ability are required. The relatively low density and low thermal conductivity compared to those of metallic materials represent a basic advantage in applications such as conveyer and running rollers, expander rollers, guide segments for rails and chains, pilot rods for hydraulic cylinders, spinning spools, measuring and feeder rolls, gears, slip bearings, ball bearings, closure gates for containers used to ship aggressive media, and electroplating clamps.

Phenolic bonded molded laminates, particularly paper-based laminates, are of advantage in situations where short runs of products are to be produced and good electrical insulating ability is required. Thus, a favorite field of application is in high-voltage technology; phenolic bonded semifinished goods are mainly used in switching gear with its couples and actuation rods, in through bearings, spacers, winding bodies, jacket tubes, spark-extinguishing chambers, explosion chamber disks, partition walls, slip plates, screening cylinders, and many other applications. Based on sheer volume, however, base materials for printed circuits (circuit boards) – as detailed above – represent the greatest percentage of industrial molded laminates. Continued development of printed circuits in such areas as epoxy binders and glass fabric has not displaced phenolic bonded material completely. On the contrary, major areas in which they are used in large quantities for economic and product-specific reasons have developed on a worldwide basis (Figs. 6.63 and 6.64).

6.1.4.8
Industrial Filter Inserts

Filter elements of fibrous materials and unfilled papers or paper-type materials such as webs – generally in the form of accordion folded tubes – are used to separate solids from liquids or gases and to remove water from fuels or mineral oils [1–3]. Specific demands are made on the structure and porosity of the papers, corresponding to the throughput, resistance (lowest possible) and retention power (highest possible) associated with their use as filtration media. Adequate strength is achieved by impregnation with alcoholic solutions of phenolic resins and drying [4]. Epoxy resins are also increasingly used for special applications. Epoxy resins can offer both certain ecological advantages (lower emissions) and technological benefits (increased productivity, higher machine part efficacy). The advantages of phenolic resins are generally to be found in superior heat and aging resistance.

The impregnated filters are used as air, oil, and fuel filters in the automotive industry [5, 6] – an area (Fig. 6.67) that consumes the major share of total production – or as hydraulic filters to clean hydraulic fluids, as general purpose filters for liquids in various industrial fields, such as the chemical industry, and in general as dust filters or filters for air conditioners.

The demands on the impregnating resin systems are quite diverse. Requirements include the necessity to achieve optimal strength, retain a certain degree of flexibility or elasticity, provide freedom from odors, and offer resistance to solvents, chemicals, and aggressive vapors and gases. Despite this, the permeability of the medium must not be adversely affected. Selected papers already exhibit special features designed to afford an optimal range of filter properties. These features relate to the selection of cellulose fibers and the type of processing on the papermaking machine.

Even the type and origin of the wood from which the fibers are obtained is a consideration. Thus, papers with low throughput resistance are obtained from fast growing conifers. Care is also taken not to defibrillate excessively the fibers of cellulosic material. In addition, the water in the paper web is not

Fig. 6.67. Selection of various automotive filters (source: Fibermark Gessner, GmbH & Co., Feldkirchen-Westerham, Germany)

removed by pressure, but by evaporation. Various additives increase the wet strength; such papers are otherwise filler-free. It is desirable that they exhibit low throughput resistance. However, the retention power can be too low for oil and hydraulic filters. In such a case, somewhat more compact products are manufactured using increased defibrillation and by removing the water under pressure during papermaking. Such grades of paper exhibit increased filtration resistance, but also offer improved separating power. Papers (webs) based on cotton and very short glass fibers can be used to achieve even better retention, but are generally somewhat higher in cost.

The customary bleached cellulose fibers and cotton fibers for special filters, are generally 4–6 mm long and 10–30 µm in thickness. Glass fibers, used for glass fiber webs, are 0.5–2 µm thick. The raw paper is soft and absorbent, and is processed by impregnation. The resins (Table 6.60) are applied by dip impregnation which involves passing the substrate material through a solution of resin adjusted to afford the desired level of resin, followed by stage drying (but not yet curing) in an appropriate oven at temperatures rising from 100 °C to 170 °C. In this process, care must be taken to ensure that the resin bath remains stable and at a constant temperature in order to prevent variation in the level of resin deposited on the filtration medium. The pre-dried, resin impregnated papers offer the advantage that they become flexible in this condition when heated from ambient to 100 °C, may thus be easily shaped (pleated and napped) in the course of further finishing operations, and retain the assumed shape after cooling [7].

The drying process is thus controlled in such a manner that the paper contains a maximum of 5–8% volatile components. This provides the flexibility required for the finishing and curing operations. Depending on the type of substrate and the required application properties, the resin level in the filter material ranges between 10% and 30%, and can reach 50% in special cases.

Table 6.60. Typical data on phenolic resin solutions in methanol used for impregnation of industrial filter papers

Composition	A[a] Modified hexa-free phenolic novolak	B Modified hexa containing phenolic novolak	C Modified phenolic resole
Nonvolatiles ISO 8618 in %	61 ± 2	65 ± 2	63 ± 2
Viscosity at 20 °C ISO 12058 in mPa·s	90 – 130	250 – 350	250 – 350
B-Time (gel time) at 130 °C ISO 8987 in min	8 – 12	8 – 12	2 – 6
Free phenol ISO 8974 in %	max. 0.5	max. 0.5	max. 1.5
Free formaldehyde ISO 9397 in %	–	–	max. 1

[a] Cured by addition of 8 parts by weight (PBW) hexamethylene tetramine per 100 PBW resin solution).

Demands for low-monomer resins are met by using methanol solutions of novolaks (e.g., resin A in Table 6.60) with free phenol levels below 0.5%, or novolak solutions already containing hexa and similarly exhibiting free phenol levels below 0.5% (e.g., resin B in Table 6.60), both of which feature good aging resistance and high levels of hydrophobicity. Solutions of resoles exhibiting low monomer levels (free phenol below 1.5%, free formaldehyde below 1%), e.g., resin C in Table 6.60, can also be used for impregnation of filter papers. Impregnation with either a novolak or a resole results in a filter paper exhibiting good mechanical strength in both the dry and wet state. The chemical resistance to fuels and lubricating oil is also good.

Reduction of the monomer levels offers the following advantages: decreased odor levels during production, storage, and during preheating, reduced soiling during the finishing operations, diminished exhaust gas treatment burden, and reduction of workplace emissions. Reduction of the monomer levels also achieves increased impregnation efficacy and allows interference-free solvent recovery.

In many cases, combinations of phenolic novolak solutions and phenolic resoles have also been accepted in practice to improve strength levels further. This permits major or minor reduction in the addition level of hexa, having a desirable effect on the process emissions.

The impregnated paper, after being cured in a subsequent processing step, also features low water absorption, high rigidity, a high level of strength, and lastly an improvement in the high temperature oxidation properties [8]. Fire retardant versions of the resins are available for some applications (for example in connection with introduction of more compact engine compartments). Such properties may be achieved by the addition of special additives [9].

For the reasons mentioned above, it is advisable to preheat the paper before it is pleated and napped, the paper web being best introduced into the pleating machine at a temperature of 60–80 °C (Fig. 6.68). Heating to higher temperatures can lead to undesirable adhesion effects during the finishing process. The pleating operation to produce accordion folded tubes (Fig. 6.69) affects the quality of the finished products to a very great extent. An uneven feed at the machine inlet or deformation of the paper can have a highly adverse effect on the separation efficiency and service life of the finished filter.

After pleating, the papers may be cured in the form of a continuous web or finished filter elements. Curing is generally performed at 160–200 °C in a throughput oven. The cured papers can be bonded to produce filter elements using products such as PVC plastisols, polyurethane adhesives, or hot melts.

6.1.4.9
Miscellaneous Impregnation Applications

The desirable impregnating properties of phenolic resins (resoles, novolaks, or mixtures of the two) dissolved in alcohol are of key technological significance for optimal saturation of paper and cotton fabric. In addition to applications such as electrical laminates, engineering laminates, industrial filters, and reinforcing fabric for abrasive wheels that have been previously noted and describ-

6.1 Permanent Bonding   255

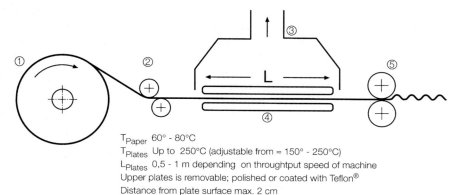

T_{Paper} 60° - 80°C
T_{Plates} Up to 250°C (adjustable from ≈ 150° - 250°C)
L_{Plates} 0,5 - 1 m depending on throughput speed of machine
Upper plates is removable; polished or coated with Teflon®
Distance from plate surface max. 2 cm

Danger of fire exists at temperatures higher than 300°C if machine stops suddenly

① Paper Roll
② Guide Rolls
③ Hood and Ventilator
④ Heating Plates/Rolls
⑤ Pleating Rolls

Fig. 6.68. Preheating of filter paper (drawing: Fibermark Gessner, GmbH & Co., Feldkirchen-Westerham, Germany)

Fig. 6.69. Accordion folded tubes (source: Fibermark Gessner, GmbH & Co., Feldkirchen-Westerham, Germany)

ed, several other uses exist and will be briefly described below. These include impregnation of paper for storage battery separators, production of paper, plastic, or metal honeycomb, gaskets for various industrial applications, filter fabric for aluminum melts, paper friction linings for segment clutches, and fishing rod prepregs. Impregnation of many different woven fabric systems is discussed in the High Performance and Advanced Composites section 6.1.5.

Separators [1–5] are used in lead-acid storage batteries to prevent direct contact of the oppositely charged lead plates. These electrically insulating paper separators, which exhibit high resistance to acids and oxidation as well as adequate mechanical strength, have been manufactured using phenolic resin impregnated cellulose paper for a long time. The separator is provided with numerous ribs on one side. The special paper is impregnated with an aqueous phenolic resin solution containing added surfactants, dried at elevated temperatures, and finally cured. Separators made with synthetic fibers such as polyacrylonitrile have also been used; the smaller pore diameter of these separators rendering them impermeable to components such as loose lead oxide particles or antimony. In recent years, the use of phenolic resin impregnated battery separators has greatly declined (particularly in the automotive sector) due to the use of modern storage batteries based on various plastics designs.

The long-familiar paper honeycomb and honeycomb structures also represent a field for use of phenolic resins as impregnating/dipping agents [6, 7]. Aramid and special cellulose-based fibers are presently the main materials used for production of honeycomb. The honeycomb is generally manufactured using the expansion process. In this method, an adhesive is first applied to the basic honeycomb material by printing the lines of adhesive being applied at right angles to the direction of product expansion. The size of the honeycomb cells can be controlled by variation of the width of the adhesive lines and the distance between these. The web of base material is then folded into sheets, which are stacked using a specific geometry.

The length of the sheets determines the width of the subsequent honeycomb block, whereas the number of sheets in the stack determines the length of the expanded block. A massive stacked block containing thousands of sheets is thus assembled. This block is cured in a heated press at elevated temperature and pressure, and after curing is pulled apart (expanded). After the blocks have been expanded they are dipped in modified phenolic resins. This provides the honeycomb with its final mechanical properties and bulk density. The mechanical properties of the material are adjusted by variation of the ratio of fiber to resin, for example from 25:50 to 50:50. Such honeycomb (Fig. 6.70) is presently in wide use, mainly for panels and construction elements in aircraft, automotive, and rail vehicle construction (Fig. 6.71).

Initially, such honeycomb and the composites produced with it were mainly used in high-tech products such as the flooring and interior paneling of aircraft; today, even passenger and freight train cars, where lightweight construction products can greatly reduce the axle load, represent areas of application. Since the cellulose fiber honeycomb can be provided with fire retardant topcoats for such applications, it is possible to produce economical, ready-to-

Fig. 6.70. Paper honeycomb for composites (photo: Euro-Composites GmbH, Echternach, Luxembourg)

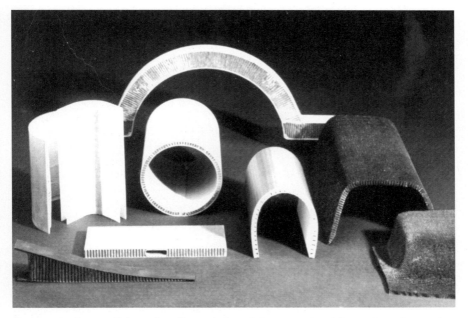

Fig. 6.71. Various cellulose-phenolic resin-based honeycombs (photo: Euro-Composites GmbH, Echternach, Luxembourg)

install construction elements for interior furnishings. Sliding doors up to 15 m in length for subway and freight train cars represent possible applications. In shipbuilding, particular attention is paid to effective fire protection. The panels used in this case can be provided with a surface skin of flame retardant material, enabling them to reach critical fire protection classes.

In the automotive industry, friction pads and braking straps (Fig. 6.72) are used in a wide variety of automatic transmissions [8, 9]. They are also used in applications such as power shift transmissions, ship winch transmissions, and motorcycle clutches. The lining materials are special types of paper impregnated with solutions of specific phenolic resins that permit adhesion to metal and provide parameters required for friction lining compounds. Special phenolic resins, which are modified with other resin systems such as epoxies and amino resins to enhance their range of properties, have been developed for impregnation of such papers.

An unusual impregnation application is production of impregnated fiberglass fabric for manufacture of filters for the foundry industry (for example to filter aluminum melts). Such impregnated fabric is generally produced on automatic equipment, which in one operation produces the woven fabric from the fiber, passes the fabric into an impregnating bath containing an aqueous phenolic resole with the lowest possible level of monomers, dries it briefly, and finally cures it in stages at temperatures ranging from 120 °C to 300 °C.

Although prepregs based on long-fiber glass fabric are generally produced using polyester and epoxy resins, phenolic resins have increased in importance due to their outstanding properties, particularly with respect to their flame resistance, low smoke development, and low cost. They exhibit excellent

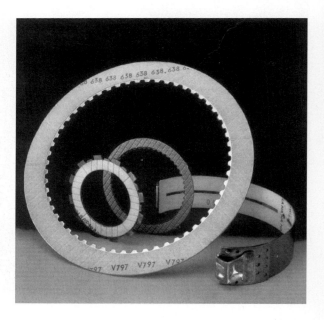

Fig. 6.72. Friction elements for use in shifting clutches for automatic transmissions (photo: Borg Warner Automotive Europe)

resistance to deformation when hot. In addition, they are less flammable than other synthetic resins (cf. High Performance and Advanced Composites section).

One of the oldest fields of application for phenolic resin-glass fabric prepregs is production of fishing rods. Special resins modified with special plasticizers are used in this case [10, 11].

6.1.5
High Performance and Advanced Composites

6.1.5.1
Introduction

High performance and advanced composite materials are fiber reinforced products that trace their origin to the early 1940s when, during the war years, lightweight materials with improved structural properties were desired.

The combination of glass fiber with a thermoset resin matrix was instrumental in launching a huge composites industry which today represents annual worldwide market of nearly 47 million tonnes. With an annual growth rate of 2 – 4%, fiber reinforced plastics are utilized in a wide spectrum of application areas ranging from transportation (Fig. 6.73), construction, marine, corrosion resistant equipment (Fig. 6.74), electrical apparatus, and aircraft to consumer/sports products [1].

Fig. 6.73. High speed train (Transrapid, Germany)

Fig. 6.74. Composite pipe

During the intervening years the use of new and improved fibers, tailored interfaces, high performance modified matrix resins, and new fabrication processes has led to the development of many highly unique and complex advanced composite systems.

Phenolic resins are used as matrix resins in the various composite market segments depending on the end use applications. The main focus of this chapter will consider the utility of phenolic resins in high performance and advanced composite market segments. As it will be discussed in the ensuing sections, a combination of attractive Fire/Smoke/Toxicity (F/S/T) characteristics, reasonably priced resin, and emerging market driven applications have been the motivating factors for the resurgence of interest in phenolic resins for fiber reinforced plastics.

6.1.5.2
Composite Market Segments

The composite industry may be arbitrarily structured in the following market segments: commodity, high performance, and advanced composites.

The commodity market segment, as the name implies, represents the general purpose, relatively cost competitive sector of the composites market where price of raw materials is the dominant factor. Phenolics participate in this market segment as molding materials (see Sect 6.1.3) as well in hand lay-up applications.

Fig. 6.75. Military composite unit

The high performance composites area combines newly developed processes with desirable product features to serve specific market areas. Either pultrusion or filament winding of glass fiber with phenolic resin matrix can be used to achieve novel Fire/Smoke/Toxicity (F/S/T) composites for oilfield applications (pipe, platform decking) or construction use in confined areas or bridges.

Advanced composite systems consist of superior performance materials composed of intermediate to high T_g matrix resins and high strength fiber (mainly carbon fiber but also glass and aramid) for critical applications. The carbonization and char forming characteristics of phenolics by controlled pyrolysis (see Sect. 6.4) are utilized in the manufacture of ablation products such as rocket nozzles, C-C brakes and C-C composites. These expensive materials reside mainly in aerospace, military (Fig. 6.75), and commercial aircraft.

Phenolic resins exhibit excellent dimensional stability with a constant use temperature range of 180–200 °C, excellent chemical, moisture, and heat resistance, and favorable F/S/T behavior. The predominant consideration in the use of phenolic resins as matrix resin in fiber reinforced composites is their unique F/S/T characteristics. Comparison studies of fiber reinforced thermosetting resin matrix systems have shown that phenolic resins are far superior to epoxy, vinyl ester, or polyester resins under fire conditions (Table 6.61). These fire conditions relate to ignitability and heat release, smoke/toxicity, and fire resistance. The propensity of phenolic resins to exhibit delayed ignitability coupled to low heat release, low smoke evolution with little or no toxic gases

Table 6.61. Comparison of the smoke density of several plastics observed during a fire (see Table 6.62)

Plastic	Smoke Density (D_n)	
	Smoldering fire	Fire
Phenolic resin	2	16
Epoxy resin	130–205	480–515
Vinyl ester resin	39	530
Polyvinyl chloride	144	364

emitted, and the capability to provide significant strength retention (to 70%) at 300 °C over a time period of 1–2 h is remarkable for an organic polymeric material. Such a material exhibits exceptional fire resistance and when incorporated into a fiber reinforced composition is used in demanding critical applications where safety of passengers during egress in case of fire is of paramount importance. Within the last two decades, the evaluation of many tests under fire conditions corroborates the superiority of phenolic resins (Table 6.62).

These fire testing data have been effective in facilitating major changes in the selection of phenolic resin reinforced fiber systems for aircraft interiors (Fig. 6.76), mass transportation, and marine/offshore areas. Controlled pyrolysis of carbon fiber/phenolic prepregs into carbon-carbon composites has for several decades been the manufacturing procedure for transforming these phenolic carbon fiber prepregs into ablation products such as rocket motor cases and rocket nozzles, and more recently into carbon-carbon brakes for military and commercial aircraft [1]. The controlled pyrolysis and char insulating characteristics, combined with chemical resistance and strength reten-

Table 6.62. Fire/smoke values of phenolic laminates[a]

Test	Values	
	Flame spread	Smoke density
Tunnel test (ASTM E-84)	5 (unpainted) 20 (painted)	5 (unpainted) 10 (painted)
OSU	Total heat release Peak heat release	26 kW/min/m^2 40 kW/m^2
Surface flammability (ASTM E-162)	Flame spread index	$I_s = 2.8$
Smoke density (ASTM E-662)	Flaming Optical density (D_M) 23	Non-flaming 7
Cone calorimeter (ASTM E1345)	Rate of heat release Effective heat of combustion	30 kW/m^2 22.3 MJ/kg
Oxygen index (ASTM D-2863)	Ignitability	>55%

[a] BP Chemicals data, 3 mm thick 35% glass hand lay-up.

Fig. 6.76. Aircraft interior (Airbus)

tion, are the motivating factors for the use of phenolic/reinforced fiber systems in oilfield applications. This area is of particular interest in an emerging market utilizing filament winding technology and pultrusion for pipes/tube inserts for oilfield use as well as secondary and tertiary structures on existing and new offshore installations [2, 3].

6.1.5.3
Fiber Reinforced Plastics

Prior to considering different processes that are used to make FRP materials, several new developments have occurred in phenolic resins, particularly those to be employed for FRP, as well as new developments in fibers.

6.1.5.4
Phenolic Resin Developments

The preferred phenolic resin for FRP is resole resin with the trend toward water-based phenolic resin rather than a solvent composition. Recent new developments consist of improved toughness/ductility provided by the introduction of an elastomeric phase [4] or silicon segment [5–7] in the resole microstructure.

Most resoles are low in free formaldehyde but have high amounts of free phenol (>5%). A recently introduced water-based resole which is environmentally and user friendly [8] possessing low formaldehyde (0.7%) and low free phenol (0.7%) is being promoted in many FRP applications.

A product known as "Dryprepreg" that is based on powder fusion of phenolic resin into woven fabric (carbon fiber, glass, aramid) is expected to provide economic advantages over existing phenolic prepregs [9]. Dryprepreg requires no refrigeration, can be stored at room temperatures (~25 °C), and maintains processability for six months.

Appending a group such as cyanate or forming a new ring structure like oxazine (see chapter on Chemistry, Reactions, Mechanisms Chapter 2) is equally beneficial since it minimizes the adverse effect of water by-product that is generated during crosslinking of typical phenolic resins and reduces void content. Phenolics modified with either cyanate or oxazine continue to exhibit favorable F/S/T behavior of the resulting cured product. Many FRP systems based on either of these modified phenolics have been reported, and the reader is urged to examine references contained in chapter on Chemistry, Reactions, Mechanisms.

6.1.5.5
Fiber Developments

In some processes such as filament winding or pultrusion, a key issue related to poor interfacial adhesion between fiber and resin was identified. This problem was resolved by several glass fiber manufacturers who developed special glasses for use in conjunction with phenolic resins [10]. Glass/phenolic composites with favorable interlaminar adhesion between fiber and matrix are obtained in filament winding and pultrusion processes using these special glasses. A recent report by Clark Schwebel Co., Anderson, SC [11] describes enhanced surface chemistry of woven glass fiber/phenolic resin composites resulting in increased laminate mechanical properties. The authors attribute the maximum amount of strength and performance of the glass phenolic composite to the optimum design of the interface between matrix and fiber surface.

6.1.5.6
Composite Fabrication Processes

Most composite processes are suited for phenolic resins. Phenolic resins can be combined with a variety of fibers such as glass, aramid, carbon fiber, or ultra high molecular weight polyethylene (UHMWPE) and transformed under appropriate processing conditions into various intermediate and/or final product forms. Various processes can be used and include impregnation (prepreg, honeycomb), filament winding, pultrusion, SMC, RTM, and hand lay up.

Resin properties and process characteristics for each of these different processes are listed in Table 6.63.

Table 6.63. Resole resin properties required for various fabrication methods

Process	Viscosity (mPa · s)	Cure Temp/°C	Thioxotropy	% Filler	Fiber size	% Fiber (wgt.)
Prepreg	1000–25,000	100–150	No	–0–	–	50–70
Ballistics	2000–6000	100–150	No	–0–	–	80–85
Honeycomb	200–1000	100–150	No	–0–	–	30–60 [a]
Filament winding	600–2000	RT-150 [b]	No	–0–	Special glass	40–70
Pultrusion	400–2000	100–150	No	Up to 20	Special glass	60–80
SMC	200–2500	150	Possibly	25–50	–	25–50
RTM	600–1000	RT-150	Possibly	–0–	–	40–60
Hand lay-up	400–800	RT	Yes	Low	–	20–40

[a] Paper also.
[b] Heated mandrel below 100 °C, postcure at 150 °C.

6.1.5.6.1
Impregnation

Prepreg

The impregnation of a woven, non-woven, or unidirectional fiber with resole results in an intermediate impregnated or "prepreg" fiber/resin system. The process involves conducting the fiber component through a bath containing resole resin (water- or solvent-based) and into a heated tower to remove solvent/water and "B stage" the resin to the intermediate phenolic prepreg. The prepreg can be tacky or dry depending on resin and treater conditions. Prepregs contain about 30–50 wt% resin and must be stored at low temperature due to the advancement of resole at room temperature.

Properties of resole resins used in the preparation of glass or carbon fiber prepregs are contained in Table 6.64 [12]. Manufacturers of phenolic resin prepregs are Hexcel, Cytec-Fiberite, Culver City Composites, and Stesalit.

Use of phenolic resin prepreg based on glass or carbon fiber as face sheets attached to honeycomb core is discussed in the honeycomb core section.

Table 6.64. Phenolic resins for prepregs [12]

Resin	A	B	C	D
Viscosity at 25 °C (mPa · s)	2000–3000	600–1000	1000–2000	10,000–20,000
Solids %	64–68	62–64	60–64	74–78
B-time at 130 °C (min)	11–15	6–8	8–12	7–11

Cargo Liner

Single or multi-ply phenolic glass prepregs can be combined with polyvinylfluoride (Tedlar) decorative film and cured into a monolithic composition which is installed as a liner in the cargo hold of commercial aircraft (Fig. 6.77). These cargo liners are based on modified phenolic resin and S-glass or aramid. These liners must possess a modest amount of impact or damage tolerance to minimize potential puncture when luggage is loaded and stored in the cargo area.

Testing of cargo liners [1] involves panels (ceiling and sidewall configuration) surviving a 5 min oil burner test with no burn through and a back side temperature of less than 204 °C.

The test consists of an 8 l/h oil burner impinging directly on the ceiling and indirectly on the sidewall panel. The burner is maintained at 927 °C with a 9 watts/cm^2 heat flux.

Besides flame containment, additional cargo liner characteristics include low weight, abrasion resistance, resistance to fastener pullout, ease of cleaning, and damage tolerance. The latter is not only important in service life but also for safety. Punctured or ruptured cargo liners are not very efficient flame barriers. Special impact tests are also mandated by aircraft manufacturers (Boeing, Airbus). Cargo liner manufacturers are M.G. Gill, Pioneer Plastics, Cytec-Fiberite, and Permali.

Ballistics

Ballistic components are relatively low resin matrix containing (less than 20% resin) composites which rely on the strength/modulus characteristics of the fiber for ballistic performance. High performance composites for ballistic applications are prepared by combining woven fabric of aramid, S-glass, or UHMW-PE with phenolic or modified phenolic resins. Usually, polyvinyl butyral is the

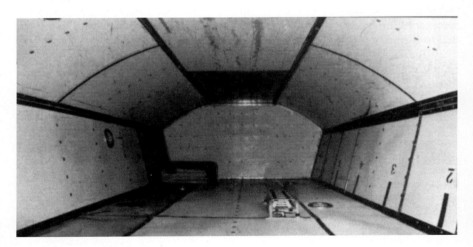

Fig. 6.77. Cargo liner (source: Permali)

Table 6.65. Phenolic resin for ballistic applications [12]

Viscosity at 25 °C (mPa·s)	3000–6000
Solids %	20–24
Shelf life at 20 °C, months	2

component in the modified phenolic resin. Key application areas consist of personal protection (helmets), land vehicles, military aircraft, and shipboard use.

Physical properties of phenolic resin for ballistics are presented in Table 6.65 [12].

The rôle of the phenolic resin in the ballistic component based on aramid or UHMWPE is primarily that of a marginal binder which easily fractures/delaminates on projectile impact and allows projectile energy to be absorbed by the high strength/modulus fiber. A different mechanism of ballistic performance occurs in S-glass/phenolic system and is related to composite compressive strength [1]. Manufacturers of ballistic components are Sioux Manufacturing, Devils Lake, ND, Owens Corning, Gentex, Carbondale, PA, Lewcott, Milbury, MA, Isola, Düren, Germany.

Carbon-Carbon Composites[1]

Phenolic prepregs consisting of carbon fibers and phenolic resin are thermally transformed and densified into carbon carbon composites (CC composites) see Sect. 6.4.3.2. The carbon fiber/phenolic prepreg is cured, carbonized (500–1000 °C) by heat/pressure, and densified by chemical vapor deposition (CVD) or chemical vapor infiltration (CVI). A key stage in the processing of these materials is the initial carbonization when the precursor is converted to a porous carbon fiber matrix system. The CC composite differs from a conventional composite due to the presence of two phases of carbon: carbon fiber reinforcement and a carbonaceous matrix. Carbon-carbon composites are used in high strength, low density, and high modulus applications where high temperature strength and high strength to weight ratio is important.

They are capable of being cycled from sub-zero to temperatures as high as 2200 °C, are lightweight (only half the weight of aluminum), and exhibit low creep at high temperatures. Temperature extremes are important for ablative use such as nose cone and rocket nozzles. On dispatch or re-entry of the Space Shuttle, the nose cone and leading edges of the Shuttle endure hot gases rushing through the nozzle/cone at high velocity, stressing and eroding nozzle walls. Carbon-carbon composites can resist high temperature erosion and abrasion without burning away.

Early evaluation of the frictional characteristics of CC composites by dynamometer measurements identified low wear rates, satisfactory frictional coefficients, and high dimensional and thermal stability. These observations

---

[1] Carbon carbon composites (CC composites) and CFC terms are used interchangeably whereby CFC signifies carbon fiber bound by carbon.

Fig. 6.78. Carbon carbon composite brakes

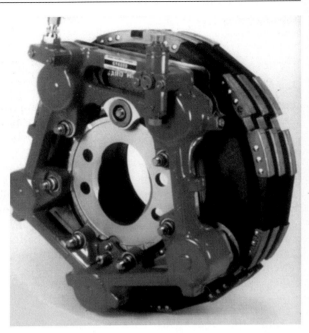

suggested wide use in aircraft and vehicle braking systems [13]. More than 80% of CC composites are used in commercial (Fig. 6.78) and military aircraft brakes. There are significant advantages associated with CC brakes. Although they may cost two to three times as much as comparably sized steel brakes, they weigh 40–50% less, remain in service longer, and handle aborted takeoffs better. Carbon carbon brakes for high speed trains and high performance automobiles, and possible use for transmission components are also under consideration. Carbon-carbon aircraft brake manufacturers are B.F. Goodrich, Allied Signal, Messier-Bugatti/SEP/Carbone, Dunlop Aerospace, and Aircraft Braking Systems Corp.

Honeycomb Core Sandwich Construction

Besides woven or unidirectional fabric, non-woven materials (including paper) can be impregnated with phenolic resin. The most attractive use of phenolic non-woven aramid fiber compositions is as honeycomb core (see also Sect. 6.1.4.5). The concept of sandwich construction evolved due to demands for greater weight reduction in military aircraft during the Second World War. It consisted of utilizing a high shear strength honeycomb core laminated with top and bottom high tensile/compressive strength face sheets into the proposed sandwich construction. Pioneering efforts by the Hexcel Corp. in the 1940s resulted in the commercialization of honeycomb core. A wide array of core materials is used, including paper, glass, aramid, carbon fiber, and aluminum. (Fig 6.79).

Expansion process

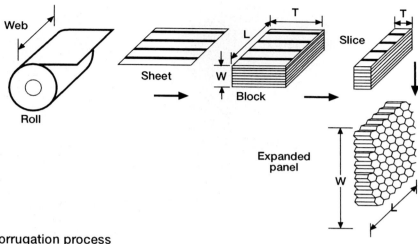

Corrugation process

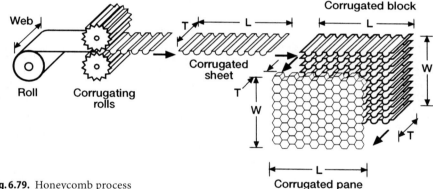

Fig. 6.79. Honeycomb process

The node adhesive and densification resin are mainly phenolic resins. Either solvent- or water-based phenolic resins are used in the manufacture of honeycomb [12]. The expansion process is the preferred process for honeycomb manufacture.

The printing of the node adhesive fixes the cell size of the honeycomb. After mechanical expansion, it is submerged into phenolic resin until the desired core density is obtained.

By utilizing epoxy/glass or epoxy/carbon fiber prepreg face sheets on aramid honeycomb, the sandwich construction provides the highest strength to weight and stiffness to weight ratios of any configuration [14]. Thus, the first applications for honeycomb composites (sandwiches) were for both fixed and rotary wing aircraft, followed rapidly by jet aircraft. Combining glass or car-

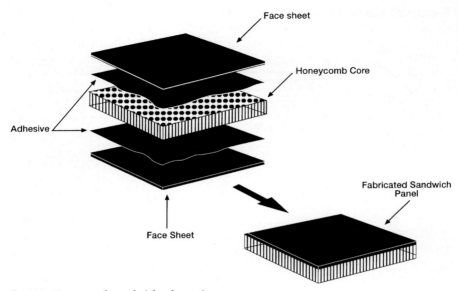

Fig. 6.80. Honeycomb sandwich schematic

bon fiber phenolic prepregs as face sheets onto honeycomb core results in sandwich construction (Fig. 6.80) for use in aircraft flooring panels, storage bins, ceilings, galleys, lavatories, and aircraft interior panels. Bonding of phenolic face sheets requires a film adhesive which was not required with an epoxy face sheet.

In the late 1980s, Bakelite AG developed a system that combined epoxy resin curing with phenolic resin condensation [15]. The objective was to combine the toughness of epoxies with the FST characteristics of phenolics into a co-cured system suitable for fiber composite structures with high strength to weight ratio and mechanical/thermal stability. The patented technology involved a single step combining epoxy prepreg with phenolic prepreg to the cured laminate. A judicious matching of polymerization rates and resin viscosity curves of the two different curing processes led to the co-cured composition [15]. Thus the toughness and the high peel strength of the epoxy segment combined with the favorable FST features of the phenolic component resulted in an attractive laminate facing for honeycomb panels. The novel co-cured system with high peel strength eliminated the use of a film adhesive. Since mid-1994 these co-cured panels have been installed as flooring panels in the Airbus 340.

Modification of phenolic microstructure for fiber reinforced composites can occur during resin manufacture or by prepreg manufacture. The controlled co-cure epoxy/phenolic conditions were extended by introducing other resin phases into the phenolic microstructure phase [4]. These studies led to a unique phenolic/elastomeric composition known as PLB (Bakelite AG designation). PLB is a recent discovery. The type and method of introduction of selected resin phases is important for enhanced improvement in mechanical properties of the resulting modified phenolic resin. Preliminary evaluation of the newly developed PLB resins has been conducted in selected aircraft interior applications. Laminate properties of PLB-14 such as flexural strength and interlaminar shear strength (Table 6.66, entries 1 and 2) compare quite favorably with similar properties of flame retardant epoxy resin and co-cured epoxy/phenolic systems. Peel strength of facings on aluminum honeycomb also compare favorably (Table 6.66, entry 3). PLB-14 exhibited 405 Newtons peel strength vs 400–410 Newtons for the epoxy systems.

The PLB-14 phenolic resin exhibits mechanical, shear, and peel strengths comparable to the other epoxy-containing systems. Yet there is no epoxy phase in PLB. Two different PLB phenolic resins (PLB-14 and PLB-15) with structurally different resin phases added to the phenolic phase were evaluated for flammability characteristics (Table 6.67). Resins PLB-14 and PLB-15 are examples of different resin phases introduced into phenolic microstructure under similar reaction conditions. The flammability results demonstrate that the resin phase of PLB-15 is superior to that of PLB-14, but both compositions are below the values of typical phenolic resin as well as the low flammability standards for aircraft interior use. Thus the PLB phenolic resin system exhibits similar mechanical strength characteristics as epoxy resins combined with outstanding FST features. These combined high performance mechanical and FST characteristics are expected to result in new and expanded opportunities for these newly developed PLB resins.

Table 6.66. Comparison of mechanical strength of PLB with epoxy containing resins

Properties	Epoxy resin Flame retardant	Epoxy/phenol resin Co-curing system	PLB phenolic Resin (PLB-14)
Flexural strength (MPa)	540	562	551
ILSS (MPa)	48	45	43
Peel strength (N)	400	410	405

Table 6.67. Flammability behavior of PLB phenolic resins

Phenolic resin systems	OSU[a]	NBS[b]	FAR[c]
Phenolic control	61/30	2.0	102
PLB–14	46.7/27.0	1.0	90
PLB–15	27.3/19.3	1.0	77

[a] Heat release KW.min/m^2; kw/m^2 MAXIMUM 65/65.
[b] Smoke density flaming mode at 4 min MAXIMUM 200.
[c] FAR 25.853 (a) burn length (mm) MAXIMUM 152.

Besides military and commercial aircraft applications, these honeycomb composites are used in marine, mass transportation, automotive, and sporting goods. Honeycomb manufacturers are Hexcel, EuroComposites, Plascore, and Aerocell.

6.1.5.6.2
Filament Winding

In the early 1960s the technique of filament winding was utilized to fabricate strong, lightweight pressure vessels for rocket motor cases and other high performance components for military and aerospace applications [16]. During the intervening years from the 1960s to the 1990s a significant shift from the military sector to the commercial area has occurred where now filament winding enjoys a dominant position in the manufacture of pressure vessels for compressed and liquefied natural gas containers, automotive drive shafts, couplings and bearings for mechanical power transmission, sports equipment such as golf shafts, tennis rackets, corrosion resistant tanks, ducts, and pipes [1]. The latter area is of particular interest in an emerging market for pipes/tube inserts for oilfield use as well as secondary and tertiary structures on existing and new offshore installations. These include sea water piping systems, gratings, handrails, ladders, and vessels and tanks where chemical resistance and F/S/T characteristics are desired [17].

Filament Winding (FW) (Figs. 6.81 and 6.82) involves the high speed precise laydown of continuous reinforcement in predescribed patterns of resin preimpregnated filament or tow on a rotating mandrel. Either wet winding or dry winding (partially cured matrix system) is carried out on the mandrel. Towpreg is used in dry winding. All fibers such as glass, aramid, and carbon fiber can be used in the FW operation – E-glass for economics and aramid or carbon fiber for special/advanced composites systems.

When the winding process is completed, the whole fixture is cured in an oven to yield the product with desired mechanical properties and structural integrity. The mandrel is removed from the cured part.

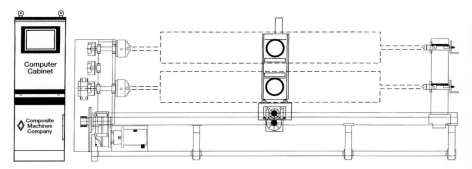

Fig. 6.81. Filament winding schematic

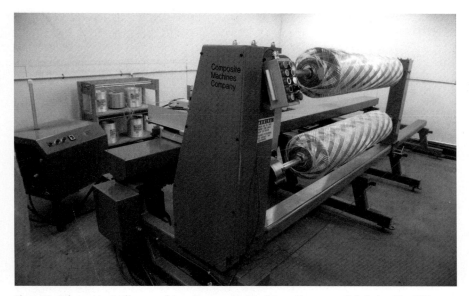

Fig. 6.82. Filament winding machine, Composite Machines Company, Salt Lake City, UT

Table 6.68. Filament winding phenolic resin [12]

Solids	72 ± 1.0
Viscosity (20 °C) (mPa · s)	400 ± 140
Free phenol %	< 4.5
Free formaldehyde %	< 0.6
Gel time (130 °C)	15 min

A recently developed low viscosity phenolic rein (Table 6.68) with low free phenol (< 4.5%) and low free formaldehyde can be formulated into novel resin systems which are readily transformed into attractive filament wound tubes [8, 9].

Curing conditions of filament wound phenolic systems require 1 h at temperatures below 90 °C followed by post cure (at 150–200 °C) for 1 h for high T_g values of 255–260 °C and significantly higher than standard epoxy or special high T_g multifunctional epoxy resins.

The use of composite materials has been steadily increasing in the offshore oil and gas industry (Fig. 6.83). The use of filament wound sea water piping systems and related vessels and tanks is expected to grow substantially due to the fire-worthiness of the phenolic FW systems. Filament wound pipe with phenolic/glass composition is being developed by Rice Engineering, Ameron, and Hunting (UK).

The recent disclosure by Folker and Friedrich of Ameron [5, 6] of high performance filament wound pipe based on a polysiloxane modified phenolic system describes composite pipe with improved impact resistance, higher hoop stress values, and improved weathering resistance (Fig. 6.84).

Fig. 6.83. Composite pipe, oil and gas industry

Polysiloxane modified phenolic resin system is filament wound into fire resistant pipe for delivering water to fight fires aboard offshore oil platforms [18]. These systems will replace steel pipes in wet and dry deluge systems on offshore oil platforms and on board ships that are subject to intense heat. Key performance features are low smoke emissions, low toxicity, and high heat resistance. One filament wound pipe composition meets International Maritime Organization (IMO) Level 3 test which requires the piping system be filled with water and subjected to heat flux of 113 kw/m² and temperature in excess of 980 °C for 30 min. Another pipe product meets jet fire test requirements according to UK Offshore Operators Association (UKOOA) guidelines for offshore use. The jet fire test replicates blowout conditions on an offshore oil platform by exposure of pipe to propane jet flame at about 1000 °C and heat flux above 300 kw/m². For jet fire certification the entire pipe composition (pipe, joints, adhesives, etc.) must withstand 5 min of exposure in dry state followed by 15 min of exposure filled with flowing water at 10 bar pressure. The resulting all-composite piping system is lighter, non-corrosive, and more temperature resistant than stainless steel, and exhibits less heat transfer throughout the pipe wall.

Fig. 6.84. Modified phenolic system for offshore pipe

6.1.5.6.3
Pultrusion

Pultrusion is a method that produces continuous sections of unidirectionally reinforced composites. Continuous fibers are "pulled" through a resin bath and then through a series of wiper rings to remove excess resin. Once in the die, material is heated, resulting in gelation and cure of the resin. Upon exiting the die, the composite is sufficiently rigid and strong to be pulled mechanically to a cutter which cuts the part to the prescribed length (Fig. 6.85).

Pultrusion is the fastest growing segment of the fiber reinforced composites industry and provides primarily thick walled structural components for marine, power transmission, civil engineering, and high-rise construction

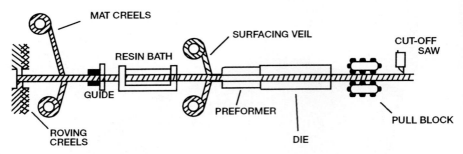

Fig. 6.85. Pultrusion process schematic

projects. Pultrusion technology has made rapid progress that not only makes composites price competitive with traditional materials like metal and concrete but also actually makes them less expensive to manufacture and assemble [19].

Worldwide volume of pultruded composites was over 150,000 tonnes in 1996. Pultrusion is becoming more sophisticated whereby parts are classified as commodity (tool handles, ladders, etc.), structural composite (civil engineering structures, structural beams, etc.), and even advanced composites (carbon fiber components for aircraft, C-C composites, and others). Most pultruded parts consist of polyester/glass with some vinyl ester/glass and small amounts of epoxy/glass (Fig. 6.86).

Strongwell (Bristol, VA) commercialized a glass/phenolic deck grating composition by pultrusion [3] (Fig. 6.87). It is durable, load bearing fiberglass grating for offshore oil and gas drilling/production platforms. The phenolic/glass pultruded product is 33% the weight of steel and exhibits enhanced

Fig. 6.86. Pultrusion operation, Strongwell Bristol, Va.

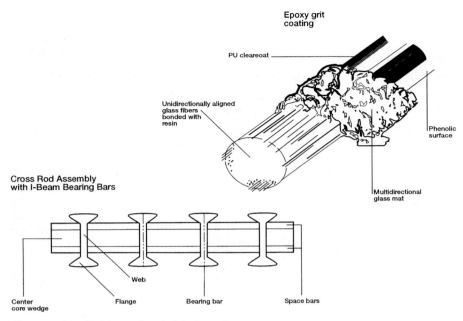

Fig. 6.87. Phenolic/glass pultruded deck grating

flame resistance, corrosion resistance, and safety properties. The Strongwell product was subjected to a 60 min fire test at temperatures exceeding 930 °C. Gratings with 120 cm span retained their structural integrity after 60 min exposure to these temperatures. The integrity of the grating is maintained without failure even after applying a 91 kg static load to the cooled grating. Measured thermal conductivity was extremely low with a 12% heat conducting rate. These favorable property characteristics suggest that the phenolic/glass composition can be used as walkways, platforms, and decking for tunnels/mass transit, aircraft, industrial/processing plants, public buildings, automotive, and marine areas.

A phenolic resin system which is "environmentally and user friendly" for pultrusion was recently reported [20]. Physical properties of the resin, PL 2499, are contained in Table 6.69. PL 2499 is low viscosity, water-based resole resin with an unusual combination of low free formaldehyde and low free phenol (< 0.7%). PL 2499 accepts a high amount of inert filler (> 20%) with a reasonably low viscosity. High filler loading does not effect the cure rate.

Table 6.69. Properties of PL 2499, Bakelite AG, Iserlohn, Germany

Solids %	75 ± 2
Viscosity (20 °C)	2000 ± 200 (mPa · s)
Free phenol %	< 0.7
Free formaldehyde %	< 0.7
Gel time (130 °C)	8.5 min

Fig. 6.88. Bridge at Bar Harbor, ME with FRP/glulam

Conducting pultrusion at temperatures of 135 °C (T_1) and 165 °C (T_2), smooth surface profiles resulted through the use of neutral catalyst with up to 10–15% filler (clay, zinc borate) being used. High strength, high T_g (> 249 °C) profiles were obtained without post cure. No F or P emissions were noticed during pultrusion.

An interesting application of phenolic pultruded profiles is the use of the profile to reinforce wood (see Wood Composites section 6.1.1). Glue laminated timber beams known as glulams are widely used for large open structures such as arches, domes, and bridges. Glulams offer many advantages over other materials such as lightweight, economical production of tapered and curved members, excellent energy absorption characteristics (seismic, damping, and acoustical responses), good chemical and corrosion resistance, better fire resistance than steel, and an attractive appearance. However, low bending stiffness and strength of glulams requires the use of glulam with large depths (1.2–1.5 m). To increase the stiffness and strength of glulam dramatically, it can be reinforced with pultruded FRP.

The pioneering work of Tingley [21] led to the development of technology known as FiRP™ glulam. It consists of pultruded material which is integrated into the glulam and allows the wood/FRP composite beam to be customized for tension and compressive strength. The wood/FRP composite is 40% lighter and said to be 25% less expensive than conventional glulam beams. FRP composites improve beam strength by as much as 85%. The FiRP glulam is generally reinforced by carbon fiber (compression strength) and aramid (tension strength) hybrids for bridge applications.

A consortium located at the University of Maine under the leadership of Professor H. Dagher and consisting of the Composites Institute of SPI, Forest Products Laboratories of US Dept. of Agriculture, and other research groups with funding from several government agencies has been established with the objective of combining glulams of lesser quality wood with pultruded glass/phenolic profile [22]. The glass reinforcement is 2 vol.% and provides an increase of 50–60% over unreinforced glulam. A 38 m long by 1.5 m wide ocean pier located at Bar Harbor, ME. (Fig. 6.88) was constructed with Maine red maple glulams reinforced with E-glass/phenolic pultruded system provided by Strongwell (Bristol, VA). The bridge was installed in 1995. Bridge properties are periodically monitored by Dagher's group and continue to exhibit favorable properties.

It was mentioned in the Wood Composites chapter that a combined pultrusion/microwave process transforms wood strands with phenolic resin adhesive into parallel strand lumber or parallam.

Thus pultrusion of phenolic FRP and wood products leads to unique reinforced materials and new market opportunities.

6.1.5.6.4
Sheet Molding Compound

Sheet molding compound (SMC) is a material based on a filled thermosetting resin (usually polyester) and a chopped or continuous strand reinforcement of glass fiber (Fig. 6.89). The SMC is formed between two pieces of carrier film. The filled polyester with catalyst, thickening agents, mold release agents, and low profile additive (LPA) is placed on the continuous carrier film. The carrier film passes under a chopper which cuts glass roving into 25.4 mm lengths.

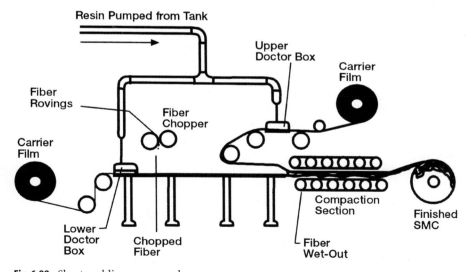

Fig. 6.89. Sheet molding compound process

Fig. 6.90. IBM think pad with fiberite SMC

After the chopped glass falls on the filled resin systems, another carrier film with a layer of filled resin combines with other containing the glass, thus enclosing or "sandwiching" the glass between the two layers. A typical SMC composition is 20–30% glass reinforcement, 40–50% filler ($CaCO_3$), the remainder being resin with thickeners, release agents, low profile additive (LPA), small amounts of catalysts, pigment, and flame retardant/UV absorbers. The biggest volume applications of polyester SMC are automotive, appliances, mass transportation, and marine. Many types of SMC compositions based on epoxy, cyanate ester, PMR-15, and phenolic have been reported [23]. Phenolic SMC is proposed as a suitable material for aircraft interiors and mass transit areas where F/S/T properties are important and expected to displace the time consuming, hand lay up materials.

Recently, a chopped carbon fiber/phenolic SMC product manufactured by Fiberite and known as Enduron 4685 was used to fabricate the base and cover of the IBM Think Pad 701 (Fig. 6.90) [24]. Key features of the CF/phenolic SMC product include excellent EMI shielding, thermal resistance (since newer computers generate more heat), and a very low shrinkage value.

6.1.5.6.5
Resin Transfer Molding

Resin transfer molding (RTM) is a "net shape" process to fabricate a fiber reinforced composite by introducing a reactive liquid resin into a woven or knitted preform contained in a closed mold. Resin is introduced via external pressure or vacuum. The resin cures at room temperature or elevated temperature by applying heat to the mold and ensuring complete cure. The advantages of RTM include low cost of FRP part fabrication because only dry fiber and resin are required, possible automation, near net shape parts, possibly low void content, and probably the only process to fabricate complex parts in one step. Disadvantages are expensive molds and questionable process control. BP [25] has developed low viscosity phenolic resin systems with delayed action catalysts for RTM applications.

6.1.5.6.6
Hand Lay Up

This method is used extensively in the marine industry for the fabrication of a relatively small number of large components. The technique involves the use of a female mold onto which a gel coat is applied. The desired structure is built up using sheets of glass fiber plies or mat followed by the application of resin with a brush or similar applicator. The method is relatively inexpensive, involving low equipment and tooling costs. Fiber weight per cent can vary from 25 to 45%. Most of the technology for the use of phenolic resins in the hand lay up process was developed by BP (UK). The facile cure of resole resins with sulfonic acid esters and more recently with cyclic phosphate esters [26] which become incorporated within the crosslinked structure represents BP technology. Hand lay up uses include railway rolling stock, cladding in buildings, underground stations, ducts in mines, and automotive heat shields [27].

6.1.5.7
Summary/Trends

Significant growth of phenolic FRP systems has occurred within the last decade due to government mandated changes, especially for aircraft interiors and some segments of mass transportation. It is anticipated that greater penetration of phenolic FRP will develop in many sectors of mass transportation such as buses, trains, ferries, and thus reduce fire threat to passengers. Similarly, the civil engineering/infrastructures and the offshore/oilfield market areas are expected to provide additional growth opportunities for phenolic FRP.

Obviously, continued improvements in phenolic resin technology, fiber enhancement through surface, interfacial treatment, and improved composite processing are "evergreen."

The future outlook in phenolic resin chemistry is quite optimistic and is supported by recent advancements in reduced resin brittleness (PLB from Bakelite AG and Ameron resins), water borne resole resins with low free phenol and formaldehyde, high T_g filament wound system (260 °C), and exceptional fire resistance of polysiloxane phenolic filament wound pipe, as well as pultruded components for offshore platforms, attractive SMC compositions (Fiberite), and a new emerging application area which uses glass/phenolic pultruded profiles to reinforce wood glulams.

6.1.6
Miscellaneous

6.1.6.1
Chemically Resistant Putties and Chemical Equipment Construction

Application of acid and heat curing binders for grouts, industrial coatings, and flooring, and in chemical equipment construction may be classified into three groups: (1) chemically resistant putties, (2) laminate coatings, and (3) solid moldings. All are produced using phenolic and/or furan resins.

Chemically resistant putties are pasty compounds that are obtained by blending the components, and are used to lay and grout acid-resistant bricks and tiles [1, 2]. Depending on the stresses to which the grouts are exposed, phenolics, furan resins, unsaturated polyester resins, or epoxies are used as binders. Only phenolic and furan resins are considered in the following discussion.

These phenolic and furan resin putties represent multicomponent systems that cure at ambient temperature. They consist of a liquid phenolic or furan resin, for example resin A or B in Table 6.70, a filler mix (putty powder), and a curing agent. The curing component can be used in liquid, paste, or powder form. In most cases, curing agents are used in a solid form that can be premixed with the filler mix to afford the "putty powder." A putty powder generally contains about 5% added curing agent. The most common curing agents are *p*-toluenesulfonic acid and sulfonic acid derivatives such as sulfamic acid or toluenesulfonyl chloride. The most common fillers are silica sand, quartz flour, barytes, pitch coke, graphite, and various fibrous materials. Before use, the putty powder and resin are blended in a high-shear mixer at a ratio of about 2:1 to 5:1 to yield a homogeneous paste of the desired con-

Table 6.70. Resin for acid-resistant and chemical equipment construction

Resin	Chemical type	Non volatiles %	Viscosity at 20 °C mPa·s ISO 12058	Density at 20 °C g/cm^3	Water %	Storage life at 20 °C months	Properties	Applications
A	Modified phenolic resole	~95	300±50	1.2	3±1	3	High acid resistance, good resistance to solvents and superior alkali resistance	Cements and laminates
B	Furan resin	~99	230±20	1.2	max. 0.5	6	High solvent, alkali and thermal resistance	Cements and laminates
C	Phenolic resole	81±2	8500±500	1.22	11±2	2	Special-purpose resin	Ramming (Haveg) mixes

Fig. 6.91. Solvent tank farm with chemically resistant tile floor

sistency. When a putty powder concentrate is used, it should be well distributed throughout the supplementary fillers before resin is added. Use of low resin viscosity material allows the production of highly filled putties.

Phenolic resin putties are particularly suitable for applications requiring resistance to strong acids and nearly all organic materials (refer to Comparisons of Thermosets chapter 5, Fig 5.7, p. 119). Even satisfactory alkali resistance for some applications can be achieved by use of modified phenolics. However, furan resin putties feature high solvent, alkali, and thermal resistance. Application examples are for use in floors (Fig. 6.91) and walls of rooms exposed to chemicals, effluent basins, catch basins, and drain gutters [3–5].

Industrial concrete and screened flooring exposed to severe chemical stress is also protected by *laminated coatings* consisting of a fibrous material, a cold curing two-component system of resin and curing agent (generally an acid chloride), and possibly further additives. Due to their good resistance to chemicals, thermoset binders including phenolic and furan resins are used in laminated coatings for surface protection. In this technique, a flexible intermediate coat is first applied to a primed substrate, and glass fiber mats or fabric then embedded in the curing agent containing phenolic or furan resin, for example, resin A or B in Table 6.70. The areas of application are similarly vessels, catch basins, effluent basins, and drain gutters. If enhanced mechanical and thermal resistance is required, the coatings are protected by layers of tiles. Glass fiber mats or fabric are used as the fibrous materials. The primed substrate or flexible intermediate layer is coated with the homogenized mix-

Fig. 6.92. Fiber-reinforced phenolic materials, in this case used for a dome consisting of 12 segments and measuring about 8000 mm in diameter for a storage tank to hold waste acid containing fractions of chlorinated hydrocarbons (Keramchemie, Siershahn, Germany)

ture of resin and curing agent, the fibrous material being embedded with rollers and then recoated with the resin/curing agent mixture. This operation may be repeated several times.

Thermoset plastics are also used in structural materials exhibiting a high level of chemical resistance for use in *chemical equipment construction* [6–8]. Areas of application are in solid moldings, composite structures, coatings, and linings with prefabricated laminate sheets. The phenolic resins used for this purpose, such as resin C in Table 6.70, are cured under heat and pressure with or without accelerators, or at ambient temperature using appropriate curing agents. Examples of applications (Fig. 6.92) are vessels, columns, and hoods made with fiber-reinforced compounds [9], and protective coatings and linings of steel vessels and parts. To produce solid moldings, the resins are blended with the fibrous fillers in a masticator to yield a homogeneous mix. The plastic, moldable mix is hammered ("rammed") or pressed into molds. It is cured in autoclaves ("Bakelisators") at temperatures up to 140 °C and pressures up to 10 bar. Addition of a curing agent is necessary when furan resins are used.

Prefabricated flexible sheets can be used to line steel vessels. Phenolic resins with graphite fillers have become established for this use. As in the case of ramming mixes, curing is carried out at elevated temperature and pressure.

6.1.6.2
Lampbase Cements ("Socket Putties")

"Socket putties" are curable, thermosetting putties/adhesive compounds used to bond the metal base to the glass body of incandescent lamps and fluorescent tubes when they are manufactured [1–3]. Bonding is carried out during fabrication of the lamp and is largely automatic. The bond must be stable over the entire service life of the lamp, and be capable of withstanding exposure to high temperatures over long periods of time.

Socket putties generally represent mixtures of phenolics and thermoplastic natural (soft) resins used as a component to provide elasticity. A standard formulation exhibits the following composition:

- 80 parts by weight mineral filler (e.g., marble flour)
- 12 parts by weight phenolic powder resin
- 8 parts by weight balsamic resin

The powdered components are homogenized and the powder mix made up to the desired pasty consistency prior to use. Technical-grade alcohol (10–15%) is generally used to form the paste. A mixture of acetone and alcohol can also be used to provide particularly good flow. Diglycol can also be used to avoid rapid drying of the putty. The paste is then injected into the base by machine, and the glass body automatically inserted. The putty is cured by exposing it to temperatures of 200–280 °C for 10–50 s.

The automatic machines generally apply the putty to the metal base in the form of a ring. About 1 g of putty is required for a standard incandescent lamp with a base 25–30 mm in diameter, and $2 \times 1-2.5$ g for a fluorescent tube. The putty-coated bases are fed to an automatic lamp making machine where the glass bulb or the tube is inserted and the putty cured. The main part of the production process is ended by soldering on the current carrying wires. Modern equipment produces more than 4000 lamps or 1500 tubes per hour. The socket putty should possess a suitable pot life for the longest possible time and exhibit high adhesion of metal to glass. The putty is foamed by slight formation of bubbles during the curing process, counteracting any shrinkage and helping to ensure that the void between the base and the glass is well sealed. The cured material should be free of stress. The required strength level, for example, in an unused incandescent lamp is set at a twist-off strength of 3.0 Nm; it should still amount to 2.5 Nm following storage for a period of 1500 h at 210 °C.

The majority of binders used are still phenolic resins (Table 6.71). Phenolic powder resins (generally combined with hexamethylene tetramine = HMTA as a curing agent) have been used as binders for lampbase putties for decades. They can include curing accelerators and natural resins as additives. Special, highly reactive phenolic novolak such as resin B in Table 6.71 are used to achieve particularly short cycle times. Such formulations contain a small amount of slow reacting soft resins. Modification with amino resins affords putties that cure to light colors.

It is useful to include additives such as silicone resins to improve the heat resistance of the phenolics when resistance to high thermal stress is required,

Table 6.71. Phenolic resins for production of lampbase cement

Resin Chemical characterization		Mixture of phenolic novolak and hexamethylene tetramine			Mixtures of phenolics modified
		A	B	C	D
Melting range (ISO 3146)	°C	95±5	90±5	95±5	–
Flow distance at 125 °C (ISO 8619)	mm	35±5	55±10	50±10	–
HMTA content (ISO 8988)	%	9.0±0.5	6.5±0.5	6.0±0.3	–
Screen analysis, fraction >0.09 mm (ISO 8620)	%	max. 2	max. 2	45±10	4±2
Bulk density (ISO 60)	g/l	360±30	360±30	490±30	750±50
Soluble in		Methanol, ethanol, isopropanol, acetone and similar			–
Application		Standard resin	Enhanced reactivity	Modified resin with high reactivity and thermal resistance	Special cement, rapid curing; particularly suitable for pilot lamps and similar

for example, in oven lamps, high-wattage lamps, projection lamps, and similar. In such cases, it is also advisable to substitute a Xylok® resin (Fig. 6.93) for the phenolic [4]. Xylok resins are the reaction products of alkyl ethers with phenol (see Chemistry, Reactions Mechanism chapter 2). They also undergo a thermosetting cure with HMTA.

Marble flour is used as the standard filler for putties; titanium dioxide is used to obtain lighter colors. Rosin, balsamic resin, shellac, or the transformation products of these natural resins are used as additives to achieve longer flow of the paste prepared from the powdered putty, and to improve the elasticity. Aside from the dispersion solvents mentioned above, trioxane can also be used. A putty prepared with this reactive diluent solidifies to form a solid mass upon cooling. The bases lined with such putties can be stored and used as required for production of lamps and tubes. However, certain industrial safety precautions such as ventilation of the workplace must be observed in working with trioxane, and the putty must be preheated to a temperature of 60–70 °C.

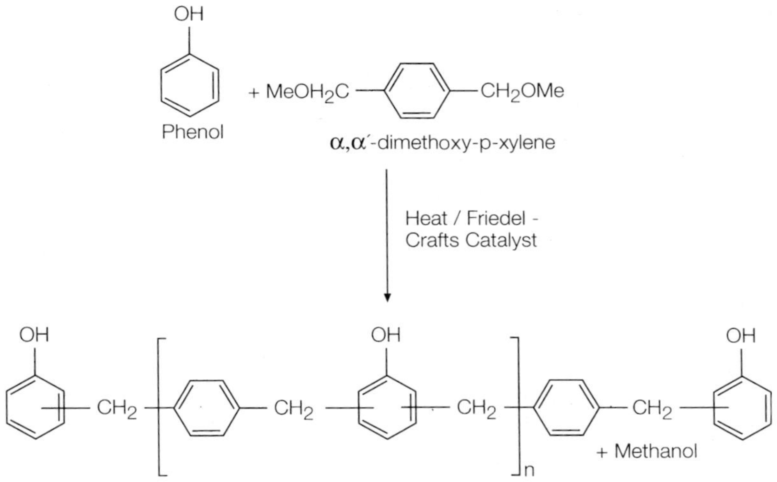

Fig. 6.93. Basic reaction in production of Xylok®

6.1.6.3
Various Applications (Brush Cements, Casting Resins, and Concrete Flow Promoters)

To complete the versatility of phenolic resin applications, three further possible uses are discussed: *brush cements, casting resins,* and *concrete flow promoters*.

Brush cements, as resins or pastes used to manufacture artists' and painters' brushes, were earlier produced using phenolic resins as the main binders. The brushes produced by this method featured excellent resistance to common chemicals and solvents [1, 2]. The bristles were bundled and coated with resin on one of the ends. The binder was cured in an oven heated to 60–70 °C, or at low temperatures following addition of acids. A disadvantage in the use of phenolic resins was occasionally the dark color or the lack of adhesion to metal when metal bases were used. For these reasons, epoxy resins are preferred binders; although more expensive, they do not exhibit the disadvantages mentioned above.

Phenolic-based *casting resins* [3] are mainly resins systems that are capable of being poured and can be cured without formation of bubbles. Depending on the application, such casting resins are produced at a molar ratio of phenol to formaldehyde of one to 1.5–3.0 using alkali metal or alkaline earth hydroxides as catalysts. The resins can be colored. This is accomplished using alcohol or phenol-soluble dyes alone or in combination with inorganic pigments. Addition of glycols to regulate the viscosity and increase the flexibility is also common. Curing of the resins following casting in molds requires more than 70–200 h at temperatures of 75–95 °C. Addition of colorless organic acids also permits curing at ambient temperature, after which the resin is post-cured at 40–50 °C.

The cured resins can be worked by machining. The final products are transparent or opaque/colored, and are used as bowling or billiard balls, or for production of costume jewelry and similar products. Special versions can also be used in pattern and mold construction.

Phenolic resin billiard balls have the advantage that they are cheaper than those made of ivory, and that they do not wobble since their center of gravity is in the center [4]. Furthermore, ivory is no longer available due to a general embargo. Synthetic resin bowling balls have replaced the earlier common wooden balls. They are much improved, their behavior more predictable, and they can be guided toward the pins when thrown, whereas balls of other materials can exhibit quite "erratic" behavior.

Concrete flow promoters are additives that considerably improve the flow of concrete when it is used (poured) and reduce the required amount of water (ratio of water to cement). Naphthalene sulfonates, melamine resin sulfonates, and, as an extender, lignin sulfonates, are used for this purpose. From the patent literature, it is known that sulfonated phenolic resins alone or in combination with the above standard products can also be used for this application [5, 6].

6.1.6.4
Phenolic Resin Fibers

Phenolic resin fibers are marketed under the trade name Kynol™, and are used alone or combined with synthetic and/or natural fibers for fire protective clothing, heat-resistant fabrics, and similar products. As asbestos has been eliminated, these fibers have become increasingly popular for applications such as fire protective clothing, particularly since they are toxicologically innocuous. They are also used for production of composites or as a starting material for fabrication of carbon fibers and carbon fiber composites materials, and particularly feature the high chemical resistance and low smoke density in case of fire, typical of phenolics (Fig. 6.94). Production of the fiber [1] is conducted by spinning of high molecular mass novolaks ("Novoloid" fibers) and subsequent curing in an aqueous formaldehyde bath, technology from the basic patents owned by the Carborundum Company, Niagara Falls, N.Y., USA [2, 3]. As a thermoset fiber and typical phenolic resin product, this material is presently marketed worldwide by Nippon Kynol, Inc. of Osaka, Japan.

Typical properties of Kynol fibers are listed in Table 6.72. Properties vary somewhat with the diameter of the fiber, and also depend to some extent on the production conditions required for special applications [4–11].

One measure of the flame resistance of a given material is its "limiting oxygen index" or LOI, which in essence states the concentration of oxygen required in the local atmosphere for continuous self-supporting combustion. The LOI of Kynol materials varies with the particular structure (fiber, felt, fabric) being tested, the test method, and apparatus, but generally ranges from 30 to 34, i.e., higher than in any of the natural organic textile fibers and in all but the most exotic man-made organic fibers.

Fig. 6.94. Smoke density of Kynol™ fibers compared to other synthetic and natural fibers (source: Kynol™ GmbH, Hamburg)

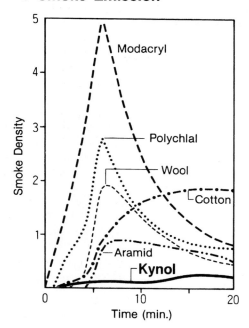

Table 6.72. Properties of Kynol™ fibers (source: Kynol™ GmbH, Hamburg)

Properties of Kynol fibers	
Color	Gold
Diameter, μ	14–33 (2~10 denier)
Fiber length, mm	1~100
Specific gravity	1.27
Tensile strength, g/d	1.3–1.8 (12~16 cN/tex)
Elongation, %	30~60
Modulus, kg/mm²	350~450 (260~350 cN/tex)
Loop strength, g/d	2.2~3.1 (19~27 cN/tex)
Knot strength, g/d	1.1~1.5 (10~13 cN/tex)
Elastic recovery, %	92~96
Moisture regain at 20 °T, 65% rh, %	6

The fibers (Fig. 6.95) and materials are highly flame resistant; in addition, they are excellent thermal insulators. However, they are not high-temperature materials in the usual sense of the term. While a 290 g/m² woven fabric will withstand an oxyacetylene flame at 250 °C for 12 s or more without breakthrough, the practical temperature limits for long-term applications are 150 °C in air and 200–250 °C in the absence of oxygen.

Fig. 6.95. Selection of various phenolic resin fiber products (photo: Kynol GmbH, Hamburg)

The fibers display excellent resistance to most chemicals and solvents. They are attacked by concentrated or hot sulfuric and nitric acids and strong bases, but are virtually unaffected by non-oxidizing acids including hydrofluoric and phosphoric acids, by dilute bases, and by organic solvents.

The novolak-based fibers are excellent thermal insulators. A nonwoven batting with a density of 0.01 g/cm^3 exhibits thermal conductivity of 0.034 kcal/m · h °C at ordinary temperatures and 0.025 kcal/m · h °C at −40 °C. The retention of properties after exposure to extremely low temperatures is excellent. The efficiency of sound absorption is high. Ultraviolet radiation, although leading to darkening of the fiber, has minimal effects on the properties, and resistance to gamma radiation is also high.

Typical recommended applications are:

1. Protective apparel (e.g., firemen's protective clothing)
2. Safety accessories such as sewn and knitted gloves, with and without extra insulation, aprons, hoods, socks, sleeves, spats, and masks
3. Flame barriers and liners for upholstered furniture
4. Miscellaneous consumer goods such as emergency bags and containers, or fire protective drop fabrics and fabrics (Fig. 6.96)

Fig. 6.96. Fire resistant clothing-based on Kynol™ fibers (photo: Kynol™ GmbH, Hamburg)

6.2
Temporary Bonding

Temporary bonding ideally provides materials with low wear coupled with high temperature resistance over an extended period of time. Pertinent applications include friction (brake and clutch) linings, abrasives (coated abrasives, grinding wheels and shapes), graphite bearings, and foundry molding materials for molds and cores.

One of the common denominators for such applications is the demand that products subject to wear and tear, such as friction linings and abrasives, be capable of reliable and lengthy use. This refers to effects such as those that arise when a friction material contacts the brake disk, brake drum, or clutch plate. In another case – in the process of grinding – the reference is to the abrasive effect that occurs when the grinding tool and the substrate come into contact. The phenomenon in both processes leads to development of relatively high temperature or even thermal shock stresses. Friction linings are thus tested with a dynamometer or in driving tests using a schedule of braking cycles or driving conditions. The products are expected and required to exhibit a very lengthy service life in spite of gradual wear.

Foundry binders must temporarily meet an extreme range of requirements with tight tolerances. The molding materials, based on silica sand or other types of sand such as zircon, chromite, or olivine, and held together with only

0.8 % to a maximum 3.0 % binder, must withstand the hot molten metal for an adequate period of time (until the melt solidifies) when the mold is poured off, and should collapse as quickly and easily as possible after the metal has solidified to keep shakeout costs low. In this case, "temporary" refers to the strength of the bond directly at the time of pouroff, when the contours and dimensional tolerances of the castings are maintained.

6.2.1
Foundry

6.2.1.1
Introduction (Economic and Technical Survey)

Shaped metal parts (Fig. 6.97) of iron, steel, and nonferrous metals (bronze, brass, aluminum, magnesium) may be produced by various methods. The most important method is by casting in which the final product is directly produced from the melt as compared to forming, staking, and sintering methods.

The annual worldwide casting production in major developed countries, for which excellent statistical information is available, presently amounts to about 71 million tonnes (Table 6.73). About 75 % of this production originates in the countries mentioned in Table 6.73. The list is led by the USA, People's Republic of China (PR China), and the Commonwealth of Independent States (CIS, the former USSR). Germany presently produces about 4 – 4.4 million tonnes of castings annually, and thus leads the list of West European countries.

Fig. 6.97. Gray iron engine blocks (source: Eisengiesserei Fritz Winter, Stadtallendorf, Germany)

Table 6.73. Castings production in various countries and worldwide in thousands of tonnes (source: Giesserei p. 44, 1997)

	Fe	Al	Nonferrous	Total	Percentage of total
USA	12,000	1650	750	14,400	20.4
PR China	10,650	550	150	11,350	16.1
CIS	9900	200	800	10,900	15.5
Japan	5700	1100	250	7050	10.0
Germany	3600	420	180	4180	6.9
India	3040	20	–	3060	4.3
Brazil	1500	100	45	1645	2.3
World	62,000	5750	2800	70,550	100

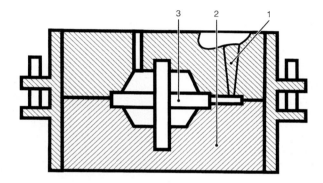

Fig. 6.98. Corebox system (1 = cope, 2 = drag, 3 = core) (from [1])

The column headed "Fe" includes gray iron and ductile iron, malleable iron and steel. The "World" annual total of 62 million tonnes in this column is composed of 65% gray iron, 21% ductile iron, 10.8% steel, and 3.2% malleable iron. Among the "Miscellaneous Nonferrous Metals", zinc alloys dominate with a share of 58% and copper alloys make up 36.5% of the total.

Casting is performed using disposable or reusable (permanent) molds. The disposable or "lost" mold used to produce most castings – particularly those made of iron – is destroyed during the casting process. The finished mold [1] may consist of one or more parts (mold sections), such as the cope, drag and the cores (Fig. 6.98). Molds and cores are generally produced from similar molding materials. In most cases, the main component of these is quartz sand (Fig. 6.99) shaped with inorganic or organic binders (Fig. 6.100). Molds are produced on molding lines, and cores using "core shooters" or "blowers" (Fig. 6.101).

The most important raw material for molds is quartz sand [2, 3]. Quartz (α-modification) is the most common modification of silica (SiO_2). When heated, α-quartz undergoes a transition to the "β-quartz" modification at temperatures above 573 °C. This reversible transition is accompanied by expansion of the crystalline lattice leading to a drop in the density from 2.65 to 2.50 g/cm³. Due to this quartz transition, "heat cracks" leading to casting de-

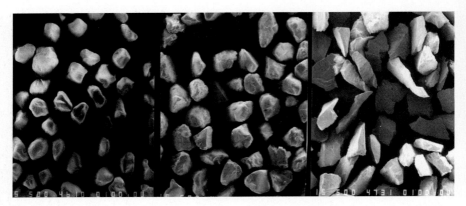

Fig. 6.99. Grain structures of quartz sand, ×50, from *left to right*: round, rounded edges, splintered (photo: Quartzwerke, Frechen)

Fig. 6.100. Scanning electron micrograph of a quartz-resin "bridge" between quartz grains, illustrating the importance of the binder (photo: RWTH, Rheinisch-Westfälische Technische Hochschule, Foundry Institute)

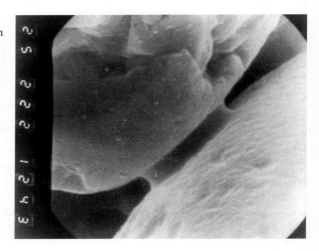

fects such as "veining" may arise when the resin-bonded molds and cores are poured off, depending on the degree of compaction of the molding material, the weight and temperature of the melt, and the crosslinking density of the cured binder.

Aside from quartz sand, chromite, olivine, or zircon sands may be used in special cases. These types of sands can be used to prevent expansion defects in molding materials exposed to particularly stringent thermal stresses. Special binder systems are generally used for specific molding processes. Molds are mainly produced using inorganic binders such as clay, in a few remaining cases cement or waterglass, and in around 2–3% of all applications organic binders such as phenolic and furan resins. Cores are prepared using molding processes involving organic binders, prominent members of which are core-making grades of phenolic resins.

Fig. 6.101. Core shooter, encapsulated to protect against emissions and permit use of gas curing processes requiring sealed systems; production of crankcase cores (photo: Loramendi, Vittoria, Spain)

In many cases, only the advances [4] which have been achieved in molding technology by use of new core/mold binders and production methods have resulted in the efficient and economical production employing a high degree of mechanization possible in modern foundries. This is illustrated by the rise in productivity at German foundries during the period from 1950–2000 (Fig. 6.102). The production per employee nearly quadrupled over these 50 years. One of the driving forces for this increase has been the meteoric rise in capacity of the automotive industry.

In an assessment of coremaking processes, the demands of which have reached a high level in recent years for ecological and economical reasons,

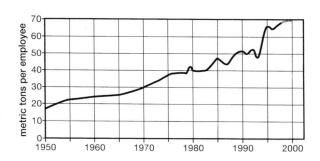

Fig. 6.102. Productivity increase in the German foundry industry from 1950 to 1997, in tonnes of production per employee (source: Verein Deutscher Giessereifachleute, Düsseldorf)

Table 6.74. History of core and mold processes using synthetic resins as binders

Year	Process	Resin
1950	Use of shell molding	Phenol novolaks
1958	Furan no-bake	Furan resins, urea resins
1962	Hot-box	Urea, furan and phenolic resins
1965	Phenolic no-bake	Phenol resols
1968	PUR-cold-box[a]	Ortho-cond. resols
1971	Furan SO_2[a]	Condensed furan resins
1978	Warm-box	Modified furan resins
1981	Warm-box-vaccum	Furan or phenol resols
1982	Ester curing no-bake	KOH-containing resols
1983	MF-cold-box[a]	KOH-containing resols
1983	Cold-box plus[a]	Ortho-cond. resols
1984	Epoxy-SO_2[a]	Modified epoxy resins
1987	Hot-box plus	Modified phenol resols
1988	Acetal (methylal)[a]	Resorcinol resins
1990	CO_2-resol cold-box[a]	Boron-containing resols

[a] Gas curing processes.

attention must presently be directed not only to the production process itself, but also to disposal of exhaust gases at the work station, minimization of emissions, and recovery of used core and mold sands. The latter is particularly prudent, since it avoids dumping and is desirable for economical and ecological reasons.

Many technical developments in foundries have only become possible due to the fact that the "art of casting" has developed into the foundry industry in its present modern form [5]. Many new core and mold production processes have been developed since the end of World War II (Table 6.74). The variety of coremaking processes is in turn basically related to six potential binder functions, which may also be analogously correlated with foundry binders (Table 6.74). The greater the extent to which a system provides a balanced mix of these (permanent, temporary, intermediate, complementary, carbon-forming and chemically reactive binding) functions, the better the system ranks in comparison to other processes.

Thus, the cores should exhibit "permanent" dimensional stability prior to the pouring operation, and feature high strength levels during intermediate handling – in transportation, storage, and insertion into the mold. The molding material should then temporarily withstand high temperatures during pouring. Its hot tensile strength over time (Fig. 6.103) should be as high as possible during the initial period following pouring, but should fall precipitously thereafter to provide good collapse properties. The complementary binding function indicates the capacity for modification and adaptation of the binder systems, and the carbon-forming function has the effect that resins produce a compatible level of carbon under pyrolysis conditions. "Compatible" means that such levels exert a positive effect on the surface of the castings, but do not contribute to undesirable "glossy carbon problems" which can considerably detract from the processing quality of the castings.

Fig. 6.103. Hot tensile strength levels of resin-bonded molding materials after 1–4 min at 650 °C or 1050 °C

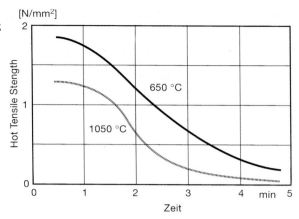

Table 6.75. Varieties of possible phenolic resin reactions and use of these in phenolic and furan resins

Phenolic resin	Reactivity	Process
Novolak	Hexa (HMTA)	Shell mold (croning)
Resole, neutral	Acid	No bake phenolic resin
	Acid, heat	Hot box phenolic resin
Resole, Ortho-condensed	Polyisocyanate	PUR cold box
Resole, alkaline	Ester, liquid	Ester no bake
	Ester, aerosol	MF-resole[a]
	Carbon dioxide	CO_2-resole[b]

[a] Methyl formate as gassing agent.
[b] Carbon dioxide as gassing agent.

The variety of coremaking processes is also related to the wide range of chemical reactivity of the resins, citing phenolics (Table 6.75) as an example. Thus, phenolic novolaks may be reacted with hexamethylene tetramine (HMTA or "hexa") at elevated temperatures. Free flowing, storable phenolic resin-coated sands are produced with hexa in the shell molding process. Phenolic resoles can react with acids at ambient temperature (no bake and hot box processes), undergo polyurethane reactions with diisocyanates (polyurethane cold box method), and be used with methyl formate or carbon dioxide for precipitation and solidification processes (methyl formate or CO_2 "resole" method).

The above comments on phenolic resins apply analogously to furan, urea, epoxy, and acrylate resins (Table 6.76). A representative process is the epoxy or furan gas curing method [6], in which the reactive resins are cured in closed systems by gassing with SO_2, which together with oxygen and water forms sulfuric acid in situ (Fig. 6.104). This process is carried out in sealed systems. The SO_2-containing exhaust is conveyed to a wet scrubber containing sodium hydroxide solution for flue gas treatment.

Table 6.76. Reactivity of furan, urea, epoxy and acrylate resins and their use in core and mold making processes

Binder type	Reactivity	Process
Furan resins	Acid	No-bake furan resin
	Acid (in situ from SO_2)	Furan-SO_2
	Acid, heat	Warm box
Urea resins	Acid	No-bake[a]
	Acid, heat	Hot box[a]
Epoxy resins	Acid (in situ from SO_2)	Epoxy-SO_2
Acrylate resins	Polymerizable	Free radical cure (FRC)[b]

[a] Generally in combination with furan or phenolic resins.
[b] Gassing with diluted SO_2 gas.

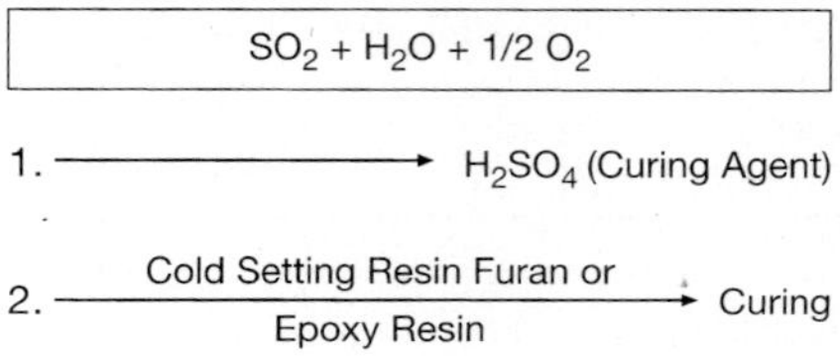

Fig. 6.104. Principle of the SO_2 process for coremaking

$$SO_2 + H_2O + 1/2\, O_2$$

1. $\longrightarrow$ H_2SO_4 (Curing Agent)

2. $\dfrac{\text{Cold Setting Resin Furan or}}{\text{Epoxy Resin}} \longrightarrow$ Curing

The different reactive properties of synthetic resins result in three main groups of coremaking – or mold fabrication – processes (Table 6.77) involving:

1. Warm and hot curing at temperatures ranging from 100 °C to 280 °C depending on the type of process.
2. Cold curing with direct addition of a curing agent at the ambient temperature prevailing in the plant during summer and winter, with possible adaptation of the resin systems for summer and winter conditions.
3. Gas curing, initiated by gaseous substances and possibly involving formation of reaction products requiring additional attention in exhaust gas disposal. Some gas curing processes must be operated at temperatures of 40–60 °C for process operations, and are thus no longer "cold" processes in a strict sense.

Statistics on binders and mold/core sand for the period 1996–1998 together with additional estimates indicate that more than 170,000 tonnes of synthetic binders (mainly phenolic and furan resins) are presently used to produce sand

Table 6.77. Classes of core molding processes (cold, hot, and gas curing)

1	Hot (180–280 °C) and Warm (100–160 °C) curing
2	Cold curing (ambient temperature)
3	Gas curing[a] – Cold (ambient temperature) and Warm (40–60 °C)

[a] One component is in gaseous form and is used as a catalyst, curing agent, solidification/hardening agent.

Table 6.78. Annual binder requirements for core and mold fabrication with chemical (synthetic) binders in the USA and Germany, 1996–1998 (source: estimated from various statistics)

			Germany	USA
A	Mold production	No bake furan resin	16,500	20,000
		No bake phenolic resin	5000 [a]	38,000 [a,b]
		No bake alkyd resin/isocyanate	–	3000
B	Hot cure core production	Shell mold (croning)	4500	17,000
		Hot/warm box	3000	14,000
C	Gas curing	PUR cold box	8000	48,000
		Silicate-CO_2 (and ester) [c]	3000	6000
		SO_2-epoxy process	1000	9000
		Resole gas curing [d]	1200	7000
		Other methods (e.g., acetal process and furan-SO_2)	300	10,000
D	Total resin requirement (tonnes)		42,500	172,000
E	Percentage of requirement (in row D) for			
	a. Mold production		50.6%	35.5%
	b. Hot cure core production		17.6%	18.0%
	c. Cold (gas) cure core production		31.8%	46.5%

[a] Including the resole/ester process.
[b] Including PUR no-bake.
[c] Mainly gas curing with CO_2 in Germany.
[d] Methyl formate or CO_2 as gassing agent.

for fabrication of molds and cores each year in the USA, and more than 40,000 tonnes in Germany (Table 6.78). The quantities of resins used for fabrication of molds and cores are nearly equal in Germany, whereas the focus in the USA according to this estimate is on binders for cores (about 65% of the total). In the USA, acid curing furan and phenolic resins, and to an increasing extent phenolic/polyurethane systems, are used in fabrication of molds. For both countries, (aqueous) alkaline resole resin systems for the ester no bake process have also been grouped under "no bake phenolic resins". The no bake alkyd resin/polyisocyanate process is mainly utilized in the USA for steel casting.

In core fabrication, resins for "cold", energy-saving gas curing processes dominate in both countries. The leader in this group is the PUR cold box process.

Internal calculations show that 80,000 tonnes of resins (phenolic novolaks, acid curing aqueous phenolic resoles, alkaline aqueous phenolic resoles, and a considerable fraction of *ortho*-condensed dissolved resoles) are used annually in the foundry industry in the USA, and 16,000 tonnes in Germany. These numbers do not include co-reactants such as acids, hexamethylene tetramine, or polyisocyanates.

Based on the figures for coremaking sand, 38% of the cores in Germany are fabricated using the PUR cold box process, 35–36% with "hot" curing processes such as the shell molding, hot box and warm box methods, and 25–26% by other gas curing processes together with a minor volume fabricated with the no bake process.

6.2.1.2
Hot Curing Processes

The shell molding and hot box processes, the warm box method, and the warm box-vacuum technique represent hot-curing processes [7]. Considering only phenolics, the resin systems used in these processes are phenolic novolak-hexa blends as well as various modified and unmodified resoles in which the modifier can represent urea resins, furan resins, or monomeric furfuryl alcohol.

6.2.1.3
The Shell Molding Process

The Croning method [8, 9], termed the shell molding or shell process in English-language foundry literature, has been applied to series production since around 1948. In this process, quartz sand is coated with a phenolic novolak-hexa film at elevated temperatures and the coated sand cured on hot pattern plates or in coreboxes to form shells or cores. Shells represent a special type of mold, and are produced nearly exclusively by the shell molding process.

The shell molding method (Fig. 6.105) may be differentiated into two basic process stages:

1. Sand conditioning, i.e., production of free flowing, storable phenolic resin coated sand by a) hot coating or b) warm coating
2. Production of shells, hollow, and solid cores on hot pattern plates or in heated coreboxes

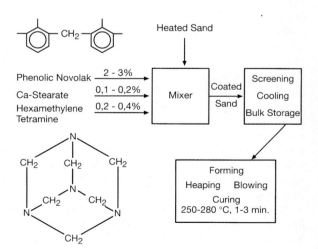

Fig. 6.105. Croning (Shell molding) process for making cores and molds (schematic description)

6.2.1.3.1
Sand Conditioning

Hot Coating

The washed and dried quartz sand is heated to 120–150 °C. The granulated resin (phenolic novolak with a melting point of around 80–100 °C, 2.0–3.5 % based on sand), which is then added to the mixers, melts and coats the grains of sand. The quantity of hexamethylene tetramine (HMTA or "hexa") required for later curing of the resin is then added in aqueous solution. The hexa level is 12–14 % based on resin. The cooling effect produced by the addition prevents premature curing of the binder. Additives such as calcium stearate are incorporated during this stage of the coating operation. The added calcium stearate promotes collapse of the sand aggregates in the mixer and serves as a parting agent in mold production. The water is driven off by blowing hot air through the mixer. After it has passed through a cooling section, the dry, free flowing sand is stored.

Warm Coating

In this process, the phenolic resins (novolaks) are used in the form of solutions. The powdered hexa is first added to sand heated to 60–90 °C in mixing equipment. After a brief mixing period, the solution of resin is metered in and the sand thus coated. At the same time, hot air is blown through the mixer to drive out the solvents and water. Calcium stearate is added shortly before the end of the mixing period. The coating process is concluded by cooling and storing the dry, free flowing molding sand.

6.2.1.3.2
Fabrication of Cores and Shells

The coated quartz sand, with a resin level (including hexa) of around 1.6–3.8 %, is capable of lengthy storage. It is converted into shells (Fig. 6.106) and cores by dumping it on heated pattern plates ("dumping method") or by shaping it under pressure ("shooting" or "blowing") in heated coreboxes on single-stage, transfer and multistage rotary shooting machines. Depending on the thickness and shape of the cores and shells, the curing periods range from 1 min to 3 min at about 250 °C. In foundry practice, curing is performed at temperatures ranging from 180 °C to 350 °C.

Shells, molds, and cores produced in this manner are not hygroscopic, and provide the casting with a high level of dimensional accuracy and surface smoothness. Steel, light metals, and copper alloys, in addition to the various types of cast iron, are cast in "shell mold packets" (assembly of shell molds and cores). The shell molding process is mainly used to produce castings for automotive, truck, and machine construction, as well as for similar uses.

The possible workplace emissions of ammonia and phenol experienced during production of cores and/or shells by the shell molding process have been alleviated by development of binder systems which are low in phenol and hexa, and thus afford a reduction in the phenol and ammonia levels to well

Fig. 6.106. Production of camshafts by the shell molding process

below the TLV limits. The free phenol levels are less than 1% in presently used phenolic novolaks. It is furthermore possible to replace hexa-containing novolak systems partially or completely by combinations of novolaks and resoles. Resoles generally exhibit lower softening points than novolaks. Care must thus be taken to avoid blocking and sticking during production and storage of shell molding sands made using such resin systems.

Plant trials have demonstrated that a significant reduction in pollutant emissions is in fact achieved when resins exhibiting low levels of free phenol are used. In these trials, an 80% reduction in the exhaust gas levels of phenols corresponded to a decrease in the level of free phenol in the binder from around 5% to less than 1%.

6.2.1.4
The Hot Box Process

The hot box process [10–13] is the method of choice for mass production of cores for gray iron, light metal, and heavy metal casting. In iron casting, unmodified and modified phenolic resoles are mainly used as binders. In light metals casting, for example, of aluminum, urea-modified furan resins are the binders of choice, and for heavy metals casting (copper, bronze) straight furan resins. In this specific coremaking technique, moist sand mix is first prepared using a liquid resin and curing agent (Fig. 6.107). This moist molding sand mix is then cured at 180–250 °C in heated coreboxes.

The hot curing with simultaneous acid catalysis employed in the hot box process affords particularly short curing cycles in series production. Complete curing of the molding sand mix in the hot coreboxes is furthermore unnecessary, since the material may be removed from the coreboxes after it has formed

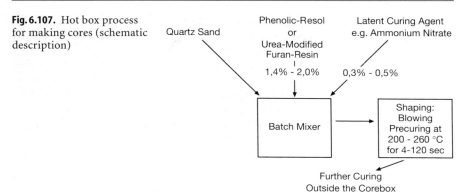

Fig. 6.107. Hot box process for making cores (schematic description)

a solid outer shell. Final, complete curing takes place outside the coreboxes. Rapid curing at elevated temperature is achieved by use of "latent" (hot) curing agents. These generally represent salts of strong inorganic and organic acids, such as ammonium nitrate coupled with specific additives, dissolved in water. The coremaking sand mix is prepared by wetting alkali-free quartz sand with the curing agent, and then coating it with the liquid phenolic resin. The moist finished molding sand mix, with a true resin level of around 1.4–2.0%, may be used for a period of 3–6 h depending on the type of resin, and is converted into cores on single-stage, transfer, or multistage rotary shooting machines (Fig. 6.108). The cores are poured off in both sand molds and (in

Fig. 6.108. View of a rotary shooting machine used to increase the productivity of hot-curing coremaking processes, for example, the hot box method

aluminum casting) permanent metal molds. Since these hot box cores exhibit particularly great dimensional stability, the nominal casting wall thicknesses may be more precisely achieved than when other methods are used. The main fields of use are in engine castings, the automotive area, and in fabrication of fittings.

A new version of the hot box method is the hot box plus process [13]. In this method, a moist sand mix is prepared with a phenolic resole and curing agent as in the standard process, but the blended curing agent additionally includes hexamethylene tetramine. The moist sand mix is then cured in the familiar manner at 180–250 °C. The molding sand properties – i.e., rapid curing, high strength levels, and good thermal properties of the cores when poured off – are comparable to those in the hot box process. However, the extremely long bench life compared to that in the conventional hot box process is a great advantage. This method can mainly be used in iron casting.

6.2.1.5
The Warm Box Process

The "warm box" method [14], in which the curing temperatures range from 130 °C to 180 °C and are thus much lower than in the hot box process, represents a variant of the latter technique. The curing periods are generally comparable to those in the hot box process. Furan resins, albeit highly modified ones, are used as binders. Phenolic resoles and novolaks are generally used as modifiers. Specific sulfonic acids and/or copper or aluminum salts of these [15] are used as (latent) curing agents. The added binder levels are lower than in the hot box process, and generally amount to 0.9–1.2% based on sand; the level of curing agent is 20–30% based on binder.

The two major advantages of this coremaking process over the hot box method are the lower energy demand and decreased pyrolytic potential, i.e., the reduction in gas evolution and emissions during casting (Fig. 6.109). The reduction in emissions is due both to the lower resin levels and to the unique structure of these binder systems. In iron casting, this method is particularly useful for production of compact cores, for example crankcase cores. However cores from warm box process are slightly higher in moisture sensitivity than those of the hot box process.

Fig. 6.109. Comparison of emissions from hot-box and warm-box cores at 800 °C

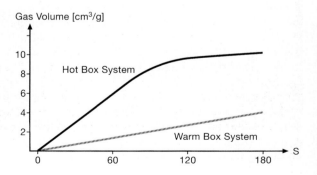

A process variant of the warm box technique is the "vacuum warm box process." This process allows cores to be produced at temperatures as low as 70–100 °C [16]. Coreboxes in which a reduced pressure can develop when a vacuum is applied are used in this method. This reduced pressure differential allows pollutants to be removed directly from the corebox and treated by conventional methods. Phenolic resoles and furan resins may be used as binders in combination with specialty curing agents. These specialty curing agents may contain materials including copper chloride. When a vacuum is applied, diluents and water immediately vaporize, causing a reduction in the pH and initiating the cure even at relatively low temperatures.

6.2.1.6
Cold Curing with Direct Addition of a Curing Agent (No Bake Process)

In the no bake process, involving cold curing with direct addition of a curing agent, cores and molds are produced from self-setting sands in cold coreboxes or molding flasks. The most common method involves the use of acid-catalyzed (phenolic and furan) resin systems which cure without application of heat. Another option consists of the use of highly alkaline binder systems – aqueous phenolic resoles containing a high level of potassium hydroxide – which are cured by addition of specific esters. A third possibility, the PUR no bake method, involves curing *ortho*-condensed resoles with polyisocyanates in the presence of amine catalysts.

6.2.1.6.1
Acid Curing

The binders are generally furan resin systems in the acid curing no bake process [17], in which the cure is brought about by addition of acids, but can also represent straight or modified phenolic resoles exhibiting the lowest possible levels of free phenol (less than 4.5%).

In the acid-catalyzed no bake process, alkali-free sand is coated with a binder system consisting of an acid-sensitive resin and a curing agent (Fig. 6.110), and is capable of undergoing cure at ambient temperature; the types of resin and curing agent are selected to match the specific production conditions as well as the requirements of special molds or cores. Blending is generally performed in continuous throughput mixers. The molding sand possesses a limited bench life, and gradually cures after being charged into the mold and compacted. The level of free formaldehyde, which should not exceed 0.5%, is presently an important factor in selection of the phenolic or furfuryl alcohol-containing binders. Furan resins, which make up the bulk (around 70–85%) of the resins used in this area, have frequently undergone price increases due to surges and fluctuations in the furfuryl alcohol prices. Thus, phenolic resoles were used as early as 1975. From the very beginning, demands to hold the levels of free phenol and formaldehyde as low as possible were proposed for these products. These demands have, in fact, been met by special developments by various manufacturers.

Fig. 6.110. No-bake acid curing process scheme

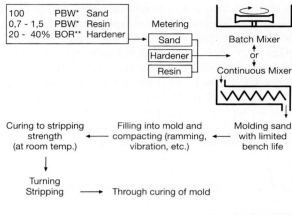

* = parts by weight
** = based on resin

The molding sands are generally prepared using the following standard formulation as a basis (PBW = Parts By Weight):
- 100 PBW quartz sand
- 0.7–1.3 PBW resin
- 0.2–0.6 PBW curing agent

The dried, alkaline-free quartz sand is wetted with the acid in a mixer, and then coated with the liquid resin. The moist molding sand can only be worked for a limited period. The length of the limited molding sand bench life, the stripping time of molds and cores, and the time required until the peak mechanical strength is reached depend on the level and type of acid used, and particularly on the temperature of the mix, in addition to the type of resin. The curing times are shortened by increased temperatures and higher acid levels, as well as by the use of stronger acids.

The most commonly used curing agents include organic sulfonic acids (mainly p-toluenesulfonic acid with sulfuric acid levels of 0.5–2%) and phosphoric acid. The phenolic resoles used in this case generally possess the following properties: low levels of free phenol and formaldehyde, low water levels, a desirable viscosity range, high curing rates upon addition of acid, and good elasticity following curing. These ensure good handling, uniform sand coating, reliable charging and compaction, remarkably little odor nuisance, and a lengthy stripping period coupled with desirable curing times. The high strength levels of the cured molding sand permit low resin metering levels and the use of up to 95% reclaimed sand.

"Reclaimed sand" is generally considered to be quartz sand from a used mold or core which has been thermally, mechanically, or chemically regenerated for reuse. The rich carbon supply and very low nitrogen level of the residual phenolic resin binders affords castings of high surface quality.

A general comparison of the use of phenolic and furan resins in this process affords the following summary. The viscosity of the furan resins is lower than that of the phenolics, and is simultaneously less temperature dependent. The reclaimed sand from used furan resin-bonded molds or cores is also generally easier to reuse. If these two points are considered when using phenolic resins (reduction of the base viscosity and thermal regeneration of reclaimed sand), significant economical and technical advantages are definitely apparent in phenolic resins, the latter particularly with respect to meeting the workplace limits for formaldehyde.

6.2.1.6.2
No Bake Curing with Added Esters

The procedure of the ester no bake process is similar to that of the acid-catalyzed no bake method with respect to the mixing and other processing operations. A highly alkaline phenolic resin generally containing up to 25% potassium hydroxide is used as a binder [18] in the ester no bake process.

In sand conditioning, 1.5–2% of the highly alkaline phenolic resole (based on sand) is first blended into the sand, followed by about 20% of an aliphatic ester (based on resin). The reaction between the specific esters or mixtures of these with the potassium hydroxide-containing resin leads to curing/setting and solidification. Depending on the type of ester or ester blend used, the stripping times may be adjusted to 5–30 min. Since the binders are free of nitrogen and sulfur, they afford foundry engineering advantages for steel, nodular iron, and aluminum casting. This process also offers the advantage of relatively low material costs, although it leads to frequently inadequate strength levels and leaves a highly alkaline sand which can only be regenerated or dumped within narrow limits. It is a poor match to a modern, environmentally acceptable fabrication concept.

6.2.1.6.3
No Bake Process with Isocyanate Curing

The polyurethane no bake method, a no bake process using polyisocyanates as curing agents [19], is chemically related to the PUR cold box process (see below). In contrast to the PUR cold box process, which is a gas curing process, a liquid catalyst (curing accelerator) is used. One reaction component is an *ortho*-condensed phenolic resole diluted with solvents, and the other is a diphenylmethane-4,4'-diisocyanate-based polyisocyanate dissolved in similar solvents. An amine incorporated into the resin is added to the molding sand as a catalyst to accelerate the cure, i.e., to facilitate rapid cure (Fig. 6.111). When the polyisocyanate comes into contact with the phenolic resin, the catalytic action of the amine initiates the actual reaction leading to polyurethane formation after a brief initial delay. Figure 6.111 illustrates this process on the basis of the compression strength relative to that in an acid-catalyzed process (with furan resin as a binder). In the case of the acid-curing molding sand, the strength level rises constantly, whereas the polyurethane reaction only proceeds spontaneously after a certain open time which provides for the shaping operation.

Fig. 6.111. Curing curves in the PUR- and furan no bake processes (comparison)

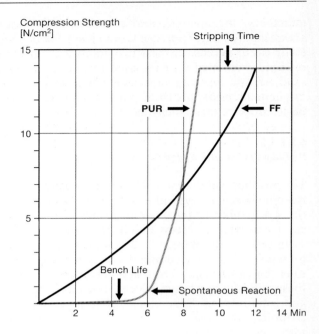

The binder system in the polyurethane no bake process is thus composed of three parts:

- Part I   *ortho*-condensed phenolic resole, 0.5–0.8% based on sand
- Part II polyisocyanate in solution, 0.5–0.8% based on sand
- Part III catalyst, about 12% based on resin

The rate of spontaneous cure can be adjusted by way of the catalyst level. In the above case involving 12% catalyst, the bench life of the molding sand is about 5 min and the stripping time about 15 min.

6.2.1.7
Gas Curing Processes Using Phenolic Resin Binders

In "gas curing" processes, one reaction component is introduced in the form of a gas, and can produce various effects/reactions (Fig. 6.112). The gas, the additional component in these processes – which are generally carried out in seal-

Fig. 6.112. Overview and principle of gas curing processes

Gas	Effect	Reaction	Process
CO_2	Precipitation	Settting	CO_2-Resole
Amine	Catalysis	PUR-Reaction	PUR-Cold-Box
Methylformate	Precipitation	Settting	MF-Cold-Box
Methylal	Formaldehyde	Reaction	Acetal Process
SO_2	Sulfuric acid	Catalysis	SO_2-Cold-Box

ed, enclosed systems to permit direct removal of the resulting exhausts – can act as a precipitating agent (to set and solidify the product), and a catalyst (such as the amines used in the polyurethane process with *ortho*-condensed phenolic resoles), can be transformed into or develop various reactive materials such as formaldehyde from methylal or formic acid from methyl formate, or – in the case of the SO_2 process – can aid in forming sulfuric acid, which in turn acts as the curing agent. In the "resole process" involving the use of highly alkaline potassium hydroxide-containing phenolic resoles which also include boron compounds (boric anhydride or borates), CO_2 similarly functions to initiate precipitation reactions.

The CO_2-silicate method, which involves the use of an inorganic sodium silicate binder, is noteworthy from an historical viewpoint. This process is very environmentally friendly and exhibits relatively low costs. On the other hand, the poor collapse of cores after casting, the moisture sensitivity of the CO_2-hardened cores, the low strength levels, and the high alkalinity of the resultant old sand represent serious disadvantages.

As a group, the gas curing processes have gained considerable significance over the past years due to efforts to save energy and costs in both short and long runs of cores. Not only are the binder levels and thus energy equivalents expended in these processes considerably reduced, but the productivity is substantially increased due to the much shorter cycle times.

6.2.1.7.1
The Polyurethane Cold Box Process

In the polyurethane cold box or amine gassing process, the amine used to gas the cores (Fig. 6.113) acts as a catalyst to accelerate the polyurethane reaction of the *ortho*-condensed phenolic resole with the polyisocyanate. The resin/binder used is produced using catalysts favoring the *ortho* orientation,

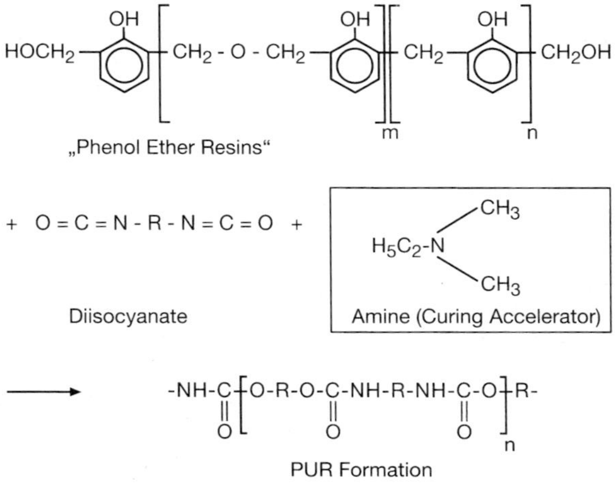

Fig. 6.113. Polyurethane-cold-box process – reaction equation

"Phenol Ether Resins"

$+ \; O=C=N-R-N=C=O \; +$

Diisocyanate

$H_5C_2-N(CH_3)_2$

Amine (Curing Accelerator)

$\longrightarrow \quad -[NH-C(=O)-O-R-O-C(=O)-NH-R-NH-C(=O)-O-R-]_n$

PUR Formation

Fig. 6.114. Polyurethane-cold-box process – scheme

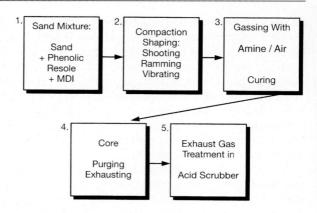

for example ones based on cobalt or zinc, and is adjusted to a specific viscosity with 30–45% of high-boiling solvents such as aromatic and aliphatic hydrocarbons, esters, ketones, and isophorone. The polyisocyanates used in the process represent straight or oligomeric diphenylmethane-4,4'-diisocyanate (abbreviated MDI), and are similarly diluted with solvents of the classes mentioned above. The overall process (Fig. 6.114) involves first introducing the coremaking sand, previously prepared by mixing quartz sand, phenolic resin, and polyisocyanate solution, into the corebox by "shooting." The core is then gassed with a mixture of amine in air or CO_2. The amine represents triethyl, dimethylisopropyl, or dimethylethylamine, and is sprayed into the corebox at levels of 0.2–1 ml/kg core weight at a pressure of 0.2–2 bar. Residual amine is purged out of the core using warm air, and treated in a scrubber containing dilute sulfuric or phosphoric acid.

The polyurethane cold box method [20], termed the Isocure™ process in the USA, has achieved considerable importance as a cold gassing method in Europe during the past 12 years. The process has been criticized due to the odor nuisance caused by the amine gassing component during production and storage of the cores. Certain technical disadvantages such as the tendency to form glossy carbon due to the high solvent levels are offset by the particularly short cycle times, good coremaking sand shootability, and relatively wide range of application which represent marked advantages of the process. Thus, this process can be used both for fabrication of cores by the corebox method, and for production of mold components in boxless casting. Figure 6.115 shows the range of mold components used for short series production of a cylinder head for racing cars.

New binders feature low levels of phenol and improved collapse in aluminum casting [21].

The cold box plus process, in which the corebox is heated to 50–70 °C [22], represents an improvement over the standard method. The purpose of the elevated corebox temperature is to evaporate solvents in the surface layers of the core, and to increase the crosslinking density and strength of the polyurethane. This affords advantages including a significant reduction in the process-specific casting defects inherent in the cold box method. A major advantage of

Fig. 6.115. Technical assembly sequence of a boxless mold: a = baseplate, b = primary core with water jacket core, c = oil space core, d = cover core (with through riser), e = finished mold ready for pouroff, with insulating plate, f = aluminum cylinder head (photo: BMW AG, Munich)

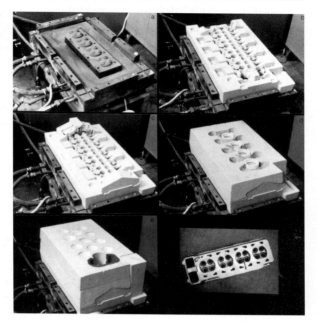

this process relative to the conventional cold box method is the reduced core-box sticking. However, slightly higher energy costs must be expected in this process, thus leading to a total expense which approaches that of the warm box process (Sect. 6.2.1.5).

6.2.1.7.2
The Methyl Formate Process

Aqueous phenolic resoles are used as binders (Fig. 6.116) and methyl formate as the gassing agent [23] in the methyl formate (MF cold box) process, which

Fig. 6.116. Methyl formate (MF) process scheme

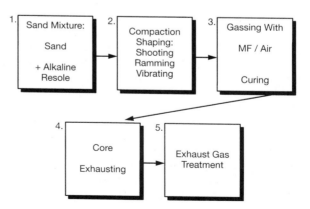

is related to the ester no bake method. The highly alkaline phenolic resoles contain up to 20% of a base, generally potassium hydroxide. When gassing occurs, the ester saponifies, leading to reactions which liberate the intermediate formic acid used to cure and neutralize the binder, and the methyl alcohol by-product. The curing and setting process is relatively complex. The strength levels of the molding materials are generally far lower than those achieved in other gassing processes. Certain applications exist in cases where these strength levels are adequate, for example, in competition to the sodium silicate process, assuming the methyl formate process does not create new environmental problems.

This process possesses the advantage that the core shooters generally need not be completely enclosed, and that partial exhausting of gases is adequate, although treatment of the exhaust gases (excess ester and methyl alcohol) becomes unavoidable when the levels exceed certain limits. The highly alkaline sand and the fact that no economical methods of exhaust gas treatment exist represent disadvantages of the process. Treatment methods such as catalytic combustion are under discussion, although these are relatively uneconomical.

6.2.1.7.3
The CO_2-Resole Process

As in the methyl formate (MF) process, this method makes use of a potassium hydroxide-containing resole. The binder differs from that used in the methyl formate process in that it contains boric acid, borates, and other additives [24]. Carbon dioxide is used as the gassing agent. When gassed, the coremaking sand mix sets due to various solidification reactions. The strength levels of the cores fabricated in this manner are identical to those in the methyl formate process. The major advantage of this process lies in the fact that no special treatment of the exhaust gases is required. The high process alkalinity again represents a disadvantage (in reclamation of used sand).

6.2.1.7.4
The Acetal Process

The acetal process is an intrinsically interesting coremaking method, although it is more of theoretical than practical utility. In this method, the binder is a resorcinol resin (polyphenol) solution, and a strong acid (mixture of sulfuric and *p*-toluenesulfonic acids) is added to the sand mix. The corebox must be held at a temperature of 35–40 °C [25]. Gassing is performed with the dimethylacetal of formaldehyde (methylal). When gassing occurs, the methylal decomposes into formaldehyde and methyl alcohol in the highly acidic medium. The formaldehyde reacts spontaneously with the resorcinol binder. This leads to a red coloration of the cores.

This process is quite interesting in theory, but presents serious problems in practice since selective heating of the corebox and the entire sand mix making up the core within it, is difficult.

6.2.1.7.5
Other Gassing Processes

Mention may be made of the epoxy or furan gas curing process using sulfur dioxide as the gassing agent. A peroxide such as methyl ethyl ketone peroxide or cumene hydroperoxide is added to the molding sand as an oxidizer. Sulfuric acid thus forms when the core is gassed with sulfur dioxide, and initiates spontaneous curing of the binder in the corebox. The binders used in this case are either acrylate-modified epoxies or low-monomer (condensed) furan resins [26]. The methods are also known as the Rütapox (or EGH) and Hardox (or FGH) processes [27, 28].

6.2.1.8
General Remarks on Core/Mold Fabrication Processes Using Phenolic Resins

Discontinuous or continuous methods of mixing may be used for core sand conditioning and applies to all the processes mentioned here. Limits are placed on the mixing procedures by the binder level, since this varies quite widely in the process-specific core sands. The mixing procedure should permit uniform distribution even of solvent-free, high-viscosity liquid binder systems on the quartz sand.

The question as to the extent of success of the cold curing and the remaining hot curing processes on the market and to whether the situation of the phenolic resins will be maintained, as well as how "good castings" will continue to be produced most economically, depends to a large extent on factors such as future environmental and workplace legislation, raw material availability, and the possibility of sand recovery and re-use. Adequate specific data on used sand utilization – sand recycling – have been published by Krapohl [29] and Boenisch [30, 31].

The influence of various fresh core sands on the strength levels of green sand reported in the cited papers is of interest. According to these data, sands from the shell molding, epoxy-SO_2, and PUR cold box processes offer benefits contributing to improvement of clay-bonded molding sands without exerting an adverse effect on the wet tensile strength of the latter. Krapohl notes that two main factors become apparent from a wide variety of individual results: interferences and the sensitivity to these. The high alkali levels of the resole processes represent such interferences. These present adverse effects both on other types of binders and even on themselves. A major factor in problem-free use of recycled (thermally or mechanically regenerated) old sand is removal of dust fractions, or stabilization of their levels. In practice, it is known that good castings of the desired quality may be produced when used sand is reasonably utilized, the used sand properties matched to the specific types of core and mold fabrication processes in use, and proper logistic procedures applied.

With respect to raw material availability, products in which renewable raw materials are used [32, 33], for example, phenolic by-products from the timber industry and cellulose production such as tannin and lignin, are doubtless of interest for the binders of the future.

6.2.2
Abrasives

6.2.2.1
Introduction (Grinding, Abrasives)

Increased process mechanization coupled with automation over the last few decades has made grinding processes a vital operation in the manufacture of numerous commercial items [1, 2].

The historical development of the abrasives industry is noteworthy by considering its evolution from a rudimentary beginning to a relatively mature industry. Presently hardly a material exists that has not been exposed to a grinding operation. It can be stated without exaggeration that the abrasives industry represents a basic building block for all other branches of industry.

Whereas natural stones were initially used as abrasives, the first synthetically bonded abrasives were reported at the beginning of the nineteenth century. Natural resins and rubber were used as binders during the following period. A patent on ceramic bonded abrasives dating from 1864 exists. Following utilization of the Baekeland patents and founding of the German Bakelite Company in 1910, the use of phenolic resins in this market area was first mentioned in 1916, and in the following decades gained steadily in importance in comparison to the use of other, conventional binders. Not only are the desirable high temperature and wear properties of phenolics of importance in such applications, but also numerous other properties such as their high resistance to aggressive chemical reagents, valuable in wet grinding.

The process of grinding represents one of the most important methods of machining, since it features high rates of material removal, the ability to hold exact tolerances and provide high dimensional accuracy, as well as excellent surface quality of the work component. It can be used even in the case of difficult-to-work materials [3]. Consequently the demands placed on the abrasive systems are very great, and include optimal grinding performance and attainment of the greatest possible precision with high economy and compliance with legal workplace and environmental requirements. These demands, as well as the wide variety of highly different materials to be worked – iron, steel, nonferrous metals, glass, wood, plastics, and others – necessitate a wide range of grinding wheels, sections, and coated abrasives as well as numerous types of grinding machines and grinding systems/processes.

Level metal surfaces result from surface grinding, whereas cylindrical surfaces are worked by cylindrical grinding. A cylindrical work piece is located between a regulating disc and the actual grinding wheel in "centerless" grinding. Diverse types of grinding machines that vary in their mechanical design and applications exist: vertical and horizontal grinders, surface, external and internal cylindrical, cut-off, thread cutting, gear cutting, tool, and guide rail grinding machines. Wood and wood surfaces are best treated with coated abrasives clamped on the surfaces of rotating discs in dish or disc sanders. In belt sanders, the continuous abrasive belt travels over two rollers, whereas the coated abrasive is wound around special rotating rollers in cylindrical sanders.

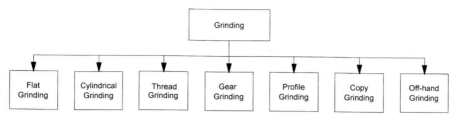

Fig. 6.117. Classification of grinding processes in DIN 8859 T 3

This brief summary only mentions a small selection of grinding methods, of which there is a greater diversity in practice.

Publications such as DIN 8859 T 3 (Fig. 6.117) summarize the above comments on grinding. In this standard, grinding processes are classified into surface grinding, cylindrical grinding, grinding of thread surfaces, grinding of gear teeth, profile grinding, copy grinding, and manual grinding [4].

All of these grinding processes may be associated with grinding wheels or bonded abrasives of different design and construction [5]. Some special cases are "wet" grinding using coolants and the particularly high-performance application of diamond wheels to grind especially hard materials.

In this field of application, the phenolic resin fulfills the function of a largely temporary binder for both bonded and coated abrasives. The term "temporary" in this case indicates a period that should be as long as possible to minimize wear to the lowest possible level. This refers in essence to the properties and high bonding strength of the resin matrix linking the abrasive grains, giving particularly great significance to the phenolic resin curing process leading to largely void and pore-free binders. Specifically, the abrasive grains should remain bonded to the matrix for the longest possible time despite thermal stresses during the grinding process. For example, in fettling castings or cutting steel, high peak temperatures of up to 800 °C (red and white heat) can occur.

6.2.2.2
Economical Significance

The use of phenolic resins for bonding has increased considerably over the last decades. In the case of grinding wheels and bonded abrasives, this use largely competes with that of ceramic bonding systems. In the case of coated abrasives, urea resins and animal glues [6] are used in addition to phenolic resins. However, phenolic resins dominate in these applications for reasons of quality.

Table 6.79, showing the growth in production of synthetic resin bonded abrasives, largely dominated by phenolic resins used as binders in the form of powders or water-borne systems, demonstrates that the competing ceramic bonding systems no longer enjoy the leading role they played in Germany even at the beginning of the 1970s. This has resulted from the increasing do-it-yourself market demand, requiring greater safety, as well as changes and evolution

Table 6.79. Production of bonded abrasives and grinding wheels in Germany from 1987 to 1995 [a]

Bonded abrasives	1987 tonnes	1991 tonnes	1995 tonnes
Ceramic bonding	15,760	15,600	17,280
Phenolic resin without fiber reinforcement	6,840	8,040	10,080
Phenolic resin with fiber reinforcement	19,315	25,170	27,140
Miscellaneous	2,830	3,260	4,300
Total Production	44,745	52,070	58,800

[a] Source: Verein Deutscher Schleifmittelwerke e.V., Oxfordstrasse 8, D-53111 Bonn, P.O.B. 2466, D-53014 Bonn.

in both the materials to be worked and the various grinding technologies and systems.

Table 6.80 shows the economical development of coated abrasives from 1987 to 1995, and demonstrates that the total production of abrasive fabric, abrasive papers, and vulcanized fiber wheels has increased by about 10% during this period. Coated abrasives bonded with phenolic resins are particularly resistant to high temperatures and achieve especially high grinding performance, particularly since multilayer designs for coated abrasives have meanwhile opened avenues for increased flexibility. Partial bonding by synthetic resins (animal glue as a base coat, phenolic resin as a sizer coat) also continues in use, although this composition is reducing in volume, and urea resins are used in commodity abrasives for economical reasons.

Table 6.80. Production of coated abrasives in Germany from 1987 to 1995 [a]

Coated abrasive	1987 1000 m^2	1991 1000 m^2	1995 1000 m^2
Abrasive paper	27,200	28,020	30,700
Waterproof abrasive paper	7,800	7,740	7,060
Abrasive cloth	18,350	20,200	22,100
Waterproof abrasive cloth	1,270	1,600	1,340
Abrasive combinations	820	780	690
Vulcanized fiber grinding wheels	1,620	1,790	1,970
Total production	57,060	60,130	63,860

[a] Source: Verein Deutscher Schleifmittelwerke e.V., Oxfordstrasse 8, D-53111 Bonn, P.O.B. 2466, D-53014 Bonn.

6.2.2.3
Grinding Wheels (Classification, Definitions)

Grinding wheels or bonded abrasives (Fig. 6.118) may be classified/named and defined according to various aspects or technical parameters:

1. Type of fabrication, for example cold, warm, or hot pressing
2. Type of abrasive, for example, refined corundum, boron nitride (CBN), diamond, or others
3. Type of specific physical property, for example, highly compacted
4. Type of bonding, for example, with phenolic resin
5. Type of use, for example, cutting, roughing, working, or polishing
6. Type of article to be ground, for example, roller grinding
7. Type of grinding process, for example, swing frame grinding

For example, the question as to whether direct curing (hot pressing) or a two-stage method (separate shaping and curing operations) is used in *fabrication* of phenolic resin bonded grinding wheels and abrasive segments is of significance, and is reflected in the density of the wheel as well as the application characteristics. The hot pressed wheels are also post cured. It should be noted that a variety of phenolic resin systems can be used in the various fabrication methods, whether these involve cold pressing followed by optimal curing of the "green" shape using specific time/temperature programs, or simply hot curing from the very beginning. The basic process conditions, affording cold or hot pressed wheels, are generally only feasible when the proper selection is made.

A second possibility of characterization is to term grinding wheels according to the *abrasive* used in them, the actual "tool" in the abrasive composite. The hardness of the abrasive "tool" largely determines the use to which the bonded abrasive is put. Thus, grinding wheels made with the particularly hard abrasives diamond or boron nitride are named after these, and called diamond or boron nitride wheels. The grinding performance can also be significantly influenced in such bonded abrasives by selection of a specially designed phe-

Fig. 6.118. Selection of various grinding wheels and bonded abrasives (photo: Tyrolit Schleifmittelwerke Swarovski K.G., Schwaz, Austria)

nolic resin or system. The expert or abrasives' user already recognizes where a bonded abrasive can be used and what materials can be worked with it (in the case of CBN and diamond mainly hard, difficult-to-work materials) when it is named according to the abrasive used, for example, in the case of "diamond" or "CBN" wheels.

Furthermore, it is also common to name grinding wheels according to specific *physical* or *chemical* properties, for example, to call them "high density" wheels, meaning that they exhibit particularly great strength due to their high level of compaction, and thus are suitable for heavy-duty work. Grinding wheels or bonded abrasives intended for use with aggressive coolants are frequently labeled or rated for "wet grinding." In principle, this means that the binder, for example, a phenolic resin, is particularly resistant to chemical attack, a property that can be achieved by use of specific resin systems and abrasive additives such as silanes.

The range of properties and thus the definition of a grinding wheel or bonded abrasive may also be classified according to the *type of bonding*. The design of the grinding wheel indicates (see below) that the binder can be inorganic in nature or represents one of the various types of organic binders. This results in grinding wheels being termed ceramically bonded wheels, synthetic resin or "Bakelit" wheels, and wheels of several other less important varieties. In the case of "Bakelit" or phenolic resin bonded wheels, it should be noted that a special group of phenolics, the "Bakelite® SP" or "Bakelite SW" resins (commercial names of Bakelite AG, Iserlohn, Germany), are used in their manufacture. Bakelit without the "e" and capitalized is the German name of the product – in English the spelling is bakelite with an "e" and not capitalized. The name of the company is with the "e" and always capitalized.

The *application* can also be reflected in the grinding wheel designation, for example in the terms "roughing wheel" or "cutting wheel," indicating that these products are used for working or separating/cutting purposes. Fettling of a work piece is also a type of "working." In the steel industry, for example, fettle grinding is used to remove scale from blocks, billets, slabs, or blooms on automatic machines. Cutting wheels, in turn, can be used to section bar steel. A grinding wheel and its use can be designated according to the type of article to be ground or the grinding process. Clear-cut differences between the two is sometimes not apparent; naming according to a combination of the two is sometimes possible. Examples are wheels for wet roller grinding or for fettling with swing frame grinders. In these examples, the articles are the steel material or the roller in the rolling mill, and the process swing frame grinding or wet grinding. These indicators of the process, or article to be ground, in turn imply specific (phenolic) resin systems that are used to produce bonded abrasives or grinding wheels of optimal efficacy in the indicated process or article-related applications.

The above remarks are designed to illustrate common definitions and simultaneously emphasize that phenolic resin bonding is of considerable importance in the design of grinding wheels (cf. next section), and that a long road of grinding wheel, grinding system, and grinding process development has led to the current technical advances. The range of current liquid and powdered phenolic resins for production of grinding wheels and bonded abrasives can be quite diversified.

6.2.2.4
Grinding Wheel Design, Bonded Abrasives (Composition and Stresses)

According to DIN 69100, the design of a grinding wheel is characterized by three basic elements among others that are shown in Fig. 6.119:

1. Grinding material (abrasive)
2. Binder that unites the abrasive grains (Figs. 6.120 and 6.121)
3. Structure dictated by the ratio of bonded abrasive grain to pores

The bonded abrasive is also characterized by specifying its hardness, defined as the resistance to separation of the abrasive grain from the bond of the binder/grain structure. DIN 69100 specifies a pertinent (Norton) hardness scale, according to which E–G represents very soft, H–K soft, L–O medium hard, P–R hard, T–W very hard, and X–Z extremely hard.

An additional characterization parameter is the structure of the bonded abrasive, 1–2 representing very compact, 3–4 compact, 5–8 medium, 9–11 open, and 12–14 very open.

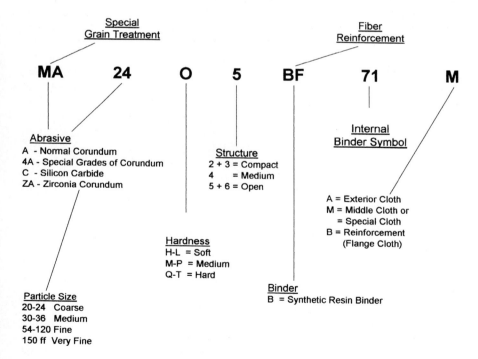

Fig. 6.119. Bonded abrasive design in DIN 69100 and [3]

Fig. 6.120. Scanning electron micrograph of a phenolic resin bridge as a bond for the abrasive grain

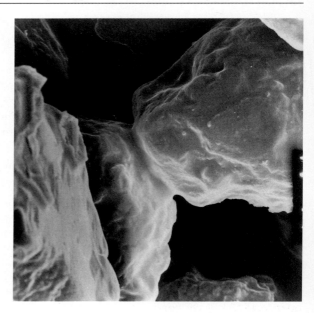

Fig. 6.121. Polished surface of a grinding wheel showing the structure (photo: Tyrolit Schleifmittelwerke Swarovski K.G., Schwaz, Austria)

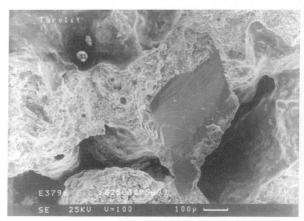

The structure of the bonded abrasive defined in this manner (ratio of abrasive grain/binder volume to pore volume) influences the bonded abrasive hardness and is far from negligible. Further pertinent parameters are the type of binder, the binder level, the type of filler, and (particularly in the case of phenolic resin bonded abrasives) the curing program whose temperature gradient partially determines the level of hardness.

Most bonded abrasives currently contain electrically fused corundum (Al_2O_3) or silicon carbide (SiC) as the abrasive grain. Crystalline aluminum oxide (α-Al_2O_3) is obtained from electrical fusion of bauxite to yield normal corundum (NC) or from purified alumina to yield refined corundum (RC) with an Al_2O_3 content of more than 99 wt%. Silicon carbide (α-SiC) is produced by heating quartz and coke together in an electrical resistance furnace at approximately 2000 °C.

Silicon carbide grains are primarily used to grind very hard, brittle materials such as stone, glass, ceramics, or tungsten carbide. In special cases, abrasive materials or abrasives such as zirconia corundum, diamond, boron carbide, and boron nitride are used to meet exceptionally high requirements.

Fillers are used in the resin matrix to increase the strength level, heat resistance, toughness, and burst resistance, as well as for other purposes. Combining abrasive and resin binder helps to achieve desirable grinding wheel properties such as cool and smooth grinding. Fillers that are currently favored are cryolite, pyrite, zinc sulfide, lithophone, calcium fluoroborate, calcium sulfate, and calcium chloride. The previously used antimony trisulfide and lead chloride are rapidly being substituted by materials such as special iron halides. Basic oxides such as calcium and magnesium oxides are also recognized as curing accelerators when phenolic resins are used as binders. However, calcium oxide is not suitable for use in wet grinding wheels because of its ability to undergo hydrolysis.

Binders are essentially classified into inorganic and organic bonding agents. Inorganic bonding is produced with ceramic, silicate, magnesite, and sintered metal (steel, bronze, or tungsten carbide powder) binders.

Aside from synthetic resins (mainly phenolics, with epoxies, polyester, or polyurethane resins to a lesser extent), rubber and shellac are used for organic bonding. Ceramics and synthetic resins represent by far the most important types of binders.

Ceramically bonded abrasives feature high natural porosity and enable cool, dry, and wet grinding. They are very brittle. They behave as rigid bodies, and follow the motions of the machine precisely during the grinding operation. Thus they are especially suited for profile and precision grinding at peripheral speeds normally ranging up to 45 m/s and in special cases up to a maximum of 80 m/s. The ceramic binder components are generally various grades of feldspar, kaolin, various types of clay, and fluxes in the form of glass frit. These are processed using the pressing or casting method. Ceramic bonding requires relatively high firing temperatures ranging between 1000 °C and 1400 °C.

In contrast, synthetic resin (specifically phenolic) bonded abrasives are considerably less sensitive to impact, shock, and lateral pressure. Their superior strength properties permit higher grinding wheel speeds and better grinding performance. They are primarily suitable for roughing and cutting wheels; their elasticity (modified phenolic binders) also renders them especially appropriate for bonding abrasives used in fine grinding and polishing operations. Synthetic resin bonding has gained in importance over the past decades (cf. "Economical Significance") and presently ranks higher than ceramic bonding in importance.

The viscoelastic properties of phenolic resin and ceramic binders are schematically compared in Fig. 6.122. The vibrational frequencies show that the bonding properties of the synthetic/phenolic resin system represent a healthy

Fig. 6.122. Comparison of ceramic and synthetic resin bonding (oscillation graphs) (from [3])

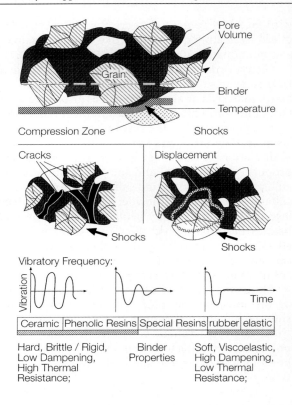

compromise between those in low dampening, brittle ceramic bonding and viscoelastic rubber bonding. Ceramically bonded abrasives tend to develop cracks when exposed to shock, whereas resin bonded materials with their elastic nature suffer at most grain displacement upon experiencing shock. The properties of the bonded materials, particularly those specific to grinding technology, may be varied by addition of fillers. The fillers are liberated by attrition of the binder framework during the grinding process due to melting or vaporization caused by the high temperatures arising in the contact zone. Formation of a lubricating film and chemical reaction with the freshly-formed metallic surface reduce the level of friction and the grinding temperature, acting to prevent structural alterations in the work piece.

The properties of grinding wheels and bonded abrasives are governed by the following factors [7, 8]: the dimensions of the grinding wheel, the type and size of the abrasive grains, the level and type of filler(s), and the strength of the (phenolic resin) binder as well as its resistance to separation of the abrasive grain from the composite during the grinding process.

Figure 6.123 illustrates the nature of the chip-cutting process that occurs during grinding on the basis of a section of the peripheral zone of a bonded abrasive [3]. The chips are formed by removal from the work piece with a variety of cutting, rake, and clearance angles at high cutting speed by the abra-

Fig. 6.123. Peripheral zone of a grinding wheel (from [3])

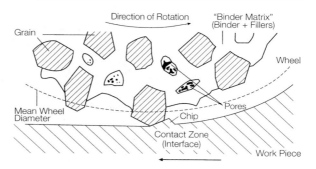

sive grains. During this procedure, very complex physical and chemical processes involving the material of the work piece and the composite of abrasive/bonding matrix take place at temperatures of around 1500 °C in the microscopically small interface between the grain and the work piece. Plastic and elastic deformations of the work piece and the bonded abrasive occur at very brief intervals, and result in chip formation, grain breakage, and degradation of the binder accompanied by generation of heat and dust. A wide variety of formulations, of which the following composition only represents an illustration [9, 10], exists to ensure that the grinding wheel or bonded abrasive (particularly when phenolic resin is used as a binder) is optimal for use in a general or specific application in the various grinding systems:

- Abrasive grain – consists of 65–95% of the total, contains natural (corundum) or synthetic materials (boron nitride)
- Binder – consists of 5–20% of the total, contains various combinations of liquid and powdered, straight or modified phenolic resins; powdered phenolic resins combined with special wetting agents such as furfural; or low melting phenolic resin combined with powdered phenolic resins
- Filler – consists of 0–20% of the total, includes mainly cryolite, pyrite, antimony oxide and zinc sulfide with a variety of other products
- Reinforcing fabric – grinding wheels may be reinforced with glass fabric impregnated with phenolic resin

6.2.2.5
Liquid and Powdered Binders (Phenolic Resins)

In manufacture of grinding wheels – as noted in the above remarks – both aqueous phenolic resoles and powdered novolak/hexa-based phenolics are used as binders for the abrasives in the system.

In this case, the liquid phenolic resins fulfill multiple functions (Table 6.81). They fix the powder resin and fillers on the abrasive grain, anchor the bonding matrix (after curing) to the abrasive grain, provide green strength to cold-pressed moldings, and influence the flow properties of the resin system during the curing process. The various aqueous phenolic resoles used as wetting agents can be highly alkaline, weakly alkaline, or neutral (for example resins

Table 6.81. Liquid phenolic resoles for production of grinding wheels and bonded abrasives [a]

No.	Liquid resin	Viscosity at 20 °C (ISO 9371)	Non-volatile (ISO 8618)	Gel time at 130 °C (ISO 8987)	Water Thinnability (ISO 8989)	Free phenol (ISO 8974)	Used to wet abrasive grain for bonded abrasives used in
1	Highly alkaline phenolic resole, low condensation degree	approx. 2500 mPa·s	approx. 80%	approx. 16 min	10:30	approx. 20%	Dry grinding, with very high strength and grinding performance
2	Highly alkaline phenolic resole, high condensation degree	approx. 800 mPa·s	approx. 77%	approx. 15 min	10:6	approx. 18%	Dry grinding, with high strength and grinding performance; good flow and storage stability of mix
3	Low-alkali phenolic resole, low condensation degree	approx. 800 mPa·s	approx. 73%	approx. 8 min	10:60	approx. 4%	Dry grinding; very good flow and storage stability of mix, for pressing with automatic machines
4	Neutral, alkali-free phenolic resole, medium condensation degree	approx. 800 mPa·s	approx. 75%	approx. 13 min	10:8	approx. 14%	Wet and dry grinding, with good performance and storage stability of mix
5	Neutral, alkali-free phenolic resole, high condensation degree	approx. 3500 mPa·s	approx. 81%	approx. 11 min	10:4	approx. 12%	Wet and dry grinding, with good performance and storage stability of mix

[a] Source: Bakelite AG, Iserlohn, Germany.

2–4 in Table 6.81), and can differ greatly from one another in their degrees of condensation. Their viscosities thus generally range around 500–4000 mPa·s at 20 °C, and the level of nonvolatile components normally varies between 70 % and 80 %. The quality of the liquid resin affects the strength level and grinding performance of the overall system, and the pourability and storage stability of the grinding wheel mix during its production. Furthermore, neutral phenolic resoles exert an influence on the wet grinding performance that is far from negligible.

Phenolic resoles exhibiting a low degree of condensation and containing a high fraction of free phenol display a particularly high powder absorption capacity, provide the mix with high adhesion and green strength, and thus afford wheels with very high strength and dry grinding performance levels. The use of highly condensed phenolic resoles (for example resin 2 in Table 6.81) affords superior processing properties with respect to the storage stability and pourability of the mix. Phenolic resoles exhibiting free phenol levels below 5 % were developed for environmental reasons and to improve the pourability and storage stability of grinding wheel mixes. However, they generally lead to a reduction of about 10 % in the grinding performance.

Furfural, which undergoes crosslinking in the presence of a calcium oxide catalyst, is best used instead of phenolic resoles as a grain wetting agent in manufacture of highly compacted grinding wheels and bonded abrasives produced using the cold and hot pressing methods. Other anhydrous reactive compounds are cresols, creosote, and anthracene oil.

The bulk of the phenolic resin bonding matrix is made up of finely milled, powdered phenolic resins that contain hexamethylene tetramine as a curing agent, and can be modified with various materials such as polyvinyl butyral, low and high molecular weight epoxy resins and rubber (resins 6 and 7 in Table 6.82). Moreover, the properties of the phenolic powder resins are significantly affected by the degree of condensation of the phenolic novolaks, i.e., their melting range and flow properties, and the fraction of hexamethylene tetramine. It can range between 4 % and 14 %. Variation of the hexa level leads to changes in the crosslinking density produced during cure, which in turn has a significant effect on the strength and hardness of the grinding wheels and bonded abrasives. For example, a softer grinding wheel formulation is required for grinding operations performed on temperature-sensitive material, where the work must be performed without excessive heat development. Use of powder resins exhibiting hexa levels below 6 % (resins 1 and 2 in Table 6.82) is desirable in such cases. In contrast, grinding operations carried out under high pressure coupled with great thermal stress require the use of bonded abrasives with a highly crosslinked bonding matrix. This is obtained by use of hexamethylene tetramine levels of 12–14 wt% (for example in resin 4 of Table 6.82).

Since the phenolic binder is intrinsically relatively brittle, elastomer-modified novolaks function as the powder resins in special fields of application, for example, to achieve particularly high burst resistance levels in high-speed grinding wheels or to withstand high thermal shock stresses. Polyvinyl bu-

Table 6.82. Powder resins for production of grinding wheels and bonded abrasives - unmodified, modified and low-dust products [a]

No.	Powder resin	Hexa level (ISO 8988)	Flow distance (ISO 8019)	Used for
1	Novolak/hexa with long flow distance	approx. 4%	60 to 80 mm	Very soft bonded abrasives, high filler levels, best for wet grinding
2	Novolak/hexa with long or short flow distance	approx. 6%	45 to 60 mm 15 to 25 mm	Soft bonded abrasives with high or low filler levels, particularly for wet grinding
3	Novolak/hexa with medium or short flow distance	approx. 9%	30 to 40 mm 15 to 20 mm	Wide range of applications (standard resin), particularly for cold and hot pressed wheels used for dry grinding
4	Novolak/hexa with long or short flow distance	approx. 14%	45 to 60 mm 15 to 25 mm	Cold and hot pressed wheels used for dry grinding; offers high thermal stability
5	Novolak/hexa, polyvinyl acetate modified, with medium or short flow distance	approx. 8%	30 to 40 mm 12 to 15 mm	Elastomeric modification of the resin bonding Matrix to achieve high burst resistance levels in dry and wet grinding
6	Novolak/hexa, epoxy resin modified, with medium or short flow distance	approx. 9%	30 to 40 mm 15 to 20 mm	Very high elastomeric modification, particularly in cold and hot pressed, glass fiber-reinforced wheels
7	Novolak/hexa, rubber modified, with medium or short flow distance	approx. 8%	20 to 30 mm 15 to 20 mm	Elastomeric modification of the resin bonding matrix, particularly in bonded abrasives used for wet grinding
8	Novolak/hexa, with antidusting additive	Various levels	Various flow distances	Major reduction in dusting

[a] Source: Bakelite AG, Iserlohn, Germany.

tyral, various types of rubber, and high molecular weight epoxy resins – including "phenoxy resins" – have been successfully used as modifiers.

Aside from phenolic resins, which afford a relatively brittle phenolic bonding matrix, other binder systems of lesser importance – such as shellac, epoxies, polyester resins, alkyds, or polyurethanes – are also used to achieve specific grinding effects, for example, in polishing.

As is typical in technology involving mixing processes, the production of grinding wheels and bonded abrasives makes use of a wide array of additives employed for purposes such as to improve the pourability and storage life of the mix, and reduce to a minimum the clumping tendency of the wet grinding

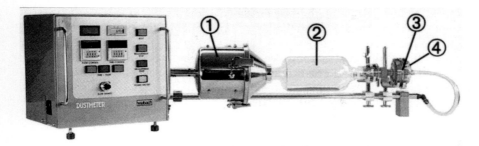

Fig. 6.124. Dustmeter used to determine the dusting properties of powdered resins (technical data: sample weight 100 g, test duration 5 min, 30 rpm, air flow 0.25 l/s, generating pot volume 2.50 l), source: Heubach Engineering GmbH, Langelsheim, Germany

wheel mix. Powdered additives and silica or derivatives, due to their uniform and low particle size, coat the granulate, thus improving the pourability and storage stability of the mix.

On the subject of minimizing dust, powder resins containing a homogeneously dispersed antidusting agent as an additive [11] were recently commercialized (for example, resin 8 in Table 6.82). When examined in a dust meter (Fig. 6.124), these new powder resins exhibit reduced levels of dust which can approach zero depending on the concentration of antidusting agent.

A new method for the production of grinding wheel mixes [12, 13] involving the use of an anhydrous resin system as a wetting agent has recently gained attention (Fig. 6.125). The abrasive grain is heated to around 130 °C, charged into a high-shear mixer, and a special, relatively low molecular weight novolak pre-heated to 100–130 °C (which behaves like a low-viscosity liquid binder at the indicated temperature) is first added in portions to wet the grain. The customary powder resins, fillers, and, if appropriate, additional hexa are then added. The novolak melt, powder resin, and fillers can then be reapplied, thus permitting formation of multiple coats. This assures uniform, coated granular particles, resulting in a storage-stable, bunkerable, and dust-free grinding wheel mix after discharge from the mixer. The advantages of this

Fig. 6.125. Flow chart of the hot coating method for production of grinding wheel mixes (from [13])

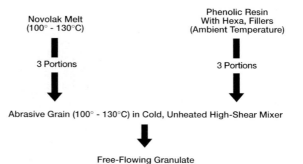

special warm coating method include a very low level of free phenol, storage-stable, dust-free mixes, a particularly homogeneous filler distribution, short mixing times, considerably shorter curing programs than those required for conventional mixes, and reduced emissions. The mixes are generally warm pressed, although cold pressing is also possible when specific pressing conditions (high compression force) are employed. The resin system and the "warm" or "hot" coating method described in this case are similar to those employed in the "warm" mixing method used to manufacture refractories (cf. section on refractories, Sect. 6.4.1).

6.2.2.6
Cold Pressed Cutting and Roughing Wheels

Phenolic resin bonded abrasives are insensitive to impact, shock, and lateral pressure. Their outstanding strength properties permit higher grinding wheel speeds and increased grinding performance. Therefore they are primarily used in production of roughing and cutting wheels, that are – as the terms imply – intended or used for cutting (Fig. 6.126) or working ("roughing", Fig. 6.127) surfaces and the like. The wide variety of demands made on bonded abrasives and their bonding matrix are material-related as well as dependent on the type of grinding and grinding conditions (pressure, speed, and temperature). The specific grinding machine and its requirements, and not least the purpose of the grinding operation, are also of importance.

In production of a cold pressing mix, the abrasive grain is first wetted and coated with a thin film of a resole-type liquid resin or, in special cases, using

Fig. 6.126. Grinding operations – cutting (photo: Tyrolit Schleifmittelwerke Swarovski K.G., Schwaz, Austria)

Fig. 6.127. Grinding operations – surface grinding (photo: Tyrolit Schleifmittelwerke Swarovski K.G., Schwaz, Austria)

furfural and/or furfuryl alcohol as a wetting agent. The powder resin is then blended into the wet abrasive grain, and fillers also added in most cases. The powdered raw materials thus "stick" to the abrasive grain. The weight ratio of liquid to powder resin depends on the particle size and distribution of the abrasive grain used, the type and level of added filler, and on the viscosity of the liquid resin or wetting agent employed. A high filler addition level requires a higher level of liquid resin. The thickness of the coating on the grains depends on the viscosity and level of liquid resin or wetting agent. The ratio of liquid to powder resin ranges from 1:2 to 1:4; when furfural or a furfural/anthracene oil blend is used, the liquid to powder ratio is 1:6–1:8. Satisfactory homogeneous mixing processes uniformly mix the entire composition without fracturing or causing attrition to the abrasive grains.

The mixing time is 2–5 min. The level of powder resin bound to the abrasive grain passes through a maximum. When the mixing time is too long, the danger of coating attrition exists and excessive heating of the mix can lead to undesirable precuring. The actual mixing process proceeds in the following manner (Fig. 6.128). The abrasive grain is pre-mixed, since it generally consists of multiple particle fractions, and the wetting agent then added. The powder resin, if appropriate in combination with the fillers, is then blended in. Near the end of the mixing process, the charge should develop into a uniform, slightly plastic but free-flowing mix that should not be excessively moist, but remains free of dust. Particularly high levels of uniformity and pourability are required for efficient processing by automatic machines. Low-alkali and alkali-free liquid resins are preferred (for example, resins 3–5 in Table 6.81).

To avoid contaminating of the mixer and lengthy cleaning operations, the work should be performed with two mixers. The abrasive grain is wet with the liquid resin in one mixer. The moist mass of abrasive grain and liquid resin is then charged into a second mixer already containing the blend of powder resin and filler. If the mix is too moist or dry, either condition may be corrected with specific additives.

Fig. 6.128. Flow chart of the fabrication process used to manufacture cold-pressed grinding wheels

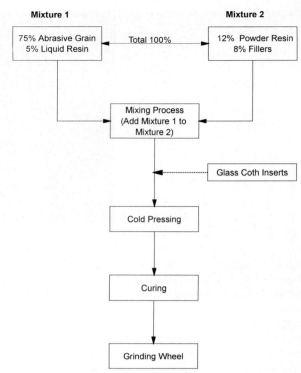

It is general practice to store the finished (screened) mix for several hours to allow it to "mature" before compression molding. However, this assumes that the mass does not tend to aggregate or solidify, as occurs with mixes using wetting agents such as furfural.

The compression molding mix is charged into appropriate molds for shaping. Special turntables are used in production of large-size wheels to ensure that the distribution of the mix in the mold is as uniform as possible. In some cases, steel rings are molded into the grinding wheels to increase their stability and strength.

Pressing is accomplished in hydraulic presses (Fig. 6.129) with a compression strength of 15–30 N/mm². The required compression strength depends on the effective plasticity of the grinding wheel mix and the desired degree of compaction. However, the compression strength should not damage the abrasive grains. The compression molding mix is normally pressed to yield a predetermined density, i.e., a specific volume for a certain type of grinding wheel. The pressing time depends on the dimensions and shape of the wheel body, the particle size and distribution of the mix, and its plasticity. The pressing time ranges from 5 s to 50 s. The green grinding wheels produced in this manner are stripped and subsequently transferred to a curing oven. The curing process [1] is most frequently carried out discontinuously in chamber ovens, less commonly as a continuous operation in tunnel ovens. It is particularly important that the green grinding wheels be properly positioned in the curing oven to

Fig. 6.129. Production of grinding wheels on hydraulic presses (photo: Krebs & Riedel, Bad Karlshafen, Germany)

ensure that air can adequately circulate between the individual grinding wheels and provide for even temperature control from all sides. The grinding wheels are placed on porous ceramic plates or on perfectly flat steel sheets. It is important that a uniform temperature be maintained. Indirectly heated curing ovens operated with gas or electricity, with automatic program controls, forced air circulation, and a fresh air feed have been used with good success for fabrication of high-quality bonded abrasives. The selection of an appropriate curing program (Fig. 6.130) depends on a variety of factors. These specifically include the wheel dimensions, grinding wheel (open or compact) structure, levels of binder and filler, and the specific properties of the binder system.

The general design of the curing programs must be such that they consider the various resin and curing parameters during the heating phase.

1. At a temperature of about 80 °C, the resin has generally been transformed into a fused mass, and the water contained in the resole is liberated when the resole begins to cure.
2. At a temperature of about 110 °C, the hexamethylene tetramine begins to decompose, initiating cure of the fused powder resin and leading to liberation of gas, particularly ammonia.
3. At temperatures up to about 180 °C, final consolidation of the structure and maximum crosslinking of the phenolic resin takes place. The bulk of the ammonia is liberated during this period. However, overcuring must be avoided since overcured wheels exhibit reduced strength levels; the final

temperature level (165–170 °C, 175–180 °C or 185–195 °C) has a considerable effect on the final properties of the bonded abrasive (hardness, toughness, brittleness).

It is possible to develop optimal curing programs permitting co-curing of resole and novolak systems from temperature-time plots of the emissions liberated during curing of bonded abrasives or grinding wheels (Fig. 6.131). Determination of the emissions liberated during curing can also lead to development of faster curing programs that are primarily applicable to thin wheels. In the case of thin wheels, it is unnecessary to provide a holding period at 80 °C; instead, the temperature may be directly increased to 110 °C in order to avoid premature through curing of the resole fraction and instead produce simultaneous curing of the resole and the novolak. As shown by Fig. 6.131, a phenol peak with a maximum at 110 °C is then obtained. The ammonia emission also shows that decomposition of the hexa proceeds parallel to the emission of phenol, a further indication that the resole and novolak cure together. The simultaneous cure of resole and novolak yields a homogeneous structure that is reflected by longer useful life and increased grinding performance levels.

After curing is complete, the wheels are gradually cooled to 50–60 °C by circulating air in the closed oven. The oven is only opened and the wheels removed after this temperature has been reached.

Standard powder resins (for example resin 3 in Table 6.82) classified according to narrow flow distance ranges as defined in ISO 8619 (15–20 mm, 20–25 mm, 25–30 mm, 30–35 mm, and 35–40 mm) have been used with particular success for production of cold pressed cutting and roughing wheels. Such standard resins, which are also marketed as dust-free versions [1, 11, 14],

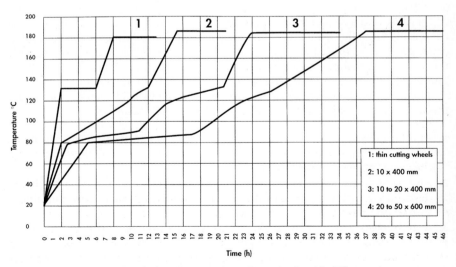

Fig. 6.130. Curing programs for production of grinding wheels with different sizes

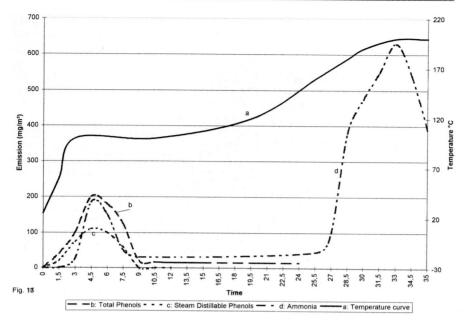

Fig. 6.131. Emission measurements during curing of grinding wheels

can be supplied to precise specifications, especially with respect to the levels of hexa (approximately 9%) and free phenol (less than 0.5%). Powder resins with hexa levels ranging between 4% and 7% are used to produce especially soft wheels. In contrast, extremely hard versions are produced with powder resins exhibiting hexa levels of 13–15% (for example resin 4 in Table 6.82).

Both alkali-containing and alkali-free liquid resoles (for example resins 4 and 5 in Table 6.81) are used to wet the abrasive grain for cold pressed grinding wheels. The latter grades are particularly satisfactory to achieve good pourability and high storage stability in mixes for processing by automatic machines (Fig. 6.132, Table 6.83).

Epoxy-modified phenolics with epoxy resin levels of up to 15% are used for cutting and roughing wheels with high burst resistance.

6.2.2.7
Production and Use of Glass Fabric Inserts

Reinforcement of bonded abrasives with wide-mesh fabric prepregs made with glass filament yarn [15, 16] has been found to be a successful method to achieve increased grinding speeds and thus improve grinding performance while simultaneously meeting safety demands, since such materials best meet the increased demands on strength. It may be noted that other materials such as bulk fiber, mats, and woven cotton, rayon, or Aramid®-based fabric can also

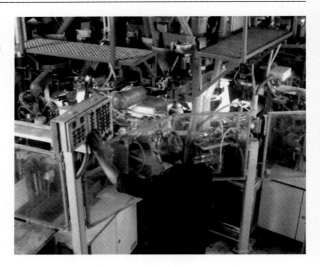

Fig. 6.132. Automatic compression molding of cutting and roughing wheels (photo: Tyrolit Schleifmittelwerke Swarovski K.G., Schwaz, Austria)

Table 6.83. Mixture for production of grinding wheels on automatic machines (from [1])

75.00 % w/w	Abrasive grain, standard particle size 30
5.00 % w/w	Low-alkali liquid resin (for example no. 3 in Table 6.81)
12.00 % w/w	Powder resin
3.40 % w/w	Pyrox, red
4.60 % w/w	Cryolite

be used as reinforcement. Weaves that have been found particularly appropriate for reinforcement (Fig. 6.133) are the linen (plain), basket, and leno weaves. The yarns used in such cases may be silanized to achieve improved bonding with the fabric. The impregnating resin (Table 6.84) should exhibit good wetting power and adhesion to the glass fabric, while simultaneously offering compatibility with the grinding wheel mix so that co-curing of the resin and the mix yields particularly high strength following the curing process. Both phenolic resoles (resin A in Table 6.84) and solutions of relatively highly condensed novolaks (resin B in Table 6.84) containing hexamethylene tetramine as a curing agent are used. Resoles generally afford high prepreg flexibility and feature good adhesion to the fabric. Novolaks increase the storage life of the glass fabric prepregs. Hexa-free phenol/novolak solutions are also used (resin C in Table 6.84) in practice, and solid novolaks (resin D in Table 6.84), with which impregnating solutions may be prepared as indicated by the suggestions in Table 6.85, similarly offered for this purpose.

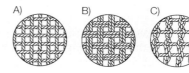

Fig. 6.133 a–c. Examples of weaves: **A** linen (plain); **B** basket; **C** leno weave

Table 6.84. Resin solutions (resoles and novolaks) for impregnation of glass fabric inserts (from [1])

Resin	Chemical characterization	Delivery form	Non-volatiles at 135 °C (ISO 8618)	Density at 20 °C mPa·s (ISO 2811)	Viscosity at 20 °C mPa·s (ISO 9371)	Gel time at 130 °C min (ISO 8987)	Flash point °C (ISO 2719)	Solvents
A	Modified phenolic-resole	70% in methanol	70±2	1.12±0.01	500±100	12±2	<16	Alcohols, ketones
B	Modified phenolic-novolak	65% in methanol	66±2	1.08±0.02	300±100	20±3	<16	Alcohols, ketones
C	Phenolic-novolak	75% in methanol	75±2	1.13±0.01	1000±200		<16	Alcohols, ketones
D	Phenolic-novolak	Granulated solid resin		Free phenol % <0.2	Viscosity at 150 °C mPa·s 400±40			Alcohols, ketones

The phenolic resin must meet various basic requirements: the curing time must be matched to the reactivity of the resins in the grinding wheel mix, "windowing", i.e., surface proliferation of the resin film between the fabric meshes, should be absent after impregnation and subsequent drying, and the prepreg should be stable in storage, remaining tack-free and flexible over a lengthy period. In the impregnating process, the glass fabric is passed through the resin solution by means of dipping, excess resin removed with rollers, and the prepreg dried to a tack-free state by steadily increasing the temperature in a drying oven. During this process, the temperature of the impregnating bath is held constant to prevent variations in the amount of resin deposited on the prepreg. After impregnation, the resin level of the fabric ranges from 28% to 35% (as required). Discs are generally die-cut out of the prepregs for use in producing grinding wheels. Several methods of recycling the considerable volume of die-cutting scraps exist [17, 18].

Cutting and roughing wheels with glass fabric inserts are fabricated in the same manner as grinding wheels without reinforcing inserts. In this case, however, it is important that the grinding wheel mix exhibits particularly good

Table 6.85. Solid resin (C and D in Table 6.84) based solutions for impregnation of glass fabric inserts (from [1])

Solution A	100 PBW[a]	Resin C from Table 6.84 (Rutaphen 1L 9264)
	10 PBW	Alcohols (methanol, ethanol)
	4 PBW	Hexamethylene tetramine
Solution B	100 PBW 1	Resin D from Table 6.84 (Rutaphen 0790–03)
	55 PBW	Alcohols (mehtanol, ethanol)
	5 PBW	Hexamethylene tetramine

[a] PBW=Parts By Weight.

Fig. 6.134. Polished section of a grinding wheel with fabric reinforcement (Photo: Tyrolit Schleifmittelwerke Swarovski K.G., Schwaz, Austria)

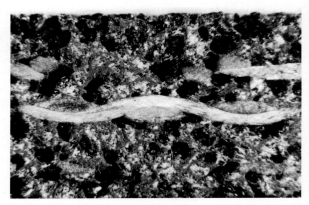

pourability and the glass fabric inserts employed are in flawless condition. The impregnating resin must again begin to flow and coalesce with the bonding resin of the grinding wheel mix at elevated temperatures, and undergo co-condensation/co-curing with the surrounding resin binder.

Excessively high compression strength should be avoided in production of cutting and roughing wheels with glass fabric inserts to prevent displacement of and damage to the fabric (Fig. 6.134).

Curing of the grinding wheels requires 13–21 h for cutting and roughing wheels of up to 10 mm thickness. The final temperature should not exceed 170–180 °C in this case. It is advisable to stack the wheels in the oven with metal sheets between the layers, and to place them under a certain pressure (with clamps or weights) to achieve sound adhesion of the fabric layers to the wheel mix. Present demands on the burst resistance of fabric-reinforced grinding wheels can correspond to operating speeds of 80–100 m/s, compared to 45–60 m/s for wheels without glass fabric inlays.

6.2.2.8
Binders for High Wet Strength Grinding Wheels and Segments

Grinding wheels or bonded abrasives used for wet grinding with coolants exhibit a drop in their bonding strength and hardness due to exposure to the aggressive cooling media. The resultant reduction in the burst resistance and grinding performance of the bonded abrasive must be reduced to a minimum to enable the bonded abrasive to meet safety demands fully and enhance its economy. In cases where the cohesive strength of phenolic resin bonded abrasives drops, particularly when alkaline grinding fluids are used, it is presumed that water molecules penetrate into the interface between the abrasive grain surface and phenolic resin binder, initially in places where defects in the form of small cleavages, voids, hairline cracks, and microscopic pores exist. Since phenolic resins, as condensation products, liberate volatile components during the curing process, the formation of such defects can be limited, but never completely eliminated. Aqueous alkaline solutions that penetrate the matrix can remove the phenolic resin binder from the surface of the abrasive grain. The adhesion between the grain and the binder is reduced in this manner.

Based on these considerations and many research projects, as well as practical experience, it is advisable to use neutral, alkali-free, specially condensed phenolic resins (for example resins 3 and 4 in Table 6.81) exhibiting good wetting power together with very desirable mixing properties to wet the abrasive grain. The powder resins used in this case are novolak/hexa mixtures (for example resins 1 and 2 in Table 6.82) exhibiting good flow and a relatively low level of hexa. For special requirements, it is advisable to use epoxy or rubber modified phenolic resins (for example resins 6 and 7 in Table 6.82) exhibiting a low level of hexa and high flow.

Pretreatment of the abrasive grain with a chemically reactive adhesion promoter such as a silane can improve the adhesion of the phenolic resin binder to the abrasive grain and thus the wet strength of the grinding wheel.

6.2.2.9
Hot Pressed, Highly Compacted Grinding Wheels (HP Wheels)

Large-surface work pieces such as steel slabs or blooms are increasingly worked in swing frame grinders (Fig. 6.135), in which the rotating wheel is pressed against the work piece with a force ranging from 20 N/mm^2 to 80 N/mm^2, and the latter is moved back and forth beneath the wheel automatically. Wheels of this type exhibit diameters of 400–900 mm; dimensions of 610x76x305 mm are frequently encountered. Such wheels must exhibit extraordinary strength, achieved by methods such as high compaction, to withstand the enormous stresses to which they are exposed. Whereas normal grinding wheels exhibit densities of 2.4–2.7 g/cm^3 up to a maximum of 2.9 g/cm^3, the density of such highly compacted wheels is about 3.1–3.5 g/cm^3. This high level of compaction can only be achieved by hot compression. The volumetric percentage of pores to grinding wheel volume should be less than 1% in such cases.

Fig. 6.135. Swing frame grinder for working steel (photo: Tyrolit Schleifmittelwerke Swarovski K.G., Schwaz, Austria)

Essentially anhydrous wetting agents, such as furfural or furfural blended with furfuryl alcohol or cresol and in some cases anthracene oil, are generally used for production of HP wheels. Aqueous resoles are unsuitable for production of HP wheels since the level of volatile components is too high.

The abrasive grain or mixture of abrasive grains is coated with the above wetting agents in the mixer as in production of cold pressed wheels, and the powder resin/filler mixture added. Powder resins exhibiting a high hexa level and short flow distance are preferred. Residual dust can be minimized by uniform addition of small amounts of anthracene oil to the mix.

The resultant mix is uniformly charged ("combed") into the mold. The mold is continuously rotated on a turntable. A less expensive mixture made with finer grades of granular material can be used at the center ring section of the wheel where steel rings are molded in to reinforce the wheel, since this section of the bonded abrasive cannot be used for grinding in any case. The wheels can be preheated. One possibility of accomplishing this is by high-frequency preheating, in which the mold together with the mix is exposed to a high-frequency field for about 10–20 min at a temperature of around 90–95 °C. Another possibility involves cold pre-pressing at a compression strength of 15–25 N/mm^2, followed by preheating in a curing oven at temperatures of 90–130 °C. The time and temperature level are determined by the heating performance of the oven and the heat capacity of the mold. As a rule, the wheels have been preheated to 80–90 °C after about 40–60 min at 120–130 °C. Following the preheating process, the mix is soft/plastic and is transferred to the hot press as quickly as possible without allowing it to cool. It is also possible to omit the prewarming step, inserting the cold-pressed green wheel into the hot compression mold and immediately placing it under the heated press.

Hot pressing is best performed in stage presses in which curing proceeds far enough that the wheel remains dimensionally stable after being stripped, and the cured zones are strong enough that trapped volatile components cause no changes such as deformation or cracking during post-curing. The pressing temperature ranges from 160 °C to 170 °C, and the pressing times are determined to be about 30–60 s per millimeter of wheel thickness.

The following guidelines are proposed for wheels of 60 mm thickness: compression strength 20–40 N/mm^2, pressing temperature 150–170 °C and pressing time 30–60 min. The wheels are then heated to a temperature of approximately 160 °C at a rate of approximately 10 °C per hour, and are held at this temperature for about 8–12 h (depending on the wheel diameter). The wheels may also be heated to 180 °C in stages to increase their heat resistance and useful life. In this case, the recommended procedure is to first allow a holding time of 2–3 h at 160 °C, then to continue heating to 180 °C and hold the wheels at this temperature for another 5–8 h.

In the last few years, the operating speeds of modern HP wheels have been increased from 60 m/s to 80 m/s, and the demands made on their burst resistance have risen sharply. It was furthermore necessary to increase the useful life of the HP wheels and thus their total performance (total number of machined work pieces per wheel). The force with which the HP wheels are pressed against the work piece also have encountered a further increase, considerably raising the level of mechanical/thermal shock. These new demands on the HP wheels required further optimization of the resin systems used in their production.

6.2.2.10
Diamond Wheels

Diamond grinding wheels are best for grinding hard metal. They are produced using hard metal, steel, bronze, and synthetic resin bonding matrices. A wide variety of options exists for production of phenolic resin bonded diamond wheels. In principle, such wheels are fabricated in the following manner. Cylindrical moldings to serve as a supporting body for the diamond-containing abrasive coating are first prepared from special phenolic resin molding compounds. Mixes of aluminum powder and powder resins – both standard and polyvinyl butyral or epoxy modified resins (for example, resins 5 and 6 in Table 6.82) – have also been successfully used for this purpose. The weight ratio of aluminum powder to powdered phenolic resin ranges between 80:20 and 85:15. The wheels are produced by hot compression molding at a pressing temperature range of 160–175 °C.

The moldings are then machined to the shaped carrier wheels required in each case. Carrier wheels produced from a mixture of aluminum and powder resin exhibit considerably better thermal conductivity than those made from a wood flour-filled phenolic resin molding compound, and thus contribute to an improvement in the useful life of the diamond wheels.

The uniformly distributed mix is charged into a compression mold heated to 170 °C. After charging, final pressure is applied after a prewarming period of about 20–30 s under a certain amount of pressure. The mold is not vented. The abrasive coating, at a thickness of 1–6 mm depending on the pertinent requirements, is then hot compression molded onto the carrier wheel.

The initial step in production of the abrasive coating mix is to prepare a premix of the powdered phenolic resin with fillers such as silicon carbide (SiC), boron carbide (B_4C) or aluminum oxide (Al_2O_3). The mixing ratio of powder resin to filler ranges between 35:65 and 90:10 as a percentage by weight, and varies according to the desired hardness and diamond grain level of the abrasive coating. The level of powder resin in the finished mix can be to up to 30% by weight. The powder resins used include polyvinylbutyral modified resins such as Rütaphen® 0309 SP or epoxy modified resins such as Rütaphen 0321 SP 01 as well as standard resins such as Rütaphen 0222 SP 04.

The diamond grain is carefully wetted with a small amount of furfural and this material blended into and intimately mixed with the premix. It is very important that extreme care be taken in this mixing process designed to achieve a completely homogeneous blend. The finished mix is then charged onto the carrier wheel in the compression mold, heated to a pressing temperature of 170–185 °C, and pressed with a force of 25–35 N/mm². The pressing time depends on the thickness of the coating and amounts to about 5–10 min. The wheel should be allowed to cool slowly in the mold.

Fig. 6.136. Selection of abrasive fabric belts (photo: Hermes Schleifmittel GmbH & Co, Hamburg)

6.2.2.11
Phenolic Resins as Binders for Coated Abrasives

Coated abrasives, in contrast to bonded abrasives, are considered to be flexible, two-dimensional grinding materials used in the form of sheets, belts (Fig 6.136), discs, and other shapes. The abrasive grain is fixed to the surface of a backing such as paper, vulcanized fiber, or fabric used as a carrier by means of a (preferably liquid) binder [19–23] that can be employed for coating, in the base or maker coat, the sizer coat, or for treatment of the backing (Fig. 6.137). Additionally some applications (e.g., for stainless steel grinding) need a second sizer coat, the so called "supersizer." The impregnation of the backing is necessary to avoid penetration of the maker coat into the fabric.

Coated abrasives thus consist of a backing, the binder and the abrasive grain. The backings used are materials such as high-quality sodium kraft paper with base weights ranging from 70 g/m² to 320 g/m² and special latex-modified, acrylate, and PVC coated papers for wet grinding, viscoelastic vulcanized fiber with a thickness of 0.4–0.8 mm and high tear and peel strength, and fabric made of cotton or synthetic fibers such as polyester with weights per unit area of 200–600 g/m². Combinations of fabric and paper or foil and paper are also used. Random polyamide fiber-based grinding mats are used as backing for coated abrasives.

The papers used for such products can be pretreated. Thus, papers used to produce water-resistant sandpaper are rendered hydrophobic and flexible by impregnating them with alkyd resin solutions, or with latex emulsions to which water-thinnable phenolic resoles have been added to improve the ther-

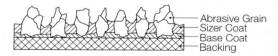

Fig. 6.137. Structure of coated abrasives (from [20])

mal resistance. Increased flexibility of a fabric-based product requires more modification of the binder. This modification can represent both internal modification of the phenolic resole and external modification by use of rubber or polymer emulsions.

When vulcanized fiber is employed as a backing, phenolic resins are almost exclusively used as binders. The diversity of different types is far smaller than in the case of fabric backings due to the small supply of vulcanized fiber.

The selection of binder depends on the application profile of the abrasive. Coated abrasives were initially and exclusively produced using animal glue as a binder. Today, synthetic resins dominate the market, particularly for high-performance products. About one third of all coated abrasives are presently manufactured using synthetic resins as the complete binders (full resin bonding), around one third using animal glue as the base coat and synthetic resin as the sizer coat (semi resin bonding), and one third still using animal glue as the exclusive binder (full glue bonding). The advantage of animal glue bonded coated abrasives is that these exhibit enhanced flexibility and do not require high temperatures to harden/dry. The advantages of the synthetic resin bonded coated abrasives are that these exhibit considerably longer performance and the grinding performance of products produced with them is very much higher. Thus, fabric and vulcanized fiber backed abrasives are nearly exclusively produced using curable resins. Aside from epoxy, urea, alkyd, and polyurethane resins, phenolics are preferred. Phenolic resins exhibit considerable advantages over the other binders, such as superior adhesion to the abrasive grain and the backing, considerably lower moisture sensitivity, and high thermal stability. The utilization of the abrasive grain performance is enhanced by the use of phenolic resins.

A wide variety of phenolic resins are available (Table 6.86). Resin selection depends on the type of backing used, the abrasive and the available coating equipment, which varies in dimensions, heat capacity, and temperature control. The holding times and throughput speeds may be adjusted by appropriate selection of a binder for the available coating equipment from those available. This relates to the different stages of reactivity and curing rates, graduated viscosity levels, concentrations of nonvolatiles, and to the various levels of residual monomers (phenol and formaldehyde).

The resins should, if possible, be stored at low (i.e. 5 °C) temperatures (Fig 6.138). Storage temperatures of 5–10 °C are recommended. The storage life of the resins is adversely affected at higher temperatures. Figure 6.138 illustrates the storage characteristics of resins A, L, and K from Table 6.86 at a temperature of 10 °C on the basis of the changes in their ISO 8989 water dilutability.

Types of abrasive grains is as follows:

1. Natural – pretreated, natural abrasives such as pumice, emery, and garnet
2. Synthetic – electrically fused corundum, zirconia corundum, or silicon carbide of various degrees of purity and particle sizes
3. Special high performance grain

Thus, particularly high-quality materials (abrasives on fabric backings) are produced using an abrasive coating of hollow Hermesit® beads, whose walls

Table 6.86. Aqueous resoles for coated abrasives [a]

Resin	Paper	Cloth	Vulcanized fiber	Non-woven	Reactivity	Non-volatiles % ISO 8618	Viscosity at 20 °C mPa·s ISO 9371	Gel time at 100 °C ISO 9396	Gel time at 130 °C ISO 9396	Free Phenol % ISO 8974	Free formaldehyde % ISO 9397
A	x		x		Moderate	79±2	3600±300	60±5	8.5±1.5	7.5±1.5	1±0.5
B	x	x			High	74±2	2800±300	26.5±3.5	7±1	6.5±1.5	<0.5
C		x			Moderate	74±2	1000±200	48±8	9±2	<5.0	<5.0
D	x		x		High	79±2	4000±300	33±5	6±1	7.5±1.5	0.4±0.2
E	x	x			High	71.5±2.5	1950±350	26.5±3.5	6.5±1.5	<2	<0.5
F[b]	x	x			Extremely high	80±2	4000±300	4±1[b]	2±1[b]	6±1	<0.8
G	x	x			Moderate	67.5±2.5	1750±250	55±10	12±2	<2.5	<0.5
H	x	x			High	70±2	2000±200	30±5	7±1	<0.9	<0.9
J				x	Moderate	72.5±2.5	2050±250	40±10	8±1	<1.7	<0.9
K	x		x		Moderate	78±3	3000±600	58±8	9±1	7.0±1.5	1±0.5
L	x	x			High	74.5±2.5	900±100	40±8	8.5±1.5	<5	0.4±0.2
M	x	x	x		High	78±3	3000±300	42±8	8±2	6±2	0.7±0.2
N	x	x			High	70±2	700±100	40±8	8±1.5	<4.5	0.3±0.1

[a] Source: Bakelite AG, Iserlohn, Germany.
[b] 100 PBW Resin F
 15 PBW Latent Hardener F_H.

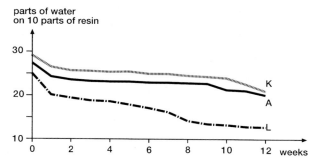

Fig. 6.138. Water tolerance of three different aqueous resoles during storage at 20 °C for 12 weeks (Source: Bakelite AG, Iserlohn, Germany)

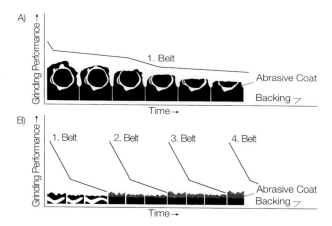

Fig. 6.139. Comparative performance of Hermisit® hollow bead grinding belts relative to conventional belts: A) slow drop in grinding performance using only one belt with Hermisit hollow beads; B) drop in grinding performance of about four conventional belts in an identical period; in both cases with simultaneous depiction of a cross section of the abrasive coat to illustrate attrition (photo: Hermes Schleifmittel GmbH & Co, Hamburg)

consist of abrasive grain and binders (Fig. 6.139). More than three times the grinding performance of conventional grinding belts can be achieved using belts produced on this basis, while the surface roughness of the machined work pieces changes only slightly.

Up to 50% inorganic fillers such as calcium carbonate in the form of chalk flour are added to the sizer coat used to bind the surface. Particularly in the case of coarse particle sizes, this reduces the shrinkage of the phenolic resin, and minimizes web warpages. In addition, fillers render the binder thixotropic to a certain extent, thus preventing dripping or uneven binder flow during the festoon drying process. Furthermore, they are beneficial by extending and improving the useful life and grinding performance of the abrasives.

Economical production of coated abrasives presently involves a considerable expenditure of capital for special coating equipment and infrastructure.

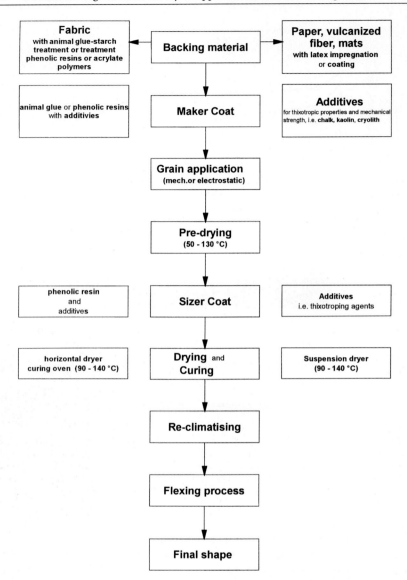

Fig. 6.140. Flow chart of the fabrication process used to manufacture coated abrasives (from [1])

Figure 6.140 shows a flow chart of the overall process and Fig. 6.141 a schematic of the fabrication process.

The webs of backing material (width up to 2 m) are unwound to a glue spreading machine (Fig. 6.142) where they are coated with a film of the appropriate binder maker coat by means of spreader rolls. The abrasive grain is then sprinkled into this or base lacquer by electrostatic scattering, in which an elec-

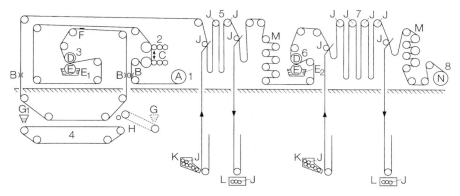

Fig. 6.141. The coating process (drawing: Hermes Schleifmittel GmbH & Co, Hamburg): (1) takeoff roll, (2) printer, (3) maker coater, (4) grain coating machine, (5) first dryer, (6) size, (7) main festoon dryer, (8) roll-up machine. A. backing roll, B. thickness measuring instrument, C. printing roll, D. applicator roll, E. pickup roll, E1. resin for first (maker) coat, E2 resin for second (size) coat, F. smoothing brush, G. grain hopper, mechanical deposition; G1. grain hopper, electrostatic deposition; H. throwing brush, J. rotating poles, K. rotating poles to supply, L. rotating poles deposit, M. tension rolls, N. jumbo roll product

Fig. 6.142. Glue spreader for production of coated adhesives (photo: Klingspor Co., Haiger Burbach)

tric field transports the abrasive grains against the force of gravity from the interior to the surface of the maker coat that "encapsulates" them. The maker coat with the abrasive grain that has been scattered into it is then dried and cured in an "intermediate rack." The festoon drying system is designed in such a manner that long loops of the backing webs are passed over rods that are drawn by a chain drive through the temperature zones of a drying/curing tunnel heated by circulating warm air (Fig. 6.143).

It is important that a holding time adequate for drying/curing of the phenolic resins is achieved. At the end of the dryer, the web is re-tensioned if required, and passed to a further roller coater or returned to the first adhesive spreader. A second layer of binder, referred to as the sizer coat, is applied to fix

Fig. 6.143. Festoon dryer for production of coated adhesives (photo: Hermes Schleifmittel GmbH & Co, Hamburg)

the abrasive grain permanently. Final curing is performed in the main drying tunnel following a curing program with temperatures between 100 °C and 140 °C. The design is similar to a festoon dryer.

Final curing in a "jumbo" roll in an oven represents another curing variant. Following final curing, the backing material is "re-acclimatized" or readjusted to an adequate moisture level. Before the abrasive web is taken up on a roll, it is subjected to a flexing operation in which the reverse side of the web is drawn over a steel roll or plate at an oblique angle under the pressure of a rubber roller; a large number of fine lateral and diagonal cracks are thus produced in the abrasive web, and provide the final product with the required flexibility. The product is then converted into finished goods (Fig. 6.144).

As in every manufacturing operation involving curable resins, productivity is an important aspect. To provide operational flexibility, resin systems exhibiting both moderate and high reactivity have been developed (cf. gel times in Table 6.86). The resins G and K listed in the indicated table may be termed moderately reactive and D, E, and H highly reactive. Furthermore the latter three resins feature a low level of monomers. Guidelines for selection of appropriate curing programs are provided for moderately reactive resin systems in Table 6.87 and for highly reactive systems in Table 6.88. The advantage of a slow curing process (refer to resin K) is that the volatile components can vaporize more easily and the microporosity of the abrasive material is relatively low. Factors such as productivity, low monomer emissions, and low microporosity must be balanced for optimum manufacturing economics.

Fig. 6.144. Finishing of coated adhesives (photo: Klingspor Co., Haiger Burbach)

Table 6.87. Curing program for moderately reactive resin systems, for example resins A and K in Table 6.86

Temperature program	Base coat		Sizer coat	
	Coarse grain size range	Medium & fine grain size range	Coarse grain size range	Medium & fine grain size range
To 75–80 °C	0.5 h	0.5 h	0.5 h	0.5 h
At 75–80 °C	0.5 h	–	1.0 h	–
At 90–95 °C	2.0 h	1.5 h	2.0 h	2.0 h
At 105 °C	0.5 h	0.5 h	1.0 h	0.5 h
At 120 °C	–	–	1.0 h	1.0 h
Total time	3.5 h	2.5 h	5.5 h	4.0 h

Table 6.88. Curing program for highly reactive resin systems, for example resins D, E, and H in Table 6.86

Temperature program	Base coat		Sizer coat	
	Coarse grain size range	Medium & fine grain size range	Coarse grain size range	Medium & fine grain size range
To 75–80 °C	0.5 h	0.5 h	0.5 h	0.5 h
At 75–80 °C	–	–	0.5 h	–
At 90–95 °C	1.5 h	1.0 h	1.5 h	1.5 h
At 105 °C	0.5 h	0.5 h	0.5 h	–
At 120 °C	–	–	1.0 h	0.5 h
Total time	2.5 h	2.0 h	4.0 h	2.5 h

Fig. 6.145. Curing program (sizer coat, coarse grain) for resoles with different reactivities

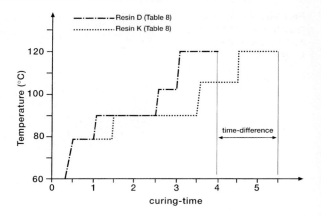

Figure 6.145 shows the difference in the curing programs used in a pilot plant for the moderately reactive resin K and the highly reactive resin D (from Table 6.86). The verified time difference is far from negligible, being around 25–30%. Highly reactive two-component systems [24] such as resin F in Table 6.86 are also used for some applications. Such highly reactive resin systems are primarily used in those instances where high-speed dryers requiring short curing times are used following the coating operation.

These resin systems represent a combination of an aqueous phenolic resole and a specific latent curing agent. The aqueous resoles themselves exhibit relatively low reactivity, and the binder systems only achieve their high rates of cure due to addition of up to 15% curing agent based on resin. Since the bench life of this binder mix (Fig. 6.146) amounts to at least 8 h, the product can be used up over the period of a shift at a coated abrasives manufacturer. Figure 6.147 shows the gel time of this two-component system as a function of the temperature and the level of curing agent, and compares the results with the corresponding figures for conventional resins. The gel times that may be achieved by this means (at 100 °C) are far shorter than those of fast curing resorcinol resin systems and considerably shorter than those of conventional, highly reactive, resorcinol-free phenolic resin binders. These coated abrasive systems can be used as base and sizer coat resins for production of coated abrasives using paper as a backing on the one hand, but on the other they are also very suitable for use as impregnating resins for fabric finishing. These systems are not suitable for production of abrasive fabric by full resin bonding or for fiber wheels, since the grinding performance levels are lower than in the case of products made using conventional, slow curing resin systems.

Specific resoles (resin C in Table 6.86) that are compatible with a wide variety of polymer emulsions and dispersions are also suitable for fabric finishing. The indicated resin C in combination with materials such as latex or PVA emulsions is also suitable for use in production of water-resistant abrasives. Various highly or moderately reactive resins – for example resins A and D in Table 6.86 – are particularly suitable in high-quality abrasive applications where phenolic resoles must exhibit high filler compatibility and cure to yield

Fig. 6.146. Viscosity at 20 °C depending on time (max. 8 h) of three different two-component systems (1 and 2 lower viscosity with F_H, 3 is resin F with F_H)

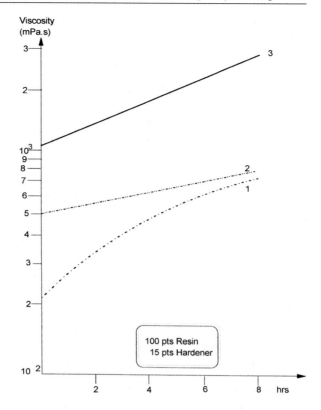

Fig. 6.147. Gel times of a two-component systems resin F and hardener F_H (Table 6.86) at 100 °C (curve 1) and 130 °C (curve 2) in comparison to the gel-time levels of resorcinol accelerated resins and standard phenolic resins (*gel-time at 100 °C)

F_H^{**} latent hardener

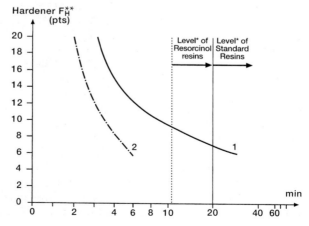

unusually viscoelastic, thermally stable products. Use of these phenolic resins assures bubble-free through curing, particularly when coarse-particle abrasives are employed, and thus has a beneficial effect on the already long useful life of the abrasive.

Products made with low-monomer resins such as C, L, and N in Table 6.86, particularly using medium and fine-grained abrasives, have been shown to exhibit a long useful life. Resins with extremely low monomer levels are suitable for use in base and sizer coats for full resin bonding in some particle size ranges. Moreover, they increase productivity appreciably.

Abrasive webs (non-woven) such as pot and pan scrubbers and scouring fabrics are closely related to coated abrasive products. In production of high-quality abrasive webs, the randomized fibers of materials such as polyamide or polyester are generally bonded using phenolic resins. This places specific demands on the phenolic resole being used, to prevent fiber damage. Phenolic resoles such as resin J in Table 6.86, that are particularly low in phenol and exhibit the high water tolerance required to enable their use in combination with latex binders, are best for this application.

As a rule, vulcanized fiber wheels (Fig. 6.148) are produced using full resin bonding. The use of specific grades of fiber is a prerequisite for production of vulcanized fiber wheels. The fiber should exhibit high transverse tear strength, high peel strength of the individual layers, and adequate elasticity. It is critical that the fiber undergoes no deformation, or only slight deformation, during the curing process, and that it does not tend to deform, much less curl, after re-acclimatization. The technical parameters of the equipment generally represent the key aspect in selection of the correct type of resin for production of vulcanized fiber wheels. Depending on the length

Fig. 6.148. Automotive repair – grinding after panel beating, vulcanized fiber disks on an electric angle grinder (photo: Hermes Schleifmittel GmbH & Co, Hamburg)

Fig. 6.149. Grinding of metal with abrasive belts, deburring of an L-profile on a backstand (photo: Hermes Schleifmittel GmbH & Co, Hamburg)

Fig. 6.150. Automotive repair – sanding of filler, strips with holes on electric orbital sander (photo: Hermes Schleifmittel GmbH & Co, Hamburg)

of the festoon, the temperatures that can be reached, and the required throughput time, phenolic resoles exhibiting various levels of reactivity and curing rates can be used, for example resin A in Table 6.86 for moderate and resin D for higher curing rates. In its main features, the production method itself corresponds to those used for other full resin bonded abrasives. The abrasive grain used for vulcanized fiber are similar to those for abrasive fabric.

The following may be noted (Figs. 6.149–6.152) as a final comment on the use of coated abrasives. Abrasive papers (strips, sheets, die-cut sections, leaf wheels, belts, broad belts) coated with electrically fused corundum, silicon carbide, or garnet grain are generally used for dry and wet sanding and polishing of wood, wooden materials, certain metals, plastics, painted surfaces, filled areas, and leather. Abrasive fabric (rolls, sheets, strips, die-cut sections, leaf wheels, belts, shells, flap wheels, small bonded abrasives) is best for dry and wet grinding of metal, low and high alloy steel, cast iron, chilled iron, nonferrous metals, ceramics, stone, glass (Fig. 6.153), and composites.

Fig. 6.151. Grinding of a square tube with a portable belt grinder (photo: Hermes Schleifmittel GmbH & Co, Hamburg)

Fig. 6.152. Removal of a stainless steel welding seam using a flap disc (photo: Hermes Schleifmittel GmbH & Co, Hamburg)

Animal glue (full bonding), animal glue with phenolic resin, animal glue with urea resin, and full bonding with phenolic resin are the types of binders used for the abrasive paper group. Although animal glue bonding of abrasive fabric is also possible, full bonding with phenolic resin is always used for heavy-duty applications. As previously noted, the latter also applies in the case of vulcanized fiber leaf wheels used for metalworking. Full bonding with phenolic resin is also encountered in fabric/paper and foil/paper combinations used in the form of sheets or belts for dry sanding of veneered wood sheets, chipboard, rigid fiberboard, and paper laminates.

Fig. 6.153. Glass grinding with abrasive belts, beveling of crystal glass rim (Photo: Hermes Schleifmittel GmbH & Co, Hamburg)

6.2.3
Friction Linings

6.2.3.1
Introduction (General Information)

Phenolic resins are used to a considerable extent in production of friction, brake and clutch linings for the automotive industry and for brake elements in rail vehicles and machines (Figs. 6.154 and 6.155).

Friction lining compounds used to manufacture disk brake pads, passenger car (Fig. 6.156) and truck drum brake linings, and clutch linings may essentially be regarded as fiber-reinforced phenolic molding compounds containing 5–35% resin. A large number of different, generally complex combinations of materials containing up to 25 or more individual components are used in these formulations. The main components are binders (preferably phenolic resins and rubber), mineral-based and synthetic organic fibers, metal fibers and chips, and organic/inorganic fillers. The required range of lining properties depends to a very great extent on the type of vehicle, the brake systems, required test schedules, mechanical conditions, and similar factors.

This results in a wide range of unmodified and modified products, novolaks and resoles, powdered and liquid resins, and resin solutions for this diverse field of phenolic resin technology [1–5]. The manufacturing processes for friction linings, which are briefly described, are as diverse as the formulation. The range of properties and behavior patterns of brake and clutch linings thus depends to a great extent on the nature and quality of the binders being used.

Fig. 6.154. Friction linings for disk and drum brakes (photo: Rütgers Automotive AG, Essen)

Fig. 6.155. Friction linings for rail vehicles (photo: Rütgers Automotive AG, Essen)

Fig. 6.156. Brake straps and linings for drum brakes (photo: Rütgers Automotive AG, Essen)

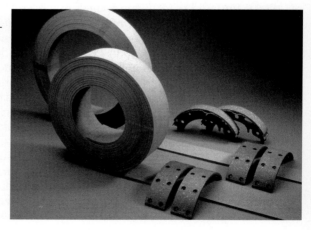

The quality of friction linings is largely assessed on the basis of the following properties: coefficient of friction, wear (including disk wear if appropriate), comfort behavior, and temperature dependence of these parameters. Phenolic resins have been widely modified to meet various frictional requirements.

The fact that the demands on friction linings – for example with respect to their service life and brake behavior – have increased considerably during the last 20 years has similarly led to comparable demands made on the binders. During the brake operation, the kinetic energy of the vehicle is largely converted into heat; peak temperatures of more than 800 °C can arise at the surface of the brake linings, depending on the stress and type of vehicle.

The main function of phenolic resins in friction linings is temporary bonding which, as defined in Sect. 6.2 (p. 291), provides the materials with the highest possible resistance to wear when these materials are subjected to abrasive and destructive use, coupled with high thermal resistance over a long interval and due to sudden heat development or auxiliary mechanical stress. Yet the other five phenolic bonding functions participate as "auxiliary functions" due to the excessive demands placed on phenolic resins as binders in friction linings.

Thus, friction linings are expected to exhibit permanent dimensional stability in the brake system, and feature high strength levels in intermediate handling (for example during fabrication and installation). The friction lining, bonded with phenolic resins, is then expected to withstand "temporarily" (the main phenolic resin bonding function) the high temperatures that develop during brake or clutch operations, temporarily in this case representing the longest possible service life. In the case of friction linings, the other auxiliary functions follow: complementary bonding function is represented by the compatibility and modification capability of phenolic resins with other thermosets or elastomers used in the lining compound, as well as with the fillers and fibers. The *carbon-forming* function of phenolics is equally important since carbonization will occur under pyrolysis conditions prevalent during the braking operation. Carbon formation plays a key role in the performance of the brake operation. Volatiles or liberated pyrolyzates can somewhat adversely affect the brake operation. The *chemically reactive* function of phenolic resins represents chemical transformations that occur with other brake lining components and improves the quality of the overall composite in the brake or clutch lining. This function also applies to post-cure and "resin rearrangement" processes that occur during the high temperature braking operation.

Thus the interrelationship of all these bonding functions and the high degree of effectiveness underscores the importance of phenolic resins as the preferred binder for these friction products for many decades. On a cost/performance basis in conjunction with bonding function, it is difficult to displace phenolic resins as the premier binder for high quality friction products.

6.2.3.2
Demands on Friction Linings

A major demand on brake and clutch linings is that they have to be effective over a wide temperature range. As is conveniently possible, the coefficient of friction [6] of brake linings should be independent of the temperature. For a long time, the index or coefficient of friction – the ratio of the frictional force to the normal force acting on a body – was considered a material constant for friction between two bodies. After friction linings were developed (for railway and automotive use) it was found that the coefficient of friction is a function of the temperature, surface velocity, and applied force. The pressure, temperature, and velocity loads are illustrated in Fig. 6.157 and represent stresses that may be expected under driving conditions. The coefficient of friction as a function of temperature resulting from speed (during braking) is affected by many factors including the organic components of the brakes and also the phenolic resin fraction.

Because of the smaller dimensions of disk brake pads compared to drum brake linings, the energy absorbed per unit area in the case of pads used in disk brakes (Fig. 6.158) is very high compared to the energy for drum brake linings (Fig. 6.159); thus, very high temperatures can arise in disk brakes. Good fading behavior, i.e., the temperature-dependent coefficient of friction remains constant to an acceptable degree, is a technical consideration. In addition, the brake lining must afford the longest possible service life, i.e., undergo low wear [7], should spare the other contact material -for example the brake disk – as much as possible, and be resistant to the effects of weather. As a function of temperature, lining wear is highly dependent on the type and level of the organic binder. The wear may be reduced by special modification of the binder.

Aside from the above considerations, the "comfort behavior" of linings nowadays represents a major factor. The term "comfort behavior" refers to the levels of noise generation (squeal) and cyclic behavior known as "judder," which should be as low as possible. The heat generated during braking can lead to momentary unevenness and deformation of the brake disk (in case of high-speed braking) that exert a perceptible effect on the driving/braking behavior. The judder caused in this manner must be compensated by the elastic behavior of the friction lining. As a general rule, judder is reduced as the modulus of elasticity decreases.

"Squeal" or generation of the other noise during braking, which is no doubt generally perceived as very unpleasant by the driver, usually occurs during low-speed braking, i.e., at low energy conversions. Brake system components capable of vibration (saddle, disk, and brake pads) are greatly involved in this phenomenon. If the friction lining acts as a dampening component in the brake system, the binder may be part of an approach designed to dampen vibration and reduce noise. The level of dampening depends to some extent on the resulting lining density which is related to the pressing operation [4]. Low compression rates afford slightly higher dampening levels. Aside from the compression rate which is purely mechanical, the proper design of phenolic resins is of considerable importance with the use of alkyl phenol-modified resins providing higher dampening levels.

Many different processes and formulations are presently used in manufacture of friction linings. Both drum linings and disk brake pads can be produced from compression molding compounds representing complex blends of numerous different components.

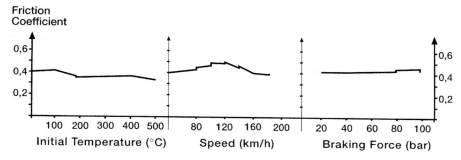

Fig. 6.157. Coefficient of friction – relationships (temperature, speed and braking force) (from [4])

Fig. 6.158. Disk brake (photo: Rütgers Automotive AG, Essen)

Fig. 6.159. Drum brake for trucks, wheel set (photo: Rütgers Automotive AG, Essen)

In development of a friction lining, the properties of many components are frequently combined on the basis of lengthy empirical test formulations. Although the effect of the binder on the final properties of the lining is limited, it is not negligible, since it depends on the overall formulation.

Aside from such considerations, the selection of phenolic resins depends on factors such as:

1. The process used to manufacture friction linings (process engineering parameters)
2. The final properties of the friction lining (material properties, purely technical parameters)
3. Environmental and workplace aspects

It is also generally impossible to predict unequivocally the effect of an isolated component on the properties of the overall system due to the complexity of brake lining systems. Repeated practice-related or direct practice trials (for example, dynamometer or practical driving tests) to examine the effects of changes in individual components are thus unavoidable.

6.2.3.3
Composition of Friction Linings

The exact compositions of industrially fabricated friction linings generally represent a portion of company-proprietary information. Due to the major efforts and difficult detailed work involved in developing them, they are, as a rule, not disclosed.

The basic design of compounds for friction linings (Table 6.89) essentially includes fibers, fillers, metals, lubricants and other additives, and binders.

Before the early 1980s, asbestos fiber was used worldwide as the *basic fiber material*. For toxicological reasons, the use of asbestos fiber was later not only limited, but finally prohibited in most countries [8]. Asbestos fiber is still used in some locations in North America (particularly Canada). However, it may be noted that asbestos fiber can generally be replaced by other fibers/fiber blends [9–11].

Asbestos fiber is problematic, since it can induce malignant cell growth when it enters the lungs by way of the respiratory tract if a certain ratio of fiber length to cross-section is present. Depending on the end use, inorganic and/or organic fibers are now generally used. Metallic fibers are also employed (Table 6.90); carbon fibers [12, 13] are primarily used in production of high-performance linings.

Table 6.89. Formulation guidelines for friction linings

		% w/w	
	1.	0–60	Fibers (organic, inorganic, metallic)
	2.	5–30	Fillers (organic or inorganic)
	3.	5–70	Metal (as powder or chips)
	4.	0–5	Lubricants and other additives
	5.	8–25	Binders (phenolic resins, rubber)

Table 6.90. Asbestos fiber substitutes (from [5])

1. Organic fibers	Aramid fibers
	Cellulose fibers
	Polyester fibers
	Polyacrylic fibers
2. Carbon fibers	Various grades
3. Mineral fibers	Glass fibers
	Slag/rock wool
	Gypsum fibers
	Ceramic fibers
	Silicate fibers
4. Metallic fibers	Steel fibers
	Brass fibers
	Copper fibers
	Zinc fibers
	Aluminum fibers

Asbestos-free linings (Fig. 6.160) based on metals can be produced, but metallic and semimetallic linings [14–16] cannot be used universally for various reasons (for example, due to corrosion). Work on replacement of asbestos fibers first concentrated on glass fiber-reinforced linings that were suitable for applications such as drum brake linings. Due to the extended range of requirements in disk brakes, which are exposed to higher stresses, it was necessary to employ a mixture of different fibers and fillers as a substitute. Depending on the composition, the sum of properties offered by asbestos-free linings can be markedly superior to that of asbestos-containing linings.

Since the quality and cost of the fibers vary greatly, the formulating task involves finding a price/performance optimized formulation that considers both demands and costs, and that fully meets the required performance profiles, for example, on a flywheel testing stand (Fig. 6.161). With reference to the fibers, the logical consequence is that mixtures are generally used. The phenolic resins that are used must exhibit good adhesion to the various types of fibers. Table 6.91 lists suggested formulations for asbestos-containing, semimetallic lining compounds, and asbestos-free friction lining compounds made with substitute fibers for manufacture of disk brake pads.

The *fillers* used are both organic and inorganic in nature. Inorganic fillers can be materials such as barytes, metallic oxides, kaolin, powdered slate, or powdered mica. One of the most common organic fillers is hardened cashew nutshell oil (CNSL) that is marketed under the names of "friction dust" or "friction particle."

Cashew nutshell oil [17] is obtained from the shells of cashew nuts. The commercial product consists of about 90% cardanol, which arises through decarboxylation of anacardic acid, and 10% cardol (Fig. 6.162). Catalytic hardening of the unsaturated side chains of cardanol or cardol leads to a polymeric product, the "friction particle" or "friction dust." The oil (CNSL) may also be used to modify phenolic resins, resoles, or novolaks that are similarly used for manufacture of friction linings.

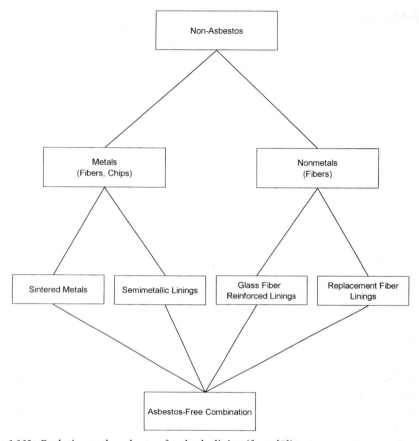

Fig. 6.160. Evolution to the asbestos-free brake lining (from [5])

Fig. 6.161. Flywheel Testing Stand for Trucks (Photo: Rütgers Automotive AG, Essen)

Table 6.91. Various friction lining mixtures (from [5])

	Containing asbestos	Semi-metallic	Asbestos substitute
Fibers	10–30	0	0–30
Metals	10–40	60–75	10–45
Lubricants	0–15	15–30	10–25
Fillers	0–30	0–5	0–20
Binders (Phenolic resins, rubber)	8–15	7–10	7–12
Friction enhancers	0–3	0–5	0–5

Fig. 6.162. Components of cashew nutshell liquid (CNSL): 90% cardanol, 10% cardol

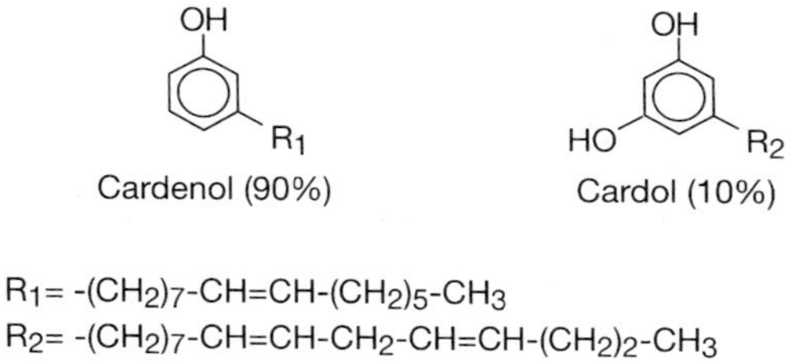

Cardenol (90%)   Cardol (10%)

$R_1 = -(CH_2)_7-CH=CH-(CH_2)_5-CH_3$
$R_2 = -(CH_2)_7-CH=CH-CH_2-CH=CH-(CH_2)_2-CH_3$

A further component of friction linings may consist of *metals* such as iron, copper, or brass in the form of powder, wool, or chips.

In addition, *lubricants* such as antimony sulfide or molybdenum sulfide are used.

Binders provide the friction lining with the required mechanical strength. Even today, phenolic resins still represent the most important binders. Rubber materials such as SBR or NBR rubber are also members of this group.

6.2.3.4
Phenolic Resins and Properties of Friction Linings

The binders used for friction linings include modified and unmodified, generally phenolic novolak/hexamethylene tetramine-based powder resins, and in the case of wet conditioning liquid resins, generally aqueous resoles. Table 6.92 provides a survey of powder resins used for manufacture of all types of friction linings. Table 6.93 lists three liquid resins that may be used both for manufacture of wet conditioned brake linings and for impregnation of clutch linings.

Phenolic resins are adapted to the different requirements by various modification and/or use of special grades exhibiting specific reactivities or viscosities. The crosslinking density is controlled by varying the hexamethylene tetramine level. At present, novolak-hexa powder resins with a hexa level of 6–14% are mainly used in the friction lining industry; the bulk of these resins exhibits a

Table 6.92. Powdered resins as binders for friction materials (examples from [1])

Modifier:	A	B	C	D	E	F	G
	no	no	no	Accelerator	NBR rubber	Phosphor, boron compounds	Epoxy resin
Flow distance at 125 °C: ISO 8619	14–20	24–30	45–55	40–50	12–16	20–30	18–28
B-transformation time at 150 °C, s. ISO 8987:	90–170	90–150	70–130	30–60	40–80	60–120	60–120
Hexa content: ISO 8988	6–7	8.7–9.3	13–14	8.5–9.5	8–9	6.3–7.3	6.3–7.3
Residual phenol% ISO 8974:	<0.5	<0.5	<0.5	<0.5	<0.5	<0.5	<0.5

Table 6.93. Liquid resins for production of friction linings

	Chemical characterization	Non-volatiles	Viscosity at 20 °C mPa·s	Gel time at 100 °C min	Gel time at 130 °C min	Free phenol	Water thinnability	Storage life at 10 °C months
H	Aqueous Phenolic Resole	67±2	350±50	45±5	7±1	8±2	100:45–55	2
J	Aqueous Phenolic Resole	71.5±2.5	1950±350	26.5±3.5	6.5±1.5	<2	>10:40	2
K	Aqueous Phenolic Resole	79±2	3600±300	60±5	8.5±1.5	7.5±1.5	10:20–30	3

hexa level of 7–10% (resins A and B in Table 6.92). Use of a low hexa level achieves relatively low hardness, which in turn leads to improved flexibility and good comfort properties. In contrast, a high level of hexa (resin C in Table 6.92) affords high hardness and high thermal resistance, but may also lead to elevated wear. The flow distance also influences the properties of friction linings. The flow distance symbolizes a combination of the curing properties and the melt viscosity. A short flow distance (such as that of resin A in Table 6.92) can lead to an open-pore structure of the linings, improving the comfort behavior.

However, the fabrication parameters must be exactly controlled and maintained when resins with short flow distances are used. Long flow distances positively affect the structure and incorporation of the fiber/filler. This also applies to the processing properties. A negative aspect is the tendency to form bubbles.

The gel or "B" times of the liquid powder or resins relate to pressing times and thus the productivity. The reactivity of the resin being used determines the productivity in fabrication of friction and clutch linings by the hot compression method to a major extent. The variety of resins exhibiting different reactivities (for example resins C and D in Table 6.92) allows optimal pressing conditions to be achieved when the reactivity is matched to the overall formulation of the friction lining and the production process. Standard reactivities (B-time around 120 s at 150 °C) allow problem-free charging of the compression mold. In addition, the tendency to form bubbles – resulting from the relatively uniform liberation of ammonia during the pressing operation – is minor. Demands for increased productivity led to the development of special powder resins (for example resin D in Table 6.92) that exhibit a markedly decreased reduction in the initiation point of the novolak-hexa reaction cure in addition to the extremely high reactivity already achieved. This occurs to increase productivity, particularly in the case of thick linings such as drum brake linings for trucks. When highly reactive resins are used, exact process control is required during fabrication.

Over the past years, the levels of free phenol in novolak-based powder resins has been markedly reduced (to below 0.5%, cf. Table 6.92). Use of phenolic resins with a low level of free phenol offers the following advantages:

1. Superior environmental compatibility
2. Reliable compliance with workplace limits (phenol)
3. Increased quality and consistency of binders

Powder resins with a free phenol level of less than 0.02% are presently also available if required to meet special requirements. Dustfree powder resins for friction materials are also possible (see Sect. 6.2.2.5, p. 326–7).

Modified resins (for example resins E, F, and G in Table 6.92) are used to achieve specific final properties in friction materials. Depending on the type of modification, these resins may chemically represent copolymers (for example, alkylphenol-modified products), homogeneous mixtures (for example, rubber-modified resins such as resin E in Table 6.92), or blends with special polymers. More "flexible" linings with good comfort properties are achieved by modification with acrylonitrile rubber. Use of chlorinated rubber as a modifier increases the thermal resistance and achieves improved fading behavior (reduces the drop in the coefficient of friction as the temperature increases) as well as good comfort properties.

Use of epoxy resins (for example in resin G, Table 6.92) affords a certain flexibility, high strength, and excellent adhesion to fibers and fillers. The adhesion properties and flexibility of the linings could similarly be influenced by incorporation of long-chain alkylphenols. Phenolic powder resins containing alkylphenols represent alternatives to CNSL-modified resins, and offer greater product consistency. Incorporation of heteroelements such as boron, phosphorus, or nitrogen (for example in resin F, Table 6.92) produces high thermal resistance. This characteristic may be utilized in applications such as manufacture of special linings for motor racing.

Aside from the above, other types of resins – such as phenoxy and melamine resins – are also used as modifiers to meet special requirements. Compared to unmodified (novolak-hexa) phenolics, elastomer-modified phenolic resins exhibit improved wear properties at a relatively low pressing force per unit of lining area and temperatures up to about 250 °C under test conditions and in practice.

Boron/phosphorus-modified resins (Fig. 6.163) generally exhibit somewhat improved thermal behavior with respect to the temperature dependence of the friction coefficient and the increase in pedal pressure required in repeated brake operations. Figure 6.164 shows a simplified presentation of the lining and disk wear in disk brake systems as a function of the binder. It

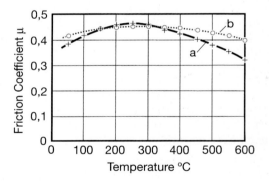

Fig. 6.163. Thermal behavior of friction linings made from (a) unmodified and (b) boron-modified phenolic resins; relationship of the friction coefficient μ and the temperature

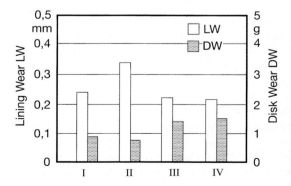

Fig. 6.164. Lining (LW) and disk wear (DW) when using variously modified phenolic resins at an operating pressure of 20 bars, I = alkylphenol resin; II = elastomer-modified phenolic resin; III = unmodified phenolic resin; IV = boron-modified phenolic resin (source: Rütgers Automotive AG, Essen)

Fig. 6.165. Selection of clutch linings (source: APTEC Reibtechnik GmbH, Leverkusen)

may be seen that the lowest lining wear in this comparison (operating pressure 20 bar) is achieved with boron/phosphorus-modified resins. It is slightly higher in the case of unmodified phenolic and alkylphenol resins, and is particularly great in elastomer-modified resins. The four resins afford exactly the reverse results with respect to wear of the metallic brake disk. The greatest disk wear is present in the case of the boron-modified phenolics, and the least in that of the elastomer-modified resins. These examples demonstrate that a large number of possibilities exist to achieve special combinations of properties by blending various resin systems. This applies equally to manufacture of brake, clutch, and machine linings, and can contribute to the high quality of these products.

6.2.3.5
Manufacture of Friction Linings

The manufacture of drum brake linings, disk brake pads, and clutch linings (Fig. 6.165) – all of which may be produced by dry or wet conditioning – is differentiated into three stages:

1. Production of the mix
2. Shaping/compression molding (hot, warm, cold)
3. Final curing (in clamps or free)

The process flow charts in Figs. 6.166–6.168 survey the manufacture of the various types of linings such as drum brake linings (Fig. 6.166), disk pads (Fig. 6.167), and clutch linings (Fig. 6.168).

These schemes, which are graphically complex, may also be briefly categorized in three processes.

6.2.3.5.1
Process 1 (Dry Mixes for Hot Pressing or Warm Shaping with Subsequent Oven Curing)

The fiber material, which has been finely divided by pretreatment, is homogeneously blended with all other components in an appropriate mixer. Be-

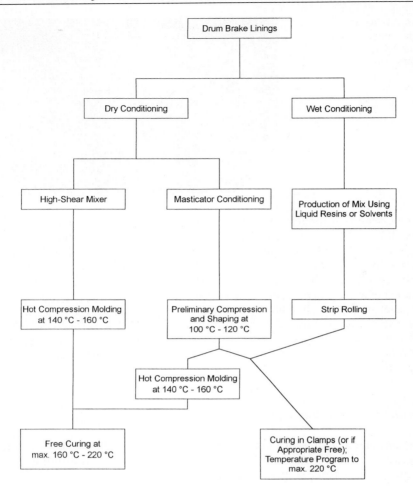

Fig. 6.166. Manufacturing process for drum brake linings

cause of its large volume, this dry mix may initially be pre-compressed at 3–10 N/mm² and ambient temperature to yield compression molding blanks.

In manufacture of drum brake liners, a considerable fraction of rubber is generally incorporated into the dry mix in the form of a milled "masterbatch" consisting of rubber, vulcanization additives, fibers, and fillers. The drum brake liners may be pre-cured by warm pressing of the compression molding blanks at 110–120 °C to yield sheets; these are then shaped to produce a curvature of the desired radius, and then completely cured in clamps in an oven. Presently, drum brake linings are also produced to a considerable extent by direct hot compression molding. Disk brake pads are produced following thorough mixing or mastication of the components by compression molding of the friction lining compounds onto cleaned, adhesive-coated metal base plates.

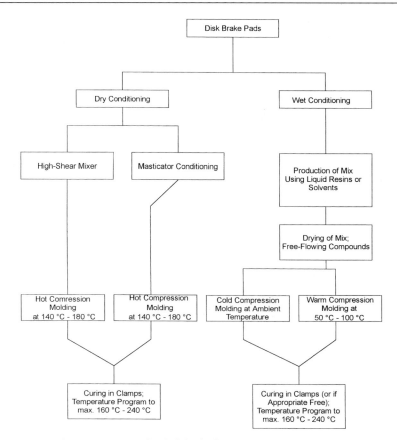

Fig. 6.167. Manufacturing process for disk brake linings

6.2.3.5.2
Process 2 (Wet Mixes for the Calender and Extrusion Processes)

In wet conditioning, mixtures of fibers, fillers and binders with added solvents, resin solution, or liquid resin are thoroughly homogenized in a ribbon blender or masticator at ambient or elevated temperatures to form a doughy mass. Masticized rubber can be blended in with solvents. The plasticized material may be shaped on calender mills, piston presses, or extruders. Following the drying process, the products are oven cured in pressure molds or hot pressed in compression molds. Machine and drum brake linings are best manufactured by this process.

6.2.3.5.3
Process 3 (Impregnation of Fiber Textiles and Yarns)

Fiber textile-based brake linings for mechanical equipment and clutch linings are manufactured from yarns using the impregnation process. The textile or

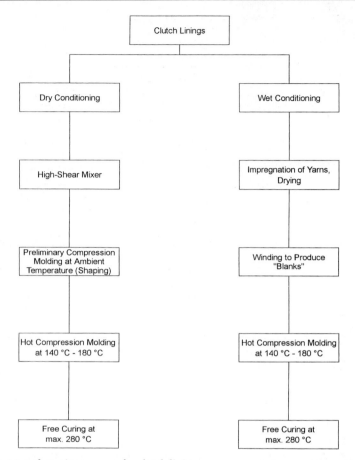

Fig. 6.168. Manufacturing process for clutch linings

yarn is impregnated with the liquid resin or resin solution (that can also contain added rubber solutions or dispersions and other special additives) using the dipping method. The material is then gradually dried at about 50–90 °C to remove solvents, and pre-cured if appropriate. The textile is shaped by a calendering operation, followed by oven curing in pressure molds or hot compression molding at 15–20 N/mm^2 and 150–170 °C. In manufacture of clutch rings by the winding process, the yarns are coiled on a special winder to form a ring, which is then hot pressed in special molds.

The pressing and curing conditions for all three processes are listed in Table 6.94.

Production costs are considerably reduced and the quality/consistency of the linings radically increased by largely automatic ("on-line") fabrication (Fig. 169) of disk brake linings [18]. On-line fabrication comprises mixing, curing, working, and coating on a uniform production line with few operating personnel.

Table 6.94. Pressing and curing conditions for brake linings

	Shaping	Shaping and curing (hot pressing)	Curing (oven, microwave, etc.)
Dry mix	cold: $P = 100-250$ N/mm^2 or at 50–80 °C	$T_A = 140$ °C – 180 °C $P_E = 20-80$ N/mm^2 Pressing time: 1/2–1 min per mm thickness, with airing periods	a) Clamped following cold shaping b) Unconstrained following hot shaping $T_A = 80$ °C $T_E = 280$ °C (max) Duration: 14–20 h
Wet mix	cold: $P = 100-250$ N/mm^2 or, at 50–80 °C	$\frac{1}{n}$	Clamped only $T_A = 80$ °C $T_E = 280$ °C (max) Duration: 14–20 h
Wet conditioned and pre-dried mix	cold: $P = 100-250$ N/mm^2 or at 50–80 °C	$T = 140$ T – 180 °C $P = 30-60$ N/mm^2 Pressing time: 1/2–1 min. per mm thickness, with airing periods	As in dry mix

P = Pressure.
P_E = Final pressure.
T = Temperature.
T_A = Initial temperature.
T_E = Final temperature.

Fig. 6.169. Automatic rotary presses (photo: Rütgers Automotive AG, Essen)

6.2.4
Auxiliaries for Petroleum and Natural Gas Production (Proppant Sands and Tensides)

Phenolic resins and their derivatives can be of use as auxiliaries in production of petroleum and natural gas. This refers both to their use as binders for sand or similar aggregates used to reinforce the bore holes and surrounding formations, and to their use in producing special tensides, for example, from alkylphenol resins, that can be employed in secondary or tertiary oil recovery [1]. Thus, phenolic resins are used for exploitation of deposits and for production of oil and gas.

The pertinent technology encompasses two important technical challenges: (1) to find the deposits and investigate their removal and (2) to produce the raw materials economically and utilize the deposits to the fullest extent possible [2]. The techniques used to exploit deposits of oil and gas are basically comparable. Deep bore holes used to find and explore for deposits are termed exploratory wells. Bore holes used to produce oil are called production wells [3]. The deepest wells can reach as far as 10,000 m into the earth. The rotary drilling method (Fig. 6.170) still remains the most important deep drilling procedure. In this method, torque is transferred from a rotary table to the drilling rod. A tool is located at the tip of the drilling rod, and is controlled

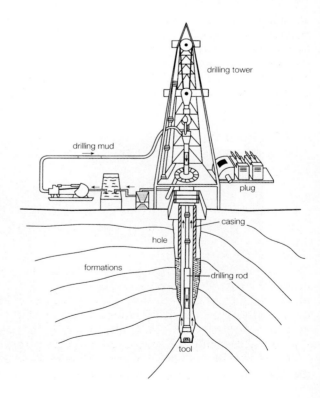

Fig. 6.170. Rotary oil well drilling rig (from [3])

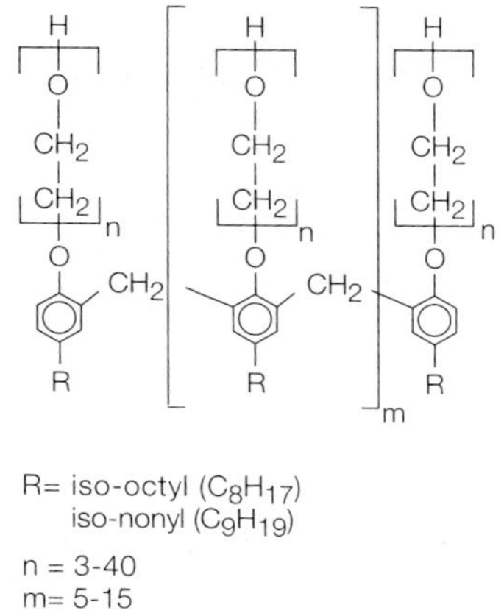

Fig. 6.171. Alkoxylated alkylphenol novolaks for use as tensides in oil production; The $(-CH_2-CH_2-O-)_n$ moiety can be replaced by $(-CH_2-CH_2-CH_2-O-)_n$

R = iso-octyl (C_8H_{17})
iso-nonyl (C_9H_{19})
n = 3-40
m = 5-15

by the torque and the pressure exerted by the mass of the drilling rod. The rock crushed by the drill is removed by flushing it through the space between the drilling rod and the wall of the bore hole. The drilling mud, a thixotropic fluid, is used to cool and lubricate the drilling tool as well as to clean the bottom of the bore hole. The mud can be reused after the drilled material has been removed by filtration. In the turbine drilling method, the tool attached to the drilling rod is rotated with the help of a turbine. The turbine is mounted in the drilling rod, and is driven by oil circulation. If petroleum is exclusively produced by its own pressure, the process is termed eruptive production (gushing). However, this generally only occurs for a short time. The oil is then removed by pumping. To further increase the oil production (secondary recovery), water is injected into the deposits, the oil displaced from these, and forced toward the production tool. The tensides are used as de-emulsifiers to separate water and oil. Ethoxylated and/or propoxylated resins based on nonyl- or isooctylphenol (Fig. 6.171) are frequently used as tensides. As the oil-bearing deposits are exploited, the amount of water in the oil/water mixture increases, and production is halted at a specific level of water dilution.

In exploitation of oil and gas wells, hydraulic fracturing represents an economic way to stimulate a well's production rate. Underground formations exhibit weaknesses, and when the wellbore is pressurized with a fluid, the wellbore formation will break apart or "fracture." As the fracture grows, it extends deep into oil and gas bearing sands. However, the fracture can close or heal unless it is filled with particles like sand or ceramic pellets. These particles are called "proppants." Proppants are essential to the success of hydraulic fracturing and today are credited with results where wells are stimulated to produce up to ten times the original production rate. Since the late 1970s, phenolic resins have been used to coat sand and ceramics for use as proppants [4–7]. Resin coated proppants not only keep the created hydraulic fractures in oil and gas formations open; they also lock the proppant in the fracture to prevent flowback.

The production of proppant sand is similar to that of coated sand used for the shell molding (in Germany "Croning") process for fabrication of molds and cores in the foundry industry (see the Foundry Binder Sect. 6.2.1.3, p. 300). The principal difference generally lies in the fact that the level of process-related precure during sand coating must be modified for the later end use of the proppant sand. In many cases it may be presumed that final curing will take place in situ in the subterranean formation, since relatively high temperatures are prevalent in most cases. On the other hand, the precure must be relatively far advanced to achieve particularly good abrasion resistance. In practice, the coating may be carried out in multiple stages (for example three coats), the initial coatings receiving an advanced level of precure and the last coat a lower level. Details are available in the extensive patent literature.

Various lubricants can be used as additives in the coated sand to increase the abrasion resistance [8]. Various types of wax, silicones, and similar materials are used as such additives. Incidentally, the brittleness of the coating can also be reduced by addition of polyvinyl butyral as a plasticizer. The strength may be increased with materials such as the conventional types of silanes also used in the foundry industry. The total resin levels (4–5%) are higher than in the shell molding process, and the addition level of hexamethylene tetramine (when a novolak is used as the binder) ranges from 13% to 17% based on novolak. A higher level of precure may be used to ensure that the proppants retain particularly good gas and liquid permeability, a property that is important for production of oil or an oil/water mixture.

Resoles are only used as coating resins to a limited extent [9]. Earlier, the resole resins were adjusted in their properties so that the natural heat present in the geological formations sufficed to cure them in situ over periods ranging from only a few days to 60 days and more. The compact, resin bonded material that resulted generally featured relatively high compression strength and permeability, properties that are of great importance for production of petroleum. The disadvantage in the use of resole resins is the limited storage life of the sand, since the resin tends to cure partially even before use, leading to a reduction in its subsequent strength.

There were initial complaints that the desired compression strength of the proppant mass was not achieved when the resole was replaced by a novolak. A possibility of improving this situation consists of using novolaks containing a large fraction (up to 90%) of o,p'- and/or o,o'-dimers [10].

The patent literature on proppants is quite extensive [11–20]. According to the patent literature, sand coated with an acid-reactive resin can also be used as proppant sand to permit curing at low temperatures. Phenol, furfuryl alcohol, formaldehyde co-condensates, and combinations of resoles with novolaks are recommended as coating materials. These are precured during the coating process by addition of latent curing agents so that a storable, dry, free-flowing material is obtained. After these proppants have been injected into the subterranean formation, where in this case relatively low temperatures prevail, they are treated with organic or inorganic acids dissolved in solvents. The desired cure then takes place even in the absence of elevated temperatures.

As indicated above, technologies involving injection of water into the oil-bearing formations have already been used for approximately 60 years in petroleum production. Simultaneous oil production and water injection, termed "secondary recovery," has meanwhile become an established procedure [21]. Secondary oil recovery techniques increase the petroleum yield to about 30–40% of total. Yields of 55–60% are only obtained in exceptional cases. Methods termed "tertiary" processes have been developed to raise the yield further, and are classified into two types: (1) application of heat [22] or (2) chemical methods [23].

Of the chemical methods, an important one is addition of tensides. The surface tension at the oil-water interface is reduced by such surfactants, and a lower residual oil saturation level is thus achieved. Such processes have been used on an industrial scale, but the economic advantage is reduced due to the high costs of investment, energy, and expensive chemicals. Thus, tertiary oil recovery processes are only of significance if there is a dire need for increased petroleum yields in times of raw material shortages.

6.3
Complementary Bonding

Phenolic, cresol, xylenol, and alkylphenol resins are combined in various ways with other thermosetting resins such as epoxies or different types of elastomeric products in the areas of coatings, adhesives, sealants, and rubber mixes. In these applications, phenolic resins act as complementary resins; their additive effects provide the overall system with specific properties including chemical resistance, weathering stability, enhanced adhesion and cohesion, and adhesion to metal. A typical example is production of protective interior can coating from combinations of phenolics with epoxies or polyester resins. Specific phenolic resins can be used as additives to increase the film hardness and to optimize the corrosion resistance results and "throwing power" of electrophoretic coatings based on various binder systems.

Complementary bonding is also important in the production of rubber mixes, particularly those used to fabricate automotive tires. Phenolic resin-hexamethylene tetramine systems provide the rubber mixtures with increased strength and abrasion resistance.

For the sake of completeness, it should also be mentioned that complementary bonding can also exist in applications involving modified phenolic resins, i.e., combination systems in which the range of properties is still largely dominated by the phenolic. Examples are rubber-modified binders for brake linings, epoxy-modified phenolics for production of high burst strength grinding wheels, and tung oil modified phenolic (impregnation) resins for cover sheets of electrical paper-based laminates to improve various physical properties.

6.3.1
Phenolic Resins for Coatings and Surface Protection

6.3.1.1
Introduction: History and Possible Applications

A direct result of the search for a synthetic substitute for natural shellac led to the discovery of phenolic resins, materials used as binders for the production of thermosetting molding compounds and in many other applications as described in Baekeland's "heat and pressure" patent [1]. Shellac, which was obtained from a secretion of the lac insect (*Kerria lacca*), was expensive and unavailable in the amounts required to produce UV resistant and glossy varnishes. An initial substitute – the phenolic resin-based "Laccain" – was developed by the German chemist C. F. Meyer as reported by Schwenk [2] and listed in Table 5.1, but did not completely meet existing demands (for example, light resistance). The search for a shellac substitute that ultimately led to the Baekeland discovery, the first completely synthetic plastic, is analogous to current efforts to substitute synthetic compounds for the natural products CNSL (cashew nutshell liquid, used as a phenolic resin modifier in various applications) and tung oils.

Some disadvantages of phenolic resins that are obstacles to their increased use as suitable coatings resins as sole binders include their limited resistance to light, brittleness in the cured state, absence of flexibility, and their taste and odor characteristics. These limit their use in the packaging area.

Distinct advantages of phenolic resins include their high level of chemical resistance in the cured state, the marked complementary bonding function, their compatibility and reactivity with other resin systems such as epoxy, melamine, alkyd, and furan resins, and the chemical bonding function, i. e., the potential chemical reactivity of the phenolic and side-chain hydroxyl groups. Examples involving this reactivity include the reaction of the phenolic hydroxyl group with epoxides, etherification of methylol groups in the side chains, polyurethane formation by reactions of nuclear or primary side-chain hydroxyl groups with polyisocyanates, and the possible production of water soluble systems by acidic derivatization [3].

Tradeoff of advantages and disadvantages results in opportunities for use of regular and modified phenolic resins (Table 6.95) in the coatings industry and surface protection areas mentioned below [4–7].

Phenolic resins are used: for *packaging enamels*, protective interior varnishes, in combination with medium to high molecular weight epoxy resins (for example the combination of Rütaphen® 9646 LB with Rütapox® 0197) or polyester resins; for hot and cold curing *anticorrosion coatings* in combination with epoxy resins or vinyl polymers; in *electrically insulating coatings* (impregnating and dynamo sheet varnishes, wire enamels) using phenolics as the sole binder or an additive, but also in combination with alkyd resins; for *oil coatings* containing binders produced by cooking alkylphenol resins with unsaturated fatty oils; for *alkylphenol resin-modified printing inks* based on natural and in some cases hydrocarbon resins; in the form of novolaks in *photosensitive coatings*; and for modern chemical and solvent resistant *powder coatings* for electrostatic

Table 6.95. Use of phenolic resins as binders or additives in the field of coatings (survey)

Applications	System	Use
Packaging varnishes	Phenolic/epoxy resins	Cans, tubes, canisters, drums
Heat-resistant anticorrosion coatings	Phenolic/epoxy resins	Corrosion protection
Cold curing systems	Phenolic resins/vinyl polymers	Wash and shop primers
Electrophoretic dip coatings	Phenolic resins as additives	Anodic or cathodic methods of vehicle body coating
Electrical impregnating varnishes	Phenolic/alkyd resins	Consolidation of coil windings
Dynamo sheet varnishes	Phenolic resins (plasticized)	Coil coating of dynamo sheet
Wire enamels	Phenolic resins as additives (to improve flow)	Wire varnishing
Photo resists	Novolaks + diazo systems	Photolithographic printing plates
Printing inks	Alkylphenol resins as additives	Uniform drying printing inks
Powder paints	Novolaks as epoxy resin curing agents	Chemical and solvent resistant systems (environmentally friendly coating methods)

powder coatings, in which novolaks can be used as epoxy resin curing agents (cf. curing of epoxy resin molding compounds, Sect. 6.1.3.11, p. 230).

6.3.1.2
Phenolic Resins as Binders in Coatings (Types of Resins and Modifications)

In most cases, however, the use of phenolic resins as coatings binders in the areas mentioned above requires internal and external modification to compensate completely or partially for the above-mentioned disadvantages and to utilize fully the attractive advantages. This involves the utilization of phenolic resin chemical reactivity, for example, by etherification, and variation of this reactivity and the molecular weight distribution, made possible by changing the ratio of phenol to formaldehyde. Phenolic coatings resins – mainly modified resoles – can be classified into eight groups (Table 6.96) on the basis of their chemical structures; the number of these groups rises considerably when the phenolic moiety is replaced by the pertinent methyl and dimethyl derivatives, and alkylphenols by *t*-butyl, octyl, nonyl, or other phenolic derivatives.

Table 6.96. Classification of phenolic coating resins according to their chemical constitution

1. Heat and acid resistant resins (general purpose resole-type resins)	5. Resoles for water thinnable paints (containing, e.g., carboxyl groups)
2. Modified resoles	6. Non-curing resins (general purpose novolaks)
3. Plasticized resoles	7. Modified novolaks
4. Etherified resoles	8. Alkylphenol resins (resoles and novolaks)

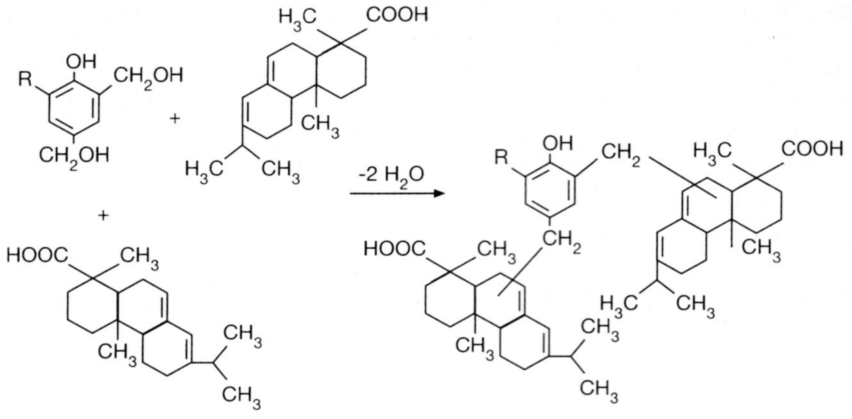

Fig. 6.172. Condensation of a phenolic resole with rosin acid

Phenolic resoles are produced using various ratios of phenol to formaldehyde, diverse catalysts, and can exist in water-borne, solid, or dissolved form. Although the reactivity is decreased by use of difunctional phenols such as o-cresol, flexibility is improved. Significant changes in the solubility (by alteration of the molecular polarity from polar to nonpolar) can be achieved by partial or complete replacement of phenol by alkylphenols such as those containing t-butyl, octyl, and nonyl groups. Etherification of the methylol groups, for example with butanol, similarly leads to changes in the reactivity, compatibility with other binders and solubility. Certain improvements in the range of properties can be achieved by such means. Chemical modification of the resoles with natural resins such as rosin affords improved solubility in nonpolar solvents and good compatibility with other coatings' raw materials. The various grades of rosin contain diterpene-type rosin acids such as abietic acid. The reaction proceeds at temperatures ranging up to a maximum of 250 °C. The main reaction involves condensation of the resoles with the rosin acid and elimination of water (Fig. 6.172).

Trifunctional phenolic resoles can undergo very dense crosslinking, after which they exhibit particularly high acid, chemical, and heat resistance. However, their resistance to alkali is very limited. Alkylphenol resoles generally represent solid products exhibiting melting points ranging from 65 °C to 100 °C, and due to their linear structure were earlier used for cooking with unsaturated oils such as tung oil, castor oil, and oiticica oil. Today, both unmodified and oil-modified alkylphenol resins are used, mainly in alkyd resin systems, as complementary resins to enhance the chemical resistance and through drying properties.

The transformation of phenol, cresol, xylenol, and alkylphenol-based resoles into their etherified counterparts in slightly acidic media leads to products that are compatible with aliphatics and aromatics, and are used in combination with epoxy, alkyd, polyvinyl, and acrylic resins for oven drying enamels with stoving temperatures ranging from 160 °C to 220 °C. "Phenolic ether" resoles are produced by etherification of the phenolic hydroxyl group [8, 9]. Grades etherified using allyl chloride are mainly used in combination with other resins for stoving coatings. Due to the blocked phenolic hydroxyl groups, these coatings feature good alkali and yellowing resistance. Water thinnable coating systems may be produced using carboxyl group-containing phenolic or Bisphenol A resoles or their salts [10, 11]. Film formation requires high stoving temperatures.

Except in powder and photosensitive coatings, straight phenolic novolaks are rarely used. Partial substitution of the phenol or cresol in phenolic novolaks by alkylphenols such as t-butylphenol improves the compatibility with low polarity solvents and enhances the oil compatibility. Alkylphenol novolaks may thus also be cooked with unsaturated oils.

6.3.1.3
Phenolic Resin Coatings Composition

A phenolic resin coatings formulation contains the following components:

1. Binders: phenolic resins alone or in combination with other synthetic resins. The binder or binder combination is designed to give the coating film good mechanical strength and provide it with chemical resistance. It should provide good adhesion to the substrate as a prerequisite for protection against corrosion. This requires prior treatment of the substrate, either by mechanical (sandblasting), physical (degreasing), or chemical means (phosphatizing).
2. Solvents: these function only as a temporary processing aid.
3. Fillers: these include both inactive materials such as silicates, sulfates, and fibers, together with active anticorrosion pigments such as chromates, and colored pigments. The purpose of fillers is to strengthen additionally the coatings layer, and if appropriate to provide it with a color effect.
4. Additives: these represent curing catalysts, flow promoters, and other materials that are added in small amounts and are not directly involved in formation of the coatings film, but they can significantly affect the resulting film properties.

6.3.1.4
Packaging Coatings (Protective Interior Coatings)

Packaging coatings (Fig. 6.173) are materials used as protective coatings or varnishes on metallic containers made of tinplate, aluminum, or black iron sheet metal used to package products such as cosmetics (spray cans), beverages, and pharmaceuticals. Protective interior coatings for food cans, tubes, spray cans, and heavy-duty packaging made of tinplate, aluminum, or sheet iron are exposed to a wide variety of mechanical and chemical stresses during manufacture and use of the packaging [12–14]. Such coatings systems are also used to coat the exteriors of containers in the packaging line.

Food can coatings [15, 16] that are applied to the sheet metal by the roller coatings method must be stack and block resistant, capable of being punched

Fig. 6.173. Various metal packages coated with interior protective varnishes

and deep drawn, sterilization resistant, and must possess the ability to withstand various ingredients. Resistance to propellants and entraining agents is additionally required in the case of interior coatings for spray cans. Tube varnishes are subjected to severe deformation, and must remain very flexible even when exposed to the ingredients. Coatings for heavy-duty packaging such as pails and drums must exhibit great shock and impact resistance. These diverse and, in some cases, conflicting requirements are best met by using binder combinations composed of relatively high molecular weight epoxy resins such as Rütapox 0197 or spezial polyesters and phenolics (Table 6.97) to afford products termed "golden varnishes" in the case of protective interior coatings. The mixing ratio of epoxy to phenolic resin is generally 55:45 to 80:20. These clear varnishes are produced by mixing 40–50% solutions of the two components at ambient temperature. Phenolic resins of low reactivity may be accelerated by using selected salts of o-toluenesulfonic acid. The varnishes may be applied by spraying them on the interior surfaces of the finished containers and stoving the coatings at 180–280 °C. They result in coatings exhibiting a golden shade whose intensity depends on the stoving conditions. If desired, the color shade of the varnish coatings can be rendered more intense by use of "tinting resins." Such resins, used as additives, are obtained by methods such as condensation of furfural with phenol.

Metal sheets or strips (in coil coatings) are continuously coated in the roller coatings process. The thickness of the coatings is generally 6–8 μm after stoving. This corresponds to an applied wet varnish coatings of 5 g/m². The desired containers may be produced from the varnished sheets or strips by shaping under pressure. The phenolic resins used in this field, such as Rütaphen 9646 LB, conform to the requirements of the U.S. Food and Drug Administration [17]. Etherified products (resins D, E, and F in Table 6.97) are used in addition to low molecular weight phenolic resoles for production of epoxy/phenolic resin combinations. Flow problems can be prevented by incorporation of additives or by pre-condensation of the resins (resin G in Table 6.97). The latter also improves the chemical resistance.

Table 6.97. Phenolic coatings resins for packaging and protective interior varnishes (selection of various types of resins; letters refer to product numbers of Bakelite AG, Iserlohn, Germany)

Type of resin (code)	Material class	Solvent	Non volatiles (%) ISO 8618	Viscosity at 20 °C (mPa · s) ISO 12058	Applications, properties
A	Cresol resole	1-Methoxy-2-propanol	44–48	approx. 350	Standard resin for protective interior varnishes
B	Phenolic resole	Aqueous	54–58	30–50	Acrylate-based, water borne systems
C	Cresol resole	Butyl glycol/butanol	51–55	approx. 1500	Standard resin for combinations with epoxy and polyester resins
D	Etherified xylenol resole	Butanol/xylene	68–72	approx. 1700	High sulfur resistance; also particularly suitable for tube varnishes (flexibility!)
E	Etherified phenolic resole	Butanol	68–72	approx. 2200	Phenol-free resin for drum coatings exhibiting high solvent resistance
F	Etherified cresol resole	Butanol	56–60	70–130	Low monomer levels
G	Epoxy-phenolic resin precondensate	Butyl glycol/1-methoxy-2-propyl acetate	40–44	approx. 1300	High flexibility, good adhesion
H	Alkylphenol resole	1-Methoxy-2-propanol	57–61	approx. 600	High flexibility

Combinations with added polyvinyl butyral are also common. This achieves excellent chemical resistance. Depending on their composition, the epoxy/phenolic or polyester/phenolic resin combinations are suitable for practically all toned, unpigmented, interior varnishes.

They are not appropriate for use in lightly pigmented interior varnishes, where polyesters or special epoxy resins are used. It is also possible to use a combination of water-borne acrylates with aqueous phenolic resoles such as resin B in Table 6.97.

6.3.1.5
Anticorrosion Primers

Anticorrosion primers include the very rapid (physically) drying, unpigmented, weldable primers based on phenolic novolaks containing small fractions of plasticizing additives such as alkyd resins or polyvinyl butyral. These primers are designed to provide protection to the interiors of tanks and storage containers, piping, and machine housings. The coating films feature good resistance to water, gasoline, and mineral oil.

Wash and *shop primers*, also termed "reaction primers," similarly represent anticorrosion primers. These are adhesion primers based on acid curing phenolic resins, flexibilizing agents, adhesion promoters, anticorrosion pigments, and phosphoric acid. The latter not only serves as a curing catalyst, but also as a phosphatizing component for the metal substrate. Wash primers were developed in an effort to protect the individual components used in shipbuilding (such as metal sheets and profiles) against corrosion in an uncomplicated manner until completion of the structure. Added film-forming agents and anticorrosion pigments enhanced the protective effect. Thus, polyvinyl butyral, phenolic resin, active pigments, and phosphoric acid in addition to the solvents still represent the basic components of these primers, even today. Methylol group-rich, acid curing types of phenolic resin are used in this application. A wash primer only offers temporary protection against corrosion. Over a period of time, the priming operation was shifted to the workshop of the component manufacturer, and the "shop" primer was developed. Its formulation resembles that of the wash primer, and it similarly offers only temporary protection against corrosion (six months to one year), but in this case represents the permanent primer coat. These air drying wash and shop primers have been further developed into economical, high quality anticorrosion primers. They are applied by spraying, including electrostatic spraying, because they dry rapidly, and can be applied "wet on wet". The coatings films are weldable, can be overcoated, are waterproof, and resist the effects of exposure to marine environments and industrial atmospheres. Shipbuilding, bridge, and general steel construction, and mechanical and automotive engineering represent the pertinent fields of application. Air drying *oil coatings* mainly used as primers and topcoat enamels in shipbuilding and hydraulic engineering may be produced on the basis of alkylphenol resins (resins O and P in Table 6.98). The alkylphenol resins are cooked with unsaturated fatty oils such as tung oil or linseed oil. The physical and chemical properties of the coatings

6.3 Complementary Bonding 381

Table 6.98. Phenolic coating resins for various applications (Bakelite AG, Iserlohn, Germany)

Type of resin (code)	Material class	Solvent	Non volatiles (%) ISO 8618	Viscosity at 20 °C (mPa · s) ISO 12058	Applications, properties
I	Modified phenolic resole	Isopropanol	48–52	ca. 500	Cold curing anticorrosion coatings, good chemical resistance
K	Phenolic resole	Butanol	62–66	ca. 4000	Hot curing, for e.g. heat exchanger coatings
L	Etherified phenolic resole	Butanol-cylohexanone	74–78	ca. 5000	EDP (electrophoretic dip coating) process, good throwing power
M	Etherified xylenol resole	Butanol	68–72	ca. 2000	Electrically insulating varnishes, in combination with alkyd resins
N	Phenolic resole	Butyl glycol/butanol	47–51	ca. 650	Dynamo sheet varnishes, in combination with polyvinyl butyral
O	Alkylphenol resole	Xylene	57–61	100–300	Additive for wire enamels
P	Alkylphenol novolak	Solid		MR[a]: 65–75 °C	Standard resin for oil coatings
Q	Alkylphenoiic resole	Solid		MR[a]: 57–67 °C	Modification of printing inks and alkyd resin coatings
R	Cresol novolak	Solid		MR[a]: 115–135 °C	Resin for photo resists
S	Modified phenolic novolak	Solid		MR[a]: 95–115 °C	Powder coatings, in combination with epoxy resins

[a] MR = melting range, ISO 3146.

Fig. 6.174. Condenser with interior tubes protected against corrosion with special-purpose, phenolic-based combination binders

films can be adjusted to produce an optimal combination for the specific application by modifying the ratio of resin to oil. The phenolic resin fraction promotes hardness and resistance to various media, and the oil fraction flexibility and adhesion. The coating films dry oxidatively (react with oxygen) and feature good water and weather resistance.

Phenolic resins have also been successfully used as binders for *heavy duty corrosion protection* (Fig. 6.174) in equipment construction [18]. The term "heavy duty corrosion protection" is only understandable when it is related to the stresses to which a protective system is exposed. In this case, particularly aggressive media such as acids, alkalis, salts, and organic solvents – but also a material such as water in the form of cooling and drinking water – render this type of special protection necessary.

Combinations with epoxy resins may also be used to enhance the chemical resistance in heavy duty corrosion protection. Straight epoxy resin coatings are generally cured using catalysts, whereas phenolic resins are cured at elevated temperatures ranging from 160 °C to 200 °C. Heat exchanger tubes, boilers, and various pieces of equipment are provided with protective phenolic resin-based coatings by centrifugal casting or spraying. The individual layers of the coatings should not exceed 40–80 µm. Solvent containing coatings systems are used that flow freely enough to allow penetration into the narrow cross-section of a heat exchanger tube, and can provide adequate wetting of the substrate. Each individual protective layer must then be precured in an oven at 160 °C, and the final coatings, up to about 250 µm thick and consisting of multiple layers, completely cured at 200 °C. When a protective coating is to be applied to the interior of piping, this operation is performed after the pipes have been welded into the system. The thin, hard, smooth protective coating with excellent adhesion for heat exchangers can be produced using coatings based on phenolics such as resin K in Table 6.98. The smooth surface is important to prevent formation of deposits or crusty materials. Deposits of materials such as lime can lead to problems such as blockage and constriction of the heat exchanger tubes, and significantly impair their efficacy. Smooth surfaces furthermore facilitate cleaning.

Electrophoretic dip coating, a method of considerable significance in the automotive industry, is carried out using water-borne, environmentally friendly coatings systems [19] containing a maximum of 5% solvent. The coatings, which have been used industrially for more than 25 years, are electrolytically applied from emulsions. When a direct current is applied to two electrodes in a dipping bath, ionized coatings particles electrophoretically migrate to the oppositely charged electrode and are deposited on substrates connected to the anode or cathode [20, 21]. Coatings systems containing "maleinate oils" such as the reaction products of maleic anhydride and linseed oil [22] as the main binder are used in the anionic method, and can be more densely crosslinked when stoved if phenolic resins are added [23]. For a long time this method was popular for priming automotive bodies.

A transition to the cathodic deposition method, that offers still more reliable corrosion protection, took place at the end of the 1970s. Epoxy resin systems that can be modified with polyesters and whose epoxide groups have been reacted with amines or ammonium salts [24] are mainly used in the cathodic EDC (electrophoretic dip coat) process. Special phenolic resins such as resin L in Table 6.98 may similarly be added in this case to improve the level of protection against corrosion and increase the film hardness of products such as primers used for automotive accessory components. Alkylphenol resins that afford increased flexibility, such as Rütaphen 9435 LA, can also be used for special parts. Special demands on electrophoretic dip coats include good deposition properties and throwing power, affording primer coats exhibiting particularly desirable properties.

6.3.1.6
Electrically Insulating Coatings

Electrically insulating coatings in which phenolic resins can be used include *electrical impregnating* and *dynamo sheet varnishes*, and *wire enamels*. Electrical impregnating varnishes are used to strengthen mechanically the windings of electric motors, which are exposed to magnetic stresses and to centrifugal forces due to rotation. The varnish simultaneously promotes dissipation of heat from the windings and provides protection against moisture and aggressive media. Phenolic resins such as resin M in Table 6.98 are used in combination with flexibilizing resins such as alkyds at ratios of phenolics to alkyds: 1:1 to 1:2 for production of impregnating varnishes of (German) Thermal Classes B and E. After the windings have been impregnated by the vacuum/pressure method, the varnish is cured at 120–140 °C over a period of 4–8 h.

Special phenolic resins such as resin N in Table 6.98 can be used as the sole binder for production of dynamo sheet varnishes. Small amounts of plasticizers improve the adhesion and die cutting properties. The varnishes are applied by roller coatings at dry coatings thicknesses of about 5–20 μm using the coil coatings process. The graphitization level of the phenolic resins used for the varnish must be particularly high when the sheet packets are welded using the Argonac process to prevent development of thick smoke.

Phenolic resins such as resin O in Table 6.98 are also used as additives in wire enamels to improve flow, increase the consolidation strength, and provide coloration. The purpose of a wire enamel is to insulate the current carrying windings in electrical machines such as electric motors, generators, and transformers from one another. The main binders for wire enamels are presently polyimides, polyamide-imides, polyester imides, polyester, polyurethanes, and oil-based enamels.

6.3.1.7
Miscellaneous Coatings Applications (Printing Inks, Alkyd Coatings, Photosensitive Coatings, Powder Coatings)

Natural resin (rosin)-based *printing ink resins* can be modified by reaction with alkylphenol resin such as Q in Table 6.98. The reaction takes place at temperatures around 250 °C. Further reactions with polyalcohols (esterification) are then carried out. The products feature good drying properties and high pigment binding levels.

Alkyd coatings can similarly be modified with alkylphenol resins. This involves either cooking the components together, or blending them at ambient temperature. This increases the chemical resistance and improves the through drying properties.

Photosensitive coatings are used with such products as novolak-diazo system-based positive photosensitive resists which are used to produce photolithographic printing plates and printed circuits for electronic components. The subject of photoresists will be discussed in more detail in Sect. 6.3.1.8. Photosensitive coatings involve the following operations:

1. Coating of a substrate surface with a layer of light-sensitive polymer
2. Heat treatment of the coated substrate at temperatures of 70–125 °C
3. Exposure through a photographic mask
4. Development by treatment with an alkaline solution
5. Possible further treatment to transfer the resultant resist structures to the substrate surface by etching or electrochemical deposition

In practice, novolaks and diazo compounds are dissolved in alcohols or glycol ethers, and subjected to reaction if appropriate. It is important that the novolak, for example, resin R in Table 6.98, exhibit uniform quality (constant specific molecular weight distribution). Depending on the required sensitivity to light, the ratio of diazo compounds to novolak ranges between 1:1.5 and 1:15. The coatings contain a variety of additives, and exhibit a solids level of around 20–30%. The substrate is coated by spraying, dipping, roller coating, or centrifugal casting. After the coating has dried, the pre-sensitized substrate surface is exposed. The unexposed areas are insoluble in alkaline medium. They exhibit hydrophobic characteristics and may be wetted with printing inks. The exposed areas are rendered soluble in alkali by chemical reaction with the diazo compound. The metallic surface is hydrophilic and is not wetted by printing inks.

As solvent-free coatings systems, *powder coatings* [25] are presently of great importance. They are generally produced by the extrusion process. In this method, the resin, curing agent, pigments/fillers, additives, and flow promoters are melted to yield a homogeneous mass at temperatures of 100–130 °C. The melt is then cooled and the extruded material reduced in size, finely milled, and separated into a specific particle size range. Electrically grounded objects are coated with the electrostatically charged powder particles. The powder coating is then crosslinked in a stoving oven at 150–220 °C for 5–20 min. Epoxy, polyester, and polyurethane systems are mainly used as binders.

Phenolic resins, for example, resin S in Table 6.98, may be used as curing agents for special-purpose epoxy powder coatings. Such coatings feature high chemical and solvent resistance as well as good flexibility and impact resistance. High-gloss versions are also possible, and powder coatings furthermore exhibit good storage stability. Applications for phenolic resin containing powder coatings exist in interior coatings for piping, in pipelines, automotive primers, the automotive accessories industry, and in certain types of packaging.

6.3.1.8
Photoresist/Imaging

6.3.1.8.1
Introduction

The use of phenolic materials in the development of an image involves the use of phenolic components that undergo a chemical or physical transformation as the image is formed. Imaging techniques that have been prominent within the last three decades and have utilized phenolic materials are photoresists, thermography or thermal recording, and carbonless copy paper or pressure sensitive recording paper. Carbonless copy paper and thermography are discussed in Sect. 6.5.2. Table 6.99 distinguishes each of these methods and identifies the type of phenolic material that participates in the formation of the image.

Table 6.99. Imaging methods using phenolic materials

Method	Chemical/physical transformations	Phenolic type	Bonding function
Photoresist	Actinic light	Cresol novolak, PHS[a]	Complementary
Carbonless copy paper	pressure, rupture charge transfer	Zinc salt of alkylphenol novolak	Chemically reactive
Thermography	Heat/charge transfer	Bis phenols, salicylic acid compounds	Chemically reactive

[a] Polyhydroxystyrene.

These varied imaging methods have had a great impact on many high technology growth industries in the 1980s and 1990s, particularly photoresists in semiconductors for computers together with thermal and pressure imaging methods for related equipment such as printers, recorders, as well as communication devices, facsimile, bar code, scanners, calculators.

6.3.1.8.2
Photoresist

In the last two decades computers have increased in memory capacity with a corresponding reduction in size. This was facilitated by technical advances that occurred in photolithographic processes used in microelectronics manufacturing. Continuous improvements in microlithography have been the predominant factor for the miniaturization of computer components. In a period of 20 years, 16 MB semiconductor memory which originally occupied a volume of 0.2 m^3 now occupies a space of 40 cm^3 or a 4000-fold reduction. Table 6.100 provides the current computer capabilities and those anticipated in the future.

The trend to reduce resist lines that form transistor circuits by roughly 10% a year results in chipmakers introducing a new generation of chips every three years with a fourfold increase in memory capacity (DRAM).

The competitive nature of the semiconductor industry is particularly keen worldwide with many nations viewing the semiconductor business as an important, high priority industry critical to national economy. For example, in 1993 electronics accounted for $340 billion in the U.S. and more that $800 billion to the global economy. Its growth rate is equally impressive in the U.S. representing 10% annually compared to GNP of about 1%, auto industry growth of 1.3%, and telecommunications of 3.7%. In 1992 the Semiconductor Industry Association (SIA) sponsored a collaborative effort among experts in industry, government, and academia to develop a common national "roadmap" for future semiconductor technology development and investment. It is known as the National Technology Roadmap for Semiconductors. The Roadmap reviewed and outlined the requirements for critical areas of semiconductor R & D, engineering, and manufacturing.

Since then, Roadmap meetings convened by SIA are held every 2–3 years.

The recently published "National Roadmap" for 1997 provides a 15 year forecast of technical challenges that must be considered to meet future

Table 6.100. Computer capabilities

Year	1997	2000	2002	2008
DRAM (MB)	64	256	1024	16[a]
Speed MHz	300	500	500	700
Resist line width (µm)	0.35	0.25	0.18	0.10

[a] GB.

demands for faster, denser, more advanced computer chips. Some salient features of the report are:

- Increased silicon wafer size from 200 mm to 300 mm by 1999 and 450 mm by 2009.
- As the industry approaches 100 nm wave length for narrow line widths, non-optical lithography based on various methods such as extreme UV, E-beam, ion beam, and X-ray must be considered.
- New fabrication plants, known as "fabs," are required with costs exceeding $1 billion per fab.
- Clean room standards for airborne particles are rated according to air quality. Class 1 chamber has 1 particle/28 l of air for newly planned fabs. Also cleanroom standards for manufacture of chemicals, gases, photoresists, solvents, etc. are necessary. Purity is critical. Solvents with less than 100 ppt impurities are common and shipped in teflon lined containers.

Recently IBM reported that copper wired chips would replace aluminum which for 30 years has been the standard for the semiconductor industry. This new development will allow chip producers to reduce voltage since copper has lower resistance than aluminum, thus reducing power consumption and cooling requirements. Further benefits of the novel IBM technology include reduction in circuitry (below 0.2 μm line width), 40% faster computer processing, and increased memory capacity.

Optical photolithography and innovative improvements have made it possible to create highly integrated circuits on silicon chips for industrial use. One of the key components is the photoresist with a total sales value in excess of $750 million worldwide. Optical photolithography involves resist exposure to actinic light using broad band, near to mid-UV region (300–450 nm) or deep UV (DUV). Exposure systems are equipped with high intensity light sources and a variety of lenses and mirrors for light collimation. The output of a Hg-Xe lamp in the 300–450 nm region has several strong peaks (g, h, i lines) which may be isolated for exposing the resist.

Resist line width and increased DRAM capacity are related as shown in Table 6.100. With continued anticipated high memory capacity and speed, it is expected that resist line width will continue to decrease to below 0.14 μm. Advances in optical photolithography are becoming more difficult due to wavelength limitation (Table 6.101).

Table 6.101. Correlation of wavelength, resist line width and resist material

Wavelength (nm)	Resist line width (μm)	Resist material
g line 436	<1.0	Cresol novolak
i line 365	0.35–0.5	Cresol novolak
DUV[a] 248	0.35	PHS+CA[b]
DUV 248	0.18–0.35	PHS+CA[b]
DUV 193	<0.18	Aliphatic polymer

[a] Deep UV.
[b] Chemical amplification.

Fig. 6.175. Procedure for positive or negative resists

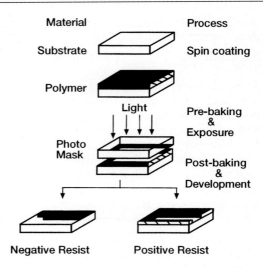

The use of deep UV or 193 nm wavelength will result in a shift from phenolic materials such as cresol novolak or PHS to aliphatic polymer (see later).

The manufacture of the semiconductor device or computer chip involves spin coating the silicon wafer with a thin coating of photosensitive polymer and exposing it to a light source through a mask. The amount of radiation exposure changes the film's chemical structure in regions exposed to light and results in differences in dissolution rate in selective solvents or developer. Dissolving either exposed or unexposed portions of the film results in positive or negative tone images (Fig. 6.175).

Photoreactive polymers, sensitive to UV light, are employed as resists-films that resist the action of etching agents.

Positive Resists

Positive resists are polymers that contain inhibitors that prevent wetting and dissolution by the developer. Specific factors are:
- Upon optical exposure the inhibitor undergoes phototransformation to an acidic material (see Fig. 6.176) and increases solubility to basic developer.
- The base developer removes the exposed areas of resist/inhibitor.

Negative Resists

Negative resists are linear polymers with photosensitive components that absorb light and promote crosslinking. Specific factors are:
- Exposed polymer is insoluble due to crosslinking.
- A developer solvent removes unexposed resin areas.

The optical photolithographic process produces a 3-D relief image in the thin film that replicates opaque and transparent areas of the mask. This image can

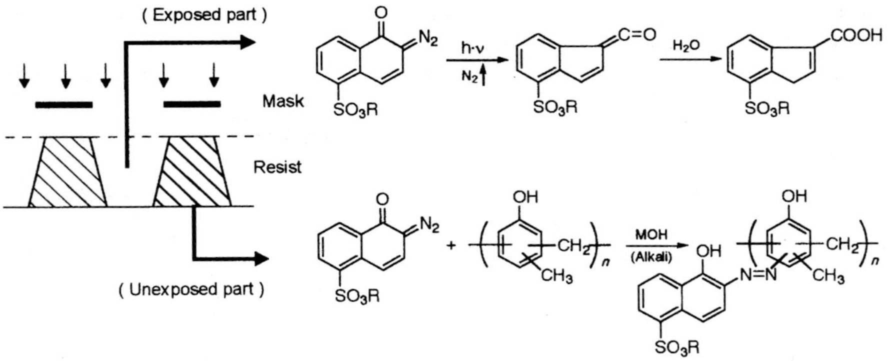

Fig. 6.176. Positive photoresist system: image formation process of novolak-quinone diazide system

be transferred to the silicon wafer by a sequence of etching, deposition, or implantation steps to the final semiconductor device.

Two key elements which influence resolution of the photolithographic process are wavelength of light and the chemical structure of the photoresist (Table 6.101). As the resist line width is reduced (Table 6.100), shorter wavelength light is necessary: (UV, 350–450 nm); DUV (190–250 nm), and possibly X-ray (IBM) and E-beam (Lucent). Intel-Motorola is investigating extreme UV (13 nm) technology. Presently the predominant technology in conventional photolithography is based on 350–450 nm (g, h, i lines) and cresol novolak resin. Incremental improvements in tool design and performance have allowed continued use of 350–450 nm to produce even smaller line widths. However greater chip capacity and computer speed characteristics are tending toward shorter wavelength (190–250 nm) or DUV [1].

Positive photoresist systems for photolithography consist of two components: an aqueous alkali soluble resin (after light exposure) and a photosensitive dissolution inhibitor. The resin is usually a cresol novolak while the photosensitive component is a hydrophobic substituted diazonaphthoquinone (DNQ) (Fig. 6.176).

The cresol novolak, soluble in a variety of organic solvents for spin coating, becomes insoluble through the incorporation of the DNQ while the excess DNQ becomes base soluble on exposure to UV light. These solubility differences result in an image that is subsequently incorporated within the device. The efforts of many investigators in U.S., Europe, and Japan provide the basis for the effective use of cresol novolak in photolithography [1–6].

The key features of cresol novolaks as initially described by Hanabata [4, 6] are polymer microstructure and molecular weight distribution (MWD).

Cresol Novolak Microstructure

Increased photoresist optical resolution occurs when cresol novolaks possess a high degree of ortho-ortho linkage to facilitate complexation with DNQ through a secondary "hydrogen-based" structural model as suggested by Zampini [7]. A hemi-calixarene structure is proposed and occurs via molecular interaction of cresol novolak phenolic hydroxyl groups and DNQ. The R group of DNQ (Fig. 6.176) is a multi-functional phenol with 1, 2, 3-trihydroxy benozophenone preferred. When "g" line radiation is used (436 nm), DNQ 5 sulfonate is recommended whereas "i" line (365 nm) requires DNQ 4 sulfonate. Favorable absorption of DNQ will occur in those wavelength regions depending on substituent location.

DNQ   5 sulfonate   DNQ   4 sulfonate

Molecular Weight and Molecular Weight Distribution

High T_g (80–150 °C) cresol novolaks are desirable as photoresist materials. Features of a suitable resist material are:
- Soluble in solvents to spin cast thin, uniform films
- Reasonable T_gs to survive processing steps
- Exhibit no flow during pattern transfer
- Possess reactive functionality to facilitate pattern differentiation after irradiation
- Have absorption characteristics for uniform imaging through resist film

The desirable high MW characteristic improves thermal flow stability and minimizes resist distortion during subsequent processing steps. Moderately polar solvents such as propylene glycol monomethyl ether (PGME), propylene glycol monomethyl ether acetate (PGMEA), ethoxy ethyl propionate (EEP), or ethyl lactate are required to dissolve cresol novolaks for spin coating. Molecular weight distribution (MWD) is equally important. Most cresol novolaks are moderately broad in MWD with M_w/M_n varying from 3 to >7. Hanabata has shown that an improved resolution image is obtained through the use of tandem (bimodal) type cresol novolak whereby medium MW material is removed leaving only high and low MW components. Recently Allen and coworkers [8, 9] conducted a more detailed examination of MWD by fractionation of commercially available cresol novolak. Cresol novolaks were ultra-fractionated using supercritical fluid technology (SCF), consisting of CO_2/cosolvents such as methanol and acetone. Various collected fractions were monodispherse ($M_w/M_n \to 1.0$) and allowed an assessment of polymer properties such as T_g, rate of dissolution, and lithographic performance. Rate of dissolution relates to how rapidly the unreacted resist can be removed by base[1]. By removing both the high MW fraction (which dissolves very slowly in base) and low MW species (has very low T_g and dissolves very rapidly in base), a cresol novolak material was obtained with satisfactory dissolution rate and a high T_g. It was identified as monodisperse middle fraction. A comparison of starting material parent novolak, monodisperse middle fraction novolak, and a composition known as tandem novolak (middle MW components removed) is shown in Table 6.102.

Each of these cresol novolak systems was evaluated in "i" line resist formulation. All three systems exhibited various deficiencies as various performance characteristics were evaluated. Allen was able to formulate a high performance resist (composition not reported) from these ultra-fractionated cresol novolak components with a balance of properties related to contrast, dissolution rate, microgrooving, and residue.

Table 6.102. Comparison of parent MW cresol novolak, monodisperse middle Fraction cresol novolak and tandem cresol novolak

Type	Parent novolak	Tandem	Middle fraction
M_n	1800	3500	1700
T_g (°C)	98	132	117
Dissolution rate (μm/min)	0.45	–	1.8

Cresol Novolak Preparation

Cresol novolaks are produced by conducting the acid catalyzed reaction between formaldehyde and one or more multi-substituted phenols (cresols, xylenols) in a polar solvent, diglyme. Oxalic acid is the preferred catalyst.

Competitive reactivities as well as positional reactivities of cresol isomers is lacking for acid catalyzed reaction of cresol with formaldehyde. Bogan [5] determined these by conducting competitive experiments consisting of equimolar amounts of phenol and cresol isomer with very low amounts of F (0.02 equivalents) under acidic conditions and then examined the monomer free polymers via NMR. The following ranking was obtained:

1. Phenol   1.00
2. *p*-Cresol   0.47 ± 0.01
3. *o*-Cresol   2.36 ± 0.05
4. *m*-Cresol   6.03 ± 1.78

---

[1] An understanding of dissolution of phenolic resist materials has been a subject of considerable discussion [1, 10, 11]. Percolation theory by Reiser and Shih [10] and a new "probabilistic model" by Willson [11] attempt to provide a better understanding of novolak dissolution and process variables.

The greater reactivity of *o*-cresol with F in comparison to phenol is surprising and is explained by a two-step mechanism of F with the cresol monomer (F addition to cresol and condensation of hydroxymethyl cresol to dimer). The addition of hydroxymethyl cresol to another cresol is approximately ten times faster as the formation of methylol from F and monomer. When competitive experiments are conducted (as in this method) to determine relative reactivities, information relates to both reaction steps (addition and condensation). Thus *o*-cresol is more than twice as reactive as phenol; *m*-cresol is six times more reactive.

Relative positional reactivities of ortho and para sites of phenol and cresols were also reported.

OH   1.0   3.7

OH, CH$_3$ 1.0   3.23

OH, CH$_3$ 1.0   0.4   1.68

Since high ortho, ortho cresol novolak microstructure is desired, it requires the copolymerzition of *m*- and *p*-cresol monomers with F. Using the same competitive experimental scheme for *m*-cresol with F, relative site reactivities for *m*-cresol were determined [12]. A lower site reactivity of *p*-cresol of 0.09 relative to *m*-cresol was reported [13, 14]. By computer simulation the *m*-, *p*-cresol novolak compositions were plotted using these measured reactivities. Comparing these plots with the synthesis of *m*-cresol homopolymer and with *m*-, *p*-cresol copolymers with 70:30 and 50:50 charged ratios, Bogan was able to show the amount of unreacted cresol at end of reaction, fraction of *m*-cresol rings in the polymer that are substituted at both the 2 and 4 positions, and, importantly, the composition of the *m*-, *p*-cresol copolymers. In Fig. 6.177,

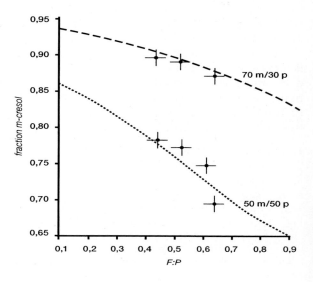

Fig. 6.177. Computer simulation (*dashed lines*) and experimental charges of *m*-, *p*-cresol/F ratios vs *m*-cresol content in the product

when 70:30 *m*-, *p*-cresol was charged in the reactor with an F:cresol ratio of 0.4–0.6, the *m*-cresol content in the copolymer was ~90% (88–90%), slightly lower than the computer simulated curve. Similarly 50:50 *m*-, *p*-cresol with the same F:cresol ratio yielded values of *m*-cresol content of 70–78% which also varied from computer simulated curve. Since *p*-cresol is exceedingly less reactive than *m*-cresol, a higher fraction of *p*-cresol in the initial charge yields not only high MWs but also more branching (2,4,6-*m*-cresol substitution pattern) at a given F: cresol ratio. As the F:cresol ratio is increased, more *p*-cresol is found in the end groups. In fact 95% end groups are *p*-cresol. *p*-Cresol is an effective chain stopper.

Other synthetic methods have been examined to prepare high ortho, ortho containing cresol novolaks. Hanabata [15] used divalent metal salts with *m*-cresol and F. The use of bis hydroxymethyl *p*-cresol has been effective in preparing alternating block copolymer resins with high ortho, ortho composition [16, 17] (Eq. 6.1); the method provides a basis for comparing high ortho, ortho block copolymer with random *m*-, *p*-cresol novolak in photoresist performance.

$$\text{(6.1)}$$

Jeffries [18] extended the block copolymer (above) method by preparing an ABA block copolymer consisting of *p*-cresol trimer as the interior (B) segment and *m*-cresol novolak as terminus A (Eq. 6.2):

$$\text{(6.2)}$$

The block copolymer exhibited similar photospeed as a 40/60 random *m*-, *p*-cresol novolak but exhibited less resolution and some scum.

The method of Casiraghi [19] was utilized to prepare high ortho, ortho *m*-cresol oligomeric novolaks [20] (Eq. 6.3):

$$\text{(6.3)}$$

55 to 75% methylene are *o,o'* bonded

The high ortho, ortho oligomer was then transformed into high MW material by reaction with F or p-cresol/F (Eq. 6.4):

$$o,o'\text{Oligomer} \xrightarrow[H^+]{CH_2O} +CH_2\text{~Oligomer~}CH_2\text{~Oligomer}+_n$$

High MW product

or

$$o,o'\text{Oligomer} + \underset{CH_3}{\underset{|}{C_6H_4}}\text{-OH} \xrightarrow[H^+]{CH_2O} \left[\text{~Oligomer~}CH_2\text{-}\underset{CH_3}{\underset{|}{C_6H_3}}\text{-OH}\right]_n \quad (6.4)$$

Photoresists based on these polymers containing ortho, ortho m-cresol intermediate provide resist profiles, acceptable photospeeds, good resolution without scum, and very good depth of focus with high image integrity.

Other aldehydes besides F have been examined in novolak preparation but have not resulted in any commercial products. Cresol novolaks with F have been the preeminent resist material from broadband actinic radiation to short wavelengths below 300 nm. Current efforts are trending to deep UV (DUV), 248 nm wavelength. Below 300 nm cresol novolaks are opaque. Improvements in the 250 nm transparency region while maintaining similar inhibitor dissolution characteristics resulted in replacing cresol novolak with poly (4 hydroxystyrene), PHS [1].

$$+CH-CH_2+_n \text{ (with } p\text{-hydroxyphenyl group)} \quad \text{"PHS"}$$

A comparison of UV absorption spectra (Fig. 6.178) of PHS and cresol novolak indicates that PHS absorbs ~60% less than cresol novolak at 250 nm. However the slow dissolution rate of PHS in alkaline media has resulted in the introduction of alkyl groups on the ring to improve the dissolution rate. Both mono and dialkyl derivatives have been successfully evaluated as matrix resins for DUV (248 nm) resist materials.

The dissolution rate of PHS was also improved by reaction of PHS with 6-hydroxymethyl-2,4-dimethyl phenol [21] (Eq. 6.5):

$$+CH\text{-}CH_2+_n\text{-}C_6H_4\text{-}OH + H_3C\text{-}C_6H_2(CH_3)(OH)\text{-}CH_2OH \rightarrow \underset{OH}{\underset{|}{C_6H_2(CH_3)_2}}\text{-}CH_2\text{-}\underset{OH}{\underset{|}{C_6H_2(CH\text{-}CH_2)_n}}\text{-}CH_2\text{-}\underset{OH}{\underset{|}{C_6H_2(CH_3)_2}} \quad (6.5)$$

Fig. 6.178. UV absorption spectrum of poly(hydroxystyrene) vs a conventional novolak [1]

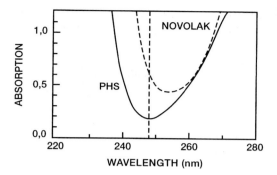

Sensitivity is a related issue which is important in the design of resist materials as shorter wavelengths are used. A recent development that has been successfully applied in providing resists with high sensitivity is chemical amplification [3, 22–25]. Chemical amplification involves the photogeneration of an acidic species that catalyzes several subsequent chemical transformations such as removal of a protecting group or crosslinking of a matrix. A chemically amplified resist consists of three components:

1. Matrix polymer
2. Photoacid generator
3. Component effecting differential solubility between exposed and unexposed areas through crosslinking or other reactions

Chemically Amplified Resists (CAR)

Components that comprise CAR type materials are:
1. Matrix resins (cresol novolaks, PHS, protected PHS and PHS copolymers)
2. Photoacid generators (onium salts, DDT and aryl bis trichloromethyl-s-triazine)
3. Crosslinking agents (melamine derivatives, benzyl alcohol derivatives)

An early example of a commercial CAR system developed by IBM [25] was the "t-BOC resist" which consisted of a mixture of poly [4-(*tert*-butyloxycarbonyl) oxy] styrene (PBOCST) and triphenylsulfonium hexafluoroantimonate, photoacid source (Eq. 6.6):

$$\begin{array}{c}\text{+CH-CH}_2\text{+}_n\\ \text{C}_6\text{H}_4\\ \text{O-C(O)-O-}t\text{-Bu}\end{array} \xrightarrow[\text{250 nm}]{\text{H}^+,\ h\nu,\ \text{Ph}_3\text{S}^+\text{SbF}_6^-} \begin{array}{c}\text{+CH-CH}_2\text{+}_n\\ \text{C}_6\text{H}_4\\ \text{OH}\end{array} + CO_2 + (H_3C)_2C=CH_2 + H^+ \qquad (6.6)$$

A small radiation dose is required to produce an image. An exposure dose of only 1–2 mJ/cm^2 was required at 250 nm as compared to classical cresol novolak/DNQ resists which need 200–300 mJ/cm^2 at 365 nm.

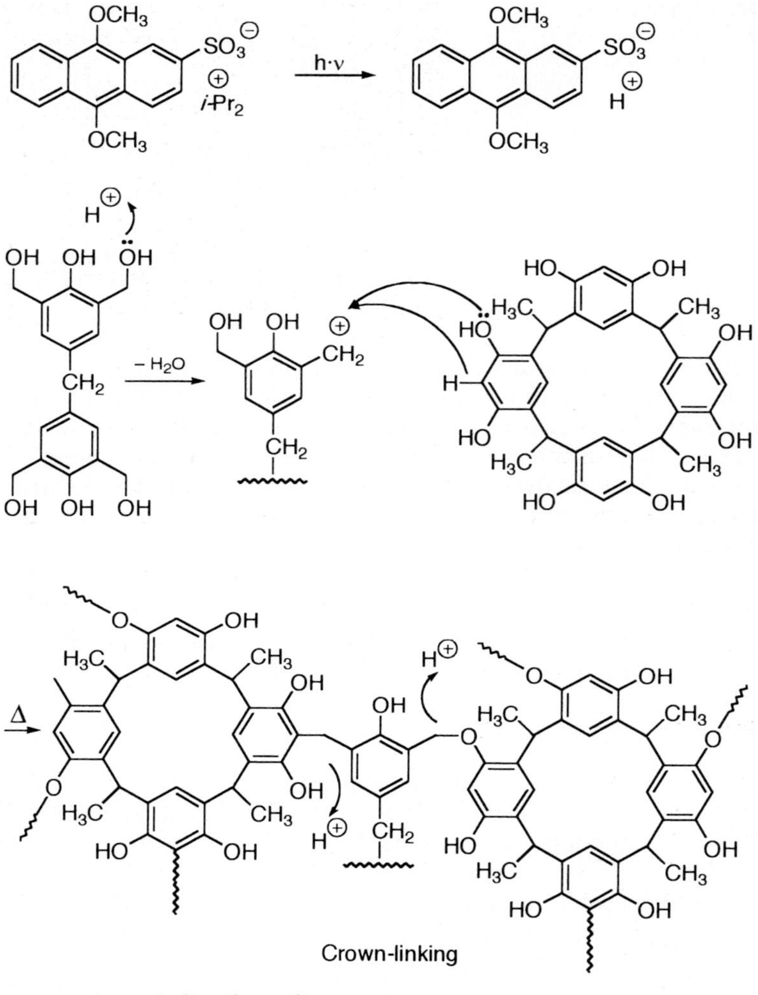

Fig. 6.179. Negative resist based on calixarene

Crosslinking agents based on methylol phenols have been prepared and evaluated by Frechet and coworkers [26, 27]. Various cresol methylols, bis phenol F tetramethylol, bis phenol A tetramethylol, among others, were shown to be effective crosslinkers of PHS in CAR systems with variable sensitivities.

Several monomethylolated, dimethylolated cresols, xylenols, and various mers [2–5] have been described by Sumitomo Chemical [28]. These methylolated materials were proposed as intermediates for DNQ compounds or for the preparation of cresol or xylenol novolak resists.

An unusual negative resist 365 nm based on a calixarene resin matrix, methylolated phenolic materials as crosslinking agent and anthracene sulfonate as photoacid generator [29] is shown in Fig. 6.179.

Fig. 6.180. High resolution 0.12 μm features [30]

For sub-0.20 μm devices, 193 nm photoresists may be required. Allen [30] recently reviewed some of the choices that the semiconductor industry must consider for narrow line widths below 0.20 μm. Aliphatic polymers are the matrix resins of choice and vary from acrylates, cyclic olefins, or hybrids. Thus as shorter wavelengths (DUV) are used, aliphatic polymers will be required (Table 6.101). Figure 6.180 shows 0.12 μm features.

Phenolic materials (cresol novolaks and PHS) will continue to be used in photolithography into the next century. As chip capacity and line widths narrow, aliphatic resin systems will be the resist resin matrix of choice.

6.3.1.8.3
Summary/Trends

Phenolic materials are important components in imaging devices that undergo either chemical or physical transformation. The use of cresol novolaks or polyhydroxystyrene resins as photoresists has transformed the computer industry into a high growth vital national industry with global impact. Computers have reduced greatly in size with a significant increase in memory capacity and speed. In achieving these computer features, resist line widths made by positive resists have continued to narrow in width and are accomplished by shorter wavelength UV.

On exposure to actinic light, DNQ undergoes a Wolff rearrangement leading to an acidic material within the cresol novolak and extractable with a base developer resulting in a positive image. High o,o- linked cresol novolaks that are based on m-, p-cresol monomers are desirable resist materials with monodisperse MWD. To achieve favorable copolymer composition cresol competitive reactivities were determined along with positional reactivities.

Random, alternating, ABA block copolymers and uncatalyzed resins are described along with resist performance.

As shorter wavelength UV (below 250 nm or deep UV) is considered, cresol novolaks are opaque at these wavelengths, and PHS with some modification is a suitable resist. Chemical amplification has been effective in using PHS materials in this wavelength region. Sub-0.20 μm resist lines may require aliphatic resin systems.

6.3.2
Phenolic Resins as Additives in the Rubber Industry

Phenolic resins are used in rubber mixes to achieve special properties [1–8]. From 1994 to 1997, sales of phenolic resins for use in the rubber industry were around 10,000–13,500 tonnes annually in Europe, excluding states of the former Soviet Union. Figure 6.181 shows that the demand rose steadily over the indicated period. This correlates with the production volume of automobiles, one of the major users of rubber (tires).

As shown in Table 6.103, the phenolic resins used in this application may be classified into three groups according to their function, type and effects:

1. *Vulcanization resins* that specifically react with rubber and act as a vulcanization agent to crosslink it
2. *Tackifying resins* that remain chemically unchanged, and during the production process provide the rubber mixes with the desired tack required for finishing
3. *Reinforcing resins* that in combination with curing agents (mainly hexamethylene tetramine or sometimes hexamethoxymethylmelamine, cf. Fig. 6.182) only undergo self condensation (no reaction with the rubber)

These additive resins are used together with a wide range of rubber materials such as natural rubber (NR), styrene-butadiene rubber (SBR), acrylonitrile-butadiene copolymers (NBR), polybutadiene (BR), polychloroprene (CR), and ethylene-propylene-diene terpolymers (EPDM). The level of activity varies

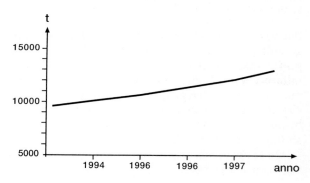

Fig. 6.181. Estimated consumption of phenolic resins in the European rubber industry (excluding states of the former Soviet Union) in tonnes, 1994–1997

Table 6.103. Functions and effects of phenolic resins in rubber mixes

Function	Type of resin	Effect
1. Vulcanization	Alkylphenol resole	Reaction with rubber
2. Tackifying	*p*-Alkylphenol novolak or *p*-alkylphenol acetylene adduct	Physical effect
3. Reinforcement	Phenolic novolak+hexa	Curing of resin during vulcanization

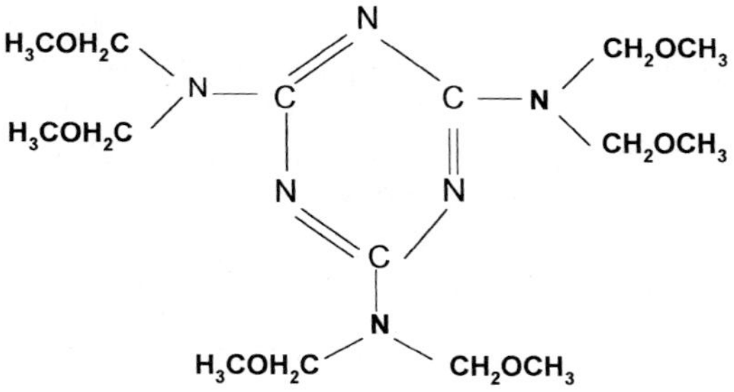

Fig. 6.182. Structural formula of hexamethoxymethylmelamine (HMMM)

with the type of rubber or elastomer. The phenolic resins are incorporated using customary equipment such as roller mills, masticators, and extruders. Solid (powder) resins with the lowest possible level of dust are used.

6.3.2.1
Vulcanization Resins

Vulcanization causes the rubber to undergo a transition from the plastic to the desired flexible state of an elastomer at the pertinent operating temperature. In this process, the individual linear or branched macromolecules of the natural or synthetic rubber materials are linked by chemical bonds, producing a large, loose molecular network. The term "vulcanization" originally only referred to crosslinking of natural rubber with sulfur, a process discovered by Goodyear in 1839 (Fig. 6.183).

In this case, three-dimensional crosslinking occurs by formation of sulfur bridges. Peroxides or other compounds may also be used for crosslinking. Even today, the use of sulfur (in combination with accelerators) is quite common.

However, it is also possible to crosslink rubber through the use of the methylol and (after cleavage) methylene ether groups of p-alkylphenol resoles (Fig. 6.184).

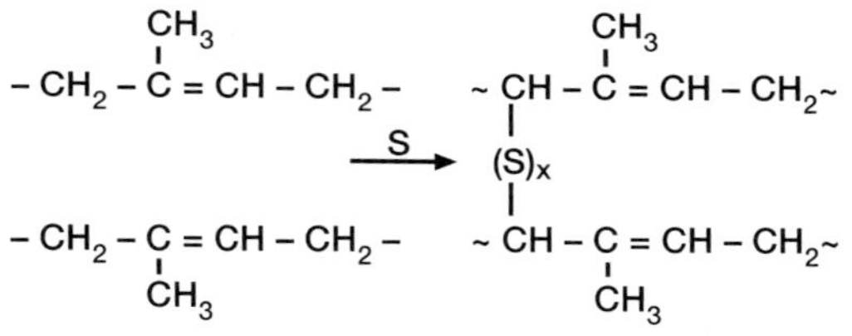

Fig. 6.183. Vulcanization with sulfur at 120–160 °C (basic reaction)

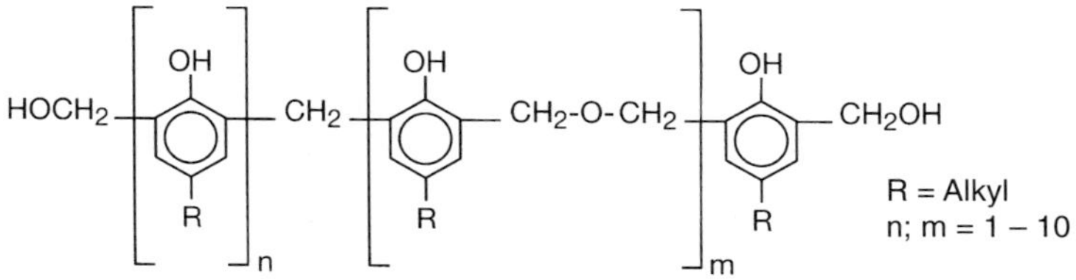

Fig. 6.184. Structural formula of a p-alkylphenol resole

Vulcanization resins are mainly used with butyl rubber (isoprene-isobutene copolymers containing relatively little isoprene). This type of rubber exhibits relatively little unsaturation. The intrinsically good oxidation resistance of butyl rubber and its relatively good heat resistance under mechanical stress are considerably enhanced by carbon-carbon bond crosslinking using phenolic resoles. It is desirable to crosslink not only butyl rubber, but also EPDM rubber with vulcanization resins. Relatively high molecular weight alkylphenol resins furthermore exhibit particularly good compatibility with rubber materials [9, 10]. Since unmodified vulcanization with resins proceeds at a relatively slow rate, it must be accelerated for an optimum rate. Lewis acids or halogen donors, such as polychloroprene or chlorosulfonated polyethylene which provide a similar effect, are used as accelerators for the crosslinking reaction. Metal salts such as $SnCl_2 \cdot 2H_2O$ accelerate the reaction to a great extent, but cause undesirable severe corrosion.

Resoles based on p-iso-octylphenol and containing 7–12% methylol groups have been found to be especially useful as vulcanization resins. The hardness and modulus of elasticity of the vulcanized materials can be adjusted by variation of the fraction of methylol groups.

Various theories exist as to how the vulcanization reaction with p-alkylphenol resoles proceeds [10, 11]. The acceleration of the reaction by acids and the fact that methylene ether groups can also react after cleavage indicates that an ionic mechanism proceeding via carbonium ions occurs [12]. However, a crosslinking mechanism that proceeds by way of the quinone methide (Fig. 6.185) is also possible [13].

Vulcanization resins are used to produce heat, steam, and oxidation resistant products. The heating tubes required in tire production represent typical examples of practical applications. The additive level of vulcanization resin is around 8–12 ph (parts per hundred of rubber).

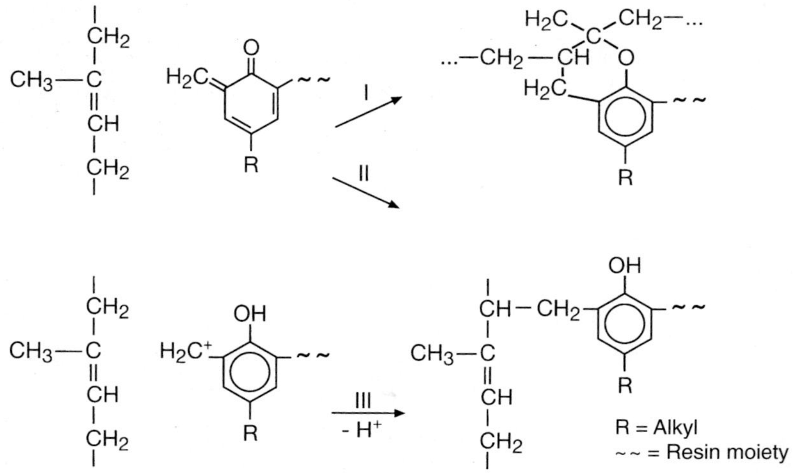

Fig. 6.185. Mechanism of resin vulcanization, possible annelation [3]

6.3.2.2
Tackifying Resins: Resins Used to Increase the Tack of Rubber Mixes

Rubber articles such as automobile tires consist of multiple layers of identical or dissimilar composition (including different types of rubber) that must adhere firmly to one another when the tire is manufactured. In automotive tires (Fig. 6.186), the tread, buffer, and sidewall must meet different types of requirements and thus generally exhibit different compositions. Rubber inner tubes and rubber rollers represent other examples of applications in which rubber mixes of various compositions are combined in a finishing operation. The individual layers making up such products should exhibit adequate contact tack when they are assembled. The synthetic rubbers SBR, NBR, and EPDM exhibit poor tack, particularly when highly modified with fillers. Such highly filled mixes render the use of tackifying resins necessary even in the case of natural rubber [14–16].

The tackifying mechanism is considered to involve polymer flow processes producing tight contact of the two surfaces to be joined and subsequent diffusion of the polymers at the surface interfaces. Both mechanisms can affect the tack, which also depends on the test conditions, the surface roughness, and on other material-specific properties (Fig. 6.187).

Fig. 6.186. Automobile tires (source: Rhein Chemie, Mannheim)

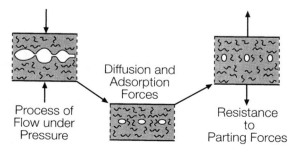

Fig. 6.187. Tackifying mechanisms (from [3])

The contact tack is tested by various means in practice. The nature of the sample surface and the preparation of the test piece affect the result to a great extent. Since the direction of the applied separation force is also significant, three different test methods are possible:

1. Separation force applied vertically
2. Peel test
3. Shear test where force is applied parallel to the contact surface

The test methods can also differ according to whether they represent dynamic determinations involving a steady increase in the separation force, or static determinations in which the force is held constant.

Optimal tack is generally achieved in practice, for example, in automotive tire production, by addition of around 2–6 parts by weight of tackifying resin. The tackifying resins must always exhibit good compatibility with the elastomers and other components of the mix, but not adversely affect the properties of the vulcanized material. No absolute result is obtained in testing of the tackifying effect or contact tack, since the efficacy is not necessarily resin-specific, but depends on many parameters. Tackifying resins may represent both natural and synthetic products such as rosin and cumarone, hydrocarbon, terpene-phenolic, and alkylphenol resins. Alkylphenol novolaks (Fig. 6.188) have also been found particularly appropriate for construction of rubber articles made up of multiple unvulcanized rubber mixes. Since the resin must be highly compatible with the rubber, alkylphenol novolaks based on *p*-iso-octylphenol are mainly used [1].

The customary addition levels are 2–4% (up to 6% in the case of EPDM). The molecular weights of the resins similarly exert an effect on the tack. The compatibility can be adjusted and the tackifying properties enhanced for special applications by using relatively long-chain or even branched alkylphenols such as isononylphenol or dodecylphenol as the phenolic component, or by co-condensation with short-chain alkylated phenols [17]. The viscosity of the mix is decreased, and thus the flow and diffusion at the interface improved by increasing the compatibility [8]. The melting points of the alkylphenol novolaks, determined by the capillary method (ISO 3146), generally range between 80 °C and 100 °C [18].

The patent literature also describes the use of co-condensates of alkylphenols with polynuclear phenols [19], trifunctional phenols [20], or compounds containing epoxy groups [21] as tackifying resins; in special cases, these afford more desirable properties than straight alkylphenol novolaks.

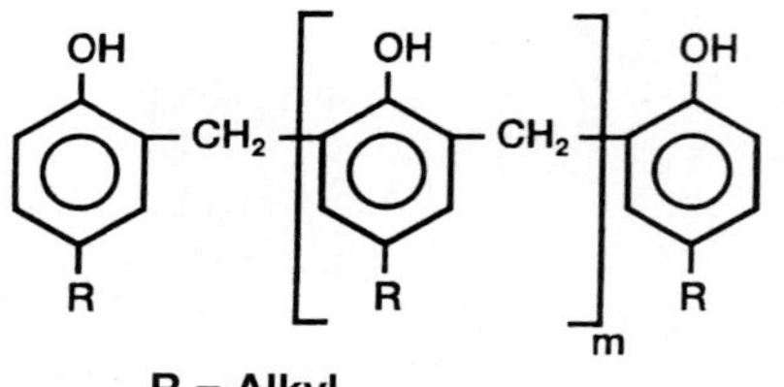

Fig. 6.188. Structural formula of a *p*-alkylphenol novolak

R = Alkyl
m = 1 − x
x = max. 12

6.3.2.3
Reinforcing Resins: Phenolic Resins Used for Reinforcement of Rubber Mixes

The term "reinforcement" denotes an increase in the hardness, tensile stress, modulus of elasticity, and viscoelastic properties. Active fillers are generally added to reinforce vulcanized materials, i.e., to increase their abrasion resistance, elongation at break, and initial tear resistance. The crosslinking density or degree of vulcanization represents a further means of influencing the hardness and stiffness. Since these methods have limitations, the search for effective reinforcing materials commenced at a relatively early date.

Phenolic resins are appropriate for this purpose, since they may be crosslinked and exhibit both high hardness and good resistance to heat, oxidation, and solvents. These are properties that represent desirable improvements in elastomers and serve to broaden their range of application to the widest possible extent. Thus, it is not surprising that the first phenolic reinforcing resins were tested as early as the end of the 1940s, and that phenolic novolaks cured with hexamethylene tetramine afforded good results. It is now known that the resins do not crosslink with the rubber, but self condense, yielding a resite network that permeates the vulcanized rubber [22].

The desired properties (hardness, tensile stress, modulus of elasticity) may thus be enhanced not only with active fillers such as carbon black, but also to a great extent using reinforcing resins [23–26]. In some areas the hexamethylene tetramine, otherwise commonly used as a crosslinking agent for the novolaks, was displaced by hexamethoxymethylmelamine and similar appropriate melamine resins for technical and industrial hygienic reasons at the end of the 1980s.

Reinforcing resins can be used in practically all types of rubber. Initial studies on resin reinforcement [3] were carried out using cashew nutshell oil-modified novolaks and nitrile rubber. When these were extended to unmodified phenolic novolaks, it was found that the compatibility of such resins with the relatively polar NBR is comparatively low. Further extensive tests [22] then demonstrated that the hexa level exerts a significant effect. It was found that the reinforcing resins (unmodified phenolic novolaks), even after curing, afford tensile strength and hardness results in the vulcanized materials that rise together with the hexa addition level and reach an optimum at a certain limit (around 8–9%). The results then fall again as the hexa addition level is further increased. In contrast to the situation in polar rubber such as NBR, no relationship of the properties to a rise in the hexa addition level is generally apparent in nonpolar rubber.

Phenolic reinforcing resins are used alone or in combination with fillers. At an addition level of 15–30 phr (parts per hundred of rubber), the resin system affords increased abrasion and initial tear resistance, reduced residual compressive set, improved heat resistance under static and dynamic load, and good solvent and oil resistance. Examples of applications include tire beads, shoe soles, conveyer belts (Fig. 6.189), drive belts, hoses, rollers, and gaskets.

Tests were also carried out to determine whether the effects on the properties of the rubber mix produced by the reinforcing resin, carbon black, and by combinations of the two exhibit interactions, and if so to what extent [23].

Fig. 6.189. Rubber conveyer belt in an underground mine

The results showed marked maxima in the levels of the parameters, and thus a certain degree of synergism, when 10–20 PBW (parts by weight) of the carbon black was replaced by reinforcing resin. The hardness of the vulcanized materials could be increased by up to 10 Shore A units when 20 PBW carbon black was replaced by reinforcing resins at high filler levels. However, this positive effect was contrasted by adverse changes in properties such as a rise in the residual compressive set at 100 °C and increased permanent deformation following dynamic stress. The practical value of such synergetic effects must therefore be established in each specific case.

Further testing involved combinations of polar and nonpolar rubber. At a constant level of reinforcing resin (phenolic novolak plus hexa) it was found that the hardness of NBR-natural rubber mixes relative to that of straight natural rubber underwent a marked increase even at relatively low NBR rubber addition levels. Similar results were observed using SBR rubber, or when resorcinol was used instead of a phenolic [27].

A special instance of reinforcing resin use is to improve the adhesion to cord. In this case, the objective is to achieve good adhesion of a rubber mix to loadbearing inserts such as steel or textile cord. The casing of an automotive tire is designed to contain the compressed air and absorb the forces that arise during use. It contains woven inserts made of cotton/polyamide, polyester, and possibly steel cord. The first resorcinol/hexa reinforcing resin systems were used together with active silica as a filler to achieve particularly good adhesion to such inserts [28]. Further development led to the use of resorcinol novolaks that offered more desirable processing properties (reduced smoking) in the rubber mix. However, the hexa can lead to corrosion of steel cord. For this reason, there has been a transition to the use of melamine resins, particularly hexamethoxymethylmelamine, as crosslinking agents [29]. The above refers to the formulation of the rubber mix; in addition, various resorcinol resin-containing latices continue to be used for pretreatment of the textile cord. About 4–8% impregnating agent (solids) is required, and the resin fraction in the latex is 15–20% (based on solids). After impregnation, the cord is heated to 120–180 °C for a period of 1–10-min. The latex produces a good bond to the rubber, and the resorcinol resin represents a particularly effective adhesion promoter. The cord is treated in dips.

6.3.3
Binders for the Adhesives Industry

Phenolic resins are not typical adhesives resins, aside from the fact that they are used in large quantities in the woodworking industry as resorcinol resin-based "glues" in glulam construction, and as "wood adhesives" for manufacture of chipboard, fiberboard and plywood sheets. The terms "binder" and "glue" or even "adhesive" overlap in such cases (see Wood Composites, Sect. 6.1.1).

However, straight phenolic resins, and to an even greater extent alkylphenol resins, are used as the main components or as additives in adhesives for a number of special applications [1–6]. Table 6.104 surveys the uses of phenolic

Table 6.104. Examples for use of phenolic resins in various adhesives mixes

Adhesives and similar products	Phenolic resins (possible, types)	Applications, properties
1. Polychloroprene basis	Alkylphenol resoles, solid and in solution	Adhesives for areas such as the shoe, automotive, furniture and building industries; good heat resistance
2. NBR basis	Cresol, xylenol, phenolic resoles, novolak/hexa powder resins	NBR adhesives with NIBR : resin ratios of 2:1 to 1.2; high plasticizer and oil resistance
3. PU basis	Modified phenolic novolak	Compatible with PU rubber to increase heat resistance
4. Rubber-metal-bonding	Phenolic resoles, phenolic novolaks, modified novolak/hexa resins	In combination with elastomers to manufacture adhesion primers for rubber-metal bonding (vulcanization)
5. Friction lining adhesives	Modified phenolic resoles	Bonding of friction linings to metal
6. Metal adhesives	Novolak/hexa powder resins	Plastisol and butyl rubber-based metal adhesives
7. Cladding adhesives	Straight and modified phenolic resoles,	Heat-curing cladding adhesives, for example ones based on acrylate or PVA emulsions
8. Rubber bonding adhesives	Alkylphenol resoles	Natural rubber-based bonding adhesives with enhanced cohesion
9. Hot melt adhesives	Alkylphenol novolaks	Tackifying resins (as additives) for EVA and PA hot melt adhesives
10. Gaskets	Phenolic resoles	Paper/cardboard-based gaskets; precipitation of resin with acid
11. Sealants	Xylenol novolaks and phenolic resoles	Adhesion promoters in NBR and Thiokol-based sealants

resins in the field of adhesives. The present consumption of phenolic resins in the European adhesives industry, excluding states of the former Soviet Union, is estimated at around 4300–4800 tonnes, about 75% of this total being used in polychloroprene adhesives.

6.3.3.1
Polychloroprene-Based Adhesives

Contact adhesives based on polychloroprene [7–10] include alkylphenol resoles (Fig. 6.190) as important additives (Table 6.105). Alkylphenol resoles, for example, those based on t-butylphenol, exhibit very good solvent compatibility. The tackifying effect of the resins arises through interactions with the elastomers. Alkylphenol resins improve the adhesion, cohesion, and particularly the thermal stability of the contact adhesives, which are applied in the customary manner (application, air dry, pressing). After this procedure, the material bond already exhibits very high initial strength, and this continues to increase. Polychloroprene adhesives are mainly used in the shoe, furniture, and automotive industries, as well as in the "do-it-yourself" area.

The types of polychloroprene employed for adhesives crystallize quite well, affording a large increase in strength. Addition of resin retards crystallization and lengthens the period ("open time") during which contact bonding occurs. This crystallization is a reversible process, i.e., the adhesive can be reactivated

Fig. 6.190. Alkylphenol resole (basic structure)

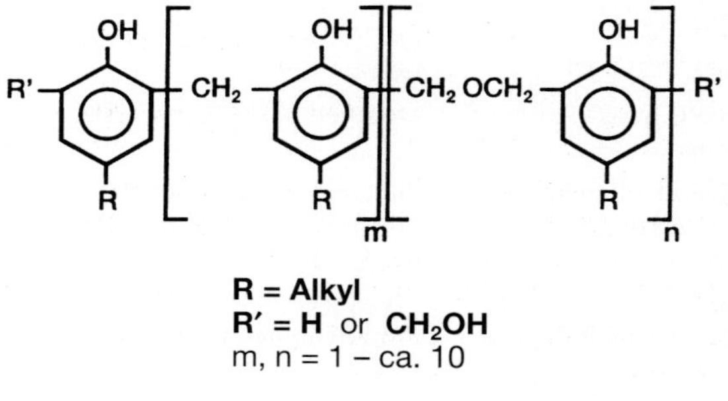

R = Alkyl
R' = H or CH_2OH
m, n = 1 – ca. 10

Table 6.105. Suggested formulation for polychloroprene contact adhesive

	Parts by weight	Raw material
	100	Polychloroprene
	40	Alkylphenol resole (e.g., Rütaphen KA 777)
	5	Magnesium oxide
	4	Zinc oxide
	2	Antioxidant
Solvents	150	Ethyl acetate
	150	Toluene
	150	Methyl ethyl ketone (MEK)
	1	Water

Fig. 6.191. Formation of a magnesium chelate with an alkylphenol resole

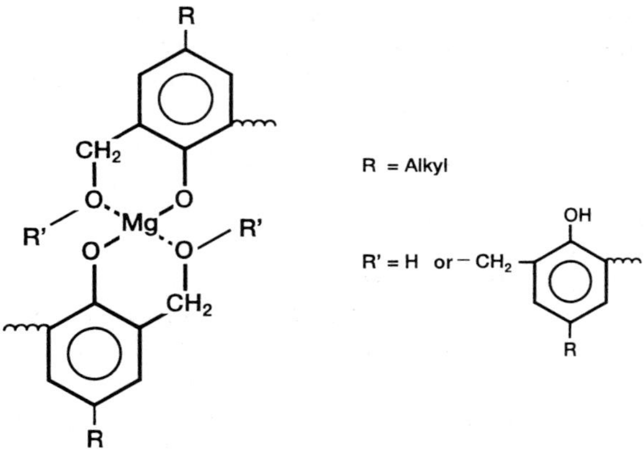

R = Alkyl

R' = H or $-CH_2-$

by application of heat. This property is utilized in the shoe industry by applying the adhesive on a single side ("one-way process"), storing the substrate temporarily after air drying the adhesive, later reactivating the adhesive by application of heat, and finally bonding the substrate (for example a shoe sole) to another by application of pressure.

The added resin provides the bond with very high internal strength that increases as the molecular mass of the resin increases, although care must always be taken to ensure that the resin remains compatible with the elastomer.

The high cohesion and thermal stability arise through formation of a chelate between the alkylphenol resoles and the magnesium oxide included in the formulation (Fig. 6.191). The chelate complexes exhibit high melting points, but good solubility. Chelate formation proceeds through the methylol and dimethylene ether groups of the "base-reactive" resin [11–16]. Chelate formation in the formulation is complete after two to three days. It is also possible to prepare the adhesive by adding magnesium oxide to the straight resin solution, and after a maturing period of about 24 h to add the chloroprene solution. This method is somewhat more involved, but permits better utilization of the resin's base reactivity, which is critical for the level of thermal stability. The open time of the adhesive can be increased as required by addition of tackifying resins such as terpene phenolics (5–20% based on alkylphenol resole), although this generally reduces the thermal stability.

The reaction with the alkylphenol resin, for example, *t*-butyl phenol resole, is affected to a great extent by the content of metallic oxide, the time, temperature, the type of solvent, and the water level. The required level of magnesium oxide is around 10% based on resin. This value is generally recommended as a minimum level for polychloroprene adhesives. Chelation proceeds most rapidly in solvents of low polarity such as toluene, hexane/ethyl acetate, and acetone. Polar solvents can retard the reaction. In polar solvents, or mixtures of these with nonpolar solvents, the reaction may be accelerated by addition of water. The water level of the resins used in such applications is generally around 0.2–0.3%. This is sufficient for acceleration of the pre-reaction. Metallic oxides other than the magnesium oxide generally used – for example, calcium and barium oxide – can also form chelate complexes exhibiting high melting points (decomposition at 250 °C), but formulations containing higher levels of these are unstable [8, 9]. The grade of magnesium oxide also influences the quality of the adhesive.

Phase separation in polychloroprene adhesives, i.e., precipitation of metallic oxide particles, can be influenced by the choice of resin [8, 13]. The mean molecular mass of the alkylphenol resoles should be around 1200–2000 to ensure phase stability of the adhesive. Commercial alkylphenol resoles recommended for use in manufacture of polychloroprene adhesives can differ in their open time and in the level of thermal stability that may be achieved with them. In general, it may be noted that although resoles with low melting points (approximately 50–60 °C) offer a long open time, they also exhibit low thermal stability. Resins with high melting points (approximately 80–90 °C) generally feature a short open time in adhesives formulations and high thermal stability of the adhesive bond. p-Phenylphenolic resoles provide extremely high thermal stability, but are very high in cost.

In addition to magnesium oxide, zinc oxide is used as a filler. Zinc oxide exhibits no tendency to form chelates, but serves to absorb the hydrogen chloride that is eliminated from the chloroprene over the course of time. Highly filled adhesives such as those for floor coverings contain magnesium or zinc carbonate, silica, quartz sand, or barytes. The antioxidant present in the formulation is designed to protect the adhesive bond against the embrittlement or softening induced by aging or simply due to oxidation-sensitive resins. Non-discoloring phenolic antioxidants [17] have been found particularly appropriate. Alkylphenol resins exert no adverse effect on the aging characteristics. The technical adhesive properties that can be achieved also depend on the type of polychloroprene. In combination with resins such as base-reactive alkylphenols, moderately highly crystalline grades of polychloroprene exhibit a lower loss of cohesive strength when crystallinity is disrupted than do highly crystalline grades [9].

Mixtures of toluene/methyl ethyl ketone/ethyl acetate and 80/100 mineral spirit are best used as a solvent. Variation of the mixing ratios permits the open time and viscosity to be adjusted within certain limits. The aliphatic hydrocarbons serve as an inexpensive diluent.

Base-reactive alkylphenol resoles may also be used as additives in polychloroprene/polyisocyanate-based two-component adhesives. An additional reaction of the phenolic hydroxyl group with the isocyanate takes place in such adhesives. In this case, crosslinking leads to an increase in the thermal stability, since the isocyanate reacts with the hydroxyl groups in both the rubber and the resin. The recommended weight ratio of isocyanate to resin to rubber is around 8:25:100, although this decreases the bench life very considerably [7, 8, 14, 15].

6.3.3.2
Nitrile Rubber Adhesives

The design of nitrile rubber adhesives is similar to that of polychloroprene materials (Table 6.106). In contrast to polychloroprene-based products, nitrile

Table 6.106. Suggested formulation for (heat curing) NBR adhesives

	Parts by weight	Raw material
	100	Nitrile rubber
	50–150	Phenolic resin (e.g., Rütaphen 9289 KP cresol resole)
	0–100	Fillers/pigments (graphite, iron oxide, calcium silicate, zinc oxide, titanium dioxide)
Solvents	300	Ethyl acetate
	100	Acetone or MEK
Plasticizer	0–20	Dioctyl or dibutyl phthalate (DOP or DBP)

rubber adhesives – which feature high plasticizer and oil resistance – are heat reactive. Depending on the required flexibility, the ratio of rubber to resin is generally 1:2–2:1. Heat-reactive, crosslinking cresol resoles or novolak/hexa systems are used as additive resins. Bonding is performed under heat and pressure following solvent removal. Solvent-free NBR adhesives are produced on masticators or roller mills. The cohesion, adhesion, and resistance to heat, oil, and solvents are increased by the added phenolic resins. Applications include PVC bonding and production of automotive seals.

Grades of rubber incorporating high levels of acrylonitrile are best for production of NBR adhesives. The phenolic resins that are employed represent resoles based on phenol, cresol, xylenol, or their co-condensates. Novolaks in combination with hexa are also customarily used as curing agents. Fillers only exhibit reinforcing properties. Plasticizers such as dibutyl or dioctyl phthalate can be added to enhance the flexibility. Processing is generally carried out at temperatures of up to 150 °C [18–20].

6.3.3.3
Polyurethane-Based Contact Adhesives

Solvent-containing, polyurethane-based contact adhesives can also be modified with phenolic resins (Table 6.107) to increase the open time and the initial, final, and hot strength levels [21]. Phenolic novolaks exhibiting a relative-

Table 6.107. Suggested (test) formulation for a polyurethane contact adhesive [21] including alkylphenol novolaks as additives (peel strength after 24 h on shoe sole rubber or plasticized PVC 1.2–1.3 N/mm^2 at ambient temperature)

Parts by weight	Raw Material
100	Dermocoll 530 elastomer
20	Alkylphenol novolak (e.g. Supraplast® D 2057 W or Rütaphen 7162 KA)
340	Acetone
130	MEK
120	Toluene

Fig. 6.192. Adhesion test of a shoe sole bonded with an alkylphenol resin containing polychloroprene adhesive

ly high melting range (90–110 °C) are generally used. Depending on the application, the rate of addition is 10–30% phenolic resin based on polyurethane. Fields of application include the shoe industry (Fig. 6.192), furniture industry, automotive construction (interior panels), and cladding of aluminum sheet, chipboard, and fiberboard with a wide variety of plastic films.

These adhesives are based on special grades of polyurethane elastomers that are used in the form of solutions. A crystallization process similar to that in polychloroprene adhesives, during which the cohesive forces responsible for the strength develop, is initiated in the drying operation. In this case as well, the rapid crystallization renders the open time excessively short. Alkylphenol novolaks are added in order to slow crystallization of the polyurethane and increase the open time of the adhesive. Depending on the elastomers and the type and level of added resin, the properties of the contact adhesive may be varied within certain limits, even when polyurethane exhibiting a strong tendency to crystallize is used. The adhesive bond normally only achieves its final strength after a period of several days, although the initial peel strength is generally sufficient to permit handling of the bonded parts.

These contact adhesives may also be used in the "one-way" process (single-side application with possible interim storage) described in the case of the polychloroprene adhesives. The phenolic resin decreases the activation temperature, for example, in the case of sole material for the shoe industry, by 15–20 °C compared to that of the unmodified adhesive.

To increase the thermal resistance, polyisocyanate crosslinking agents may also be incorporated into the adhesives batch. In this case, the phenolic resins are also cured via their hydroxyl groups.

6.3.3.4
Rubber/Metal Bonding

Bonding of metal to rubber is of great importance in various applications [4, 21]. In contrast to the situation in adhesive bonding, rubber/metal bonding involves vulcanization of an unvulcanized rubber mix onto a metallic sub-

Fig. 6.193. Engine suspension (example of rubber/metal bonding)

strate that has been pretreated with a bonding agent. This permits considerably higher adhesive strength levels to be achieved [22]. The bonding agents are vulcanizable formulations generally based on chloroprene rubber, to which novolak/hexa-type phenolic resins or resoles containing reactive methylol groups are added. It is also common to incorporate epoxy resins.

Typical examples of rubber/metal bonding are found in rubber springs, bearings, engine suspensions (Fig. 6.193), shock absorbers, tread padding for tracked vehicles, linings for pipes, pumps and fittings, engine supports, sealing components, and bridge supports.

A two-coat system including a primer is used when an extreme range of requirements must be met (Fig. 6.194 B), for example, when resistance to hot

Fig. 6.194. Rubber/metal bonding scheme in [21] (A = without primer, B = with primer)

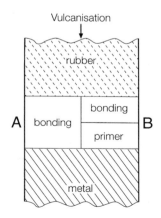

water is necessary. Phenolic resoles or combinations of these with novolaks represent an important component of these adhesion primers, which also include elastomers (nitrile-butadiene or polychloroprene rubber) as film-forming agents.

The resins are also crosslinked when the elastomers are vulcanized. As a result, the reactivity of the resins must be matched to the vulcanization process. Diffusion processes at the interfaces are of particular significance for development of a good bond, i.e., layers are formed in which components of the elastomer and the thermosetting binder crosslink with each other.

The process as such may be described in the following manner:

1. The metallic surface is phosphated or roughened
2. A bonding agent, and if appropriate a primer as well, is applied by brushing, spraying, roller coating, or blade spreading; the solvent is evaporated at ambient temperature or at 60–80 °C until the coating is tack-free
3. The rubber mix to be vulcanized is applied by pressing or transfer molding process. Vulcanization takes place at 130–200 °C, depending on the quantity and grade of the rubber

With regard to the properties of the finished articles, it is presumed that in case of optimal adhesion, the elastomeric material fractures under tensile stress, and that this fracture does not take place in the bonded surface. This circumstance is termed "100 % rubber fracture."

6.3.3.5
Friction Lining Adhesives

Although bonding of friction linings to metal backings was previously accomplished using rivets, a significant portion is presently adhesive bonded. Phenolic resins modified with vinyl polymers to enhance the adhesion are used for this purpose.

Bonding of friction linings to the metallic backing material – adhesive bonding of a thermoset molding to metal – offers significant advantages over attachment by riveting [21] for the following reasons: 1) the thickness of the lining can be utilized to a greater extent since no rivets interfere; 2) the lack of rivet holes increases the active braking surface; and 3) thinner linings can be used for clutches. Current asbestos-free linings are particularly suited for adhesive bonding; their strength is further stabilized by the bonding process, and the simplified assembly and longer service life of the linings reduces costs.

However, the advantages mentioned above presuppose that the adhesive bonds remain both completely reliable throughout the entire service life of the linings and thermally stable even under varying stresses in automotive operation, and that the adhesives can be processed by highly efficient, uncomplicated methods. Phenolic resoles modified with products such as vinyl polymers to eliminate the brittleness of the adhesive film and enhance adhesion to the materials being bonded are best used as friction lining adhesives. Such adhesives are generally supplied in the form of solutions. The composition of the adhesives can vary when they are adapted for use with specific grades of linings and types of process operations in the bonding shop. Thus, the viscosity of the adhesives solution is adjusted for the pertinent method of application and the characteristics of the linings.

Adhesive films based on phenolic resoles, polyvinyl formal or butyral and nitrile rubber are also used, in some cases reinforced with weblike carrier materials based on glass, synthetic, or linen fibers. Such phenolic resole and vinyl polymer-based adhesives are also used in applications such as production of lightweight structures in aircraft construction, large-scale construction elements made of composites, for cladding work, and production of honeycomb structures [23] (see High Performance and Advanced Composites, Sect. 6.1.5).

Process stages are recommended when bonding linings for drum brakes:

1. The substrate surfaces are prepared for bonding by sandblasting or grinding. The lining material is available in the production area, and is in a condition appropriate for bonding (surfaces roughened).
2. In series production, the adhesive is applied by rollers or extruders. Clutch sections are also coated by screen printing. The thickness of the coating ranges from 70 μm to 200 μm, and depends on the type of adhesive and grade of lining. It is useful to coat both bonding surfaces (lining and metal).
3. If adhesive films are not used, the solvent in the layers of adhesive must be removed as completely as possible. This occurs during storage at ambient temperature or is accomplished by drying at 50–70 °C in a chamber or tunnel oven. The parts can then be bonded immediately, or after a lengthy period of storage.
4. The linings and brake shoes are pressed together using special clamping jigs, and the adhesive cured at maximum temperatures of 180–220 °C in a chamber or continuous throughput oven. The holding time in the oven depends on the reactivity of the adhesive and the required peak temperature. It is generally adequate if a maximum temperature of 180–220 °C is used in the contact layer and persists over a period of 5–10 min.

Disk brake linings are bonded to the pads in a manner similar to those employed in the process used to bond drum brake linings.

In practice, the adhesive bond is subjected to various tests to assure a lining assembly quality adequate for automotive use. For example, it is customary to shear off the lining or apply force to separate it from the backing in special test equipment, measuring the force required to accomplish this. Both the failure appearance and the measured strength levels are of significance. Aside from cold shearing trials, tests of the adhesion strength at elevated temperatures, for example 250–300 °C, are of particular significance for assessment of adhesive bonds in automotive operation. In this case, it is possible to heat the parts to the pertinent test temperature, and then to shear off the lining while hot or after cooling. Thermal shock tests, in which the test article undergoes multiple heating and cooling cycles, represent particularly stringent trials [21].

6.3.3.6
Other Applications of Phenolic Resins in Adhesives Formulations and Sealants

6.3.3.6.1
Metal Bonding

In vehicle construction (automobiles, rail vehicles), structural materials are bonded with adhesives based on plastisols and butyl rubber (including reactive grades), for example, in elastic bonding of sheet metal parts with profiles. Since the adhesives in this case function to some degree as sealants, they can also be classified as such. A novolak/hexa powder resin with a hexa level below 7–8% is added at a rate of 3–5% to increase the adhesive strength. Typical applications for such adhesives include their use in bonding sheet metal structures for hoods, roofs, and doors.

6.3.3.6.2
Cladding Adhesives

Phenolic resin impregnated rigid paper (cf. chapter on electrical laminates) is a base material for production of copper-clad boards for printed circuits and decorative laminates with metal surfaces. In decorative sheets, untreated or anodized aluminum foils are also used in addition to copper foils. The bond between the rigid paper and metal foil can also be produced using vinyl (for example, polyvinyl butyral) modified phenolic resoles. The condition of the foil surface is critical for proper bonding between the metal foil and the adhesive.

Whereas commercial copper foils exhibit a surface with good adhesion and suitable for bonding by treatment of one side, other metal foils must be subjected to a mechanical or chemical treatment.

Such treatment includes degreasing as well as mechanical or chemical roughening of the surface. The foils are coated by uniformly applying solutions of the modified phenolic resoles. The drying process is controlled to ensure that the adhesive is evenly distributed over the foil and no bubbles develop on the surface. A prerequisite for the required solder bath resistance of a copper foil employed for cladding a base material used for printed circuits is a volatile component level below 0.5 wt%. The coated foils are bonded to the backing material under pressure in stage presses at 140–160 °C to yield a composite sheet. Heat curing cladding adhesives based on acrylate or PVAc emulsions may also be modified with phenolic resoles such as Rütaphen 6245 KP.

6.3.3.6.3
Natural Rubber-Based Bonding Adhesives

Bonding adhesives are permanently tacky materials that are applied to backings such as paper, fabric, or foils, and bond to the surfaces of many materials even under slight pressure. The adhesive mix, which is applied in the form of a solution, emulsion, or melt, consists largely of natural rubber materials, resins, fillers, and plasticizers. Bonding adhesives resist short-term stress well, but tend to "creep" under permanent stress, particularly at elevated temperatures. For this reason, adhesive mixes containing natural rubber are modified with alkylphenol resoles to increase their cohesion, especially at elevated temperatures. In this case, vulcanization of the rubber takes place at elevated temperatures. Applications include adhesive tape for use in painting automobiles. NBR rubber-based adhesive tapes can similarly be modified with phenolic resins.

6.3.3.6.4
Hot Melts

Ethylene/vinyl acetate and polyamide polymers are used as base materials for hot melts used in the paper packaging industry, furniture industry (bonding of edge and surface foils), and in assembly bonding (electronics, automobiles, commercial vehicles, and aircraft), with hydrocarbon resins generally being used as tackifiers. Fairly long-chain alkylphenol novolaks such as Rutaphen 9485 KA are used to achieve certain properties and improve adhesion.

These products are manufactured by melting the components together in heated equipment (stirring vessels, masticators, or extruders) and then discharging the mix in the form of pastils, granulate, blocks, or rods.

6.3.3.6.5
Gaskets

Slurries of fibrous materials such as cellulose, minerals, fiberglass, or Aramid with binders such as latex combined with fillers in water may be used to produce gaskets. The material is deposited on a screen belt or circular screen by means of a vacuum. It is dried at around 180–200 °C. Phenolic resoles may be included as additives to increase the thermal stability. Applications for such gaskets are in the chemical and automotive industries, and in mechanical engineering. Such gaskets may also be manufactured from special types of paper impregnated with alcoholic solutions of phenolic resoles (cf. sections on impregnation, Sect. 6.1.4.9 and 6.1.5.6.1, p. 254 pp and 265 pp).

6.3.3.6.6
Sealants

Compounds that remain plastic or elastic are termed sealants. Sealants designed to fill expansion and connecting joints are composed of fillers, binders, and special additives. Thiokol or NBR sealants contain about 1–8% phenolic resin based on the binder fraction as an adhesion promoter. Materials including xylenol novolaks, for example, Rütaphen 0782 KP, are used as adhesion promoters in NBR-based sealants. Specially modified phenolic resoles such as Rütaphen 0784 KP are particularly suitable for use in Thiokol sealants.

6.4
Intermediate and Carbon-Forming Bonding

Intermediate bonding generally relates to the production of moldings that are to be transformed into another class of product, for example, by thermal treatment, where the dimensions of the product correspond to those of the original. A particularly good example is production of "glassy carbon" (Sect. 6.4.4) crucibles (Fig. 6.195). Thus, "green" (uncured) crucibles are produced from filler-free phenolic resins, fully cured, and subjected to a lengthy pyrolysis process that finally leads to glassy carbon crucibles exhibiting outstanding chemical resistance. As expected, elimination of pyrolysates causes shrinkage of the crucible, with its volume reducing by about 50%. However, the original, desired geometric ratios of diameter to footprint, height to footprint, and height to diameter must and are retained in this process.

In their function as an intermediate binder, phenolic resins are similarly of considerable importance in production of refractories such as magnesia or dolomite bricks. The green bricks are initially only cured, and are later carbonized (the binder "cracked") by separate pyrolysis or use in a steel mill, a process that again can lead to changes in the volume and density. Ideally, the dimensional relationships of length to width to height should also remain constant in this case. In principle, the same applies to production of green grinding wheels by cold compression molding (see Abrasives Sect. 6.2.2). The uncured resin or mixture of liquid and powder resin provide the cold pressed green wheel with intermediate dimensions that ensure the required dimensional relationships in the final product even subsequent to the cleavage reactions that proceed during the curing process.

Due to their crosslinked structure with a nominal carbon content of around 80%, phenolic resins are capable of facile carbonization in the cured state, i.e., they exhibit a further bonding function that can be transformed into a carbon bond under certain physical conditions. This capability is utilized in applications such as production of graphite materials and refractories. Resin systems that afford a high carbon yield and simultaneously exhibit good mixing properties (in cold and warm mixing processes) and excellent compatibility with fillers and graphite are generally best suited for such uses.

Section 6.4.1 will deal with the pyrolysis of phenolic resins, since this process – involving factors including the carbon yield, oxygen resistance of the

Fig. 6.195. Production of glassy carbon crucibles (original molding and pyrolyzed final product of glassy carbon)

polymeric carbon, and evolution of volatile decomposition products – is of great prominence in production of the carbon-containing products mainly considered in the pertinent chapter. The significance of carbon formation in determining the quality of phenolic resin-bonded brake linings is also quite vital, since the binders pyrolyze at the point the lining and the brake disk or drum come into contact during the braking operation. Carbon and pyrolysis gases also develop when phenolic-bonded foundry molding materials are poured off, and exert special effects on the casting operation and casting quality.

6.4.1
Pyrolysis of Phenolics

Cleavage by-products and (where appropriate) solvents are liberated when phenolics are cured at temperatures to 200 °C. Optimal three-dimensional crosslinking of the resins to yield a resite is important to provide a favorable starting site for carbon formation during pyrolysis.

Theoretical considerations of resite segments and knowledge from detailed pyrolysis studies [1–6] show that a theoretical residual carbon yield of 55–70 % may be expected in pyrolysis, based on an 80 % carbon level in an ideal resite segment. The variables that are involved are the crosslinking density, the size of the molecules after curing, foreign inclusions (trace elements), and the number of carbon-containing compounds that are eliminated. For structural reasons, the carbon yields in other types of thermosetting resins are lower (Table 6.108).

Various ranges can be established in a simplified consideration of resite pyrolysis (Fig. 6.196). Curve 1 in Fig. 6.196 represents degradation of the resite and creation of the carbon structure by tracing the development of density and strength [7, 8] while curve 2 describes the emissions. The density and strength levels (Fig. 6.197) are reduced as the temperature rises to about 600 °C, then increase as the temperature continues to rise to about 800 °C, leading to formation of "polymeric carbon" at the end of the ring forming reactions.

The structure passes through two phases in the transition from point A to point C in Fig. 6.196. A phase of falling density, in which the volume initially remains constant, is located between points A and B; an increase in the density and strength with a simultaneous reduction in the volume occurs between points B and C, and only degassing phenomena, that have little or no effect on

Table 6.108. Carbon content and theoretical carbon yield		Carbon Content (%) (theoretical)	Carbon-yield (%) (possible level)
	PF	<80	55–70
	FF	<75	50–60
	EP	<75	25–35
	UP	<60	15–25

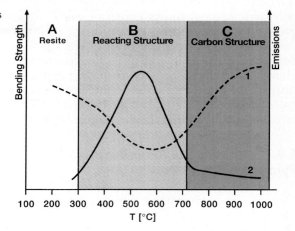

Fig. 6.196. Simplified pyrolysis scheme, density and strength (1), emissions (2)

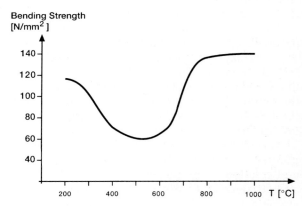

Fig. 6.197. Bending strength vs temperature, determined by thermal treatment of phenolic resin films at temperatures of approximately 300–900 °C (from [7])

the density and strength levels, can still be observed at temperatures of 800 °C and above. Reasonable estimation of the applicability of methods used to determine the carbonization residue requires comparison of the methodology and the pyrolysis process. On the basis of such considerations, pyrolysis may best be evaluated using thermogravimetric analysis at temperatures to 1000 °C. Complete incineration is subsequently carried out to determine the true carbon content of the carbonization residue.

Using special test equipment (Fig. 6.198), various resin systems (after being cured at temperatures to 200 °C) can be evaluated for their possible carbon yields [9] under an argon atmosphere at temperatures to 1000 °C. It has also been found that pertinent information regarding the oxygen sensitivity of a resin system can be determined when parallel tests are conducted under an atmosphere containing 1.5% added oxygen. Differences in the methane, hydrogen, water, and carbon monoxide levels together with major differences in the carbon yields have been observed. Aromatic, high molecular mass cleavage products are conveyed to a separate testing system, and the structures of the pyrolysis residues visualized with scanning electron microscopy.

6.4 Intermediate and Carbon-Forming Bonding

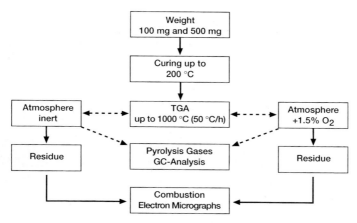

Fig. 6.198. INASMET test system for measuring C-yield of phenolic resin systems under inert and oxygen containing conditions (source: Bakelite AG, Iserlohn, Germany)

Such tests have shown that the effect of the heating rate – whether heating is carried out over 22, 29, or 37 h – is relatively minor. The sample size has a greater effect. The carbon yield can increase by up to a third for larger sample sizes. This indicates that the greater volume leads to a higher level of carbonization of the outwardly diffusing organic components. When the test is conducted in an atmosphere containing 1.5 % oxygen, this phenomenon disappears partially or completely, since in this case a certain amount of combustion dominates. To some extent, the knowledge gained in such tests can be extrapolated to end use applications.

Hexamethylene tetramine is generally used to "harden" phenolic novolaks. In general, carbon residue or "polymeric carbon" is similarly obtained when hexa (HMTA) free novolak systems are pyrolyzed. Special tests [10] have demonstrated that the pyrolysis residues of HMTA-free systems are highly dependent on the mean molecular mass of the pertinent novolak (Fig. 6.199). The carbon yield is only around 20 % at a low molecular mass such as 400, whereas carbon yields of about 50 % are achieved when the molecular mass increases to around 1000.

Fig. 6.199. Carbon yields from HMTA-free phenolic novolaks vs molecular mass

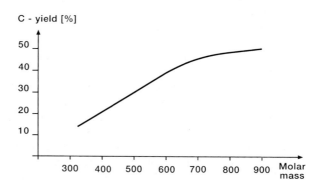

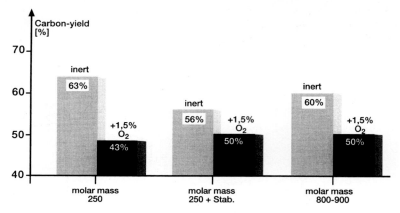

Fig. 6.200. Carbon yields in pyrolysis of cured novolaks with different mean molecular masses (in inert and oxygen containing conditions)

In addition, it was found that relatively high carbon yields (over 60%) are obtained in pyrolysis of very low molecular mass novolaks cured with 10–14% HMTA (Fig. 6.200, molecular mass 250), but these drop sharply in the presence of oxygen – a process that can be retarded by the addition of phenolic antioxidant additives (middle bars in Fig. 6.200). The latter effect can be explained by assuming active participation of the phenolic additives to form extended ranges of uniform structure. This structure is then less susceptible to attack by oxygen. Carbon yields of up to 60%, that only fall by about 10% (absolute) under the effect of oxygen, are achieved in the case of novolaks that exhibit molecular masses of 600–900 and are cured with 10–14% HMTA (right group of bars in Fig. 6.200).

This can indicate that optimal crosslinking of these novolaks affords enhanced oxidation stability. However, it is also possible that the increased oxidation stability is completely or partially associated with the nitrogen level, since the carbonization residues are not free of nitrogen derived from curing with HMTA. Work by Böhm and Mang [11] has shown that a significant effect on the oxidation stability can be achieved by incorporation of nitrogen in amorphous carbon. Sugar was carbonized in this study. Such carbonization was also performed using sugar that had previously been treated with specific amines, and resulted in polymeric carbon containing up to 5% nitrogen. This treatment increased the resistance to oxygen by several powers of ten. The effects of boron and phosphorus as oxygen stabilizers are well known [12, 13]. However, direct incorporation of boron or phosphorus in the phenolic resins generally afforded no significant stabilizing effect.

When pyrolyzed under an inert atmosphere, solid resoles afford a carbon yield of 66–70%; these values drop by about one third in the presence of oxygen. This behavior may be associated with the fact that oxygen-containing groups such as ether bridges are still present in the resite structure and partly induce the greater degradation. In the case of aqueous resoles condensed under alkaline conditions, the carbon yields obtained from the cured resins

Fig. 6.201. Effect of oxygen on carbon yields in pyrolysis of furan resins

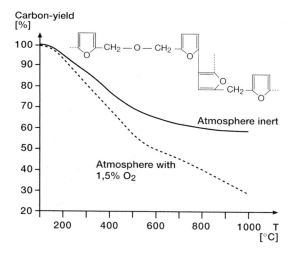

are around 50–65%, but drop by about one third under the effect of oxygen. The presence of alkaline catalysts has an effect in this instance.

Phenolics were also compared with furan resins. Furan resins exhibit an even greater tendency to oxidize during pyrolysis (Fig. 6.201). For example, the original carbon yield of nearly 55% falls to around 31% in a non-inert atmosphere. This is presumably associated with the greater instability of the partly linear, relatively oxygen-rich structure, which is partly crosslinked with double bonds and is less uniform than that of a phenolic resite.

The effect of oxygen on pyrolysis of phenolics is also apparent in the structure of the carbon components observed in scanning electron micrographic analysis. An oxidative atmosphere generally causes increased growth of carbon whiskers from the gaseous phase at the surface of the carbon. The heating rate has a certain significance when pyrolysis is carried out under a slightly oxidizing atmosphere (1.5% oxygen). The whiskers are large and numerous at a low heating rate (Fig. 6.202). When the heating rate is higher (Fig. 6.203), no

Fig. 6.202. Scanning electron micrograph of carbon from pyrolysis of a novolak-hmta resin in an oxygen-containing atmosphere, slow heating rate

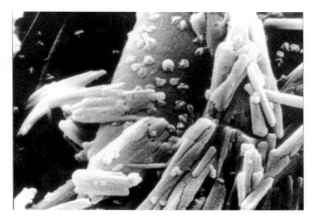

Fig. 6.203. Scanning electron micrograph of carbon from pyrolysis of a novolak-hmta resin in an oxygen-containing atmosphere, fast heating rate

Fig. 6.204. Specific surface area of resin-derived carbon by TGA

Carbon from	Atmosphere	Specific Surface (m^2/g)	
		Inert	1,5% O_2
1. Cured resols		< 1	< 10
2. High crosslinked novolaks		< 1	< 6
3. Uncrosslinked novolaks		< 10	150-180

whiskers are formed, and a "cauliflower" structure results. On the whole, far less material is also liberated. These observations can be explained by the relationships expressed in the Boudouard equilibrium ($2CO \leftrightarrow C+CO_2$).

On the whole, studies have shown that the carbon yield and quality depend on the type of resin (phenolics generally better than furan resins), the crosslinking density of the resins, the pyrolytic atmosphere (inert or oxygen containing), and trace elements.

The specific surface area of the carbon formed from phenolics (Fig. 6.204) is also affected to a considerable extent by the surrounding atmosphere. The following trends are nonetheless apparent:

- Cured aqueous phenolic resoles result in a specific surface area less than 1 m^2/g when pyrolyzed in an inert atmosphere. Values which can be up to ten times higher occur in the presence of oxygen.
- The specific surface area of the pyrolysis residue from an unmodified novolak greatly depends on the crosslinking density and the molecular mass. After treatment in an inert atmosphere, specific surface areas of less than 1 m^2/g are measured at higher molecular masses when the level of added hexamethylene tetramine is also high.

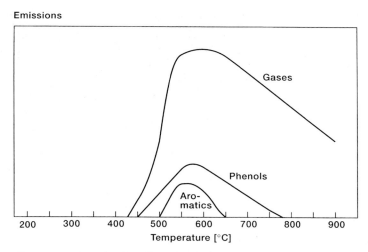

Fig. 6.205. Cleavage products in phenolic resin pyrolysis

- In an oxygen-containing atmosphere, novolaks of various crosslinking densities in some cases yielded considerably higher values to less than 6 while uncured low molecular-mass novolaks extend into the range of 150 m²/g to 180 m²/g when pyrolyzed.

These results are of general significance since the oxidative stability depends on the magnitude of the specific internal surface area. Reduction of the specific surface area leads to improved oxidative stability.

The cleavage products which arise during pyrolysis of phenolic resins as the temperature increases from 350 °C to a maximum of about 450 °C are mainly hydrogen, water, carbon monoxide, low molecular mass hydrocarbons, and a certain fraction of phenolic derivatives and aromatic hydrocarbons (Fig. 6.205).

The composition of the cleavage products changes significantly in the presence of oxygen. In pyrolysis tests, higher molecular mass pyrolysis products such as derivatives of benzene and phenol as well as polycyclic aromatics may be condensed in a cold trap and separated using various analytical methods. Such experiments may be carried out as comparative tests by including conventional carbon-forming agents such as pitch (Table 6.109) or special chemical by-product binders (e.g., Rauxolit, VfT, Castrop-Rauxel, Germany).

The results can be summarized as follows:

- Benzene derivatives, and phenol derivatives such as xylene and cresols, arise during pyrolysis of cured phenolic resins.
- Naphthalene, phenanthrene, anthracene, and similar compounds are only produced at very low levels.
- Benz[a]pyrene, acenaphthene, benz[a]anthracene, chrysene, benzofluoranthene, and other polycyclic aromatics were only observed during pyrolysis of pitch.

Table 6.109. Pyrolysis products of organic binders in an inert atmosphere

Concentration %	Phenolic resin	Rauxolit[a]	Pitch
Toluol	–	0.04	0.19
XY101	0.03	–	0.43
Phenol	7.4	0.76	0.56
Naphthalene	1.27	0.04	0.71
Fluorene	0.04	0.24	0.32
Phenanthrene	0.03	0.34	0.87
Anthracene	0.02	0.07	0.23
Fluoranthene	0.1	0.88	1.37
Pyrene	–	0.74	0.83
Benz(a)pyrene	–	–	0.77
Acenaphthene	–	0.08	0.4
Benz(a)anthracene	–	0.13	1.02
Crysene	–	0.09	0.77
Benz(b)fluoranthene	–	–	0.87
Benz(k)fluoranthene	–	–	0.58

[a] Rauxolit is a special binder for refractories based on a by-product of the chemical industry.

- Fluoranthene and pyrene (structures I and II respectively in Fig. 6.206) can occur at levels comparable to those experienced in pyrolysis of pitch when novolaks are pyrolyzed in the uncured state, since "loose structures" (structure III in Fig. 6.206) are possible in that case. Only minimal levels of fluoranthene or pyrene arise during pyrolysis when the precursor products are adequately crosslinked.
- Total emissions of organic cleavage products are considerably less than in the case of conventional carbon-forming agents such as pitch.

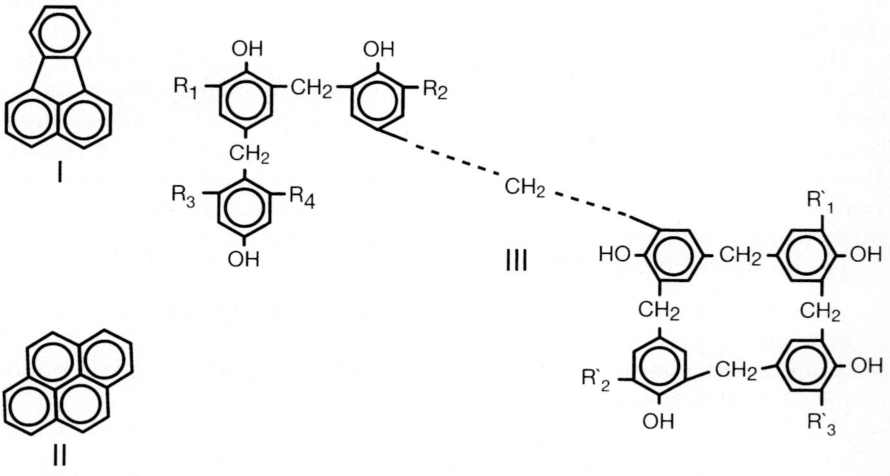

Fig. 6.206. Formulas of I = fluoranthene, II = pyrene, III = parts of the resite structure

To summarize, phenolic resins are generally quite well suited for use as carbon-forming agents. It is important to select the most suitable resin for the application, and to employ optimal curing and pyrolysis conditions. The emissions which arise during pyrolysis relatively are uncomplicated, and there is little tendency for formation of polycyclic hydrocarbons, the pyrolysates generated also depending on the lattice structure of the cured resin.

6.4.2
Phenolic Resins as Binders for Refractories

6.4.2.1
Definitions

Refractories are materials that are used at temperatures exceeding 1000 °C (Table 6.110). The softening points of these materials are generally considerably higher than the operating temperature. The terms "refractory," "highly refractory," and "extremely refractory" derive from the appropriate level of softening temperature (Table 6.111). Certain construction materials of the refractories industry do not reach the softening temperature of 1500 °C stipulated for "refractory" products, but otherwise satisfy the range of requirements for thermal insulation. Such materials are then termed "fire resistant."

Refractories may be classified [1–6] according to their chemical composition, their shape (shaped or unshaped), and according to the process or tech-

Table 6.110. Classification of refractories (possibilities of differentiation)

Classification	Groups	Applications (Examples)
1. Classification according to chemical composition (ISO 1109)	High-alumina	High-temperature furnaces, products for steel casting
	Fireclay	Firing construction, blast furnaces, ladles
	German brick	Furnace construction
	Silica	Air heaters, coking furnaces
	Basic materials, e.g. magnesite	Steel melting furnaces, containers for secondary metallurgy
	Zircon silicate, SiC, etc.	Glass melting furnaces, blast furnaces, nonferrous metals industry
2. Classification according to shape	Shaped	Bricks, blocks, spouts, plugs
	Unshaped	Ramming mixes, runner mixes, mortars, gunning mixes
3. Classification according to manufacturing process (ISO 2246)	1. Bonding <150 °C	Refer to Fig. 6.207
	2. Bonding 150–800 °C	
	3. Bonding >800 °C	

Table 6.111. Classification of refractories according to their softening temperatures

Designation	Softening temperature
Fire resistant	≥1000–1500 °C
Refractory	≥1500–1770 °C
Highly refractory	1770–2000 °C
Extremely refractory	≥2000 °C

nology used to manufacture them. ISO Standard 1109 classifies refractories according to their chemical/mineral composition, for example, as high-alumina, Al_2O_3 containing products (corundum and bauxite), silicate materials (containing more than 93% SiO_2), fireclay products, German brick products (containing 85–93% SiO_2), basic materials (magnesia, magnesia-carbon, forsterite, dolomite, dolomite-carbon, and others), and special products such as carbon materials, nitrides, borides, and others. Refractory construction materials are further classified into shaped products such as bricks and shapes, or unshaped mixes such as mortars or ramming, trough, and taphole mixes.

ISO Standard 2246 describes a classification according to the manufacturing process associated with the pertinent type of binder (Table 6.112). This standard describes unfired products (manufactured at temperatures below 150 °C), thermally treated products (produced at temperatures ranging from 150–200 °C), and fired or fusion cast products (manufactured at temperatures above 800 °C). At temperatures below 150 °C, products such as clay, sodium silicate, cement, or organic materials such as molasses or lignin sulfonate may be used as binders. Products that are thermally treated at 150–200 °C may similarly be bonded by inorganic chemical (phosphates, sulfates), hydraulic (cement), or organic chemical means (binders such as tar, tar pitch, or synthetic resins such as phenolics).

In the area of organic binders, tar pitch was earlier used to a major extent as a starting material to provide satisfactory carbon formation and achieve

Table 6.112. Classification of refractories according to their manufacturing process (ISO 2246); classification of binders

Bonding temperature	Product	Binders (examples)
1. Below 150 °C	Unfired products	Clay Phosphates, magnesium sulfate Cement Lignin sulfonate, molasses Other organic binders
2. 150 to 800 °C	Thermally treated products	Phosphates, sulfates Cement Tar, pitch, and synthetic resins (formation of bonding carbon)
3. Above 800 °C	Fired or fused products	Bonded by firing or fusing followed by solidification

Table 6.113. Production of refractories (selection) (from [3])

Country		Production in millions of tonnes			Monetary value in billions of DM
		Shaped	Unshaped	Total	
European Union (EU)	1988	3.1	1.8	4.9	
	1993	2.7	1.5	4.2	approx. 5.3
German share[a] of total	1988	1.07	0.73	1.8	
	1993	0.98	0.54	1.52	1.93
Japan	1988	0.95	0.8	1.75	–
	1993	0.72	0.78	1.50	–
USA	1989	1.33	1.18	2.5	
	1993	1.27	1.0	2.3	approx. 2.9
Russia [2]	1991	3.41	1.71	5.12	–
	1994	3.11	1.56	4.67	

[a] Representing approximately 14,000 tonnes of binders in Germany; of this approximately 8000–9000 tonnes tar pitch and approximately 5000–6000 tonnes phenolic resins.

properties such as particularly high slag resistance; however, for more than 25 years, synthetic resins such as phenolics and furan resins have advanced in worldwide acceptance as binders and carbon-forming agents for refractories to provide further quality improvement and particularly for reasons of environmental impact and toxicity [6–10]. Products derived from tar contain carcinogenic polycyclic hydrocarbons such as benz[a]pyrene.

6.4.2.2
Economic Importance and Applications

Worldwide refractories production presently amounts to around 22–24 million tonnes [3]. The production of refractories has declined over the past two decades (Table 6.113). This has mainly resulted from technological changes in the consuming industries, and from improvements in refractories leading to longer service lives. As shown in Table 6.113, worldwide production declined by around 11–12% over the period from 1988 to 1993/1994.

The main user of refractories (Table 6.114), accounting for about 70% of the total, is the iron and steel industry. In this field, synthetic resins (phenolics and furan resins) are presently in common use as binders for shaped products such as magnesia, dolomite, bauxite, and andalusite-based bricks, for isostatically shaped products, and unshaped mixes such as those for blast furnace tapholes, troughs, and tundish liners, and as impregnating agents for the vacuum/pressure process. The scheme in Fig. 6.207 surveys the uses of organically/phenolic resin bonded products in steelmaking, i.e., in the area of the blast furnace, during further transportation of the steel to and from the converter, and in continuous steel production.

Table 6.114. Refractory users and specific refractory consumption (from [3])

Users (totals worldwide)	Share of refractory consumption in %	Specific refractory consumption in kg of refractory per tonne of product
Iron and steel industry	approx. 70	18–25[a]
Nonferrous metals industry (Al, Cu, Ni, Pb, Sn, Zn)	2–3	6–9
Glass industry	3–4	5–8
Cement and lime industry	4–7	0.6–1.2
Ceramics	4–7	–
Chemistry, petrochemistry	2–4	–
Others[b]	approx. 10	–

[a] With large regional variations, for example extremes of 10–15 kg refractory per tonne in Japan, France and Germany compared to 70–75 kg in Russia [2].
[b] This includes the refractories industry, coal gasification, thermal power stations, thermal waste processing, electric storage heaters.

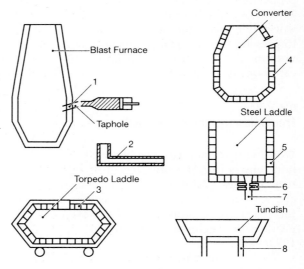

Fig. 6.207. Illustration of high quality refractories for iron and steel making (1 = taphole mass, 2 = ramming mass for trough liners, 3 = bauxite- and andalusite-bricks, 4 = magnesia-carbon-bricks, 5 = dolomite carbon bricks, 6 = slide gates, 7 = shroud, 8 = submerged nozzle)

Phenolic resin bonded bricks [11] offer four fundamental advantages:

1. Manufacture and use is environmentally friendly
2. Manufacture saves energy when the cold mixing process is used
3. The products do not pass through a plastic phase when heated, as do those bonded with materials such as pitch
4. Higher levels of carbon (more graphite or carbon black) can be used, thus achieving high wear and slag resistance

Fig. 6.208. Isostatically shaped products (high quality refractories)

Fig. 6.209. Use of phenolic bonded ramming mixes: production of blast furnace troughs for discharge of the melt

Isostatically pressed products (Fig. 6.208) may also be markedly improved with respect to their service life by use of phenolic resin binders, particularly when compared to that of ceramically bonded shapes (increased thermal shock resistance). Novolak/hexa powder resins in combination with furfural, or special liquid modified phenolic resoles without furfural are used as binders in this case.

Unshaped mixes for blast furnace troughs (Fig. 6.209) are used in the form of either plastic ramming mixes containing resin solutions as binders, or dry vibratory compacted mixes with powder resins as binders, and afford particularly high wear resistance, as has been frequently demonstrated in practice.

Phenolic resins exhibiting low reactivity but a high residual carbon level following carbonization are used in manufacture of blast furnace taphole mixes (also refer to Fig. 6.207), ensuring satisfactory setting of the blast furnace tapholes and thus a high level of safety, while simultaneously affording extremely short plugging times. The reduced reactivity of these resins prevents premature hardening of the mixes in the plugging machine.

6.4.2.3
Environmental Impact in Use of Phenolic Resins

More than ever, the emission behavior (Fig. 6.210) is presently a prime concern in selection of a binder for refractories [12]. In the case of synthetic resin binders, and particularly phenolics, the levels of monomers and odoriferous cleavage and pyrolysis products were reduced significantly years ago by optimization of existing and development of new resin systems [13]. As a result of these extensive development efforts, products exhibiting very low monomer levels are presently available on the market.

Reliable workplace exposure determinations show that modern, i.e., low-monomer synthetic resin binders may be used and handled without problems from a technical or industrial medical hazard standpoint. In contrast to the situation in the case of conventional tar and pitch binders, the emissions from phenolic and furan resins exposed to pyrolytic conditions are limited to a few easily measured components due to the regular, predictable structures of the resins.

Phenolic and furan resins contain no polycyclic aromatics such as benz[*a*]-pyrene. There is also no significant tendency to form such compounds during pyrolysis (Table 6.115). The pyrolysis products from phenolic resins and other binders listed in Table 6.115 were determined using the INASMET method as

Fig. 6.210. Examples for uses of phenolic resins in production of refractories: monobloc stopper plugs made by isostatic compression (*left*) and graphite crucibles manufactured by the spinning process (*right*)

Table 6.115. Pyrolysis products (guide components) from phenolic resins, pitch substitutes ("chemical binders") and tar pitch (inert atmosphere) (from [13])

Concentration in %	Phenolic resin	Chemical binder	Pitch
Toluene	–	0.04	0.19
Xylene	0.03	–	0.43
Phenols	7.40	0.76	0.56
Naphthalene	1.27	0.04	0.71
Fluorene	0.04	0.24	0.32
Phenanthrene	0.03	0.34	0.87
Anthracene	0.02	0.07	0.23
Fluoranthene	0.10	0.88	1.37
Pyrene	–	0.74	0.83
Benz[a]pyrene	–	–	0.77
Acenaphthene	–	0.08	0.40
Benz[a]anthracene	–	0.13	1.02
Chrysene	–	0.09	0.77
Benz[b]fluoranthene	–	–	0.87
Benz[k]fluoranthene	–	–	0.58

already described in Sect. 6.4.1. The pyrolyses were carried out under an inert atmosphere in this case. When the determination is performed in the presence of 1.5 vol.% added oxygen, the fraction of phenols liberated in the phenolic resin pyrolysis is reduced to more than 90%, and the levels of the other listed pyrolysis products largely fall to below the limit of detection. The fractions of polycyclic aromatics liberated during pyrolysis of the "chemical binder" Rauxolit and tar pitch only change to an insignificant extent when pyrolysis is conducted in the presence of the indicated volumetric concentration of oxygen.

Measurements made in practice under the difficult analytical conditions of a steelmaking plant also exist (Fig. 6.211), and demonstrate the composition of the pyrolysis gases liberated when steel ladles are heated up. The emissions from ladles lined exclusively with either tar pitch or phenolic resin bonded dolomite bricks were compared. The results of these tests demonstrated that the emission potential of tar pitch is higher than that of phenolic resin by a factor of around 1000 even under the very favorable combustion conditions experienced in optimized ladle firing. One of the reasons for this remarkable dif-

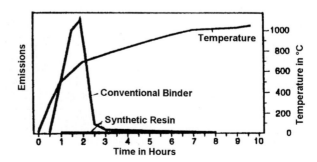

Fig. 6.211. Emissions ("organic carbon") of carbon-containing ladle linings made using conventional (tar pitch) and synthetic resin (phenolic) binders – measurements in practice

ference is the fact that phenolic resin pyrolysates undergo more facile thermal oxidation, even at a relatively low atmospheric oxygen level. This observation also agrees with the previously mentioned results of laboratory measurements.

6.4.2.4
Shaped Products

When phenolic resins are used to produce shaped refractories, the resin performs two main functions:
1. It acts as a binder in the shaping process.
2. It provides a source of carbon due to its favorable carbonization properties at temperatures between 300° and 1000 °C.

The selection of a specific phenolic resin depends on the processing method [14, 15], since the various types of resins may be associated with different processing methods (Table 6.116). Figure 6.212 shows a flow chart illustrating production of shaped refractories. The main raw materials aside from silicon carbide, silicates and other mineral products are carbon in the form of graphite and carbon black, and magnesia, dolomite, bauxite, and andalusite (Table 6.117). However, the deposits of these products throughout the world vary in quality. Particular attention must be paid to the level of calcium oxide when liquid phenolic resins are to be used (anhydrous binders possibly being required).

Table 6.116. Phenolic resins for refractories: binders, mixing processes and applications

Applications	Cold-mixing	Warm-mixing
Magnesia and Magnesia-Carbon bricks	Resols Novolak solutions	Novolak Novolak solution
Dolomite bricks	Novolak solutions	Novolak
Bauxite and Andalusite bricks	Resols Novolak solutions	Novolak –
Isostatically formed products	Resols Novolak solutions	Novolak
Sliding gates	Resols Novolak solutions	–
Graphite crucibles	Resols Novolaks	Novolak solutions
Blast furnace taphole mixes	Resols Novolak solutions mixture	–
Trough mixes	Powdered resins Novolak solutions	–
Ramming mixes	Powdered resins	–
Impregnated, ceramically bonded products	Resols Novolak solutions	Novolak

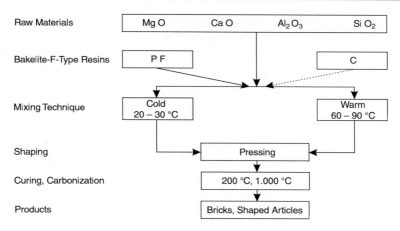

Fig. 6.212. Production of refractories, schematic flow chart

Table 6.117. Raw materials for refractories (selection)

Raw material	Places where found	Formula	Melting temperature
Magnesia	Austria, Turkey, China, Australia, Canada, Brazil	>90% MgO, Fe_2O_3, CaO, SiO_2, Al_2O_3	2800 °C
Dolomite	Germany, UK, USA, Italy, India	60% MgO + 40% CaO	2450 °C
Bauxite	Guyana, China, Brazil	Al_2O_3, SiO_2, >88% Al_2O_3, SiO_2, Fe_2O_3, TiO_2	2000 °C
Andalusite	South Africa	Al_2O_3 SiO_2, 55–60% Al_2O_3, Fe_2O_3, SiO_2, TiO_2	1800 °C

In case of more stringent demands, particularly in the necessary crude iron treatments (desiliconization, dephosphorization, or desulfurization) and for steelmaking equipment, high-alumina (greater than 45% Al_2O_3) products with a graphite level of up to 20% and bonded with pitch or – as is currently more and more the case – phenolic resin can be used to meet the mechanical and chemical/metallurgic stresses. The addition of antioxidants counteracts oxidative stresses due to air, carbon dioxide, or iron oxide.

The mixing process may be carried out at room temperature or at elevated temperatures. At present, shaping is mainly performed in controlled hydraulic presses (Fig. 6.213). These can be continuously adjusted to match structural requirements by means of special electronic process controls. Thus, parameters such as the height and shape of bricks can be continuously monitored and corrected during the pressing operation. Depending on the tonnage of the press and hydraulic drive, the applied pressure can range from 80 N/mm² to 200 N/mm². The isostatic compression method, a special case that is considered below, is carried out using molds of flexible materials such as plastics or

Fig. 6.213. Hydraulic press with two molds used for manufacture of refractory bricks (from [6]) (source: Didier-Werke, Wiesbaden)

rubber. In this method, the mold is placed in an autoclave and subjected to an even pressure applied through a liquid. The pressurized liquid produces an even compression force in all directions. The production stage designated as "curing, carbonization" in Fig. 6.212 can be conducted using modern ovens, dryers, or burners and lasts over a period of hours to several days. The time depends on the size of the parts, the cleavage products liberated during cure, or binder decomposition (carbonization), the required grade, and other factors related to process engineering and application technology. The expected material shrinkage is compensated in advance. The parts are fired in annular, tunnel (Fig. 6.214), and other types of ovens. These systems can be operated with all types of fuels, or electrically heated.

6.4.2.5
Manufacture of Bricks

The first stage in manufacture of bricks (Fig. 6.215) based on andalusite, bauxite, dolomite, magnesia, and similar materials consists of a cold or warm mixing operation. In the cold mixing method, the mixing operation is performed at room temperature. This is obviously the most desirable method from the standpoint of energy costs.

Fig. 6.214. Tunnel oven for curing phenolic bonded refractories (from [6]) (source: Didier-Werke, Wiesbaden)

Fig. 6.215. Phenolic resin bonded refractory bricks (converter, taphole and ladle bricks)

The refractory components (Table 6.117) and certain additives such as carbon black, graphite and possibly metal (Si, Al) powder are intimately blended with a 3–5% binder fraction. A recommended order of mixing is first to wet the coarse grain with most of the binder, and then blend in the fine-grained fractions together with the remaining binder. The added graphite produces a proportionate increase in the thermal shock resistance and improvement in the slag resistance. The metallic additives are designed to reduce the susceptibility to oxidation of the carbon, particularly the polymeric carbon derived from the synthetic resin binder.

Liquid resole-based resins containing various levels of free phenol (resins A, B, and C in Table 6.118) and novolak solutions (resin D in Table 6.118) have been used with great success for production of magnesia-carbon bricks (Tables 6.118 and 6.119), including those with high levels of graphite. Hexa (10–14% based on resin solids) is added when novolak resins are used.

The bonding and adhesive power of such mixes, and thus their compressibility and the green strength of shapes made from them, may be further improved by addition of a novolak/hexa-based powder resin such as H, J, or K

Table 6.118. Phenolic resins for production of magnesia-carbon, dolomite, dolomite-carbon, bauxite, and andalusite-based bricks

Resin	Mixing method	Type	Delivery form	Non-volatiles % ISO 8618	Free Phenol % ISO 8974	Viscosity at 20 °C MPa · s ISO 12058	Viscosity at 100 °C MPa · s ISO 12058	Water level %	Use(s)
A	Cold	Phenolic resole	Liquid	80±2	approx. 15	7000±1000	–	max. 4	MgO-C
B	Cold	Phenolic resole	Liquid	79±2.5	9.0±1.0	600±100	–	3.0	MgO-C
C	Cold	Phenolic resole	Liquid	73.5±1.5	max. 4.9	1100±100	–	max. 4	MgO-C
D	Cold	Phenolic novolak	Solution	70±2	max. 0.5	2000±150	–	max. 2	MgO-C Bauxite, Andalusite
E	Warm	Phenolic novolak	Melt	–	max. 0.5	–	650±150	max. 0.3	All-purpose
F	Cold	Phenolic novolak	Solution	61±1.5	max. 0.5	2000±150	–	max. 0.3	Dolomite
G	Cold	Phenolic resole	Liquid	73±1	max. 4.9	750±150	–	–	Bauxite

Table 6.119. Powder resins for various uses (refractory manufacture)

Resin	Type	Delivery form	Residual phenol % ISO 8974	Flow distance at 125 °C mm ISO 8619	Use(s)
H	Phenolic novolak with hexa	Powder	max. 0.5	35 ± 5	Bricks (additive)
J	Phenolic novolak with hexa	Powder	max. 0.5	50 ± 5	Bricks, ramming mixes
K	Phenolic novolak with hexa	Powder	max. 0.5	70 ± 10	Ramming mixes
L	Modified, hexa-free phenolic novolak	Powder	max. 0.8	90 ± 15	Trough mixes

in Table 6.119. The moisture level of the mix is furthermore reduced by partial replacement of the liquid resin by the powder resin.

Bauxite and andalusite bricks exhibiting high green strength may be produced using liquid resoles such as resin G in Table 6.118, or novolak solutions (resin D in Table 6.118) in combination with powder resins (resins H and J in Table 6.119).

When dolomite is used as a refractory component, the presence of water must be avoided completely. Hydration and consequent decomposition of the crystalline structure will occur due to the calcium oxide content. Appropriate binders should not contain water, nor should they form water during the curing process. Resin solutions that employ ethylene glycol as a solvent are similarly unacceptable. Novolak solutions containing less than 0.3% water, such as resin D in Table 6.118, are therefore used to manufacture dolomite or dolomite-carbon bricks using the cold mixing method. A common, but relatively expensive, solvent for that is anhydrous furfuryl alcohol. Certain esters are similarly suitable as solvents, but in general are relatively rarely used due to their emission and odor properties. Powdered hexa is blended into the novolak solutions during the mixing operation, or is introduced by addition of a powder resin with a high hexa content.

A further binder option used in the cold mixing method consists of acid-sensitive phenolic or furan resins with special curing systems. However, these binder combinations can only be used when the refractory raw materials do not react with acids or acidic compounds. Understandably, such mixes exhibit only limited bench lives.

All mixtures prepared using the cold mixing method may be shaped by compression at room temperature.

In the warm mixing process, the mixing operation – as in processing of tar pitch – occurs at elevated temperatures (above 70 °C). In this case, solvent-free novolaks such as resin E in Table 6.118 are particularly useful as binders. In manufacture of dolomite bricks, this method offers the special advantage that the molten resin is practically anhydrous and liberates no water during the curing reaction. The hexa curing agent is blended at temperatures ranging from 70 °C to 85 °C, care being taken that the temperature does not exceed 85 °C for any lengthy period to avoid premature decomposition of the hexa. After the blending operation, the mix is generally pressed in a warm condition to ensure an adequate level of compression.

All shapes produced using the cold and warm mixing processes are cured by heating to temperatures ranging from 180 °C to 200 °C, the temperature being gradually increased (10 °C per hour) from 80 °C to 180 °C. Volatile components including water (from resoles), the monomeric components phenol and formaldehyde, and when hexa is used as a curing agent (for novolaks) ammonia are liberated during the curing process. These products are conveyed to an afterburner.

6.4.2.6
Isostatically Pressed Products for Continuous Casting

Numerous products for use in continuous casting (Figs. 6.216 and 6.217) may be produced using the isostatic compression method (Fig. 6.218). These include shrouds (Fig. 6.208), submerged nozzles, and monoblock stoppers (Fig. 6.210, left).

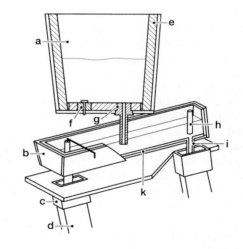

Fig. 6.216. Illustration of high quality refractories in a two-strand continuous caster (a = steel casting ladle, b = tundish, c = mold, d = steel stand, e = ladle lining, f = gas purge set, g = ladle sliding gate with phroud, h = monobloc plug, i = submerged nozzle, k = tundish lining) (source: Didier-Werke, Wiesbaden)

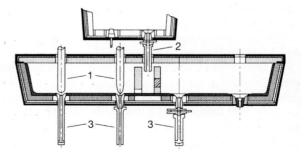

Fig. 6.217. Isostatically formed Refractories for the tundish of a continous caster (1 = monobloc stopper, 2 = ladle shroud, 3 = submerged nozzle) (source: Didier-Werke, Wiesbaden)

Fig. 6.218. Isostatic press for manufacturing premium continuous casting components such as monobloc stoppers, submerged nozzles and shrouds (source: Didier-Werke, Wiesbaden, Germany)

The mixtures for such products are mainly produced by the cold mixing process. A synthetic resin-based method so common it may already be termed "conventional" is the use of phenolic novolak/hexa powder resins in combination with furfural. Due to the toxicity of furfural, this method is increasingly being replaced by those using furfural-free, environmentally friendly phenolic resin systems.

An important point in the manufacture of isostatically pressed products is ensuring that mixing of the oxide-based refractory components with graphite and other additives as well as 10–20% of added binder yields a fine-grained, homogeneous granulate that may be charged into the isostatic compression mold without separation occurring.

Satisfactory granulation of the mix is generally achieved when aqueous resoles such as resin M in Table 6.120 are used. The mix must be dried to a certain extent to produce a free-flowing, storable granulate. However, this postdrying stage may be omitted when special additives that produce drying of the granulate during the mixing process itself are included. Another option is the use of novolak solutions that require addition of hexamethylenetetramine for curing and similarly afford dry granulates when special additives are incorporated during the mixing operation.

Following the isostatic pressing operation at room temperature, the green shapes are cured by slowly increasing the temperature to 80–180 °C at a rate of 10–30 °C per hour, followed by carbonization in the absence of air at temperatures ranging up to about 800 °C.

Table 6.120. Phenolic resins for use in various refractories (shaped and unshaped products)

Resin	Type	Delivery form	Non-volatiles % ISO 8618	Free phenol % ISO 8974	Viscosity at 20 °C mPa · s ISO 12058	Use(s)
M	Phenolic resole	Liquid	77.5 ± 1.5	max. 4.9	2600 ± 200	Isostatically shaped products
N	Phenolic novolak	Solution	74 ± 2	4.3 ± 0.3	4500 ± 500[a]	Isostatically shaped products and slide gate plates
O	Phenolic resole	Liquid	76.5 ± 1.5	20 ± 2	850 ± 50	Graphite crucibles
P	Phenolic resole	Liquid	75 ± 1.5	13 ± 2	1000 ± 200	Graphite crucibles
Q	Phenolic resole	Liquid	80.5 ± 2	max. 4.5	7000 ± 1000	Graphite crucibles
R	Phenolic resole	Liquid	80.5 ± 1.5	21 ± 3	2600 ± 200	Taphole mixes
S	Modified phenolic resole	Solution	79 ± 2	8 ± 2	4750 ± 250	Taphole mixes
T	Phenolic novolak	Solution	83 ± 3	max. 0.5	7500 ± 500	Taphole mixes
U	Phenolic novolak	Solution	69 ± 2	max. 0.5	5500 ± 500	Trough mixes

[a] Viscosity at 25 °C.

6.4.2.7
Slide Gates, Graphite Crucibles, Insulating Plates

Slide gates (Fig. 6.219), for example, those used to close steel casting ladles (Figs. 6.207 and 6.216), are mainly fabricated using the cold mixing method, and can be bonded using phenolic resoles in combination with powder resins. The novolak solutions recommended for the cold mixing method may also be used (with added hexa) for this purpose.

In this case as well, a certain amount of powder resin may be added to reduce the volatiles level and increase the carbon yield. Powder resins containing both antidusting and release agents have been developed for this purpose, and reduce the tendency of the mixes to stick during pressing. After the products have been shaped by pressing at room temperature, they are cured and subsequently pyrolyzed to carbonize the binder in a manner analogous to that described in the case of isostatically pressed shapes.

Two modes of operation, the isostatic pressing method and the spinning process, are mainly employed to manufacture graphite crucibles (Fig. 6.210, right). As previously mentioned, the isostatic pressing method requires the use of a free-flowing, homogeneous granulate that can be produced using phenolic resoles such as resins M and Q in Table 6.120, or novolak solutions in combination with hexa as a curing agent. In the spinning process, a crucible mold rotating around a conical cylinder spins out a highly plastic, doughy mix to the upper edge of the mold.

In addition to the cold mixing method, the warm mixing process using appropriate novolak solutions is recommended for production of such mixes. Drying, curing, and pyrolysis of the crucibles require slow increases in the temperature to avoid cracking or other defects due to sudden evolution of excessive levels of volatile components.

Isothermal insulating plates are used both for thermal insulation of permanent molds in steel mills, and to line cold distribution troughs for continuous casting. The effective thermal insulation holds the melt at a sufficiently high temperature for a certain period of time. These insulating plates consist of 95% inorganic and organic fillers, and a 5% binder fraction in the form of novolak/hexa-based powder resins. The resins provide the shapes with high mechanical strength and good moisture resistance. Fillers such as sand, waste fiber, mineral/rock wool, and waste paper, together with about eight times their volume of water, are finely divided and transformed into a pulpy sus-

Fig. 6.219. Slide gate plate for a continuous casting line

pension in appropriate mixers. These mixes are dewatered and shaped in screen molds by applying pressure using a vacuum, dried in an oven at 80–100 °C, and finally cured by increasing the temperature to 180–200 °C.

6.4.2.8
Unshaped Refractory Mixes

The fields of application and uses of unshaped mixes are quite varied, and range from blast furnace taphole and trough mixes as well as free-flowing tundish mixes through mortar and grouting mixes to vibratory, coating, gunning, and repair mixes. Consequently, the demands made on the binder properties also differ widely.

Novolak solutions such as resin T in Table 6.120, aqueous resoles such as resin R in Table 6.120, and combinations of the two may be used for production of tar pitch-free taphole mixes. In this case, the novolak solution is used without added hexa to avoid premature curing of taphole mixes in the plugging machine due to the effects of elevated temperature.

The phenolic resole R in Table 6.120 can also be used as a sole binder, and features very high binding and adhesive power, low reactivity and a very low (3–4%) water level. Resin solutions representing a combination of resole and novolak (resin S in Table 6.120) are also available.

Special plasticizing agents (plasticizers, waxes, oils) are used in addition to the binder to prevent premature curing and improve the plasticity of taphole mixes [16–18].

Phenolic resins are already used as binders in production of (dry or plastic) mixes for the runners used to convey the flow of steel from the blast furnace (Fig. 6.207). Novolak solutions using added hexa as a curing agent, for example, resin U in Table 6.120, have been found particularly appropriate for the plastic mixes, and provide the rammed trough wall with excellent green strength due to the very high compressibility of the mixes made with them.

Dry, free-flowing vibratory compacting mixes are increasingly used in current practice, and are expected to meet the following demands: they should be low dusting or dust-free, have low odor during the curing process, should provide high strength after curing, and high wear resistance in use. Hexa-free powder resins (resin L in Table 6.119) are particularly suitable for this application. These binders liberate no ammonia during the curing process, and may thus be considered essentially free of odor. Backing mixes are used for certain applications, and are rammed behind patterns to form shapes.

The previously described powder resins (H–L in Table 6.119) are best suited for this purpose, i.e., resins with both high and low levels of hexa are used, and after curing provide the rammed wall with a high level of strength, permitting easy stripping of the patterns.

6.4.2.9
Impregnation of Refractories

Introduction of carbon into the pores of refractories such as ceramically bonded products has been found to be a particularly useful method of improving

Fig. 6.220. The pressure/vacuum process for impregnation of refractories (e.g., sliding gate plates) or carbon materials with liquid phenolic resins

1. Evacuation   2. Aspirate resin by vacuum   3. Impregnation by pressure

the quality, particularly with respect to the slag resistance [19]. A relatively simple means of introducing carbon into shapes is to impregnate them with phenolic resins using the cold or warm impregnating method, and then to carbonize the binder. Use is made of the vacuum/pressure process (Fig. 6.220). This method is quite familiar from the technology of carbon and graphite materials used to manufacture equipment components and tubes. Special phenolic resole and novolak/hexa solutions (resins V, W, and X in Table 6.121) are available for room temperature impregnation. These resins particularly feature low viscosity and good wetting and impregnating properties. Special temperature programs (Fig. 6.221) permit curing of the resins without exudation of the binder at the surface of the shape part. Impregnating procedure is already being used, i.e., for the improvement of slide gates and shaped articles [20]. Also impregnated bricks are used for test runs.

Table 6.121. Phenolic resins for impregnation of refractories

Resin	Type	Delivery form	Nonvolatiles %	Free phenol %	Viscosity at 20 °C mPa · s
V	Phenolic resole	Liquid	73 + 1.5	7.5 ± 1	575 ± 75
W	Phenolic novolak with hexa	Solution	60 ± 1.5	max. 0.5	630 ± 50
X	Phenolic novolak with hexa	Solution	72 + 2	max. 0.5	630 ± 50

Fig. 6.221. Curing program for resin impregnated refractories (source: Bakelite AG, Iserlohn, Germany)

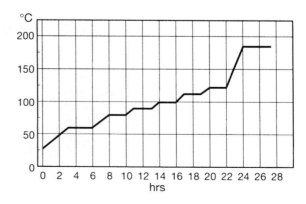

6.4.3
Carbon and Graphite Materials

6.4.3.1
Introduction (General Information)

Carbon and graphite are ceramic engineering materials that are highly suitable for use in equipment construction [1] due to their high level of corrosion resistance, and desirable chemical and physical properties. Such materials are particularly appropriate for components that come into contact with corrosive substances such as hydrochloric, hydrofluoric, and other acids in equipment such as columns (Fig. 6.222), plates, and packing used in operations including distillation and rectification. Thus, carbon and graphite materials are widely used in production of chemical equipment, electrode material, graphite electrodes for steel smelting, and in many other applications. These materials can generally be produced from coke, carbon black, and graphite-based mixtures with carbonizable binders.

For many years, these materials have been produced using phenolics such as powder resins, aqueous resoles, and novolak/resole-based resin solutions as binders, impregnating agents, and carbon donors. In this application, phenolics generally compete with conventional binders based on tar pitch and similar residue-derived products [2–8].

The base mixture generally represents a compound used to produce moldings that are commonly carbonized at 800–1200 °C in a reducing atmosphere. In special cases, heat treatment is carried out at temperatures up to 2600–3000 °C. The phenolic resin undergoes the decomposition and reorien-

Fig. 6.222. CFC packing for separating columns (photo: SGL Carbon International, Meitingen)

Fig. 6.223. Scanning electron micrograph of the fractured surface of a CFC composite article with a carbonized resin matrix (photo: SGL Carbon Group, Meitingen)

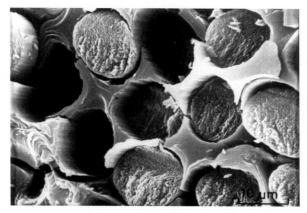

tation processes described in the chapter on pyrolysis to form polymeric carbon, and thus acts as a carbon donor. This means that the carbon or graphite particles are bonded together to form a homogeneous composition. The resultant system is generally porous, exhibiting a mean porosity of 20–30% that can affect the range of engineering properties of the materials.

Depending on the process and parameter conditions, heat treatment (termed "carbonization" or "cracking") of the phenolic resin binder affords a carbon yield of 40–60% based on the original organic starting material [9]. The polymeric carbon yield has a major effect on the quality of the carbon-based material. Porous carbon and graphite articles may be treated with impregnating resins. This reduces the porosity and increases the strength of the carbon or graphite articles. Aside from phenolic resins, the patent literature describes the use of furfural-tar mixtures [10] and polyurethane prepolymers [11] as impregnating agents.

CFC (Carbon Fiber Composite) or Carbon Carbon Composite (see page 267) represents a relatively modern carbon-based engineering material. CFC is a fiber composite made up of a carbon matrix and reinforcing fibers of the same material [12] (see Sect. 6.1.5). Modification of the type of fibers, their orientation, the fiber fraction, and the production method permits relatively wide variation of the range of properties of these high-quality materials. These materials consist of continuous phases of fibrous or compact carbon, and carbon in crystalline and vitreous form. CFC products are generally produced using crystalline carbon fibers and an amorphous carbon matrix (Fig. 6.223).

As may be inferred from this general description, various aspects apply to the use of phenolic resins in production of carbon and graphite materials:

1. In the form of binders and adhesives, they act to connect and bond the individual carbon particles.
2. Phenolic resins render it possible to produce the desired shape by compression or extrusion processes. In some applications, their purpose is to achieve adequate green strength in the moldings and high strength in the finished product following the curing process.

3. Liquid phenolic resins can be used as impregnating agents for porous carbon and graphite articles to reduce their permeability and increase the strength of the material.
4. As in production of glassy carbon, they act as an additional carbon donor in other carbon-based materials, for example, in treatment of graphite.

6.4.3.2
Phenolic Resins as Binders for Carbon and Graphite Engineering Materials (Including CFC)

Aqueous phenolic resoles are commonly used as binders. The coal, coke, or graphite raw materials of various particle size distributions are intimately mixed with the liquid phenolic resin. As in production of refractories, powder resins may also be incorporated to enhance the green strength or increase the carbonization residue. After the product has been shaped by compression, extrusion, or isostatic pressing at ambient or elevated (60–80 °C) temperature, it is heat-cured by slowly increasing the temperature to 180–200 °C in stages according to a defined program. When powder resins are mainly used, it is also possible to shape the products directly by hot compression molding at temperatures ranging from 160 °C to 180 °C. Depending on the end use, the cured moldings or pressings are used immediately, or are subsequently heated to high temperatures in a reducing atmosphere to carbonize the binder.

The premium chemical and physical properties of carbon-based materials, particularly their good thermal conductivity and high temperature resistance, render them useful not only in conventional applications such as electrode materials, carbon brushes, slip-ring seals, and bearing materials, but also increasingly in modern technologies such as rocket engineering, for nuclear reactors and in production of fuel elements, for highly stressed heat exchangers (Figs. 6.224–6.226), and in desulfurization of flue gas. In keeping with the rising demands on such applications, the phenolic resins used in them have been adapted to new operating techniques and modified to provide the desired ranges of material properties over the past years.

Phenolic resins of various compositions – aqueous phenolic resoles as well as resin solutions generally based on novolaks – are used for production of CFC materials. It is also possible to use low-melting novolaks applied by the melt coating process [12] at temperatures ranging from 90 °C to 120 °C in combination with added hexamethylene tetramine.

Depending on the starting materials and manufacturing process, CFC materials can offer a variety of properties and application opportunities. The production process is outlined in Fig. 6.227. Various grades of carbon fiber are used. Polyacrylonitrile-based carbon fibers are mainly used for textile intermediates such as staple fiber, yarns, woven tapes, and various weaves of broadcloth. The function of a component is already taken into consideration during its manufacture. This means, for instance, that a housing is produced with laminated roving fabric, a bolt cut from a semifinished piece made with staple fiber fabric, and a pump shaft fabricated from a tube produced with windings of roving fabric. The fact that phenolic resins afford higher carbon

Fig. 6.224. cross-section through a fiber jacketed tubular heat exchanger (photo: SGL Carbon Group, Meitingen)

Fig. 6.225. Partial view of a Diabon® HF 1 tubular bundle (photo: SGL Carbon Group, Meitingen)

yields than epoxies, polyester resins, and similar carbon donors has also been demonstrated in production of CFC. In addition, phenolic resins generally represent a less costly alternative than the other synthetic resins.

Depending on the orientation of the fibers within the composite, changes in its external dimensions may be observed. A sheet made with roving fabric (Fig. 6.228) experiences no dimensional changes along the fibers arranged in

Fig. 6.226. Carbon fiber reinforcement of a tube base (photo: SGL Carbon Group, Meitingen)

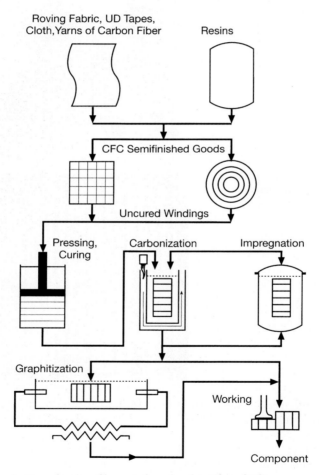

Fig. 6.227. Production of carbon fiber-reinforced carbon (from [12])

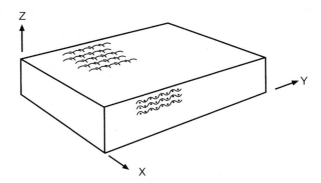

Fig. 6.228. Model of a level CFC sheet of roving fabric (from [12])

Fig. 6.229. CFC heating element (photo: SGL Carbon Group, Meitingen)

the x and y planes. However, shrinkage vertical to the plane of the fabric insert may be observed.

Massive articles with multidimensional fiber reinforcement are obtained using special weaving and knitting techniques producing an arrangement of carbon fibers that extends into three-dimensional space in all directions [13]. Such a three-dimensional, fiber-reinforced object is subsequently impregnated with binders. Major CFC applications include aircraft brakes and furnace construction (Fig. 6.229). Other fields of application are rocket technology and pattern making. All components are capable of use at temperatures up to 2500 °C. This underscores the utility of these engineering materials.

Table 6.122. Typical Data of Impregnating Resins (A = Resole, B = Novolak)

		Resin A	Resin B
Viscosity at 20 °C (method Hoeppler ISO 9371)	mPa · s	1000 ± 200	630 ± 50
Gel time at 130 °C (ISO 9396)	min s	11.30 ± 2.30	8.55
Non-volatile at 135 °C (ISO 8618)	%	75.5 ± 1.5	60 + 1.5
Residual phenol (ISO 8974)	%	6.5 ± 1.5	approx. 0.5
Water content K. F. Fischer (ISO 760)	%	14.5 ± 2.5	approx. 0.4

6.4.3.3
Phenolic Resins as Impregnating Agents for Porous Carbon or Graphite Articles

Liquid resins (aqueous phenolic resoles) or novolak-based resin solutions with added hexa are used to impregnate carbon-based materials (Table 6.122). These resins feature particularly good impregnating abilities and high carbon yields. The two different impregnating resins listed in Table 6.122 can lead to similar impregnation results. The purpose of impregnating porous carbon and graphite materials is to reduce their permeability and increase their strength.

Impregnation is carried out in a vacuum-pressure vessel (also refer to Sect. 6.4.2.9) that is pressurized at up to 10 bars after the impregnating agent has been charged. The process conditions, including the viscosity of the impregnating agent and the impregnating time, are largely determined by the range of dimensions exhibited by the pores of the material. The impregnated articles are heated to 150–220 °C to cure the impregnating agent. Since part of the impregnating agent escapes from the porous system during this treatment, possibly rendering it impossible to achieve the desired effect, curing can also be carried out under pressure [14]. Large-scale impregnating systems operate essentially automatically [15]. The resin-impregnated carbon and graphite materials are largely impermeable to fluids, and they exhibit strength levels that are generally 50% higher than those of non-impregnated materials. The impregnated carbon-based materials find a wide variety of uses [16–23]. It is important that (liquid) resins capable of homogeneously filling the pores be used; not only the resin as such, but also the volatile substances that are liberated during curing and are captured within the resin exert a significant effect on pore filling (cf. Fig. 6.230). The impermeability required for some applications, such as high-temperature nuclear reactors and heat exchangers, is only achieved by repeated impregnation and pyrolytic treatment. Since the diameters of the porous channels decrease with every cycle, it is advisable to reduce the viscosity of the impregnating agent with each cycle.

6.4.3.4
Electrode Production

The use of phenolic resins as binders for carbon and graphite materials is considerably more expensive than that of the coal tar pitch frequently used for this purpose. The most economical raw materials – including carbon black, coke,

Fig. 6.230. Microsection of impregnated carbon article (dark spots are trapped gas bubbles from the curing reactions) (photo: SGL, Meitingen)

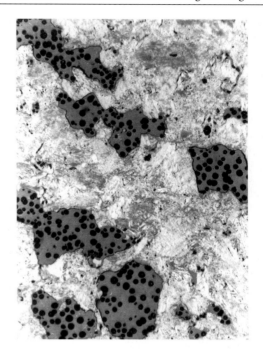

tar, pitch, and some grades of natural graphite – are thus used for large-volume commercial products such as carbon anodes for aluminum production or graphite electrodes for steel smelting. Since the carbon-based products manufactured using these economical raw materials cost approximately the same as phenolic resins, the latter are generally not used for production of such mass articles.

For environmental reasons, for example, in connection with the production of electrodes, and for reasons of cost and quality (specifically to reduce the consumption of anodes still further), efforts have been made to replace partially and improve the conventional pitch binders by including other carbon donors such as phenolic resins, or to upgrade the electrodes by surface treatment. At the present state of knowledge, the following may be noted with respect to this development: as preliminary trials have shown, partial or complete replacement of the pitch binder by phenolic resins is in principle possible, special novolak solutions being particularly suitable for this purpose [24].

In principle, it is also possible to use resoles for this purpose, although these must be free of alkalis and alkaline earths. The use of resoles offers the advantages that the cold mixing method could be used, and that carbonization would be complete at about 850 °C. Emissions of aromatics fall to 20% or less of those experienced with pitch binders. Polycyclic aromatics only occur in traces in the pyrolysis emissions, if any are present. The fraction of binder may be decreased somewhat below the level used in the case of pitch binders; it is possible that a slight change in the particle size distribution of the selected coke may be necessary, and special attention must be paid to the mixing sequence, as in the manufacture of refractories.

6.4.4
Glassy Carbon (Polymeric Carbon)

The engineering material *glassy carbon* is obtained by pyrolysis of ultrapure resins. Glassy carbon is a distinctive pyrolysis product of cured phenolic (or in some cases furan) resins, and exhibits a highly refined range of properties including thermal resistance to temperatures of up to 3000 °C, a low specific gravity, high alkali resistance, gas permeability, and biocompatibility, the latter allowing it to be used for implants in the human body. In the very precisely controlled pyrolysis process, liquid resins such as hexa-containing phenolic novolaks or phenolic resoles in furfuryl alcohol are cast in molds or solid resins (filler-free molding compounds) shaped by pressing or transfer molding. The castings or moldings are then cured, and finally subjected to special pyrolysis to obtain pure carbon [1–7].

The manufacturing process used to produce glassy carbon may be broken down into the following stages:

1. Shaping of the liquid phenolic resins (or of a filler-free phenolic novolak/hexa injection molding compound)
2. Curing and post-curing
3. Solid phase pyrolysis at temperatures up to 1000 °C, or if appropriate up to 3000 °C for high-temperature treatment

Phenolic resins are particularly well suited for this application since they afford a high (60–70%) yield of carbon on pyrolysis, based on the cured starting material. In general, highly crosslinked aromatic polymers such as polyphenyls, polyimides, and epoxy formulations can also be considered. Typical methods in phenolic resin processing such as casting, transfer molding, injection molding, centrifugal casting, and hot pressing – identify phenolic resins as a convenient and versatile precursor to glassy carbon [8].

After it is shaped and cured, the resin object may be machined. The molding is pyrolyzed at temperatures below 600 °C at a heating rate of 1–5 °C per hour. Considerable loss in mass and shrinkage (see Fig 6.195, p. 416) takes place over this range of temperature (Fig. 6.231). The dimensional change (linear shrinkage) in unpyrolyzed shaped articles can amount to around 25%; the objects undergo about 5% expansion during further high-temperature treatment at temperatures up to 1000 °C or 3000 °C (Fig. 6.232). The rate at which the temperature is increased during pyrolysis depends on the speed at which the pyrolysis products (cf. section on pyrolysis of phenolic resins, Sect. 6.4.1, p. 417 pp) diffuse out of the mass. Several weeks are required in the case of wall thicknesses amounting to several millimeters.

The lengthy period of heat treatment represents the reason why the wall thickness of glassy carbon articles is generally limited to 4 mm. Amorphous (polymeric) carbon is formed during pyrolysis; in contrast to graphite, this material consists of hexagonal microcrystalline layers that are not aligned in a

Fig. 6.231. Loss in mass and shrinkage during pyrolysis (from [2])

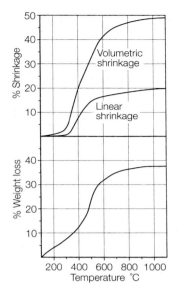

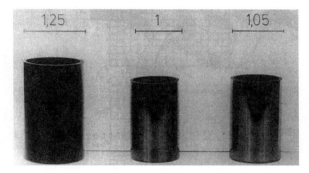

Fig. 6.232. Dimensional change of moldings during thermal treatment (1.25 resin matrix, 1 Sigradur® K, 1.05 Sigradur G) (from [2])

regular manner over wide areas, but form a polymer-like nodular structure (Fig. 6.233) and are thus amorphous as a whole. Due to this structural arrangement, the apparent density is only 1.5 g/cm³, and glassy carbon features relatively low electrical and thermal conductivity.

Glassy carbon is macroscopically free of pores. However, voids exist between the hexagonal graphite nodules or graphite layers, and can exhibit pore diameters of 1–3 nm as in the case of glasses. Glassy carbon exhibits extraordinarily high resistance to corrosion by acidic and alkaline reagents and melts, and is only attacked by oxygen and oxidizing substances at temperatures above 550 °C.

The higher the temperature at which heat treatment is carried out, the greater the resistance to oxidizing reagents such as nitric acid, perchloric acid, or oxygen. The applications of glassy (polymeric) carbon follow from its properties. This engineering material is resistant to corrosion, impermeable to gases and liquids, and resistant to temperatures up to 550 °C in air and up to more than 3000 °C in vacuum or in the presence of inert gas. This tough, hard material is

Fig. 6.233. Structural model of glassy carbon by Jenkins and Kawamura (from [8])

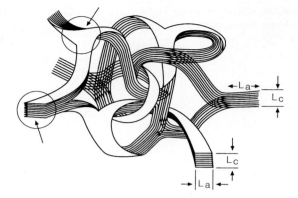

resistant to thermal shock, exhibits high permanent strength, an excellent surface quality, high chemical purity, and is resistant/inert in contact with biological systems. The main applications are in laboratory and chemical engineering equipment as well as medical and dental inserts (Fig. 6.234). Thus, glassy carbon is well suited for construction of laboratory equipment such as crucibles (Fig. 6.235), tubes, and test tubes; other applications include casting molds, implants, electrodes for electrochemical processes, and heart pacemakers [9].

Fig. 6.234. Glassy carbon crucible for dental work (photo: HTW, Meitingen)

Fig. 6.235. Glassy carbon for various applications such as crucibles and beakers used in laboratory work (photo: SGL, Meitingen)

Pyrolysis of phenolic resin foam (see Insulation chapter, Sect. 6.1.2.2) yields porous glassy carbon that can be an ideal material for insulation at temperatures up to more than 3000 °C [10]. Fibrous glassy carbon may be obtained by pyrolysis of phenolic resin fibers [11]. Such fibers can be used for products such as heat-resistant protective clothing [12].

6.5
The Chemically Reactive Bonding Function

6.5.1
Summary: Epoxidation, Alkoxylation, Polyurethane

The chemically reactive bonding function (Fig. 6.236) allows phenolic resins to be transformed into compounds of other important synthetic classes and enables products with other functional groups to be produced from phenolics (see Chemistry, Reactions, Mechanisms chapter 2). For example, the phenolic hydroxyl groups of phenolic resins can be ethoxylated. Raw materials for tensides used in tertiary petroleum recovery are obtained in this manner. Epoxidized novolaks are important raw materials used to manufacture epoxy injection molding compounds used to encapsulate components such as transistors, microprocessors, capacitors, and diodes, or to produce high-performance commutators. Moreover, *ortho*-condensed phenolic resoles can be reacted with 4,4'-diphenylmethane diisocyanate to yield polyurethane systems whose additive curing mechanism in their use as foundry binders (PUR "cold box" gas curing process) is spontaneously accelerated by the presence of gaseous amines. These three application systems, all of which make use of the chemically reactive bonding function, are described in the chapters on additives

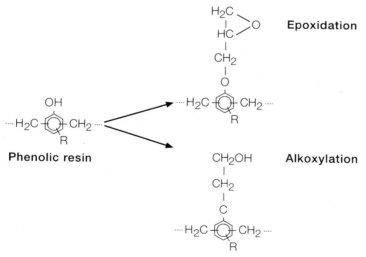

Fig. 6.236. Epoxidation – alkoxylation

in petroleum production (tensides) (Sect. 6.2.4), molding compounds, Sect. 6.1.3.11, and foundry binders Sect. 6.2.1.7.1. Another chemical end use is the application of p-alkylphenol novolaks as "acidic" dye developers in production of carbonless copy paper and thermography. Both techniques are similar mechanistically in the development of an image or color formation.

6.5.2
Alkylphenol Resins as Dye Developers for Carbonless Copy Paper

Alkylphenol resins based on tertiary butyl, octyl, nonyl, or phenylphenol can be used as developers for carbonless copy paper. Due to the large number of 2,2'-methylene bridges, the acidity of these alkylphenol resins is higher than in unsubstituted novolaks. In this application, the phenolic resins possess a chemically reactive bonding function in their use as acidic components [1–5].

Carbonless copy paper generally consists of two or more sheets of paper (Fig. 6.237). Carbonless copy paper is the standard type of a paper in which the reverse side of the paper substrate can be treated with a "CB" coating containing one or more microencapsulated dye precursors. Two dye precursors, for example crystal violet lactone and N-benzoylleucomethylene blue, are generally used in most commercial papers [6]. These dye precursors are dissolved in high-boiling hydrocarbons. Crystal violet lactone develops a blue shade within seconds, but this is not very persistent; on the other hand, the developing time of N-benzoylleucomethylene blue is much longer, but this compound affords a considerably more permanent color shade. The front side of the paper substrate is coated with a "CF" layer containing one or more dye developers. Both the dye precursor and the dye developer are transparent in

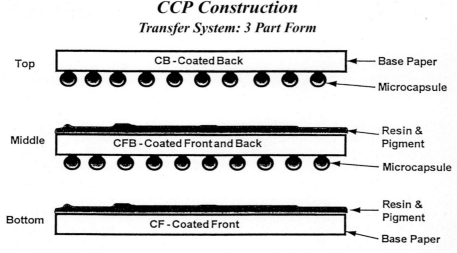

Fig. 6.237. Carbonless copy paper construction: transfer system, three-part form (photo: Schenectady International Inc.).

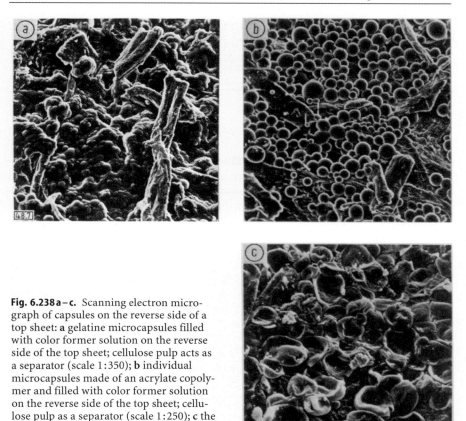

Fig. 6.238a–c. Scanning electron micrograph of capsules on the reverse side of a top sheet: **a** gelatine microcapsules filled with color former solution on the reverse side of the top sheet; cellulose pulp acts as a separator (scale 1:350); **b** individual microcapsules made of an acrylate copolymer and filled with color former solution on the reverse side of the top sheet; cellulose pulp as a separator (scale 1:250); **c** the capsules shown in the center following the writing process (scale 1:500) (from [6])

the formulation present in the coatings on the reverse side of the first sheet and front side of the next. This applies until the CB and CF coatings are brought into contact and pressure applied, for example by a typewriter, which ruptures the microcapsules in the CB layer and liberates the dye precursor (Fig. 6.238). The dye precursor then comes into contact with the CF coating, and reacts with the dye developer to form the image [7–12].

Self-contained CCP papers (Fig. 6.239), in which only one side of the paper substrate is coated, are presently commercially available. In this case, the microcapsules containing the dye precursor are also ruptured when pressure is applied, and the dye developer, for example, an alkylphenol novolak, reacts with this to form a copy image (Fig. 6.240). Self-contained paper also relies on a chemical, color developing reaction between a microencapsulated leuco dye and a dye developer, which is usually an acidic substance. It differs from the transfer type paper in that the microencapsulated dye and dye developer are both coated on the same surface.

CCP Construction
Self-Contained

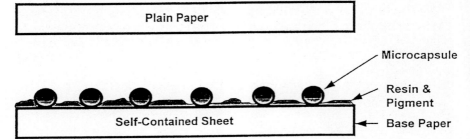

Fig. 6.239. Carbonless (self-contained) copy paper construction (photo: Schenectady International Inc.)

Fig. 6.240. Scanning electron micrograph of the dot of an "i" produced by pressure on carbonless copy paper (photo: BASF, Ludwigshafen, Germany)

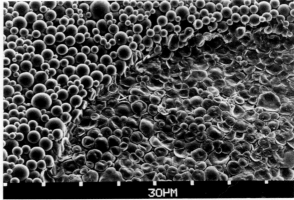

The back of a self-contained sheet can also be coated with a layer of microcapsules (CB), thus allowing the sheet to be used with additional sheets of CF or CFB paper in a multi-part form.

Self-contained paper is most often used in applications where the top sheet, or the original, is plain paper such as security paper for banks.

Microcapsules of gelatin or synthetic polymers serve the primary mechanical function of isolating the color formers (also called "leuco dyes" or dye precursors) from the color developer before being used.

Current [2] dye developers in use (Table 6.123) include:

1. *Acidic clays* such as high-montmorillonite grades of bentonite, in which the color-developing activity of the clay is often activated by treating it with a mineral acid.
2. *Synthetic phenolic resins* that are specifically designed for carbonless copy paper applications, and can represent non-chelated or chelated types. Among the earliest resins (novolaks) used for this purpose were those

based on *p*-phenylphenol. These were an improvement over the use of acidic clays with respect to the moisture sensitivity, and afforded an image with improved fading resistance (especially in the case of crystal violet). The high cost of *p*-phenylphenol and the quest for even better reactivity and image fading resistance led to the development of alkylphenol resins complexed or chelated with zinc. Current versions of these products are available in a pre-dispersed, fine particle form ready to use in a CF coating formulation.

3. *Synthetic, non-polymeric compounds based on salicylic acid.* As in the case of the novel phenolic resins described above, the salicylic acid functional group is reacted with zinc to form a stable chelate or complex.

Due to the advantages of the alkylphenol resin-based dye developers (cf. Table 6.123), these have displaced the acidic clays in the United States and Southeast Asia, and to some extent in Japan and Latin America as well. The advantages of the salicylates over phenolic resins are not so great as to justify the higher price for all applications. Thus, the main markets for zinc alkyl-salicylates used in carbonless copy paper are presently in Japan and to a limited extent in Europe.

Table 6.123. Comparison of types of dye developers for carbonless paper (from [2–4])

Property	Acid clays	Phenolic resins	Zinc salicyliates
Coating rheology	Fair to poor	Good	Good
Coat weight	High (5–6 gsm)	Low (3–4 gsm)	Lowest (theoretically)
Coating types	Aqueous	Aqueous, solvent, print-on	Aqueous, solvent, print-on
Color development	Adsorption	Adsorption & solution	Adsorption & solution
Moisture resistance	Sensitive	Not sensitive	Not sensitive
Fade resistance	Fair to poor	Good	Fast
Image speed	Fast	Medium to fast	Fast
Background yellowing	None	Slight	Slight to none
Abrasiveness	Medium to high	Low	High

6.5.2.1
Thermography

Thermography imaging processes were initiated in the 1950s and pioneered by the efforts of Ricoh of Japan. As it was mentioned earlier, the same imaging characteristics apply in thermography as carbonless copy paper or pressure sensitive recording. Printing paper is coated with both a leuco dye (chromogenic dye) and an acid developer such as diphenols, salicylic acid derivatives. Color development occurs by the application of heat from a print head that can be a simple stylus to a solid state shaped silicon element. The desired reaction takes place preferably at 65–120 °C. Papers that are able to record at lower temperatures are known as high sensitivity papers and contain developers which possess lower melting points.

Uses include computer printers/word processors, X-Y plotters, printing calculators/adding machines, medical/scientific recorders, fax (facsimile) machines, tag, label, and barcode printers.

6.5.3
Antioxidants/Stabilizers

6.5.3.1
Introduction

The attractive polymer additives market is valued at $16 billion (1996) worldwide with a favorable global outlook. Asia is currently leading the way in consumption and growth. Global growth is expected to be 3–5% annually into the next century. The antioxidants/stabilizer segment of the additives market represents about 20% of the total market or over $3 billion worldwide with growth projected at 4–5% annually. The burgeoning growth of thermoplastic resins, rubber compounds, and petroleum products since the 1940s has been responsible for the myriad types of antioxidants/stabilizers that are available today. The utility of phenolics as antioxidants resides mainly with hindered monomeric phenols or polyhydroxy aromatic compounds and rarely resins.

The pervasiveness of atmospheric oxygen is responsible for the facile oxidative degradation of organic compounds. Hence materials must be stabilized against this intrusion of oxygen. Materials such as polymeric products, rubber compounds, petroleum products, and foods/beverages require stabilization.

6.5.3.2
Types of Antioxidants/Stabilizers

These stabilizers, commonly referred to as antioxidants [1] can be naturally occurring products or synthetic materials.

6.5.3.2.1
Natural Antioxidants

β-carotene
α-tocopherol

6.5.3.2.2
Synthetic Antioxidants

- Hindered phenols
- Secondary aromatic amines
- Special sulfide esters

- Trivalent phosphorus compounds: phosphites based on phenolic derivatives
- Hindered amine light stabilizers (HALS)
- Metal dithiocarbamates/thiophosphates

6.5.3.3
Classification of Antioxidants/Stabilizers

A further classification of antioxidants distinguishes them as primary and secondary stabilizers [2].

6.5.3.3.1
Primary Antioxidants

Primary antioxidants are capable of functioning alone and consist of hindered phenols, aromatic amines, HALS, and metal dithiocarbamates/thiophosphates.

6.5.3.3.2
Secondary Antioxidants

Secondary stabilizers are phosphites and sulfide esters and are combined with primary antioxidants. Antioxidants based on phenolic derivatives (hindered phenols and phosphites with phenolic substituents) maintain a significant share of the antioxidant market.

Zweifel recently reviewed most antioxidants/stabilizers in a 1997 publication [3].

6.5.3.4
Mechanism of Oxidation

The mechanism by which oxidation (autoxidation) of an uninhibited organic material (RH) occurs is by a free radical chain reaction:

1. Initiation:

$$RH \rightarrow R\cdot, ROO\cdot, HO\cdot$$
$$ROOH \rightarrow RO\cdot + \cdot OH$$
$$2\,ROOH \rightarrow RO\cdot + ROO\cdot + H_2O$$
$$ROOR \rightarrow 2\,RO\cdot$$

This occurs with oxygen sensitive materials such as unsaturated products (fats, oils, rubber) or during thermally stressing polymers (compounding, extrusion, molding) or UV/light induced stresses.

2. Propagation:

$$R\cdot + O_2 \rightarrow ROO\cdot$$
$$ROO\cdot + RH \rightarrow ROOH + R\cdot$$

The rate of oxidation follows allyl H > tertiary H > secondary H > primary H

3. Termination:

    2 R· → R–R
    ROO· + R· → ROOR
    2 ROO· → other products

Termination occurs when O_2 concentration is very low, such as diffusion of O_2 in thick molded parts or when radicals combine resulting in crosslinking or disproportionate into low MW fragments.

Use of radical scavengers or H donating antioxidants (AH) such as hindered phenols or secondary aromatic amines inhibits oxidation by reacting with peroxy radicals where rate of AH is much faster than the reaction of ROO· with organic materials.

6.5.3.5
Effect of Antioxidants/Stabilizers

Various antioxidants are effective in terminating autoxidation.

6.5.3.5.1
Hindered Phenols and Secondary Aromatic Amines

These materials (AH) are hydrogen donating antioxidants and terminate radical chain reaction by:

    ROO· + AH → ROOH + A·

where A· is

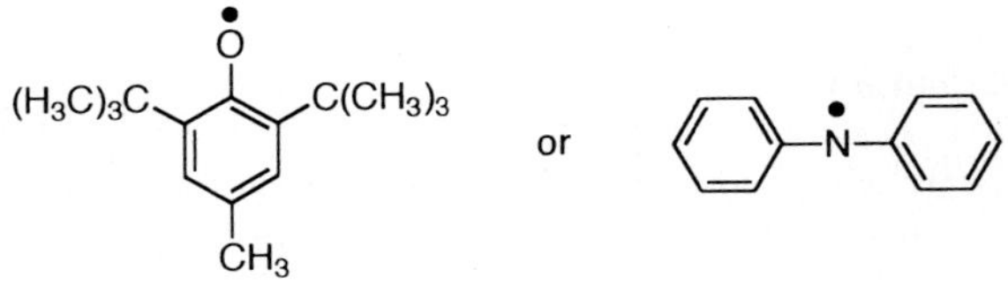

Reaction of AH with ROO· is much faster than reaction of ROO· with RH.

6.5.3.5.2
Trivalent Phosphorus Compounds

Phosphorus compounds such as phosphites or phosphonites react with peroxide derivatives:

   $(RO)_3P + ROOH → (RO)_3PO + ROH$

6.5.3.5.3
Hindered Amines (HALS)

These are classified as light stabilizers rather than antioxidants and are derivatives of 2,2,6,6-tetramethyl piperidine (**I**)

HALS are highly effective stabilizers for polymers and function in the following manner:

$$>NH + O_2 \rightarrow >NO\cdot$$
$$>NO\cdot + R\cdot \rightarrow >NOR$$
$$>NOR + ROO\cdot \rightarrow >NO + R=O + ROH$$

HALS terminates propagating reactions by reacting with alkyl and peroxy radicals.

Antioxidants are used in low amounts, ~0.01–0.05 wt%, in saturated thermoplastic polymers while higher concentrations, 0.5–2 wt%, are required for unsaturated elastomers and blends with polybutadiene (ABS, etc.).

6.5.3.6
Market Areas

6.5.3.6.1
Food/Beverages

Antioxidants traditionally used in food products containing fats and oils to prevent spoilage or rancidity due to oxidation consist of naturally occurring α-tocopherol, ATP, **II**, as well as four phenolic substituted products: BHA (butylated hydroxy anisole) **III**, BHT (butylated hydroxy toluene or ditertiary butyl cresol) **IV**, TBHQ (*tert*-butyl hydroquinone) **V**, and PG (propyl gallate) **VI**:

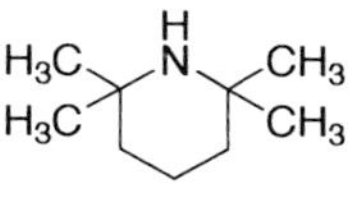

I

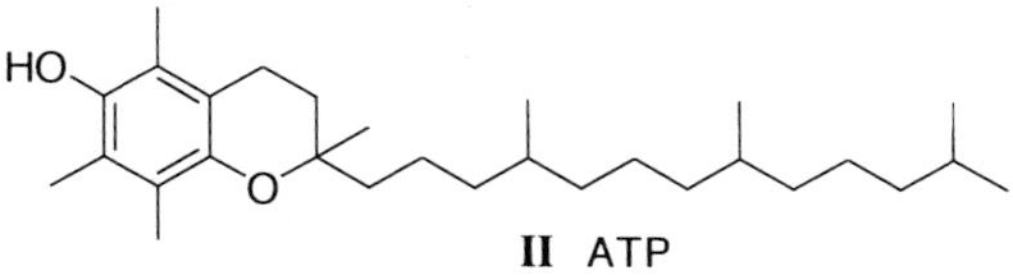

II ATP

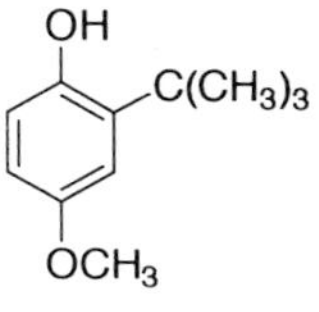

III BHA

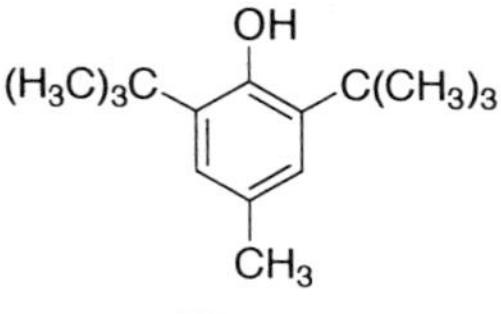

IV BHT

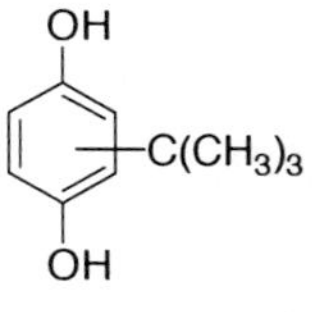

V TBHQ

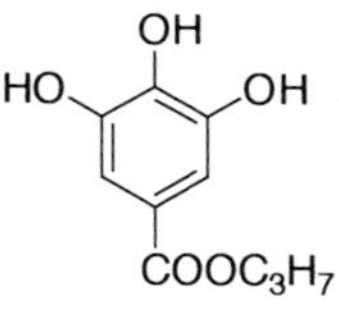

VI PG

Antioxidants are FDA approved and are used at less than 200 ppm.

6.5.3.6.2
Petroleum Products

Fuels

Fuels contain some unsaturated hydrocarbons and undergo oxidation on storage. Continuing oxidation leads to color development followed by the formation of gums and residues. The latter are extremely troublesome to fuel supplying components like carburetors. Hindered phenols such as BHT and 2,6-ditertiary butyl phenol are used at low levels, 5–10 ppm, and provide excellent storage stability.

Lubricating Oils

These materials are used in jet engines, turbines, crankcases of automobiles, etc. and are stabilized with hindered phenols, or secondary aromatic amines or hindered phenols combined with metal dithiophosphates.

Computer simulation studies [4] have been developed to model the relationship of the base oil composition (hexadecane/tetralin-model oils) and antioxidant consumption through the inhibition of oxidation.

6.5.3.6.3
Rubber Compounds

Unsaturated materials (elastomers, rubbers) are more easily oxidized, and hence a higher concentration of antioxidant is required to minimize oxidation. Hindered phenols combined with phosphites or aromatic amines are suitable stabilizer systems.

6.5.3.6.4
Polymeric Materials

Most commercially available organic polymers are susceptible to oxidative degradation during processing and the lifetime of the fabricated object. The introduction of antioxidants/stabilizers allows these polymeric materials to retain their physical properties and ensure reasonable service life during storage and ultimate use. The explosive growth of thermoplastic materials from the 1940s to the present has fostered the large stabilizer market of hindered phenols, alkylaryl amines, phosphites, thioesters, acids, quinolines, alkylated phenols, and other antioxidants. It has evolved into a "quality additive" market area.

Structures of some recent hindered phenols are shown:

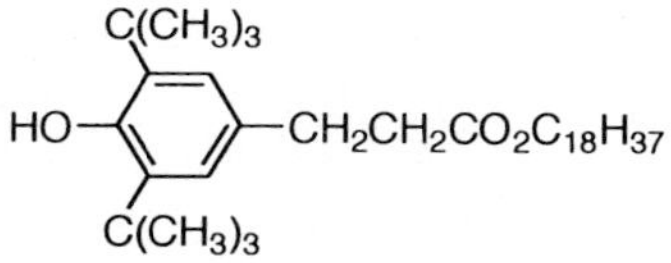

VII

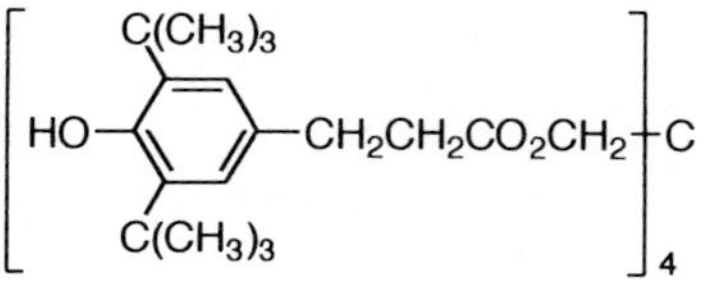

VIII

[Structure IX shown with R groups]

Studies [3] have shown that **VIII** significantly outperforms both BHT (**IV**) and **VII**, providing excellent long term thermal stability to polypropylene homopolymer at 135 °C.

Synergistic effects or combining primary and secondary stabilizers, are noted between hindered phenols and thioether esters. By incorporating a thioether substituent into a phenolic compound (**X**), intermolecular cooperative effect of phenol with the thioether substituent in the ortho position provided extended processing of polybutadiene in a Brabender kneader at 160 °C [3].

[Structures X and XI shown]

Similarly synergist effects (primary combined with secondary) occur when hindered phenols are combined with aryl phosphites. Thus hindered phenol **VIII** with phosphite **XI** provided moderately constant melt flow of polypropylene homopolymer after five consecutive extrusions.

A new secondary antioxidant **XII** of the phosphite family was recently introduced [5]. The novel phosphite exhibits high resistance to hydrolysis and less yellowing in polyolefins when combined with hindered phenol. The presence of amino ester group facilitates metal deactivation of potentially harmful metal ionic polyolefin catalysts. It is effective for polyolefins, olefin copolymers, polycarbonates, and polyamides.

Novel butyl ethyl propanediol component introduced into phosphites as secondary antioxidants provides superior hydrocarbon solubility and improved handling (**XIII**) [6].

[Structures XII and XIII shown]

UOP Guided Wave Systems (El Dorado Hills, CA.), through their fiber optic "in line" process capabilities, have developed a UV fiber optic spectroscopy method for online monitoring of hindered phenols and phosphites in polypropylene. The method provides successful measurement of these antioxidants even in the presence of potential interferents [7].

6.5.3.7
New Developments

Factors which must be considered in the development of improved and new antioxidants are low volatility, low extractability, low color development, thermal/UV stability, non-toxicity, and when necessary, FDA approval.

6.5.3.7.1
Polymer Bound

Stabilizers which are highly soluble in resins and exhibit very low extractibility are attached to polymers. Several of these "polymer bound" stabilizers are being commercialized. Polysiloxane bound UV stabilizers and phenolic antioxidants are reported [8] to be highly compatible with polypropylene and highly resistant to extraction.

6.5.3.7.2
Benzotriazole Types

A different method requires the use of a reactive UV stabilizer that copolymerizes with a variety of vinyl monomers and does not contribute to coloration of the resulting polymer [9].

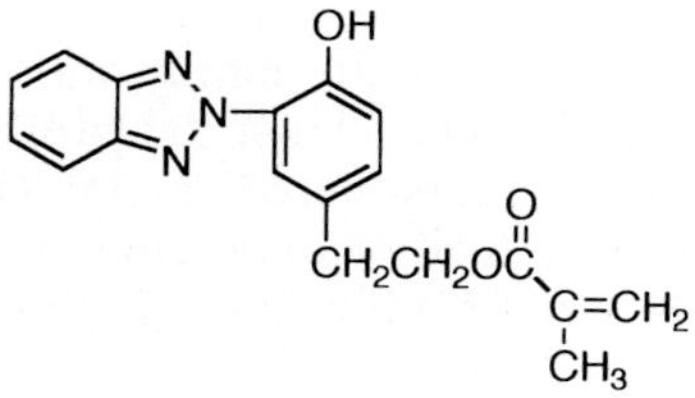

XIV

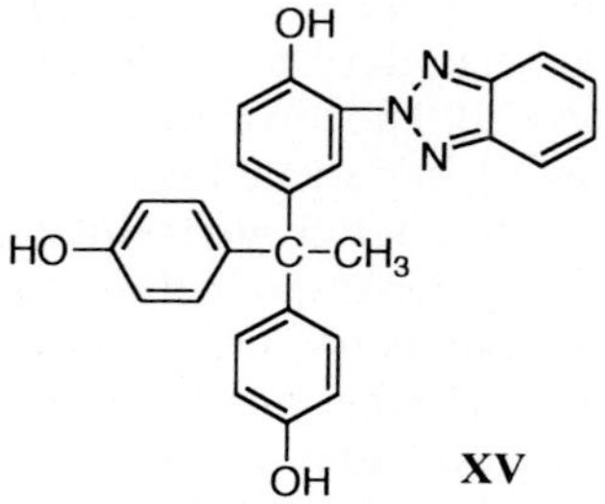

XV

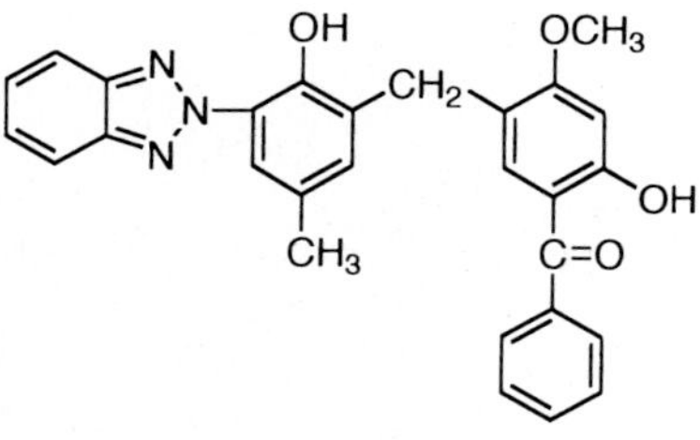

XVI

The benzotriazole (**XIV**) on copolymerization is nonvolatile, nonmigratory, and non-blooming.

Another benzotriazole material (**XV**) proposed as a UV light stabilizer exhibits extremely high temperature stability (1 wt% loss at 324 °C) [10]. **XV** is non-migrating with long term stability, and developed for high performance engineering thermoplastics such as polysulfone, polycarbonate, etc.

Coupling the benzotriazole functionality with benzophenone has resulted in a high temperature UV stabilizer (**XVI**) [11].

6.5.3.7.3
Triazine

An added feature of new UV stabilizers developed for engineered thermoplastics such as polyesters, polyamides, polycarbonates, and polyacetals is their high extinction coefficients in the spectral range up to 400 nm and their absorbtion of the UV portion of sunlight. Triazine compound **XVII** exhibits significantly longer life time in engineered thermoplastics than 2-(2-hydroxyphenyl) benzotriazole and 2-hydroxy benzophenone.

XVII

6.5.3.7.4
Lactone/Hydroxylamine

Recent efforts involving non-phenolic derivatives consist of improved stabilizers based on lactone and hydroxylamine functionalities. Both are reported to be effective scavengers of alkyl radicals and prevent autoxidation at an early stage.

6.5.3.8
Summary/Trends

Susceptibility of organic compounds to atmospheric oxygen requires stabilization to counteract autoxidation. Hindered phenols are known to inhibit the free radical chain reaction occurring during autoxidation. As a primary antioxidant or combined with secondary antioxidant (phosphite or metal dithiophosphates), hindered phenols maintain a large segment of the antioxidant/stabilizer market.

New types with phenolic substituents are polymer bound, benzotriazole, and triazine types. Non-phenolic types include lactones and hydroxylamines.

Continuing research in developing antioxidants for a range of polymers from the commodity types (polyolefins, vinyl polymers, etc.) as well as the newly commercial engineering resins will result in continued improvements in antioxidants in addition to new materials. These anticipated future developments are expected to improve performance and productivity in polypropylene, demanding polyethylene applications, and other resin systems that require high temperature processing, high shear rates, high melt viscosity, or long melt residence times.

References to Chapter 6

1. Knop A, Pilato LA (1985) Phenolic resins. Springer, Berlin Heidelberg New York
2. Gardziella A, Haub HG (1988) Phenolharze. In: Kunststoff-Handbuch, vol. 10, Duroplaste. Carl Hanser, Munich, pp 12–40
3. Adolphs P, Giebeler E, Stäglich P (1987) In: Houben-Weyl (eds) Methoden der organischen Chemie, 4th edn, vol E20, part 3. Thieme, Stuttgart, pp 1974–1810
4. Ruisinger K (1989) Magazin Sammeln (Oct 1989)
5. Gardziella A, Müller R (1990) Phenolic resins. Kunststoffe, Issue 10
6. Gardziella A, Adolphs P (1992) Phenolic resins (PF). Kunststoffe, Issue 15
7. Behring M (1998) Gift im Auto. Auto-Bild No. 41 (Oct 10, 1988)
8. Bakelite AG (1992) Rutaphen-Phenolic Resins, Guide Product Range Application, Company Publication, September 1992, Bakelite AG, Iserlohn, Germany

References to Sect. 6.1.1

1. Pizzi A (1994) Advanced wood adhesives technology. Dekker, New York
2. Pizzi A, Mittal KL (eds) (1994) Handbook of adhesive technology. Dekker, New York
3. Gardner DJ, Wolcott MP, Wilson L, Huang Y, Carpenter M (1995) Wood Adhesives 1995, Symposium June 29–30, Portland, OR, p 29
4. Scheike M, Dunky MM, Resch H (1995) Wood Adhesives 1995, Symposium June 29–30, Portland, p 43
5. Shaler S, Keane D, Wang H, Mott L (1998) Forest Products Society Annual Meeting, June 21–24, Merida, Yucatan, Mexico, Session 2, paper 1
6. Maloney TM (1996) Forest Prod J 46(2): 19
7. Spelter H, McKeever D, Durbak I (1997) Review of wood-based panel sector in U.S. and Canada. US Department of Agriculture, FPL-GTR-99 June
8. Sellers T Jr (1985) Plywood and adhesives technology. Dekker, New York
9. PS1–95 Construction and Industrial Plywood, Sept 7, 1995 American Plywood Association, Tacoma WA
10. Technical information related to liquid resins fo plywood, OSB, particleboard, fiberboard (1998) Bakelite AG, Duisburg, Germany
11. Kim MG, Nieh WL, Sellers T Jr, Wilson WW (1992) Ind Eng Chem Res 31:973
12. Kim MG, Nieh WL-S, Meacham RM (1991) Ind Eng Chem Res 30:798
13. Coggeshall B (1995) Wood Adhesives 1995, Symposium June 29–30, Portland, OR, P186
14. Tomita B, Matsuzaki T (1985) Ind Eng Chem Prod Res Dev 24:1
15. Tomita B, Hse C-Y (1992) J Poly Sci Part A, Poly Chem 30:1615
16. Tomita B, Hse C-Y (1993) Mokuzai Gahkaishi 39:1276
17. Tomita B, Ohyama M, Doi K, Hse C-H (1994) Mokuzai Gahkaishi 40:170
18. Tomita B, Ohyama M, Hse C-H (1994) Holz 48:522

19. Hse C-Y, Xia Z-Y, Tomita B (1994) Holz 48:527
20. Ohiyama M, Tomita B, Hse C-Y (1995) Holz 49:87
21. Sidhu A, Steiner P (1995) Wood Adhesives 1995, Symposium June 29–30, Portland, OR, p 203
22. Higuchi M, Roh JK, Tajima S, Irita H, Honda T, Sakata I (1994) Proceeding from the Adhesives and Bonded Wood Symposium, Forest Products Society, Madison, WI, pp 429–449
23. Ellis S, Steiner P (1991) Wood & Fiber Science 23(1):85
24. Ellis S, Steiner P (1992) Forest Prod J 42(1):8
25. Ellis S (1993) Forest Prod 43(2):66
26. Ellis S (1996) Forest Prod 46(9):69
27. Ellis S (1995) Wood & Fiber Science 27(1):79
28. Kamke F, Saunders HG (1996) Forest Prod J 46 (6):63
29. Anderson AW, Troughton GE (1996) Forest Prod 46(10):72
30. Steiner PR, Troughton GE, Andersen (1993) AW Forest Prod 43(10):29
31. Wolf F (1997) 31st Particleboard/Composite Materials Symposium, April 8–10, Washington State University, Pullman, WA, p 16
32. Broline BM, Holloway TC, Moriarity CJ (1995) Wood Adhesives 1995, Symposium June 29–30, Portland, OR, p 97
33. Nagy E (1995) Wood Adhesives 1995, Symposium June 29–30, Portland, OR, p 105
34. Chapman KM (1997) CA 126:265, 283W
35. Maylor R (1995) Wood Adhesives 1995, Symposium June 29–30, Portland, OR, p 115
36. Park B-D, Riedl B (1997) 31st Particleboard/Composite Materials Symposium, April 8–10, Washington State University, Pullman, WA
37. Guss LM (1995) Wood Adhesives 1995, Symposium June 29–30, Portland, OR, p 21
38. Stahl U (1997) 31st Particleboard/Composite Materials Symposium, April 8–10, Washington State University, Pullman, WA, paper 22
39. Churchland MT (1995) Ceram Trans 59:63
40. Geimer RL, Christiansen AW (1996) Forest Prod J 46(11/12):67
41. Wang XM, Reidl B, Geimer RL, Christiansen AW (1996) Wood Sci Technol 30(6):423; (1997) C.A. 126:909, 28f
42. Glasser WG, Sarkinen S (eds) (1989) Lignin: Properties and Materials, ACS Symposium Series 397. American Chemical Society, Washington, DC
43. Seller T Jr (1995) Lignin – adhesive research for wood composites. Forest Products Laboratory, Mississippi State, MS
44. McDonough TJ (1993) TAPPI J 76(8):186
45. Faix O, Argyropoulos DS, Robert D, Neirinck V (1994) Holz 48:387
46. Argyropoulos DS (1994) J Wood Chem Tecknol 14(1):65
47. Rodriquez F, Gilarranz MA, Oliet M, Tijero J, Barbadillo P (1997) C.A., 126:265, 316j
48. Lora JH et al. (1991) Symposium sponsored by USDA Forest Service, Seattle, WA, November 19–21
49. Lora JH, Senyo WC, Creamer AW, Wu CE (1996) Forest Prod J 46(6):73
50. Mott L (1997) Particleboard/Composite Materials Symposium, 8–10 April, Washington State University, Pullman, WA
51. Dressler H (1994) Resorcinol: its uses and derivatives. Plenum Press, New York
52. Werstler DD (1986) Polymer 27:757
53. Kim MG, Amos LW, Barnes EE (1993) J Poly Sci Part A Poly Chem 31:1871
54. Christjansen P, Köösel A, Suurpere A (1997) Poly Eng & Sci 37(6):928
55. Scopelitis E, Pizzi A (1993) J Appl Poly Sci 47:351
56. Tingley DA (1996) U.S. 5,547,729; (1996) 5,565,257
57. Gardner DJ, Davalos JF, Munipalle UM (1994) Forest Prod J 44(5):62
58. Vick CB, Richter K, River RH (1996) U.S. 5,543,487
59. Vick CB (1995) Wood Adhesives 1995, Symposium 29–30 June, Portland, OR, p 47
60. Vick CB, Okkonen EA (1997) Forest Prod J 47(3):71
61. Qiao P, Davalos JF, Trimble B, Bender R, Dailey H Jr (1998) International Composites Expo'98, 19–21 January, Nashville, TN, session 22A

References to Sect. 6.1.2.1 and 6.1.2.2

1. Finnegan JA, Gersh BP, Lang JB, Levy R (1995) In: Kroschwitz JI (ed) Kirk Othmer encyclopedia of chemical technology, vol 4. Wiley, New York, p 644+
2. Mohr JG, Rowe WP (1978) Glass fiber. Van Nostrand Reinhold, New York
3. Tomita B, Matsuzaki T (1985) Ind Eng Chem Prod Res Dev 24:1
4. Scopelitis E, Pizzi A (1993) J Appl Poly Sci 47:351
5. Technical information related to liquid resins for glass fiber, mineral wool, floral foam, and orthopedic foam (1998) Bakelite AG, Duisburg, Germany
6. Knop A, Pilato LA (1985) Phenolic resins: chemistry, applications and performance, future directions. Springer, Berlin Heidelberg New York, p 224+
7. C.A. (1995) 122:189,885u; (1994) 120:136,048a; (1994) 120:78,708c; (1993) 119:182,376 h
8. C.A. (1995) 123:342,868 h
9. Owens Corning (1989) U.S. Pat 4,883,824; U.S. Pat 4,945,077 (1990)
10. U.S. Pat 5,489,619 (1996)
11. C.A. (1995) 123:341,697q
12. U.S. Pat 5,432,207 (1995); U.S. Pat 5,514,725 (1996)
13. U.S. Pat 5,407,963 (1995)
14. C.A. (1994) 121:84839
15. U.S. Pat 4,575,521 (1986)
16. C.A. (1995) 123:289,084w
17. U.S. Pat 5,444,098
18. C.A. (1982) 96:144112t
19. Owens Corning (1995) U.S. Pat 5,441,992
20. C.A. (1995) 123:259112m
21. C.A. (1993) 119:140,840z; (1994) 121:110,609c; (1995) 122:135,592c, 189,691c, 189,980; (1995) 123:289,152s
22. C.A. (1995) 122:316,451b
23. C.A. (1993) 119:140,729t, 204,755k, 227,189t; (1994) 120:56,155t; (1995) 122:83,355f, 83,356g; (1997) 126 252,210v
24. C.A. (1994) 120:78,707b, 165,866w
25. C.A. (1994) 121:84,839h
26. C.A. (1993) 119:119,060k
27. Cleary TG, Quintiere JG (1991) NISTIR 4664, U.S. Dept. of Commerce, NIST, Gaithersburg, MD
28. Stevens MG, Voruganti V, Rose R (1996) Proc Int Conf Fire Saf 21:245
29. Grand AF, Bundick JR (1998) Fire and Materials '98. 5th Int'l Conference and Exhibition, 23–24 February, San Antonio, TX
30. Grand AF, Weil ED (1998) 43rd Int'l SAMPE Symposium & Exhibition, 31 May, 4 June, Anaheim, California
31. Bomberg MT (1993) Cellular Polymers, paper 24
32. Bomberg MT, Kumaren MK (1995) Cell Polym III, 3rd Int Conf, paper 1
33. Booth JR, Grimes JT (1993) Therm Cond 22:783
34. Teichner SJ (1991) Chem Tech, June, p 372
35. Pekala RW, Alviso CT, Nielson JK, Tran TD, Reynolds GAM, Dresselhaus MS (1995) Mat Res Soc Sym Proc, 393, 413
36. Rettelbach T, Ebert H-P, Capa R, Fricke J, Alviso CT, Pekala RW (1996) Therm Cond 23:407
37. C & EN (1995) 13 February, p 64
38. Aviation Week & Space Tech (1995) August 14, 1995, p 58

References to Sect. 6.1.2.3

1. Eisele D (1988) Textilvliesstoffe aus Duroplaste. In: Becker/Braun (eds) Kunststoffhandbuch, vol 10. Carl Hanser, Munich Vienna, pp 763–775
2. Gardziella A (1996) Thermosets – status report on current technology. Address at the International Plastics Congress '96, 17–18 April, Süddeutsches Kunststoffzentrum, Würzburg, Germany (reprint pp F1–F14)
3. Gardziella A (1992) Materials Symposium on Phenolic Resin-Bonded Textile Mats in Automotive Interiors (Phenolharzgebundenes Textilvlies im Kfz-Innenraum), Automobiltechnische Zeitung ATZ 94 1:17–19
4. Eisele D, Gardziella A, Marutzky B, Müller F-J (1992) Duroplastisches Textilvliesmaterial: bewährt und verbessert. Kunststoffe 3:229–234
5. Müller F-J (1996) Phenolharzgebundene Faservlies-Werkstoffe. Seminar in Haus der Technik, Essen, Germany, 21–22 May
6. Eisele D (1990) Textilvlies-Flachware und Formteile für die Automobilausstattung. Textil Praxis International 16(10):1057–1063
7. Eisele D (1990) Textilvlies-Flachware und Formteile für die Automobilausstattung. Address at Technical Symposium Phenolharze – vielfältig in der Anwendung, 16–17 May, Süddeutsches Kunststoffzentrum (reprint pp 57–70)
8. Bakelite AG (1992) Binders for the manufacture of textile and natural fibre felts. Bakelite AG Company Booklet, Iserlohn, Sept
9. Rütgers AG (1988) EP 0 254 807
10. Rütgers AG (1993) EP 0 540 836
11. Rütgers AG (1994) EP 0 595 003
12. Matec Holding AG (1990) EP 0 375 618
13. Eisele D (1993) Faserhaltige Bauteile für die Automobilausstattung. J. Borgers Booklet, Bocholt, 1 July
14. J. Borgers GmbH (1993) Triflex für den Automobilbau. J. Borgers GmbH & Co. KG Company Booklet, Bocholt
15. Eisele D (1987) Geruch und fogging von Automobilinnenausstattungsmaterialien. Melliand Textilberichte 3:206–215
16. Kalweit M (1996) Phenoplastgebundene Faservlies-Werkstoffe. Seminar in Haus der Technik, Essen, Germany, 21–22 May

References to Sect. 6.1.3

1. Baekeland LH (1908) DRP 23,3803. Heat and pressure patent
2. Becker W, Braun D, Woebken W (1988) Duroplaste. In: Kunststoff-Handbuch, vol 10. Carl Hanser, Munich Vienna
3. Schönthaler W, August H, Bauer W, Decker K-H, Ehnert GP, Graf W, Harms W, Huster FJ, Mattelé P, Morawetz G (1988) Formmassen aus duroplastischen Harzen. In: Kunststoff-Handbuch, vol 10. Carl Hanser, Munich Vienna, pp 195–384
4. Schönthaler W (1991) Duroplaste: Zukunft von Anfang an, Technische Vereinigung, Würzburg
5. Knop A, Pilato LA (1985) Molding compounds. In: Phenolic resins, chemistry, applications and performance, future directions. Springer, Berlin Heidelberg New York, pp 156–211
6. Schönthaler W (1998) In: Kunststoff-Handbuch, vol 10. Carl Hanser, Munich Vienna Phenolharz-formmassen, pp 207–265
7. Ehrenstein GW, Bittmann E (1997) Duroplaste, Aushärtung, Prüfung, Eigenschaften. Carl Hanser, Munich
8. Saechtling HJ, Oberbach K (1995) Kunststoff-Taschenbuch, 26th edn. Carl Hanser, Munich
9. Braun U (1994) Duroplastische Formmassen – Grundlagen und neue Produkte. In: Niemann K, Schröder K (1994) Spritzgiessen von Duroplasten. Hüthig, Heidelberg

10. Saechtling HJ, Oberbach K (1995) Härtbare Formmassen. In: Kunststoff-Taschenbuch, 26th edn. Carl Hanser, Munich, pp 537–565
11. Bachmann A, Müller K (1973) Pressmassen. In: Phenolplaste. VEB Deutscher Verlag der Grundstoffindustrie, Leipzig, pp 100–161
12. Gardziella A (1995) Härtbare Harze und Formmassen. Kunststoffe 12(85):2184–2186
13. Gardziella A (1996/97) Duroplaste-Statusreport zum Stand der Technik, 224–227. Kunststoff-Kautschuk-Produkte, Jahreshandbuch 1996/7. Hoppenstedt, Darmstadt
14. Gardziella A (1996) Duroplaste – Statusreport zum Stand der Technik. Address at the Kunststoffe '96 Congress, in Würzburg (Organizers: Süddeutsches Kunststoffzentrum in Würzburg, Carl Hanser, Munich, and KI Verlagsgesellsin Bad Homburg)
15. Hansen A (1993) Duroplastische Bindemittelsysteme, Chemie und allgemeine Anwendungen. In: Anwendungen von duroplastischen Kunststoffen im Automobilbau. Proceedings of SKZ Symposium, 23–24 September, Würzburg, pp 9–26
16. Bakelite AG (1992) Formmassen Lieferprogramm, Eigenschaften und Anwendungen. Company Publication, Bakelite AG, Iserlohn, Germany
17. Bakelite AG (1996) Moulding compounds delivery program, characteristics and properties. Company Publication, Bakelite AG, Iserlohn, Germany
18. Kraemer M (1995) Härtbare Formmassen. Kunststoffe 10:1635–1636
19. Schönthaler W, Schirber H (1990) Phenolharze vielfältig in der Anwendung. SKZ Symposium, Erlangen, 16–17 May
20. Gardziella A, Schirber H (1993) Anwendungen von duroplastischen Kunststoffen im Automobilbau. SKZ Symposium, 23–24 Sept Würzburg
21. Gardziella A, Baumgärtel K (1995) Duroplastische Werkstoffe aktuell. SKZ Symposium, 26–27 April, Würzburg
22. Gardziella A, Baumgärtel K (1997) Duroplastische Werkstoffe in der Elektronik und Elektroindustrie. SKZ Symposium, 4–5 June, Würzburg
23. Jellinek K, Schönthaler W, Niemann K (1983) Neuere Entwicklungen bei härtbaren Formmassen. Chem Ing Techn 55(1):30–38
24. Domininghaus (1988) Füll- und Verstärkerstoffe sowie andere Zusatzstoffe. In: Kunststoff-Handbuch, vol 10. Carl Hanser, Munich Vienna, pp 157–194
25. Gardziella A (1963) Sonderpressmassen auf dem Phenolharzsektor. Industrieanzeiger 13, Issue on Kunststoffe und Kunststoffverarbeitung 1. Verlag W. Giradet, 12 February, Essen
26. Gardziella A (1965) Wirkung von Füllstoffen auf die Eigenschaften von Phenolharz-formmassen. Kunststoffe 55(1):20–25
27. Wiegand H, Müller K (1969) Einfluss der Formmasse-Zusammensetzung auf die Formstoffeigenschaften von Phenoplasten. Kunststoff-Rundschau 12:716–720
28. Schulz H, Godawa KH (1972) Galvanisierbare duroplastische Formteile. Kunststoffe 62(6):402–406
29. Braun U (1995) Hochtemperaturbeständige Formmassen. Proceedings of SKZ Symposium, 26–27 April, Würzburg, pp 41–56
30. Breuer H, Dupp G, Schmitz J, Tullmann R (1990) Einheitliche Werkstoff-Datenbank-eine Idee setzt sich durch. Kunststoffe 80(11):1289–1294
31. Bakelite AG (1995) CAMPUS before world-wide-spread. Technical Information Brochure, Bakelite AG, Iserlohn, Germany
32. (1996) Molding compounds overview, properties according to new ISO-Norms. Company Publication, Bakelite AG, Iserlohn, Germany
33. Saechtling HJ, Oberbach K (1995) Verarbeitung Duroplaste. In: Kunststoff-Taschenbuch, 26th edn. Carl Hanser, Munich, pp 265–276
34. Leitfaden-Formmassen der Bakelite AG (1997) Company Publication, Bakelite AG, Iserlohn, Germany
35. Schönthaler W (1970) Wichtige Verarbeitungseigenschaften und ihre Beschreibung. VDI Verarbeitung härtbarer Formmassen Symposium, Düsseldorf, Germany, November 1970
36. Bauer W (1974) Fortschritte bei der Verarbeitung duromerer Formmassen. Kunststoff-Berater 4:206–208

37. Derek H (1986) Pressen und Spritzgiessen – Wege der Duroplastverarbeitung. Kunststoffe 76(8):659–662
38. Fischbach G (1988) Prozessführung beim Spritzgiessen härtbarer Formmassen. Dissertation at the RWTH, Aachen
39. Bauer W (1988) Grundsätzliches zur Auswahl und Verarbeitbarkeit von Formmassen. In: Kunststoff-Handbuch, vol 10. Carl Hanser, Munich Vienna, pp 198–206
40. Danne W (1994) Der Verarbeitungsprozess. In: Niemann K, Schröder K (1994) Spritzgiessen von Duroplasten. Hüthig, Heidelberg
41. Kleinemeier B, Menges G (1978) Mechanical properties of selected injection molded thermosets. Polymer Engineering and Science 18(13):996–1000
42. Brandau E (1993) Duroplastwerkstoffe, Technologie, Prüfung, Anwendung. Verlag Chemie, Weinheim
43. Müller K (1988) Vergleichbarkeit der Basiseigenschaften von Duroplasten und Thermoplasten. International Thermoset Congress, March, Bad Mergentheim
44. Niemann K, Schröder K (1994) Spritzgiessen von Duroplasten. Hüthig, Heidelberg
45. Casellas A (1994) Duroplastwerkzeuge. In: Niemann K, Schröder K (1994) Spritzgiessen von Duroplasten. Hüthig, Heidelberg
46. Adolf J (1994) Formteilgestaltung duroplastischer Artikel. In: Niemann K, Schröder K (1994) Spritzgiessen von Duroplasten. Hüthig, Heidelberg
47. Thienel P, Kürten C (1994) Kaltkanaltechnik. In: Niemann K, Schröder K (1994) Spritzgiessen von Duroplasten. Hüthig, Heidelberg
48. Schröder K (1994) Spritzgiessmaschinen, Pressautomaten, Steuerungen und Peripherie. In: Niemann K, Schröder K (1994) Spritzgiessen von Duroplasten. Hüthig, Heidelberg
49. Thienel P, Hoster B (1994) Ermittlung der Füllbildkonstruktion härtbarer Formmassen mit einem Sichtwerkzeug. In: Niemann K, Schröder K (1994) Spritzgiessen von Duroplasten. Hüthig, Heidelberg
50. Bielfeld B, Hadeball H, Lohmann H, Schönthaler W (1988) Das Verarbeiten von Duroplasten auf Schnecken-Spritzguss-Maschinen. Kunststoffe 3, 4 and 5, pp 177, 263, 346 ff
51. Rothe J (1978) Formgebung von Duromeren in Spritzgiess-Maschinen. VDI Wirtschaftliche Herstellung von Duromer-Formteilen Symposium, Symposium Booklet, Mannheim, p 95
52. Wendt G, Schönthaler W, Niemann K (1969) Planung und Herstellung von Werkzeugen für das Spritzgiessen von Duroplasten. Kunststoffe 59(4):211ff
53. Niemann K, Danne W, Tüscher T (1981) Spritzgiessen duroplastischer Präzisionsteile mit Vortrocknen. Plastverarbeiter 32(9):1200
54. Menges G, Schmidt L, Schultheis S, Wortberg J (1977) Berechnung des Füllvorganges beim Spritzgiessen von vernetzenden Formmassen. Plastverarbeiter 28(8):401–404
55. Michael KP (1987) Vermeidung der Gratbildung beim Spritzgiessen von duroplastischen Formmassen, IKV of the RWTH. Aachen, Communication, p 128
56. Niemann K (1997) Kaltkanaltechnik – Stand und Einsatzmöglichkeit bei Duroplasten. SKZ Angussminimiertes Spritzgiessen" Symposium, Booklet, pp 25–34B
57. Schönthaler W, Niemann K (1976) Druck- und Temperatur-Messungen in Duroplast-Spritzgiess-Werkzeugen. Kunststoffe 68(2):66–70
58. Schönthaler W, Niemann K (1981) Optimiertes Spritzgiessen von Duroplasten, Möglichkeiten einer Prozess-Steuerung. Kunststoffe 71(5):348–351
59. Rüschenbaum I, Sievert E (1987) Herstellen gratfreier Duroplast-Formteile durch das Gegendruck-Verfahren. Diplomarbeit D 8717 at the Märkische Fachhochschule of Iserlohn
60. Braun U, Danne W, Schönthaler W (1987) Angussloses Spritzprägen in der Duroplastverarbeitung. Kunststoffe 77(1):27–29
61. Schröder K (1994) Qualitätsüberwachungssysteme. In: Niemann K, Schröder K (1994) Spritzgiessen von Duroplasten. Hüthig, Heidelberg
62. Saechtling HJ, Oberbach K (1995) Normung und Gütesicherung. In: Kunststoff-Taschenbuch, 26th edn. Carl Hanser, Munich, pp 751–771

63. Benfer W, Schulze-Kadelbach R (1983) Fliessverhalten von Duroplasten. Final Report, IKV, Aachen
64. Niemann K (1994) Anwendungsbeispiele duroplastischer Formmassen und Formteile. In: Niemann K, Schröder K (1994) Spritzgiessen von Duroplasten. Hüthig, Heidelberg
65. Bakelite AG (1996) Sonderformmassen Kommutatoren. Company Publication, Bakelite AG, Iserlohn
66. Bachhuber K (1983) Formmassen zur Herstellung von Kommutatoren. From Ref. 21, paper VI/1-9
67. Braun (1988) Formmassen für Kommutatoren. In: Kunststoff-Handbuch, vol 10. Carl Hanser, Munich Vienna, pp N1-10
68. Braun U (1993) Härtbare Formmassen für moderne Anwendungen im Automobilbau. In: Anwendungen von duroplastischen Kunststoffen im Automobilbau. Proceedings of SKZ Symposium, 23-24 Sept, Würzburg, pp 59-80
69. Gardziella A, Schönthaler W (1981) Technische Phenolharze und härtbare Formmassen im Automobilbau. Kunststoffe 71(3):159-166
70. Pahl H (1993) Anforderungen an duroplastische Formmassen für Funktionsteile im Kfz-Motorraum am Fallbeispiel – Steuerbalken des Bremskraftverstärkers. In: Anwendungen von duroplastischen Kunststoffen im Automobilbau. Proceedings of SKZ Symposium, 23-24 Sept, Würzburg, pp 43-58
71. Schuchert M, Dekovaky T, Eyarer P (1991) Energie- und Rohstoffaufwand zur Herstellung von Werkstoffen: Vergleichende Betrachtung der Werkstoffgruppen Stahl, Aluminium und Kunststoff. Colloquium Booklet, 12th Stuttgart Plastics Colloquium
72. Bayerl H (1997) Vergleich von duroplastischen und thermoplastischen Isolierstoffen. In: Proceedings of SKZ Symposium, 4-5 June, Würzburg, pp K1-19
73. Heinrich E (1997) Granulat-Formmassen. In: Proceedings of SKZ Symposium, 4-5 June, Würzburg, pp M1-14
74. Müller K (1988) Vergleichbarkeit der Basiseigenschaften von Duroplasten und Thermoplasten. International Thermoset Congress, March, Bad Mergentheim, Germany
75. Braun U (1993) Hochtemperaturbeständige Formmassen. In: Anwendungen von duroplastischen Kunststoffen im Automobilbau. Proceedings of SKZ Symposium, 23-24 September, Würzburg, pp 41-56
76. Günther P (1995) Hochtemperaturbeständige Phenolharzformmassen. Proceedings of SKZ Symposium, 26-27 April, Würzburg, pp VII.1-10
77. Ehrentraut P, Schnabel P (1986) Hochtemperaturbeständige duroplastische Kunststoffe, Eigenschaften, Einsatzgebiete und Verarbeitung. Plastverarbeiter, Issue 3
78. Morawetz G (1988) Epoxidharzformmassen für die Umhüllung elektronischer Bausteine. In: Kunststoff-Handbuch, vol 10. Carl Hanser, Munich Vienna, pp 361-367
79. Messing M (1995) Epoxidierte Novolake und zugehörige Novolakhärter zur Herstellung von Elektronikformmassen. Proceedings of SKZ Symposium, 26-27 April, Würzburg, pp 243-258
80. Morawetz G (1988) Epoxidharzformmassen für allgemeine Anwendung. In: Kunststoff-Handbuch, vol 10. Carl Hanser, Munich Vienna, pp 338-360

References to Sect. 6.1.4.1-6.1.4.7

1. Becker GW, Braun D (1988) Duroplaste. In: Kunststoff Handbuch, vol 10. Carl Hanser, Munich Vienna
2. Knop A, Pilato LA (1985) Phenolic resins, chemistry, application and performance. Springer, Berlin Heidelberg New York, pp 230 ff
3. Zehrfeld J (1988) Kern- und Faserverbundwerkstoffe. In: Kunststoff Handbuch, vol 10. Carl Hanser, Munich Vienna, pp 490-508
4. Bakelite AG (1996) Resin systems for fibre composites. Company Publication, Bakelite AG, Duisburg, Germany, June

5. Böttcher A (1995) Phenolharze für Faserverbundwerkstoffe, Einsatz im Transportsektor. In: Proceedings of SKZ Duroplastische Werkstoffe aktuell Symposium, 26–27 April, Würzburg
6. Stesalit AG (1995) StesaPreg und StesaTape zur Herstellung von Hochleistungsfaserverbundwerkstoffen. Company Publication, Stesalit AG, Zullwill, Schwitzerland
7. Böttcher A (1995) Zähelastisch und Flammfest. Kunststoffe, 8:1142–1144
8. Pilato LA, Michno MJ (1994) Advanced composite materials. Springer, Berlin Heidelberg New York
9. Herrmann G (1982) Handbuch der Leiterplattentechnik, 2nd edn. Eugen G. Lenze, Saulgau
10. Borchard K (1988) Duroplastische Tafeln, Rohre und Profile. In: Kunststoff Handbuch, vol 10. Carl Hanser, Munich Vienna, pp 385–450
11. Bakelite AG (1996) Resins and curing agents for manufacture of printed circuit boards. Company Publication, Bakelite AG, Duisburg, Germany, June
12. Bakelite AG, October 1993.Rutaphen resins for impregnation. In: Rutaphen phenolic resins guide, product range, application, pp 50–63
13. Bakelite AG (1992) Rutaphen phenolic resins, bonding agents for paper laminates. Bakelite AG, Iserlohn, Germany, Sept
14. Isola AG (1995) Schichtpreßstoffe. Technical Information Brochure of Isola AG, Düren, Germany, Oct
15. CompaPublication (1989) Beschichtungs- und Imprägnier-Anlagen für Papier-Basismaterialien. CompaPublication, Caratsch AG, Brenngarten, Switzerland
16. Deutsche Babcock-VITS-Maschinenbau (1988) Impregnating plants. Company Publication, Deutsche Babcock-VITS-Maschinenbau, Langenfeld, Germany
17. Deutsche Babcock-VITS-Maschinenbau (1988) Impregnating plants vertical. Company Publication, Deutsche Babcock-VITS-Maschinenbau, Langenfeld, Germany
18. G. Siempelkamp (1995) Press lines and finishing lines for the production of technical laminates. Technical Bulletin, G. Siempelkamp & Co
19. Ullmann F (1983) Enzyklopädie der technischen Chemie, 4th edn, vol 23, p 747 („Vulkanfiber"), Chemie, Weinheim
20. Isola AG (1997) Basismaterialien für gedruckte Schaltungen. Company Publication, Isola AG, Düren, Germany, Oct
21. Shell International (1991) EP 0 476 752 A1

References to Sect. 6.1.4.8

1. Plohnke K (1988) Technische Filtereinsätze. In: Kunststoff-Handbuch, vol 10, Duroplaste, Carl Hanser, Munich Vienna, pp 1069–1072
2. Purchas DB (1977) Solid/liquid separation equipment scale-up. Uplands Press, Croydon, UK
3. Warring RH (1981) Filters and filtration handbook. Trade and Technical Press, Morden, UK
4. Bakelite AG (1996) Phenolic resins as binders in filter engineering (German). Company Publication, Bakelite AG, Iserlohn Germany, June
5. Fibermark Gessner (1996) Newsletters News and Facts – Kfz-Filterpapier. Fibermark Gessner GmbH, Feldkirchen-Westerham, Germany
6. Fibermark Gessner (1996) Heavy duty automotive filter media for innovative filter technology (German). Fibermark Gessner GmbH, Feldkirchen-Westerham, Germany
7. Erdmannsdörfer H (1971) MTZ, Motortechnische Zeitschrift 32(4):123–131
8. Bakelite AG (1992) Industrial filter inserts. In: Rutaphen phenolic resins, guide product range application. Company Publication, Bakelite AG, Iserlohn, Germany, p 59
9. Degussa (1977) DE-C 2 756 973

References to Sect. 6.1.4.9

1. Bakelite AG (1992) Bakelite Resin B 428 [German]. Company Publication, Bakelite AG
2. Behrendt D (1976) Galvanische Elemente, Primär- und Sekundärelemente. In: Ullmann's Enzyklopädie der Technischen Chemie, vol 12, 4th edn. Urban und Schwarzenberg, Munich Berlin, pp 73–111
3. GRACE GmbH (1992) DARAK DARAMIC, Produktbeschreibung Separatoren. Company Publication, GRACE GmbH, Norderstedt 1, Germany
4. Müller R (1988) Batterieseparatoren. In: Kunststoffhandbuch, vol 10, Duroplaste. Carl Hanser, Munich Vienna, pp 1072–1075
5. Bakelite AG (1993) Battery separators. In: Rutaphen phenolic resins, guide, product range, application, p 60. Bakelite AG, Iserlohn, Germany
6. Menei L, Willkomm H (1996) Wabenkonstruktionen: Auch dreidimensional in Form. Schweizer Maschinenmarkt 19
7. Euro-Composites GmbH (1996) Euro-Composites – die Zukunft leicht gemacht. Company Publication, Euro-Composites GmbH, Echternach, Luxembourg
8. Borg Warner Automotive (1993) Reibelemente. Company Publication, Borg Warner Automotive, Heidelberg, Germany
9. Lanzerath G, Patzer H (1986) Sychronizer blocker ring with organic lining. SAE Technical Paper Series, International Congress and Exposition, 24–28 February, Detroit, MI, USA
10. Bakelite AG (1993) Angelruten-Prepregs. In: Rutaphen phenolic resins, guide, product range, application. Company Publication, Bakelite AG, Iserlohn, Germany, October, p 62
11. Rosato DV, Grove CS Jr (1964) Filament winding. Interscience, New York

References to Sect. 6.1.5

1. Pilato LA, Michno MJ (1994) Advanced composite materials. Springer, Berlin Heidelberg New York
2. Barnes FJ (1996) SAMPE J 32(2):12
3. Composites Technology (1996) Jan/Feb p 48
4. Böttcher A, Pilato LA (1996) Proc Int'l SAMPE Tech Conf 28:1353
5. Folker JL, Friedrich RS (1998) Proc Int'l Composites Espo '97, 27 January 1997, Nashville, TN, session 22 A; also Chem Abs 128:24,052
6. Chem Abs (1997) 127:66360
7. Qureshi S(1997) Georgia Pacific (unpublished results)
8. Böttcher A, Pilato LA (1997) Int'l SAMPE Sym & Exh 42:336
9. Bakelite AG (1996) (unpublished results)
10. Composites Technology (1996) Jan/Feb, p 31
11. Brown GL Jr, Creech B (1997) Proc Int'l Composites Expo '97 27 Jan, Nashville, TN session 4-C
12. Technical information related to liquid resins for prepegs, ballistics, filament winding, pultrusion (1998) Bakelite AG Duisburg, Germany
13. Schmidt DL, Davidson KE, Theibert LS (1996) SAMPE J 32(4):44
14. Hexcel Honeycomb Literature (1996) Dublin, California
15. Jellinek K, Meier B, Zehrfeld J (1987) Bakelite Patent EP 0242512
16. Peters ST, Humphry WD, Foral RF (1991) Filament winding composite structure fabrication. SAMPE, Covina, CA
17. Knop A, Pilato LA (1985) Phenolic resins. Springer, Berlin Heidelberg New York
18. Composites Technology (1967) Nov/Dec, p 16
19. Goldsworthy WB (1995) Plast Tech, Mar, p 36
20. Böttcher A, Pilato LA, Klett MW (1997) Int'l SAMPE Tech Conf 29:635
21. Tingley DA (1994) U.S. Patent 5,362,545; (1996) 5,547,729; (1997) 5,641,553; (1997) 5,648,138

22. Dagher H, Shaler S, Abdel-Magdid B, Landis E (1998) Int'l Composites Expo '98, 18-21 January, Nashville, TN session 22D
23. Plastics Technology (1993) July, p 30
24. High-Performance Composites (1997) 5(1):23
25. Clark NJ (1994) BP literature
26. U.S. Patent 5,378,793 (1994)
27. Orpin MR (1993) Cruise & Ferry '93, London

References to Sect. 6.1.6.1

1. Subat G (1988) Chemischer Apparatebau und chemikalienbeständige Kitte. In: Handbuch der Kunststoffe, vol 10, Duropplaste. Carl Hanser, Munich Vienna, pp 1083-1088
2. Fries A (1997) Chemikalienbeständige Beschichtungen auf Basis duroplastischer Harze, SKZ Symposium, Würzburg, 4-5 June, pp XII 1-17
3. Bakelite AG (1993) Rutaphen resins for acid-resistant constructions. In: Rutaphen phenolic resins guide product range application. Bakelite AG, Iserlohn, Germany, Oct, pp 123-124
4. Bakelite AG (1992) Rutaphen resins for acid-resistant constructions. Technical Information Brochure, Bakelite AG, Iserlohn, Germany, Sept
5. Bakelite AG (1997) Binder systems for the building industry. Bakelite AG, Iserlohn, Germany, Jan
6. Bureick G (1984) Anwendungsbeispiele auf dem Gebiet der GFK-Konstruktionen. Talk held at the Haus der Technik, Oct, Essen, Germany
7. Rotluff H (1980) Auskleidungen mit Bahnen aus Graphit-Phenolharz in Oberflächenschutz mit organischen Werkstoffen im Behälter-, Apparate- und Rohrleitungsbau. In: Series Ingenieurwissen. VDI, Düsseldorf
8. Falcke FK (1985) Handbook of acid proof construction. Chemie, Weinheim
9. Gibbesch B, Schedlitz D (1996) Fiber-Reinforced Phenol Resins for Plant Construction, Kunststoffe 84 (6) 773

References to Sect. 6.1.6.2

1. Bakelite AG (1992) Rutaphen resins for the productioin of socket putties [German] Company Publication, Bakelite AG, Iserlohn
2. Kwasniok A (1988) Lampbase cements. In: Plastic-handbook [German], vol 10, Thermosets. Carl Hanser, Munich, pp 1092-1094
3. Bakelite AG (1993) Socket putties. In: Rutaphen phenolic resins; guide, product range, application. Company Publication, Bakelite AG, Iserlohn, Germany, Oct, pp 125-126
4. Harris GJ, Edwards AG, Coxon G (1976) Chemica Petrochemica VI(7-8):403

References to Sect. 6.1.6.3

1. Pallapies E (1988) Pinsel- und Bürstenharze (Resins for artists' brushes and paintbrushes), Kunststoff-Handbuch, vol 10, Duroplasts (Thermosetting resins). Carl Hanser, Munich, pp 1095-1097
2. Bock E (1983) Pinsel und Bürsten. Selbstverlag der Bürsten- und Pinselindustrie
3. Koch O (1988) PF-Harze für Edelkunstharze, Modell- und Werkzeugharze, Kunststoff-Handbuch, vol 10, Duroplaste. Carl Hanser, Munich, pp 540-547
4. Edelkunstharze (1989) Company Publication, Raschig AG, Ludwigshafen, Germany
5. Sandler SR, Karo W (1977) Polymer syntheses, vol II. Academic Press, New York p 57
6. Harris TG, Neville HA (1953) J Polym Sci 4:673

References to Sect 6.1.6.4

1. Nippon Kynol (1996) Kynol Novoloid fibres. Company Publication, Nippon Kynol, Inc., Osaka, Japan
2. U.S. Pat. 3,650,102 (1972)
3. U.S. Pat. 3,723,588 (1973)
4. Hayes JS Jr (1996) Novoloid fibres. In: Kirk-Othmer (eds) Encyclopedia of chemical technology, vol 16, 3rd edn, pp 125–138
5. Economy J, Wohrer L (1971) Phenolic fibres. In: Bikales NM (ed) Encyclopedia of polymer science and technology, vol 15. Wiley-Interscience, New York, pp 370–373
6. Economy J, Wohrer LC, Frechette FJ, Lei GY (1973) Appl Polym Symp 21:81
7. Economy J (1978) Phenolic fibres. In: Lewin M, Atlas SM, Pearce EM (eds) Flame-retardant polymer materials, vol 2. Plenum Press, New York, pp 210–219
8. Conley RT, Quinn DF (1975) Retardation of combustion of phenolic, urea-formaldehyde, epoxy, and related resin systems. In: Bikales NM (ed) Encyclopedia of polymer science and technology, vol 1. Wiley-Interscience, New York, pp 339–344
9. Dynatech R/D Company (1976) Report to American Kynol, 26 April
10. Batha HD, Hazelet GJ (1978) U.S. Pat 4,076,692, assigned to American Kynol, Inc., 28 February
11. Industrial Health Foundation (1972) Report to the Carborundum Company, Sept 1972

References to Sect. 6.2 and 6.2.1

1. Bindernagel E (1983) Molding sands and molding processes in foundry engineering [German]. Giesserei-Verlag, Düsseldorf, p 2
2. Weiss R (1984) Raw materials for molds, their occurrence, properties, testing and opportunities for use [German]. VDG, no 11. Giesserei-Verlag, Düsseldorf
3. Weiss R (1982) Silicon dioxide [German]. In: Ohmann's Enzyklopädie der Technischen Chemie, vol 21. Chemie, Weinheim
4. Cobos L, Gardziella A, Kwasniok A (1995) Coremaking process comparisons – recent studies [German]. Giesserei-Erfahrungsaustausch 6:219–227
5. Bakelite AG (1993) Rütaphen phenolic resins guide, product range, applications. Bakelite AG, Gennaer Strasse 2–4, Iserlohn (Germany), Oct, pp 65ff
6. Gardziella A, Kwasniok A (1984) The SO_2 coremaking process: current technology, provisional appraisal and continued development [German]. Giesserei 10:393–403
7. Gardziella A (1988) Foundry auxiliaries based on reactive resins [German]. In: Becker/Braun (eds) Kunststoffhandbuch, vol 10 Duroplaste (sects 12.2.1–12.3.6). Carl Hanser, Munich – Vienna, pp 931–984
8. Robert SL (1971) Practice of the shell molding process [German]. Giesserei, Düsseldorf
9. Berndt H (1981) Molding sands and technology of the shell molding process [German]. VDG-Taschenbuch No 9, Giesserei, Düsseldorf
10. Jasson MP (1961) Core production in hot boxes at the Renault Works. British Foundryman 54(7):309–319
11. Gardziella A (1981) The hot box process [German]. In: Handbuch der Fertigungstechnik, vol 1, Urform. Carl Hanser, Munich Vienna, pp 358–372
12. Gardziella A (1988) Phenolic, furan and amino resins for the hot box process [German]. In: Kunststoffhandbuch, vol 10, Duroplaste. Carl Hanser, Munich Vienna, pp 950–960
13. Rütgers AG (1982) The hot box plus process. EP 03 16 517
14. Gardziella A (1988) Foundry auxiliaries based on reactive resins [German]. In: Becker/Braun (eds) Kunststoffhandbuch, vol 10 Duroplaste. Carl Hanser, Munich Vienna, pp 958–960
15. Quaker Oats Chemicals (1981) The warm box process. US PS 43 83 098
16. IQU (1981) The vacuum warm box process. EP 00 44 739 A1
17. Bakelite AG (1993) The no bake process. In: Bakelite AG Company Publication – Rütaphen phenolic resins guide, product range, applications, Oct, pp 72–76

18. Borden Ltd. (1983) EP 00 85 512
19. Kögler H, Scholich K (1978) Experience with an anhydrous, rapidly curing binder system for mold and core sands [German]. Giesserei 65(5):101–105
20. Ashland Inc (1967) DBP 1 583 521
21. Rütgers AG (1989) EP 0 362 486
22. Boenisch D (1983) DE-OS 33 42 225 A1
23. Borden Ltd. (1983) EP 00 86 615
24. Foseco International (1987) The Ecolotec process, EP 03 23 096 B1
25. Hüttenes Albertus (1986) The red set process, EP 02 90 551 B1
26. Gardziella A, Kwasniok A (1988) The gas curing epoxy process established in practice. Giesserei 13:401–409
27. SAPIC (1972) DE-PS 22 39 835
28. Ashland Inc (1983) EP 01 02 208
29. Krapohl HP (1991) Gas curing foundry sand cores – a comparison of properties from coremaking to sand regeneration [German]. Dissertation at the Foundry Institute of the RWTH Aachen
30. Boenisch D (1991) Recycling of used core materials, pt 1. Cold regeneration of resin-bonded old sands [German]. Giesserei 78(21):733–744
31. Boenisch D (1992) Recycling of used core materials, pt 2. Core waste – a versatile material [German]. Giesserei 79(11):428–435
32. Rütgers AG (1992) Modification of furan resins with lignin, US PS 52 88 774
33. K. Kurple (1989) PCT-WO 89/07 497

References to Sect. 6.2.2

1. Bakelite AG (1997) Resins for abrasive producer. Company Publication, Bakelite AG, Iserlohn, Germany, Feb
2. Verlag Chemie (1981) Schleifen und Schleifmittel. In: Enzyklopädie der technischen Chemie, 4th edn,, vol 20, pp 449–455. Verlag Chemie, Weinheim
3. Selgrad V (1990) Grobschleifen mit phenolharzgebundenen Schleifscheiben. Talk at Phenolharze – vielfältig in der Anwendung, Technical Symposium of the Süddeutsches Kunststoffzentrum, Würzburg in Erlangen, May 16–17
4. Verlag Moderne Industrie (1980) Schleifen. In: Enzyklopädie Naturwissenschaft und Technik. Verlag Moderne Industrie, Landsberg am Lech, pp 3813–3824
5. Spur G, Stöferle Th (1980) Handbuch der Fertigungstechnik, vol 3, Spanen. Carl Hanser Verlag, Munich/Vienna
6. Gardziella A (1988) Schleifmittel und Reibbeläge mit duroplastischen Bindemitteln. In: Becker/Braun (eds) Kunststoff-Handbuch, vol 10, Duroplaste. Carl Hanser Verlag, Munich/Vienna, pp 891–893
7. Stade G (1962) Technologie des Schleifens. Carl Hanser Verlag, Munich, p 24
8. Riege W (1981) Schleifen und Trennschleifen in der Gussputzerei. VDG Taschenbuch, no 8. Giesserei-Verlag GmbH, Düsseldorf, p 132
9. Bakelite AG (1993) Rutaphen phenolic resins for abrasives. In: Rutaphen phenolic resins; guide product range application. Company Publication of Bakelite AG, Oct, pp 95–101
10. Schwieger K-H, Coselli K (1988) Schleifscheiben und Schleifkörper. In: Kunststoff-Handbuch, vol 10, Duroplaste. Carl Hanser Verlag, Munich/Vienna, pp 894–908
11. Rütgerswerke AG (1993) EP-B 05 95 003; (1993) US-A 5,399,606
12. Rütgerswerke AG (1987) EP-B 02 48 980; (1989) US-A 4,918,116
13. Gardziella A, Müller R (1990) Phenolharze. Kunststoffe 10:1174
14. Bakelite AG (1992) Rutaphen resins for grinding wheels. Company Publication, Bakelite AG, Iserlohn, Germany
15. Duda H (1988) Glasgewebeeinlagen. In: Kunststoff-Handbuch, vol 10, Duroplaste. Carl Hanser Verlag, Munich/Vienna, pp 908–910

16. Bakelite AG (1992) Phenolic resins as binders for fiberglass mesh reinforcement of grinding wheels. Company Publication, Bakelite AG, Iserlohn, Germany
17. Bakelite AG (1992) DE-C 42 11 446
18. Rütgerswerke AG (1992) DE-C 42 11 445
19. Bakelite AG (1992) Rutaphen resins for coated abrasives. Company Publication, Bakelite AG, Iserlohn, Germany
20. Adolphs P, Dietrich J (1988) Bindemittel für Schleifmittel auf Unterlage. In: Kunststoff-Handbuch, vol 10, Duroplaste. Carl Hanser Verlag, Munich/Vienna, pp 910–920
21. C. Klingspor GmbH (1981) Bandschleifen, Theorie und Praxis. Technical Bulletin 12/1981
22. Hermes Schleifmittel (1998) Schleifmittel. Technical Bulletin, Hermes Schleifmittel GmbH & Co, Hamburg
23. Hermes Schleifmittel (1998) Schleifwerkzeuge. Company Publication, Hermes Schleifmittel GmbH & Co, Hamburg
24. Rütgerswerke AG (1984) EP-B 01 39 309; (1985) US-A 4,587,291

References to Sect. 6.2.3

1. Bakelite AG (1996) Harze für die Reibbelagsindustrie. Company Publication, June
2. Gardziella A, Schönthaler W (1981) Technische Phenolharze und härtbare Formmassen im Automobilbau. Kunststoffe 71(3):159–196
3. Adolphs P, Paul H-G (1988) Bindemittel für Reibbeläge. In: Becker/Braun (eds) Kunststoff-Handbuch, vol 10, Duroplaste. Carl Hanser, Munich Vienna, pp 921–929
4. Paul H-G (1990) Phenolharzgebundene, asbestfreie Reibbeläge. In: Phenolharze - vielfältig in der Anwendung, Tech Symposium of the Süddeutsches Kunststoffzentrum, Würzburg, May 16–17 (Reprint), pp 86–107
5. Eckart A, Goldbach D (1992) Phenolharzgebundene Hochleistungsbremsbeläge für den Rennsportbereich und andere Hochleistungsanwendungen. In: Proceedings of Technical Symposium Anwendungen von duroplastischen Kunststoffen im Automobilbau, Süddeutsches Kunststoffzentrum, Würzburg, Sept 23–24 (Reprint) pp 134–156
6. Conbomb CA (1985) Memoires de Mathematique et de Physique de l'Académie des Sciences 10:161–331
7. Uetz H, Wellinger K, Gürlegik M (1968) Gleitverschleißuntersuchungen an Metallen und nicht metallischen Werkstoffen. Wear 11:173–199
8. Kamper P (1984) Bremsen ohne Asbest. VDI-Nachrichten 48:19
9. Rütgers AG (1983) US-PS 4,373,038
10. Rütgers AG (1983) US-PS 4,384,054
11. Rütgers AG (1981) US-PS 4,273,699
12. Ho TL, Kennedy FE, Peterson MB (1979) Evaluation of aircraft brake materials. American Society of Lubrication Engineers 22:71
13. Bill RC (1978) Friction and wear of carbon-graphite materials for high energy brakes. American Society of Lubrication Engineers 21:268
14. Liu T, Rhee SK (1978) Wear 76:213
15. Newman LB (1978) Friction materials, recent advantages. Noyes Delta, Parkridge, New Jersey, USA
16. Schumann R (1981) Antriebstechnik 20(9):383
17. Gardziella A, Haub HG (1988) Phenolharze. In: Becker/Braun (eds) Kunststoff Handbuch, vol 10, Duroplaste. Carl Hanser Verlag, Munich – Vienna, pp 17–18
18. Rütgers Automotive AG (1995) Company Publication, Essen, Germany

References to Sect. 6.2.4

1. Reismüller H (1976) Erdöl und Erdgas, Gewinnung. In: Ullmann's Enzyklopädie der Technik, vol 11. Ausgabe, pp 19–39
2. Mayer-Gürr (1973) Erdöl- und Erdgasgewinnung in der Bundesrepublik Deutschland, ÖI-Zeitschrift für die Mineralölwirtschaft 1
3. Schmidt KH, Romey I (1981) Kohle Erdöl Erdgas, Chemie und Technik. Vogel-Verlag, Würzburg, pp 31 ff
4. Santrol Products (1983) US PS 4,527,627
5. Dow Chemical (1998) US PS 3,857,444
6. Exxon Production Research (1998) US PS 3,929,191
7. Acme Resins (1988) US PS 4,785,884
8. Acme Resins (1982) US PS 4,439,489
9. Exxon Production Research (1974) US PS 3,929,191
10. Univar(1981) US PS 4,413,931
11. Santrol Products (1985) US PS 4,518,039
12. Santrol Products (1985) US PS 4,553,596
13. Santrol Products (1989) US PS 4,888,240
14. Santrol Products (1986) US PS 4,585,064
15. Santrol Products (1986) US PS 4,597,991
16. Santrol Products (1988) US PS 4,717,594
17. Santrol Products (1988) US PS 4,732,920
18. Santrol Products(1995) US PS 5,422,183
19. Santrol Products (1997) US PS 6,597,784
20. Sinclair AR (1983) Improved well stimulation with resin-coated proppants. Paper presented at a Symposium of AIME, Oklahoma City, Feb 27
21. Smith CHR (1966) Mechanics of secondary oil recovery. Reinold, New York
22. Bleakley WB (1974) Oil and Gas Journal 72:69–78
23. Noran D (1975) Oil and Gas Journal 73:77–81

References to Sect. 6.3 – 6.3.1.7

1. Baekeland LH (1908) Heat and pressure. Pat DRP 233,803
2. Schwenk E (1983) 80 Jahre Kunstharze. Company Publication, Hoechst AG, Frankfurt/Main, Germany
3. Vianova (1956) DE-PS 113,775
4. Bakelite AG (1996) Binders for the lacquer industry. Company Publication, Bakelite AG, Iserlohn, Germany, June
5. Kwasniok A (1996) Phenolic resins for surface protection [German]. Kunststoffhandbuch, vol 10, Duroplaste. Carl Hanser Verlag, Munich – Vienna, pp 986–994
6. Kwasniok A, Schröter St (1996) Phenolharze. In: Kittel H Lehrbuch der Lacke und Beschichtungen (Textbook of coatings and Coating), vol II, 2nd edn. Wissenschaftliche Verlagsanstalt, Stuttgart
7. Bakelite AG (1993) Rütaphen phenolic resins, guide: product range, application. Company Publication, Bakelite AG, Iserlohn, Germany, Oct, pp 106–111
8. Vianova (1984) EP 0,122,424
9. Vianova (1984) EP 0,122,425
10. Reichhold-Albert-Chemie AG (1966) DE-PS 1,519,329
11. Vianova (1969) DE-AS 1,965,669
12. Read RT, Holt JC (1975) JOCCA 51:58
13. Marsal P (1978) Neue Verpackung (NEUVAU), 31
14. Goldschmidt AE (1984) In: Knappe's Glasurit-Handbuch, Lacke und Farben, 11th edn. Curt R. Vincentz Verlag, Hanover Germany
15. Fink H (1970) Plaste und Kautschuk, p 677
16. Immout Corp USA (1984) DE-OS 3,446,178

17. Code of Federal Regulations, Title 21, Foods and drugs; sect 175.300, Resinous and polymeric coatings, p 496
18. Menzel H (1988) Heavy-duty corrosion protection [German]. Kunststoffhandbuch, vol 10, Duroplaste. Carl Hanser Verlag, Munich Vienna, pp 1029–1035
19. Streitberger HJ (1988) Electrophoretic dip coats [German]. Kunststoffhandbuch, vol 10, Duroplaste. Carl Hanser Verlag, Munich Vienna, pp 1023–1029
20. Pouchol J (1983) 15th AFTV Congress, Cannes; (1984) 58th World Surface Coatings Abstracts 716, 503
21. Keneth K, Earhart A (1972) Coatings and varnish production H1:35
22. PPG Ind. (1962) FR-PS 13941 007
23. Vianova (1968) DE-OS 1,929,523
24. Seyfert KH (1988) Powder coatings [German], Kunststoffhandbuch, vol 10, Duroplaste. Carl Hanser Verlag, Munich Vienna, pp 1047–1058
25. PPG Ind. (1975) US-PS 3,883,483

References to Sect. 6.3.1.8

1. Dammel RR (1992) Diazonaphthoquinones-based resists. SPIE tutorial text series, vol 11. Bellingham, WA
2. Wallroff GM, Allen RD, Hinsberg WD, Simpson LL, Kunz RR (1993) Chem Tech April, p 22
3. Reichmanis E, Novembre AE (1993) Annu Rev Mater Sci 23:11
4. Hanabata M, Uetani Y, Furuta A (1988) Proc. SPIE, 920, 349; Hanabata M, Furuta A (1991) J Vac Sci Technol A9(2):254
5. Bogan LE Jr (1991) Macromolecules 24(17):4807
6. Hanabata M (1994) Adv Mat Optics & Elect 4:73
7. Zampini A, Templeton MK, Szmanda CR (1987) Proc SPIE 771:136
8. Allen RD, Chen KJR, Gallagher-Wetmore PM (1995) SPIE 2438:250
9. Tsiartos PC, Simpson LL, Qin A, Willson CG, Allen RD, Krukonis VJ, Gallagher-Wetmore PM (1995) SPIE 2438:261
10. Shih H-Y, Reiser A (1996) Macromolecules 29:2082
11. Tsiartas PC, Flanagan LW, Henderson CL, Hinsberg WD, Sanchez IC, Bonnecaze RT, Willson CG (1997) Macromolecules 30:4656
12. Bogan LE Jr, Wolk SK (1992) Macromolecules 25:161
13. Bogan LE Jr (1992) Macromolecules 25:196
14. Bogan LE Jr (1993) Proc SPIE 1925:564
15. Hanabata M, Furuta A, Uemura Y (1986) Proc SPIE 631:76
16. Zampini A, Fischer RL, Wickman JB (1989) Proc SPIE 1086:85
17. Zampini A, Turci P, Cernigliaro GJ, Sandford HF, Swanson GJ, Meister CC, Sinta R (1990) Proc SPIE 1262:501
18. Jeffries A, Honda K, Blakeney AJ, Tadros S U.S. 5,302,688
19. Casiraghi A, Cornia M, Ricci G, Balduzzi G, Casnati G, Andreeti GD (1981) Makromol Chem 182(11):2973
20. Sizensky JJ, Sarubbi TR, Toukhy MA (1995) U.S. 5,413,894
21. Honda K (1994) U.S. 5,340,687
22. Frechet JMJ (1992) Pure & Appl Chem 64(9):1239
23. Reichmanis E, Houlihan FM, Nalamasu O, Neenan TX (1994) Adv Mat Optics & Elect 4:83
24. Yoshida M, Frechet JMJ (1994 Polymer 35(1):5
25. MacDonald SA, Willson CG, Frechet JMJ (1994) Accounts of Chem Res 27(6):151
26. Lee SM, Frechet JMJ, Willson CG (1994) Macromolecules 27:5154
27. Lee SM, Frechet JMJ (1994) Macromolecules 27:5160
28. Chem Abs (1997) 127, 33,973p; 33,974g; 33,975r; 33,976a; 33,977t; 33,979v; 33,980p; 33,981q; 33,982r; 33,984t; 33,985u
29. Ueda M, Takahishi D, Nakayama T, Haba O (1997) PME Preprints ACS Las Vegas Fall 1997 Meeting, p455
30. Allen RD (1997) Semiconductor International, Sept, p 73

References to Sect. 6.3.2

1. Kwasniok A ()1988 Kautschuk-Additivharze. In: Kunststoffhandbuch, vol 10, Duroplaste. Carl Hanser, Munich Vienna, pp 1075–1080
2. Bakelite AG (1993) Rutaphen resins for the rubber industry. In: Rutaphen phenolic resins guide, product range, application. Company Publication, Bakelite AG, Iserlohn, Germany, Oct, pp 119–121
3. Sattelmeyer R (1990) Phenolharzanwendungen in der Kautschukindustrie. Talk at Phenolharze – vielfältig in der Anwendung, Technical Symposium of the Süddeutsches Kunststoffzentrum, Würzburg (SKZ) in Erlangen, May
4. Giller A (1971) Gummi-Asbest-Kunststoffe 24:405
5. Giller A (1976) Gummi-Asbest-Kunststoffe 29:766
6. Giller A (1971) Über die Wirkung von Phenolharzen in Kautschuk. GAK 5:405–409, 450
7. Kempermann Th (1973) Trends im Einsatz von Kautschuk-Chemikalien in der Gummi-Industrie, Kautschuk + Gummi, Kunststoffe 4:234–246
8. Fries H (1985) Entwicklung des Einsatzes von Harzen in der Gummi-Industrie, GAK 9:454–460
9. Giller A (1961) Kautschuk + Gummi 14, WT 201
10. Giller A (1966) Kautschuk + Gummi, Kunststoffe 19:188
11. Christov D, Nenov N, Busowa Z (1977) Angewandte MakromolChem 65:49, 63
12. Giller A (1966) Kautschuk + Gummi, Kunststoffe 19:188
13. Lattimer RP, Kinsey RA, Layer RW, Rhee CK (1989) Rubber Chem Technol 62:107
14. Skewis JD (1983) Am Chem Soc Rubber Division, 124th Meeting, Oct
15. Rhee CK, Andries JC (1982) Gummi-Asbest-Kunststoffe 35:185
16. Esch E, Fries H, Dahl H, Kempermann Th (1977) Kautschuk + Gummi, Kunststoffe 30:524
17. Bayer AG (1978) EP 0,001,580A1
18. Kwasniok A, Schröter St (1995) Phenolharze. In: Klebharze. R Hinterwaldner, Munich, pp 130–131
19. Bayer AG (1977) DE 2,746,138
20. Hoechst AG (1975) DE 2,537,656
21. Schenectady Chemicals, Inc. (1975) DE 2,537,718
22. Giller A (1969) Reichholdt-Albert-Nachrichten 2:18
23. Leicht E, Sattelmeyer R (1987) Kautschuk + Gummi, Kunststoffe 40:126
24. Fries H, Esch E, Kempermann Th (1979) Kautschuk + Gummi, Kunststoffe 32:860
25. Nieberle J, Paulus G, Queins H, Schöppl H (1989) Kautschuk + Gummi, Kunststoffe 39:108
26. Borowith J, Kosfeld R (1981) Angewandte Makromol Chem 100:23
27. Hoechst AG (1986) DE 3,604,744
28. Degussa AG (1970) Technical Reports of Degussa AG, Haftverfahren, Publication G-17-6805
29. Bayer AG (1968) Cohedur, Company Publication, Bayer AG

References to Sect. 6.3.3

1. Kwasniok A, Schröter St (1994) Phenolharze. In: Klebharze. R Hinterwaldner Verlag, Munich, pp 117–138
2. Bakelite AG (1997) Binders for the adhesive industry. Bakelite AG, Iserlohn, Germany, Mar
3. Bakelite AG (1993) Rutaphen resins for the adhesive industry. In: Rutaphen phenolic resins: guide, product range, application. Bakelite AG, Iserlohn, Germany, pp 114–118
4. Kwasniok A (1988) Klebstoff-Additivharze. In: Kunststoffhandbuch, vol 10, Duroplaste. Carl Hanser Verlag, Munich Vienna , pp 1080–1083
5. Verlag Chemie (1978) Ullmann's Enzyclopädie der technischen Chemie, 4th edn, vol 14. Verlag Chemie, Weinheim, pp 227–268
6. Knop A, Pilato LA (1985) Phenolic resins in rubbers and adhesives. In: Phenolic resins. Springer, Berlin Heidelberg New York, pp 288–298

7. Hultzsch K (1971) Kunstharze in Elastomer-Klebmassen. Farbe und Lack 12:1165–1172
8. Schunck E (1969) Der Einfluß von Kunstharzen auf die Eigenschaften Polychloropren-Klebstoffen. Reichhold-Albert-Nachrichten 2:34–40
9. Schunck E (1979) Die Wirkung von Kunstharzen auf Polychloropren-Typen mit verschiedener Kristallisationsneigung in Klebstoffmischungen. Adhäsion 3:33–36
10. Landau M, Müller H, Rohleder U (1980) Polychloroprenklebstoffe. Adhäsion 3:64–70
11. Martin RA (1966) An analysis of prereaction, zinc oxide and magnesium oxide on neoprene phenolic adhesives. Adhesives Age, May, pp 28–30
12. Hitachi Chemical Co (1967) DE-OS 1 719 145
13. Garrett RR, Lawrence RD (1966) Beeinflussung der Phasenbildung in Neoprene-Lösungsmittelklebern durch bestimmte Faktoren. Adhäsion 7/8:296–299
14. Dollhausen M (1982) 25 Jahre Baypren – 25 Jahre Fortschritt auf dem Polychloropren-Klebstoffen-Gebiet. Adhäsion 10:23–27
15. Zwetkoff P, Eltimova R (1974) Die Vernetzung von Zweikomponenten-Polychloropren-Klebstoffen. Adhäsion 6:166–168
16. Bakelite AG (1977) Binders for the adhesive industry. Company Publication, Bakelite AG, Iserlohn, Germany, Mar
17. Bayer AG (1972) Baypren. Publication KA 30020, Bayer AG, Leverkusen, Germany
18. Carr AF (1973) Adhäsion 11:334
19. Carr AF (1973) Klebstoffe aus Nitrilkautschuk. Coating 11:334–338
20. Straschill M (1973) Klebstoffe aus künstlichem Gummi. Seifen-Öle-Fette-Wachse 23:665–669
21. Seidel H (1990) Ausgewählte Anwendungen von Phenolharzen in der Klebstoffindustrie Talk at Phenolharze – vielfältig in der Anwendung, Technical Symposium of the Süddeutsches Kunststoffzentrum, Würzburg in Erlangen, May 16–17, 1990, Symposium Reprints, pp 149–160
22. Luck H (1971) Adhäsion 15(7):235
23. Holz E (1968) Technische Phenolharze. In: Vieweg R, Becker E (eds) Kunststoffhandbuch, vol 10, Duroplaste. Carl Hanser Verlag, Munich, p 79

References to Sect. 6.4 and 6.4.1

1. Gardziella A, Suren J, Belsue M (1992) Carbon from phenolic resins: carbon yield and volatile components. Proceedings of 5th International Carbon Conference (Carbon '92), Essen, June 22–26, pp 69–71
2. Schwieger KH (1988) Carbonisierung von Phenolharzen. In: Kunststoff-Handbuch, vol 10, Duroplaste. Carl Hanser, Munich Vienna New York, pp 1096–1099
3. Gardziella A, Schwieger KH (1988) Curable synthetic resins for the manufacture of refractory products [German]. Fachberichte Hüttenpraxis Metallverarbeitung 6:566
4. Gardziella A (1986) Phenolic resins, general economic account and application possibilities, especially as binders with high carbon yield. 4th International Carbon Symposium (Carbon '86), Baden-Baden, June 30 – July 4
5. Gardziella A, Suren J, Wandschneider P (1989) Impregnation with phenolic resins and possibilities of improving the quality of ceramically bonded refractory products. Aachen Proceedings, Oct 12–13, pp 57–63
6. Gardziella A, Suren J (1977) Phenolic resins as impregnating agents for refractories – present state of development. Proceedings of the Unified International Technical Conference on Refractories (UNITECR), vol II, New Orleans, LA (USA), Nov, pp 975–998
7. Lausevic Z, Marinkovic S (1986) Mechanical properties and chemistry of carbonisation of phenol formaldehyde resins. Carbon 24(5):575–580
8. Jenkins GM, Kawamura K (1976) Polymeric carbons-carbon fibre, glass and char. Cambridge University Press, Cambridge, UK

9. Gardziella A, Solozabal R, Suren J, Wiesche V in der (1991) Synthetic resins as carbon forming agents for various refractories (carbon yields, analytic methods, structures and emissions). Preprints, UNITECR Congress, Aachen, pp 260–264
10. Gardziella A, Suren J (1988) Selected criteria for phenolic resins as binders and carbon-forming agents in refractory products. Fachberichte Hüttenpraxis und Metallverarbeitung, issue 4. Sprechersaal-Verlag, Coburg, pp 308–310
11. Mang D, Boehm HP (1992) Inhibiting effect of incorporated nitrogen on the oxidation of microcrystalline carbons. Carbon 30(3)
12. Jones LE, Thrower PA (1991) Influence of boron on carbon fiber microstructure, physical properties and oxidation behaviour. Carbon 29(2):251–269
13. McKee DW, Spiro CL, Lamby EJ (1984) The inhibition of graphite oxidation by phosphorus additives. Carbon 22(3):285–290

References to Sect. 6.4.2

1. Koltermann M (1979) Company Brochure, Radex-Rundschau, Issue 4
2. Koltermann M (1982) Stahl und Eisen 102(5):197–199
3. Routschka G (1996) Taschenbuch Feuerfestwerkstoffe. Vulkan-Verlag, Essen
4. Gardziella A, Schwieger K-H (1984) Härtbare Kunstharze zur Herstellung von feuerfesten Erzeugnissen – Möglichkeiten und Eigenschaften, Fachberichte Hüttenpraxis Metallweiterverarbeitung 6. Sprechsaal Verlag, Coburg
5. Majdic A (1975) Überblick über die feuerfesten Baustoffe, Schriften der Gesellschaft deutscher Metallhütten- und Bergleute, Issue 29
6. Didier AG (1990) Company Brochure, Feuerfeste Werkstoffe und ihre Merkmale (Refractory materials and their properties), May
7. Gardziella A (1986) Phenolic resins, general economic account and application possibilities, especially as binders with high carbon yield. 4th International Carbon Symposium, Baden-Baden, June 30 – July 4
8. Gardziella A, Solozubal R, Suren J, in der Wiesche V (1991) Synthetic Resins as Carbon Forming Agents for Various Refractories, Proceedings of UNITECR'91, Unified International Technical Conferences on Refractories, Aachen, Germany October 1991, pp 260–4
9. Schwieger K-H (1988) Refractories. In: Plastics Handbook, vol 10, Thermosetting plastics. Carl Hanser Verlag, Munich Vienna New York
10. Gardziella A, Suren J (1993) Kunstharzgebundene Feuerfesterzeugnisse -weltweite Erfahrungen im Stahlwerk. Stahl und Eisen 6:75–78
11. Suren J (1990) Refractories, a modern application of phenolic resins used as binding and impregnating products. Paper at Phenolic Resins, Manifold in Application, SKZ Symposium, Erlangen, May 16–17
12. Müller FJ (1993) Umweltverträglichkeit. Stahl und Eisen 6:77
13. Gardziella A, Müller F-J (1990) Umweltrelevanz und Anwendungsfunktionen der Phenolharze. Kunststoffe 4:510–514
14. Bakelite AG (1993) Rütaphen phenolic resins; guide, product range, application. Company Brochure, Bakelite AG, Iserlohn, Oct, pp 92–94
15. Bakelite AG (1992) Bakelite Harze für feuerfeste Erzeugnisse. Company Brochure, Bakelite AG, Iserlohn, Sept
16. Rütgerswerke AG (1977) DE-C 2 723 792
17. Rütgerswerke AG (1986) DE-A 3 637 720
18. Rütgerswerke AG (1987) EP-A 0 269 788
19. Gardziella A, Suren J, Wandschneider P (1989) Impregnation with phenolic resins and possibilities of improving the quality of ceramically bonded refractory products. Aachen Proceedings 11: special issue of Interceram, 32nd International Colloquium on Refractories, Oct 12–13. Verlag Schmid Publication, Freiburg
20. Gardziella A, Suren J (1997) Phenolic resins as impregnating agents for refractories – present state of development. Proceedings of UNITECR, Unified International Technical Conference on Refractories, New Orleans LA, USA, Nov, pp 975–998

References to Sect. 6.4.3

1. Härtel G, Künzel J, Hardörfer F (1995) Füllkörper aus Kohlenstoff. Chemie Technik 9 44 ff
2. Wege E (1988) Kohlenstoff-und Graphitwerkstoffe. In: Kunststoff-Handbuch, vol 10, Duroplaste. Carl Hanser, Munich
3. Verlag Chemie (1977) Werkstoffe aus Kohlenstoff und Graphit. In: Ullmanns Enzyklopädie der technischen Chemie, 4th edn, vol 14. Verlag Chemie, Weinheim New York, pp 600 ff
4. Knop A, Pilato LA (1985) Phenolic resins. Springer, Berlin Heidelberg New York, pp 156–160
5. Gardziella A (1986) Phenolic resins, general economic account and application possibilities, especially as binders with high carbon yield. Carbon 86, 4th International Carbon Symposium, Baden-Baden, June 30 – July 4
6. Gardziella A, Suren J (1988) Selection criteria for phenolic resins as binders and carbon-forming agents in refractory products. Sprechsaal Verlag, Coburg. Fachberichte Hüttenpraxis Metallweiterverarbeitung 4:308–310
7. Bakelite AG (1992) Rutaphen resins for carbon and graphite materials [German]. Company Publication, Bakelite AG, Iserlohn, Germany, Sept
8. Gardziella A, Suren J, Belsue M (1992) Carbon from phenolic resins: carbon yield and volatile components – recent studies. Carbon '92, 5th International Carbon Symposium, Essen, June 22–26, pp 69–71
9. Vohler O, Wege E (1983) Formprodukte aus Kohlenstoff. In: Winnacker/Küchler (eds) Chemische Technologie, vol 3. Carl Hanser, Munich, p 278
10. Gardziella A (1988) Kohlenstoff aus Phenolharzen-Grundlagen und Anwendungsmöglichkeiten. Conference of the Carbon Work Group of the German Ceramics Society (Deutschen Keramischen Gesellschaft), Frankfurt/Main, April 15
11. Union Carbide (1964) US-PS 3 375 132
12. Mitsubishi (1975) DE-PS 2 527 923
13. Rütgers AG (1987) EP 0 248 980
14. Neumann U (1990) CFC-Werkstoffe, hergestellt unter Verwendung von Phenolharzen. Phenolharze – vielfältig in der Anwendung, Technical Symposium of the Süddeutsches Kunststoffzentrum, Würzburg, May 16–17, pp 182–206
15. Schunk & Ebe (1977) DE-AS 2 732 553
16. Nikku Industrie (1976) US-PS 4 103 046
17. SGL Carbon Group (1995) Apparate und Anlagen aus Diabon®. Company Publication, SGL Carbon Group, Meitingen, Germany, May
18. SGL Carbon Group (1995) Kolonnen und Kolonnenbauteile aus Diabon. Company Publication, SGL Carbon Group, Meitingen, Germany, April
19. SGL Carbon Group (1995) Diabon F Plattenwärmeaustauscher. Company Publication, SGL Carbon Group, Meitingen, Germany, Sept
20. SGL Carbon Group (1995) Diabon Chlorwasserstoff Syntheseanlagen. Company Publication, SGL Carbon Group, Meitingen, Germany, June
21. SGL Carbon Group (1995) Diabon Kammerwärmeaustauscher Baureihen KD92, KFD 93. Company Publication, SGL Carbon Group, Meitingen, Germany, March
22. Härtel G, Künzel J, Bernt D (1995) Optimale Auslegung von Quenchen. Chemie Technik 2:22 ff
23. SGL Carbon Group (1995) Diabon Rohrbündelwärmeaustauscher. Company Publication, SGL Carbon Group, Meitingen, Germany, March
24. Bakelite AG (1988–1992) Unpublished results of Bakelite AG, Iserlohn, Germany

References to Sect. 6.4.4

1. Wege E (1988) Glaskohlenstoff (Polymerkohlenstoff). In: Kunststoffhandbuch, vol 10, Duroplaste. Carl Hanser Verlag, Munich, pp 1090–1092
2. Dübgen R (1990) Glaskohlenstoff – vom Duroplast zum keramischen Hochleistungswerkstoff. Phenolharze – vielfältig in der Anwendung, Technical Symposium of the Süddeutsches Kunststoffzentrum, Würzburg in Erlangen, May, pp 209–225

3. Bakelite AG (1993) Rutaphen resins for carbon and graphite materials In: Rutaphen phenolic resins – guide, product range, application. Bakelite AG, Iserlohn, Germany, Oct, pp 92–94
4. Bakelite AG (1992) Rutaphen resins for carbon and graphite materials. Company Publication, Bakelite AG, Iserlohn,Germany
5. Verlag Chemie (1977) Ullmann's encyclopedia of industrial chemistry [German], 4th edn, vol 14. Verlag Chemie, Weinheim New York, pp 600ff
6. General Electric (1959) GB-PS 921,236
7. Plessey (1960) GB-PS 956,452
8. Jenkins GM, Kawamura K (1976) Polymeric carbons. Cambridge University Press, Cambridge, UK
9. Siemens AG (1976) Sigri Elektrographit DE-AS 2,613,072
10. Union Carbide Corp. (1962) US-PS 3,302,999
11. The Carborundum Company (1969) DE-OS 1,948,415
12. The Carborundum Company (1972) DE-OS 2,246,572

References to Sect. 6.5.1 and 6.5.2

1. Wiley (1988) Carbonless copy paper. In: Encyclopedia of polymer science and engineering, vol 11. Wiley, New York, pp 78–79
2. Schenectady (1995) Carbonless copy paper, technology and products. Schenectady International
3. Sullivan J (1998) Carbonless technology. Schenectady International, Schenectady, NY, USA
4. Schenectady (1998) Business imaging products. Schenectady International, Schenectady, NY, USA
5. The Mead Corporation (1977) DE 2,719,938 C2
6. Sliwka W (1978) Mikrokapseln. In: Ullmanns Encyklopädie der technischen Chemie, 4th edn, vol 16. Chemie, Weinheim New York, pp 675–682
7. National Cash Register (1973) US 3,723,156
8. National Cash Register (1973) US 3,732,120
9. National Cash Register (1973) US 3,737,410
10. The Mead Corp (1977) US 4,025,490
11. National Cash Register (1979) US 4,165,102
12. Mead Corporation (1977) US PS 4,025,490

References to Sect. 6.5.3

1. Daxter M (1992) Antioxidants. In: Kirk-Othmer encyclopedia of chemical technology, vol 3, 45th edn. JI Kroschwitz, Wilby, NY, p 424+
2. Klemchuk PP (1985) Antioxidants. In: Ullman's encyclopedia of industrial chemistry, vol A3. VCH, Weinheim, p 91+
3. Zweifel H (1997) Macromol Symp 115:181
4. Igarashi J, Yoshida T (1995) Lubri Sci 7(2):107
5. C & EN (1994) C & EN, Sept 19, p 35
6. Plastics Technology (1995) Nov, p 21; (1995) C.A. 122 189,378f, 189379g, 189380a
7. Foulk SJ, Miller KL (1993) Proc SPIE-Int Soc Oct Eng 2069:85
8. Plastics Technology (1997) April, p 26
9. Noramco, New Brunswick, NJ 08901, Norbloc 7966 (1996)
10. Hoechst Celanese, Dallas, TX 75234, THPE-BZT (1997)
11. Fairmount Chemical (1984) U.S. 5,362,881
12. C.A. (1993) 118 214256a

CHAPTER 7

Chemical, Physical and Application Technology Parameters of Phenolic Resins

7.1 Introduction (Purpose and Objective)

Testing or identification of phenolic resins in principle differentiates these according to quality guidelines – whether a stipulated specification is met – or by their classification and suitability for certain uses or in specific fields of application. The former type of testing is carried out in modern quality and/or production control facilities at the manufacturer and the customer. Testing in the areas of application technology, research, and development may be seen as a determination of parameters according to application-specific aspects.

Such product testing should be highly practice-oriented, since the test cascade ends with, or centers directly on, further industrial processing. For this reason, the selection of resin or binder parameters to be monitored by constant quality control – whether in received goods or final product testing – is made to ensure that these are matched as closely as possible to the application and further processing needs (shown in the form of a diagram in Fig. 7.1).

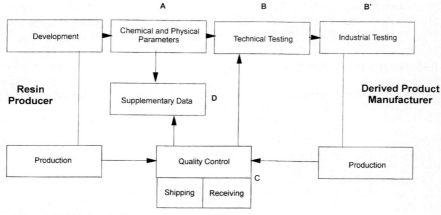

A = Chemical + Physical Data (Wide Range)
B = Practice-Relevant Application Testing
B' = Industrial Testing
C = Selection of A
D = Supplementary Testing

Fig. 7.1. "Test cascade" for quality and suitability testing of phenolic resins

7.1 Introduction (Purpose and Objective)

Nowadays, such testing must additionally be considered within the guidelines of quality assurance (QA) certification as specified in ISO 9001 and subsequent standards. The significance of quality control (Fig. 7.1) is particularly great in "just-in-time" deliveries where the customer – in agreement with the resin supplier – completely omits received goods quality control, or only carries out point analyses or determinations on randomly selected samples.

The final consequence is that chemical and physical tests of a resin should only be carried out using reliable and standardized test methods. "Reliable and standardized" means that a procedure has been tested and verified by means such as cooperative testing in multiple laboratories (best done internationally, for example under the auspices of the International Standards Organization = ISO), and that its reproducibility and comparability have been estimated by mathematical error analysis.

Standards should best be operator-independent. The fact must be considered that in-house methods or individual standards for phenolic resin test procedures have meanwhile existed in Germany, and similarly in the USA, Great Britain, Japan, and other countries for nearly 90 years (since Baekeland's discoveries and inventions), but that a centralized selection of these under DIN and their translation into ISO and recently into CEN (for Europe) standards has only taken place within the past two decades. It is by no means self-evident that such standards are presently accepted without alterations by either the phenolic resin manufacturers or their customers. This is a gradual process, although it is promoted by progressive ISO 9001 ff certification.

Supplementary procedures based on modern analytical methods exist in addition to the common chemical and physical test methods listed mainly in ISO 10082. In this case, we refer to detailed tests for quality determination and substance identification. Structural analysis of phenolic resins must be considered separately, since other – generally research-related – standards apply in that case. Finally, application technology tests matched to the practical use join the array of procedures. Table 7.1 provides an overview of this situation.

Taken together, this means that application technology tests must be used in addition to chemical and physical quality testing both to support application-related development of a new product, and to monitor the properties of sales products.

Table 7.1. Classification of test methods for phenolic resins

Test group	Basis	Examples	Users
1 Standard chemical and physical testing	ISO 10082	Viscosity by various methods	Development and production
2 Quality control	ISO 10082	Höppler viscosity	Production control
3 Supplementary tests	Modern analysis	Gel chromatography	Development and application research
4 Application testing	Specific methods	Impregnation	Application research

7.2
Standard Chemical and Physical Tests

In some cases, the tests carried out in phenolic resin chemistry and in production/quality monitoring are based on long-standing procedures. In addition, modern methods have been adopted or basically familiar test procedures adapted to conform to the findings of modern analytical technology; three basic principles have been observed in this development:

1. The test procedure should be operator-independent (a requirement that cannot always be met) and highly reproducible, i. e., verified by cooperative testing (Round robin testing).
2. The test technique should be rationalized to the greatest possible extent to avoid burdening the product with unnecessary costs in the case of quality monitoring.
3. The test data should ideally bear a relationship to the later use (Table 7.2).

Methods such as fully automatic viscosity measurement, automatic operator-independent gel time determination, automatic potentiometric titration and fully automatic gas chromatographic determination of residual phenol have thus been introduced. Efforts are also in progress to replace the mechanical, point screening technique for determination of the particle size distribution of powder resins by test methods based on the use of laser technology.

Table 7.2. Relationships between test methods and applications (examples)

Test method	Field of application	Effects in which area
Gel time	Coated abrasives, laminates etc.	Adjustment of curing conditions
Nonvolatiles	Coatings resins, laminates	Concentration, final properties
Water miscibility	Coated abrasives	Processing and viscosity adjustment
Electrical conductivity	Electrical laminates	Disturbances due to electrolytes
Residual monomers	All applications	Environment and workplace
Thermogravimetry	Refractories, carbon materials	Carbon level in final product
Viscosity	Foundry, refractories processing	Product transfer, mixing for
Screen analysis	Powder resins for various applications	Uniformity in processing
Specific surface area	Powder resins for grinding wheels	Absorption capacity for liquid resin
Flow distance	Friction linings, grinding wheels	Melting and curing behavior
Moisture determination (K. Fischer method)	Powder resins for various applications	Storage life, processing

Table 7.3. Physical methods for liquid and solid (powdered) phenolic resins

Resin (solid or liquid)	ISO Standard	Test method
Solid	ISO 3146 A	Melting range, capillary method
Powder	ISO 8620	Sieve analysis of powdered resins
Powder	ISO 60	Determination of apparent density
Liquid	ISO 2811	Density (pyrometer method)
Liquid	ISO 3675	Density (hydrometer method)
Liquid	ISO 9371	Determination of viscosity (4 methods)
Liquid	ISO 8975	pH-value
Liquid	ISO 9944	Electrical conductivity
Liquid	ISO 8989	Miscibility in water

Table 7.4. Chemical methods for liquid and solid phenolic resins

Resin (solid or liquid)	ISO standard	Test method
Solid (powder)	ISO 8619	Flow distance on a glass plate
Solid and liquid	ISO 8987	Reactivity on a B-transformation test plate
Liquid	ISO 9396	Geltime under specific conditions
Liquid and so rid	ISO 11 409	Temperatures of reactions by DSC
Liquid	ISO 9771	Acid reactivity of phenolic resins
Liquid	ISO 8618	Non-volatile matter
Liquid and solid	ISO 11 402	Free formaldehyde (3 methods)
Solid (powder)	ISO 8988	Hexamethylenetetramine content
Solid and liquid	ISO 760	Water content (K. Fischer)
Solid and liquid	ISO 3451-1, A	Ash content

The present basis for standard chemical and physical tests is mainly represented by the ISO standards listed in overview standard ISO 10 082 (Tables 7.3 and 7.4).

7.3 Description of Physical and Chemical Test Methods (ISO 10 082) and Their Significance [1, 2]

ISO 10 082 provides an overview of test methods used to investigate phenolic resins. The ISO regularly revises this standard at five-year intervals. Apart from various other national standards, the basis for ISO 10 082 was mainly DIN 16 916, Parts 1 and 2, the first complete standard on test methods for determination of phenolic resin parameters prepared in 1979 by the DIN standards committee. The ISO standard added procedures for use of liquid and gas chromatography as well as differential scanning calorimetry (DSC). The application of laser technology to determination of the particle size distribution will similarly be mentioned there in the future.

ISO 10 082 includes brief descriptions to provide an overview of the methods mentioned below.

7.3.1
ISO 3146, Melting Range, Melting Behavior

In ISO 3146, Method A (capillary method), a sample of resin is heated at a defined rate in a capillary and its physical behavior observed. Since phenolic resins represent mixtures of generally oligomeric condensates of phenol and formaldehyde rather than chemical compounds in the usual sense, they generally exhibit a wide melting range when heated, from the point at which they soften to that at which they melt, rather than a precisely defined "melting point." This melting range can thus bracket the temperatures from the sintering (or bonding) point up to the beginning of the molten state. Definition of the visible change in state is difficult at times. In general, the first visible change is defined as the "sintering point." In the molten state (at the "melting point"), the phenolic resin is either completely liquid, becomes translucent, or separates from the capillary wall. Due to these circumstances, application of this method requires experience and a knowledge of the product. The melting range of powder resins is affected by their moisture level. It is thus advisable to agree on drying conditions (for example 48 h drying over P_2O_5) for comparative determinations.

There has been no lack of experiments directed toward determination of the melting behavior in an operator-independent manner. Substitution of the capillary method by thermoanalytic measurements (DTA, DSC) is only possible to a limited extent since the latter are relatively time-consuming. In addition, the glass transition and softening ranges frequently overlap when DTA or DSC are used, precluding a clear result in such cases as well. "Softening point" determinations with photometric detection may be used for plant measurements, but must generally be considered highly product-specific. The same applies to "sticking point" determinations on the Kofler apparatus. On the other hand, it has been repeatedly found that "capillary melting point" determinations exhibit adequate reproducibility if carried out by experienced personnel.

7.3.2
ISO 8620, Screen Analysis with the Air Jet Screen and Particle Size Analysis as Specified by ISO 13 320 (Laser Method)

In principle, the air jet screen (Fig. 7.2) represents a closed container in which the powdered resin is placed on a screen and is exposed to a jet of air that emerges from a rotating nozzle located beneath the screen. The air is drawn through the screen. Common screen mesh sizes are 45 and 90 µm. The parameters affecting the determination – aside from the screen mesh size – are the level of reduced pressure and the duration of screening.

Only defined particle sizes (for example, "percentage larger than 45 µm") can be determined with the air jet screen. This type of analysis represents a point determination and the single result affords no information on the "particle size distribution." Approximations of the particle size distribution can

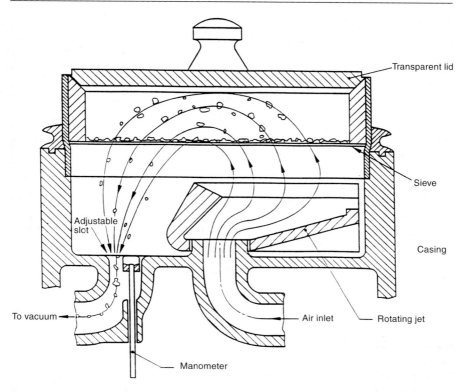

Fig. 7.2. ISO 8620-compliant alpine air jet screen (principle)

only be achieved by analogous determinations using multiple screen mesh sizes.

The laser technique (ISO 13 320) meanwhile makes it possible to determine the entire particle size distribution in dispersed wet or dry particle mixtures, thus providing a complete overview of all particle sizes and the magnitudes of their fractions. Application of the ISO 13 320 procedure [3] to novolak-HMTA powder resins by measurement in an aqueous dispersion does not afford a particle size distribution comparable with that obtained with mechanical screen analysis, since the HMTA is dissolved. However, it is possible to determine the true particle size distribution of powdered phenolic resins by dry dispersion analysis using modern instruments based on laser technology [4]. Dry dispersion is carried out using regulated inputs of compressed air and particles. In this procedure, the sample is dispersed and transported into the dry measuring cell by a stream of air. The measurement is highly efficient, and can be used in quality control for particle sizes of approximately 0.7 µm and above (Fig. 7.3).

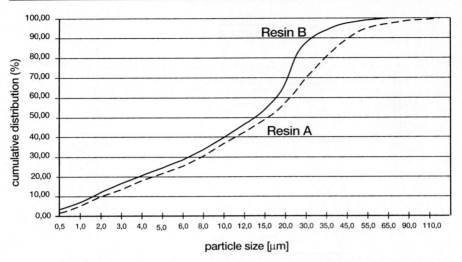

Fig. 7.3. Determination of the particle size distribution of two powder resins by the laser technique (a = coarse milling, b = fine milling)

7.3.3
ISO 60, Bulk Density

Determination of the bulk density is particularly used for volume characterization of the unmodified and modified, milled phenolic resins (with added hexa) used to manufacture friction linings, grinding wheels, refractories, and phenolic resin bonded textile felts (mainly cotton-based materials). Particularly in the case of bonded textile felts, the fabrication lines – for example, those using the scatter method – require that the volume of powder necessary to achieve a uniform resin level – for example, 30% – in the cured sheet goods or special moldings for interior use in automotive construction be as constant as possible.

The following method is used to determine the bulk density. A funnel is positioned 20–30 mm above a defined receiver (such as a 100 ml graduated cylinder cut off at the 100 ml mark). The outlet of the powder-filled funnel is initially sealed. The seal is then removed, allowing the powder to flow evenly into the receiver. Excess powder is scraped off the top of the receiver. The contents of the receiver are then weighed.

The "prior history" of the sample must be taken into account in this determination method. The finely milled material could have been compacted (undesirable for this measurement) by storage and transportation. The prior compaction can be largely eliminated by screening the sample prior to the determination.

7.3.4
ISO 2811 and ISO 3675, Determination of Density

The pycnometer (ISO 2811) or hydrometer method (ISO 3675) may be used to determine the density of a liquid phenolic resins (in g/cm^3) required for various calculations, such as the weight of a defined volume. ISO 2811 may be used for all liquid resins, whereas ISO 3675 is only applicable to resin with a viscosity of less than 1 Pa·s.

7.3.5
ISO 2555, ISO 3219, and ISO 12058, Determination of Viscosity

Three methods are used to determine the viscosity (particularly of solutions, but also of melts). It is possible to perform the measurement using a viscometer at a defined velocity gradient, for example, a plate-and-cone viscometer (ISO 3219). Such viscometers – particularly the "ICI Viscometer" in Fig. 7.4 – have been found ideal for rapid (direct) determinations of solution viscosities and of melt viscosities at elevated temperatures in the production plant.

Determinations with the Brookfield viscometer (ISO 2555) or Höppler falling ball viscometer (ISO 12058) are used as comparative measurement pro-

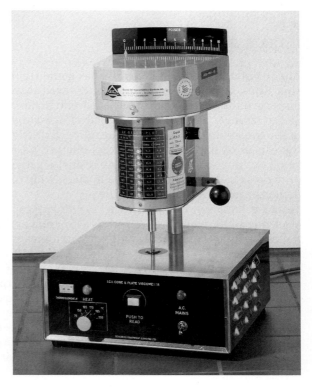

Fig. 7.4. ICI Cone-and-plate viscometer for rapid viscosity determination (photo: Bakelite AG, Iserlohn, Germany)

cedures. The Brookfield rotary viscometer measures the rotation speed of a defined spindle in the liquid. In the Höppler falling ball viscometer, the rate at which a ball falls in the liquid test resin is determined. The Brookfield rotary viscometer is mainly used (at a measuring temperature of 25 °C) in the pertinent areas of U.S. industry. Modern falling-ball viscometers operate fully automatically, and are operator-independent. They are fully computerized and are mainly used for efficient operation in situations where plant operation requires that large numbers of measurements be carried out.

The reasons for the wide variety of methods for determination of the viscosity of liquid resins – although the "essence" of standardization is to reduce the number of methods for a determination to a single one – are historical in nature, but also related to specific applications. As a side note, it may be remarked that the somewhat "unscientific" flow cup method (DIN 53211), in which the flow rate of a resin solution out of a standardized cup is determined, is still used in certain industrial areas, for example, by the paint and varnish industry. This method has acquired certain "established rights" since it is uncomplicated and much experience has been gained with it.

Since liquid phenolic resins and resin solutions frequently represent non-Newtonian liquids, the same method should, if possible, be used for comparative measurements and the test method (and temperature) to be used for the viscosity should be defined in specifications to avoid discrepancies.

7.3.6
ISO 8975, Determination of pH

The pH of a liquid resin generally has some significance for further processing. Thus, neutral liquid resins are best used as wetting agents in production of wet grinding wheels since the resin bond in the cured grinding shape is subject to stronger attack by alkali (catalyst in resin production) in the presence of the coolant liquid used in wet grinding. Excessively low pH values, or better the causes of these, can produce premature post-condensation in phenolic resoles or phenolic-modified furan resins, negatively affecting the storage life of the resins. It would be possible to cite a wide range of further examples to demonstrate the significance of the pH.

The principle of ISO 8975 is that the difference in potential between a glass electrode and a reference electrode is a function of the solution pH. The potential differential is measured using a potentiometer. The fact that the pH determined with a glass electrode represents a measure of the activity rather than the concentration of hydrogen ions is noted in ISO 8975. In dilute aqueous solutions, the activity coefficient is nearly equal to one, and the difference between the concentration and activity is small. Liquid phenolic resoles and organic solutions of phenolic resins generally contain little water. The difference between the hydrogen ion activity and concentration can be considerable in such cases. No exact relationship between the pH and the alkali content exists in the case of alkali-containing phenolic resins (due to phenolate formation).

7.3.7
ISO 9944, Determination of Electrical Conductivity

The electrical conductivity detects the presence of electrolytes, which should be avoided in impregnating resins for paper base electrical laminate. Conversely, a certain degree of conductivity can be desirable when electrostatics would not develop in an application.

The principle of the ISO 9944 determination of electrical conductivity may be described as follows. A mixture of acetone and water is added to the resin solution. After any possible precipitate has settled, the conductivity of the suspension above the resin is measured.

7.3.8
ISO 8989, Water Miscibility

The water miscibility of aqueous phenolic resins (resoles) can represent an important parameter for various reasons. In some processing methods, for example, in the base and size coats of coated abrasives, the resins are diluted with water. Water miscibility is also a measure of the compatibility of aqueous resins with aqueous dispersions (for example, PVA emulsions or latices). Moreover, the storage life may be inferred from the change in water miscibility during storage under certain conditions of temperature (measured over time).

ISO 8989 is applied by determining the quantity of water (as a mass percentage) required to produce turbidity (generally at a test temperature of $23 \pm 1\,°C$) when diluted in the resin. Addition of water is continued until persistent turbidity is observed after stirring the mixture for approximately 30 s.

7.3.9
ISO 8819, Determination of the Flow Distance

The "flow distance" at 125 °C is a measure of the flow properties of a powder resin, and provides an understanding regarding the curing characteristics as well as the impregnating and melt properties. This method was originally developed for use in grinding wheel fabrication (in-house method of Norton Co.), and the result also represents an important parameter in production of brake linings. In this procedure, the resin melts (thus including the melt viscosity of the resin in the test), and the flow of the melt is terminated as the phenolic resin system is fully cured.

Tablets (generally weighing 0.5 g and measuring 12.5 mm in diameter) are prepared under defined conditions of the test. The tablets are placed on a preheated glass plate, which is then set on a tilting stand in a drying oven at a temperature of 125 °C. The tablets remain in a horizontal position in the oven for 3 min, and are then tilted at an angle (60°) for 20 min. The resin flows down the glass plate (Fig. 7.5). The flow distance that results is measured in millimeters (for example, 15–20 mm is a "short" and 50–60 mm a "long" flow). With

Fig. 7.5. Schematic flow distance device according to ISO 8619

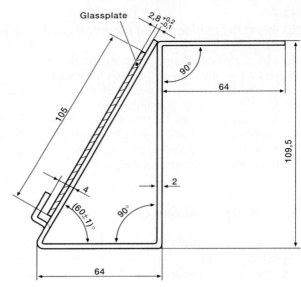

some experience, these data can be related to the production parameters and if appropriate to the expected final properties of the grinding wheel or the brake lining. The method is highly reproducible and also capable of comparison under constant test conditions, but does not represent a substitute for measurements of the melt viscosity or curing characteristics if these data are required for optimization or evaluation of the production process. The melting range of a powder resin only correlates with the flow distance when the content of hexa and the percentage of residual phenol in the resins being compared are at the same level in the comparison samples.

The "flow time" during which the melted resin moves a specific distance (in seconds) is measured when the above flow distance method is carried out in the absence of a curing agent (novolak powder resin without hexa). This method is frequently used as a rapid procedure for determination of the melt behavior of novolaks in production of molding compounds or coating of quartz sand for the foundry industry. Assuming pertinent product experience, this can be used as a rapid method for evaluation of the melt viscosity, but does not replace the scientifically founded procedure using a cone-and-plate or rotary viscometer.

7.3.10
ISO 8987, ISO 9396, and ISO 11409, Measurement of the Curing Characteristics

The curing behavior of a phenolic resin represents an important processing parameter and provides information on whether a resin system (resole or novolak/hexa) cures rapidly or reacts according to a slower curing mechanism.

Fig. 7.6. Testing of the "B" time (ISO 8987, method A). This method requires experience, but is then quite reproducible and is applicable to many phenolic resins (photo: Bakelite AG, Iserlohn, Germany)

The procedures described in ISO 8987 and ISO 9396 determine the reaction time that a phenolic resin requires for conversion to the "B stage" at a specified temperature. These methods do not describe the transition to the completely crosslinked "C stage". In ISO 8987, the "B time" is determined on a hot plate at a defined temperature. Two types of hot plates are described: (1) a plate with depressions in the form of spherical segments (Method A), and (2) a plate without depressions (Method B). In Method A (Fig. 7.6), the time at which threads of the melt break (at a measurement temperature of 130 °C, 150 °C, or 160 °C) is determined using a glass rod (with which the melt is stirred), and in Method B with a spatula (with which the melt is spread out in segments on the flat hot plate). Methods A and B require practice and the results are affected by the individual, i.e., are not operator-independent.

Measurement of the gel time at a specified temperature using an automatic instrument, as described in ISO 9396, is an operator-independent procedure (Fig. 7.7). This method is mainly applicable to liquid resins, and is generally carried out at 100 °C or 130 °C (important for liquid resins used in production of coated abrasives or impregnating resins for paper base electrical laminate). Some time after initiation of the polycondensation reaction, the phenolic

Fig. 7.7. Automatic testing of the gel time (ISO 9396)

resins reach a critical point at which the viscosity undergoes a sudden rise. The automatic instrument determines the time required to reach this point. In the instrument, the resin is contained in a thermostatted test tube in which an agitator moves up and down. The end point (gelation) is reached when the viscosity of the sample rises so high that the test tube is raised together with the agitator and switches off the running stop clock (which then indicates the gel time).

In ISO 11409, the reaction enthalpies and reaction temperatures at the beginning, midpoint, and end of curing at a specified rate of heating are determined by means of DSC (Differential Scanning Calorimetry). The heat flow (quantity of heat) toward a sample located in a small, tightly sealed steel container is measured as a function of the time or temperature. The heating rate is usually $5 \pm 1\,°C/min$, but the measurement can also be performed under isothermal conditions. The heat differential that exists between the sample container in which the substance is located and a "neutral" (sample-free) reference system is measured and indicated. The percentage conversion at rising temperatures and thus parameters such as differences in the curing rates of comparison samples can also be determined using this method.

7.3.11
ISO 9771, Acid Reactivity of Liquid Phenolic Resins

Liquid, aqueous phenolic resoles or furfuryl alcohol-containing phenolic resoles (used as foundry binders or in acid-resistant construction or for the manufacture of phenolic foam) can react and cure by means of acidic catalysts (acidic salts or acids). The quantities of heat liberated in this process can be considerable, and can heat the sample to high temperatures. ISO 9771 describes measurement of the peak temperature that develops during curing of a phenolic resin mixed with an acid as a curing catalyst, and the time required to reach it. The resultant data indicate the reactivity of the resin and can be used for comparative determinations.

7.3.12
ISO 8618, Determination of Nonvolatile Components

The nonvolatiles in a waterborne or dissolved resin are those components that are effectively incorporated into the subsequent application. The volatile substances in a resin are either solvents (including water), monomeric components that do not undergo reaction under the prevailing conditions, or elimination products of the curing reaction such as the water and formaldehyde produced when resoles cure. The volatile components can exert a negative effect on the properties of a final product, for example, by causing porosity or cracking of moldings, which must be compensated by appropriate measures during processing. This description indicates that the nonvolatiles determination must be viewed as a complex procedure resulting in the parameters to be observed in ISO 8618.

In ISO 8618, the liquid phenolic resin (aqueous resole or resin solution) is weighed into a shallow dish of specified dimensions. The nonvolatile fraction is the percentage residue obtained when the volatile components are evaporated under specific conditions of temperature and time. Cooperative testing has shown that drying ovens whose interiors exhibit little or no temperature variations may generally be used to achieve comparable results. Modern drying equipment maintains defined conditions of temperature and air flows. A test temperature of 135 °C is generally suggested. A special procedure at 150 °C is recommended for specific (impregnating) resins.

7.3.13
ISO 8974, Residual Phenol, Gas Chromatographic Determination

The determination of residual phenol has gained considerable importance since many phenolic resin producers began to reduce the level of residual (free) phenol in numerous types of resins by significant amounts (to less than 0.2 % in novolaks and less than 4 % in resoles). Gas chromatography, a modern and efficient method of determination that is operator-independent, can also be performed using automatic equipment, and offers large-scale producers the possibility of carrying out a large number of determinations within a short period of time. It is excellent for this purpose.

In ISO 8974, a sample is dissolved in an appropriate solvent (generally acetone) and the level of phenol determined by gas chromatography using m-cresol as an internal standard. Capillary columns, for example, a 25 m capillary column with an interior diameter of 0.32 mm and coated with OV-1701 silicone, are used presently. Helium is used as a carrier gas in the chromatograph. The amount of the m-cresol reference substance in the standard acetone solutions depends on the expected level of phenol. Highly alkaline resins are neutralized prior to the determination to prevent formation of phenolates.

7.3.13.1
DIN 16916-02-L 1 and DIN 38409-16, Conventional Methods for Determination of Residual Phenol

For reasons of completeness, the Koppeschaar wet analytical method and the colorimetric method (for residual phenol levels below 1 %) for determination of residual phenol will also be described below. The free phenol is separated by steam distillation in both procedures; alkaline resins must be neutralized prior to this separation.

In the *Koppeschaar method* (no limit to residual phenol level), the phenol is determined by bromination with a bromide/bromate solution to yield tribromophenol. The reagent solution liberates a certain amount of bromine when it is acidified:

$$BrO_3^- + 5\,Br^- + 6\,H^+ \rightarrow 3\,H_2O + 3\,Br_2$$

The excess reagent after addition of the aqueous phenol solution and bromination of the phenol liberates iodine when potassium iodide is added (producing a blue color with starch), and the iodine is back-titrated with sodium thiosulfate solution. In the *colorimetric method* of DIN 38,409-16 (for phenol levels below 1%), the phenol separated by steam distillation is oxidatively transformed into a quinonimine dye with 4-aminoantipyrine and potassium hexacyanoferrate (III) solution, and colorimetrically determined (extinction against a blank solution) in a photometer.

7.3.14
ISO 11 402, Determination of Free Formaldehyde in Phenolic Resins and Co-Condensates

The requirement that condensation resins produced with formaldehyde bear hazards markings and be classified necessitates differentiated methods (Table 7.5) for exact determination of the free formaldehyde level.

The principle by which free formaldehyde is determined in unmodified phenolic and furan resins and their mixtures is the reaction of *hydroxylamine hydrochloride* with formaldehyde (ISO 9397, cf. Table 7.5):

$$CH_2O + NH_2OH \cdot HCl \rightarrow H_2O + CH_2NOH + HCl$$

The hydrochloric acid formed in the reaction is potentiometrically determined with reagent-grade NaOH solution. ISO 9397 cannot be used if the resins have been modified with urea and/or melamine resins, since the methylol groups of the amino resins can partially hydrolyze.

In the *sulfite method*, free formaldehyde and formaldehyde hemiacetals are converted into hydroxymethane sulfonate by an excess of sodium sulfite at 0 °C. The excess sodium sulfite is titrated with iodine solution. The resultant hydroxymethane sulfonate is decomposed with sodium carbonate solution, and the sodium sulfite liberated in the reaction titrated with iodine solution:

$$HOCH_2-SO_2Na + Na_2CO_3 \rightarrow CH_2O + Na_2SO_3 + NaHCO_3$$

$$Na_2SO_3 + I_2 + H_2O \rightarrow Na_2SO_4 + 2 HI$$

Table 7.5. Free formaldehyde level in condensation resins (selection of procedures, ISO 11 402)

Procedure	Suitable for testing of
Hydroxylamine hydrochloride procedure	Phenolic resins, furan resins[a] (unmodified with urea or melamine resin)
Sulfite procedure	Urea resins, melamine resins, furan resins, urea-melamine resins, furan-urea resins
KCN procedure	Melamine-phenolic resins, urea-phenolic resins, urea-melamine-phenolic resins

[a] Unmodified furan resins.

The free formaldehyde in phenolic-amino resin co-condensates is mainly determined by the "*KCN method.*" In this method, free formaldehyde is reacted with an excess of potassium cyanide:

a. $CH_2O + (excess)\ KCN \rightarrow NC-CH_2-OK$

b. $2\,KCN + Hg(NO_3)_2 \rightarrow 2\,KNO_3 + Hg(CN)_2$

The excess of potassium cyanide is back-titrated with mercury (II) nitrate solution using diphenylcarbazone as an indicator.

7.3.15
ISO 8988, Hexamethylene Tetramine (HMTA) Content

This method is mainly used for novolak-based phenolic powder resins containing hexamethylene tetramine as a curing agent for the novolak. ISO 8988 describes two procedures, the *Kjeldahl method* and the *perchloric acid method*. The Kjeldahl method is inapplicable if other nitrogen-containing components (for example amino resins) are present in the phenolic. The perchloric acid method is only applicable if the resin contains no alkaline or acidic additives. In doubtful cases, these methods are only applicable when a knowledge of the base formulation exists.

In the Kjeldahl method, HMTA is converted into ammonium sulfate by hot degradation with concentrated sulfuric acid in the presence of an added catalyst mix, for example a mixture of 97 g Na_2SO_4, 1.5 g $CuSO_4 \cdot 5\,H_2O$ and 1.5 g selenium. Ammonia is liberated by addition of sodium hydroxide, and is (steam) distilled into a receiver containing hydrochloric acid. The excess hydrochloric acid is in turn back-titrated with NaOH:

$(NH_4)_2SO_4 + 2\,NaOH \rightarrow Na_2SO_4 + 2\,NH_3 + 2\,H_2O$

$NH_3 + HCl \rightarrow NH_4Cl$

The *Dumas nitrogen determination* used in elemental analysis can also replace the Kjeldahl nitrogen determination. In the Dumas method, a 10–30 mg sample of resin mixed with copper oxide dust is incinerated under a stream of pure oxygen in an automatic combustion system. The combustion gases are carried over copper oxide and copper by a flow of helium, and the nitrogen determined using a thermal conductivity detector. The entire procedure is automatic. This method also determines total nitrogen.

In the perchloric acid method, the powder resin is dissolved in an appropriate solvent, and the HMTA content determined by direct titration of the tertiary amine functional group with perchloric acid.

7.4
Miscellaneous Chemical Test Methods Used for Analysis of Phenolic Resins

ISO 10082 recommends that the *water content* of phenolic resins be determined using the Karl Fischer method of ISO 780. Modified Karl Fischer reagents, for example pyridine-free materials and reagents that are insensitive to the presence of aldehydes and ketones, may also be used for determination of water in phenolic resins. The *ash content* is determined as specified in ISO 3451/1 A. In this method, the ash is determined by direct incineration, i.e., by combustion of the organic fraction of the product and further heat treatment of the inorganic residue. The *alkali number* and *alkali content* may be determined by various titrimetric methods.

7.4.1
ISO 11401, Liquid Chromatography for Separation of Phenolic Resins

Liquid chromatography is generally used in research for structural elucidation of phenolic resins (resoles and novolaks), but in some cases is also used for routine analytical testing, for example, to characterize and compare production batches, or to check the level of residual phenol. The ISO thus standardized this method.

ISO 11401 describes three procedures [5]:

A. Gel permeation chromatography (GPC), also known as gel chromatography
B. High-performance liquid chromatography (HPLC) on polar columns
C. High-performance liquid chromatography (HPLC) on nonpolar columns

A phenolic resin can be separated by molecular size with the aid of Method A (gel chromatography). The free phenol and the sum of the dihydroxydiphenylmethanes (in novolaks) as well as various methylolphenols (in resoles) are quantitatively separated in this procedure; resin fractions of higher molecular mass are only incompletely separated due to the large number of isomers (see Structure, Chapter 4).

Methods B and C (high-performance liquid chromatography) separate mixtures by the molecular mass and polarity of the compounds. The effects of molecular mass dominate on polar separation columns (Method B), and those of polarity on nonpolar columns (Method C). These methods also permit quantitative determination of individual resin components of low molecular mass. Due to the different solubility characteristics of the resins, Method B is better suited for novolaks, and Method C for resoles.

High-performance liquid chromatography affords chromatograms of the entire range of resin components. In the case of novolaks, it is possible to identify isomeric mixtures despite the large number of isomers of compounds with more than four aromatic rings by performing simultaneous measurements at multiple wavelengths using an array detector.

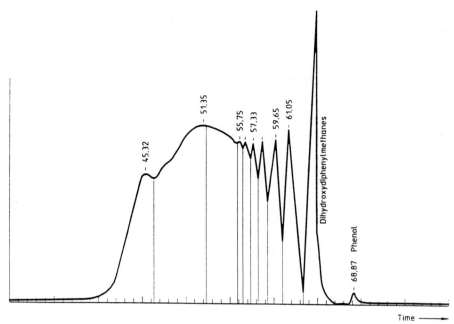

Fig. 7.8. Example of a GPC phenolic novolak chromatogram [5]

In Method A, the phenolic resin is dissolved in an appropriate solvent and the molecular mass distribution (Fig. 7.8) determined by separation on a column packed with polymer gel exhibiting pores of various diameters. In Method B, the resin solution is separated by passage through a polar column. The eluent is a mixture of solvents run with a "concentration gradient" of the individual components. Novolaks (Fig. 7.9) and resoles that are soluble in tetrahydrofuran can be analyzed in this manner. In Method C, the phenolic resin is dissolved in tetrahydrofuran and separated by passage through a nonpolar column (Fig. 7.10). As in Method B, the eluent is an appropriate mixture of solvents run with a concentration gradient.

7.4.2
Thin-Layer Chromatography (Company-Specific Method)

In the 1950s, separations of products (for example, separation of the individual low molecular mass components of an oligomeric mixture in aqueous phenolic resoles) or of components resulting from oxidative cleavage of cured phenolic resins were carried out by paper chromatography. This method was simplified by use of thin-layer chromatography (TLC), in which the "migration" of individual components along a silica gel-coated glass plate is visualized.

Although these "historical" methods were very useful in the past, they have meanwhile been displaced by gas chromatography, gel chromatography, high-

506   7 Chemical, Physical and Application Technology Parameters of Phenolic Resins

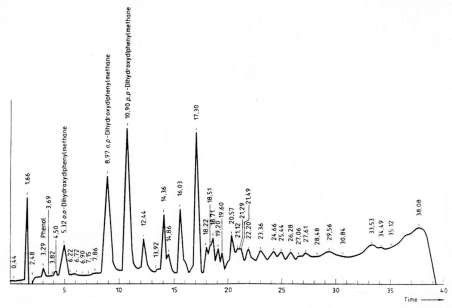

Fig. 7.9. Example of an HPLC phenolic novolak chromatogram produced using a polar column [5]

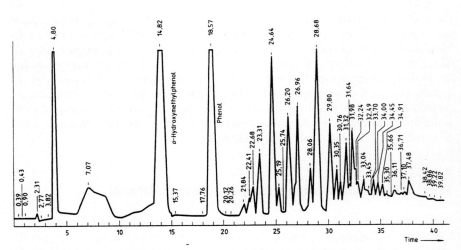

Fig. 7.10. Example of an HPLC phenolic resole chromatogram produced using a nonpolar column [5]

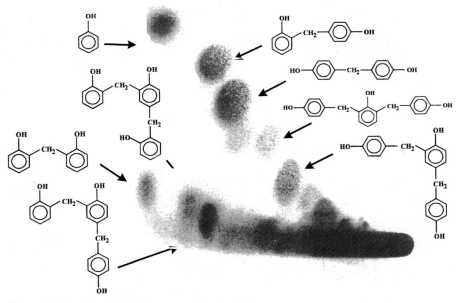

Fig. 7.11. Thin-layer chromatogram of the low molecular mass components of a phenolic novolak (TLC plate 200×200 mm, Silica Gel 60 bonded with gypsum; developing solvents direction 1: chloroform/methanol 90:10; direction 2: toluene/MEK/DEA 75:75:20; development with diazotized p-nitraniline)

performance liquid chromatography, and other modern methods of structural analysis.

In plant analytical practice, however, it has been found that TLC to this day remains well suited for rapid (qualitative and quantitative) preliminary testing of commercially manufactured phenolic resins to obtain basic data on their quality. Figure 7.11 shows a thin-layer chromatogram of the low molecular mass components in a standard novolak. Relative conclusions as to the degree of condensation may be drawn from this chromatogram, in which compounds with two and three aromatic rings are separated. Figure 7.12 shows a chromatogram of the mono-, di-, and trimethylolphenol mixture in a low molecular mass phenolic resole, and is similarly suitable to provide basic production information.

7.5
Application Technology Testing

Practice-oriented application testing plays a significant part in the "test cascade" used to determine the parameters and suitability of a resin, as well as in its development, continued development, and improvement of its properties (Position B in Fig. 7.1). In some cases, such tests are also carried out (either sporadically or over a certain period of time) on samples from running pro-

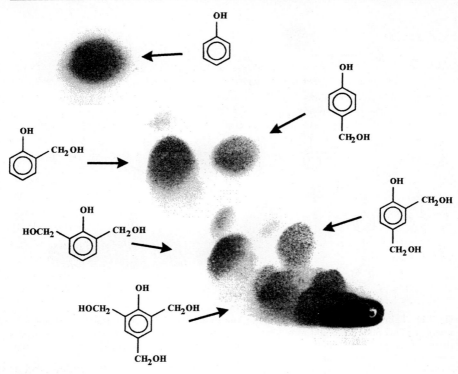

Fig. 7.12. Thin-layer chromatogram of a low molecular mass phenolic resole (TLC conditions as in Fig. 7.11)

duction, for example, on phenolic or furan resin-based foundry binders for the no-bake process. Such application tests generally deal with the phenolic resin as a binder for use in production of articles in a specific application, and consider the function and effects on quality in the final product. The tests that are performed represent chemical and physical trials designed to conform to the final product and its application.

For technical, economical, and equipment-related reasons, it is not always possible to test in a manner as closely related to practice as would be the case in actual industrial use in all fields of application (Position B' in Fig. 7.1). In such a case, however, the application tests are performed in a manner that ensures that the main technical requirements can be covered and a prediction can be made as to whether the product is suitable for industrial use or, if not, what defects it exhibits. Major application tests for specific fields are described below.

7.5.1
Refractories

The binder is generally carbonized when these products (for example, refractory bricks for steel production) are used. It is therefore necessary to determine the level of the carbon yield that may be expected following carbonization, since this magnitude affects the quality of the products. The carbon yield is generally determined by TGA (Thermal Gravimetric Analysis) with a thermobalance (at temperatures up to 1000 °C) in an inert or partially oxidative atmosphere. In the case of phenolic resins, the yield is generally 50–70%. The procedure can be carried out by subjecting the resin system to a specific curing program prior to thermoanalysis (cf. Sect. 6.4.1 Pyrolysis of Phenolics).

Further possibilities for application technology testing are offered by magnesia or dolomite mixtures with graphite, where the phenolic resin is used as a binder; test pieces (for example, cylindrical shapes) are prepared from this mixture and appropriately cured (using specific curing programs), after which physical tests are performed on the test pieces. A similar procedure is followed in testing the impregnation capacity of a refractory and the cure of the impregnated product (cf. Sect. 6.4.2 Phenolic Resins as Binders for Refractories) [6].

7.5.2
Foundry Binders

Foundry binders are generally tested [7] by preparing practice-related (wet or dry) mold or coremaking sand mixes and determining properties such as the flowability, shaping ability, bake-on temperature, peelback behavior, and bench life. Test pieces are produced by means of the various (hot, cold, or gas curing) mold and coremaking processes. The processes (for example, the gas curing method) are usually carried out in a manner corresponding to that in the foundry industry (Fig. 7.13). The mechanical strength levels, thermal (hot distortion or thermoshock) properties, and the levels of gas evolution at ele-

Fig. 7.13. Foundry pilot plant (core shooter, gassing system, scrubber) for practical testing of foundry binders (photo: Bakelite AG, Iserlohn, Germany)

vated temperatures (matched to the casting conditions) can be determined with the test specimens. "Break-down" that permit conclusions regarding the later behavior of the poured-off sand when the cast articles are shaken or vibrated out of the mold (shakeout) are performed in the case of aluminium casting (melt temperature approximately 700 °C, internal temperature of cores generally not higher than 400 °C). Such break-down tests can be performed by simple means on a laboratory shaker or using vibration or screening equipment.

7.5.3
Abrasives (Grinding Wheels, Abrasive Shapes, and Coated Abrasives)

In the case of the liquid wetting resins and powder resins used in production of grinding wheels, it is important that the mixing behavior as well as the powder absorption capacity of the liquid resins used to wet the grain be tested. Moreover, cold or hot pressed and oven-cured test pieces may be used to determine the burst strength, the wet strength (effect of coolant in wet grinding), and further physical properties. In the case of coated abrasives, abrasive paper may be produced in the laboratory and properties such as the flow and curing properties of the liquid resins determined during this process.

7.5.4
Friction Linings

Physical determinations of properties such as the strength levels are similarly carried out on test pieces in this application; in this case, the test pieces are produced from friction lining mixes corresponding to those used in practice. Moreover, the thermal stability of the friction lining mixes can be determined using thermoanalysis. It is also possible to carry out still more intensive testing on a friction lining test stand using specific test programs (Fig. 7.14); in this case, the resin manufacturer must weigh the advantages and disadvantages of installing such a friction lining test stand in-house or contracting the work at an outside testing facility. Parameters such as the coefficient of friction and wear of friction segments can be measured with simple formulations using such test equipment, and it is possible to determine the effects of various grades of resins.

7.5.5
Impregnating Resins for Industrial and Paper Base Electrical Laminates

Application testing of these resins is matched to the fabrication process used for paper base electrical laminates. This includes impregnation of paper on a pilot plant line. If desired, copper-clad sheets can be produced from the impregnated paper on a hot press. Physical data such as the punching properties, surface resistance, specific volume resistance, electrical dissipation factor, flammability, water uptake, copper adhesion, solder bath behavior,

Fig. 7.14. Brake lining test stand for semi-industrial tests and quality assurance

mechanical strength levels, and corrosion properties are determined with the sheets. The tests can also be extended to include etching of the copper foil (suitability test for production of printed circuits).

7.5.6
Resins for the Coatings Industry

Phenolic coating resins are tested for compatibility with solvents, drying oils, and other types of resins (such as epoxies); standard formulations are produced and the coatings application properties examined. The flow and reactivity represent important coating properties. Generally tested coatings film properties include the hardness, adhesion, deep-drawing capability, sterilization resistance, chemical resistance, and water and weather resistance.

7.5.7
Resins for the Adhesives Area

Alkylphenol resins are used to produce products such as polychloroprene and nitrile rubber adhesives. Properties such as the reactivity with magnesium oxide are of significance in these applications. Determination of the major adhesives properties, such as the "open time," and of the tensile, peel, and shear strengths of standard adhesive bonds are of importance in their use as adhesives.

7.5.8
Resins for the Rubber Industry

In the rubber industry, phenolics (unmodified and alkylphenol resins) are used as reinforcing or tackifying resins. It is thus advisable to produce rubber mixes and vulcanisates using the specific resins for use in preliminary testing. The hardness, strength levels, elongation, and aging are tested in the vulcanized products.

7.5.9
Resins for Textile Felt Production and Related Applications

Phenolic powder resins are used for purposes including production of industrial textile felts for applications in the automotive, household, and building areas. Preliminary trials include the emission tests already discussed in the chapter on textile felts as well as tests for flame resistance of the final products. Laboratory facilities for production of textile mats are on the market. Even without these facilities, it is possible to produce a random blend of resins and fibers in a simplified manner to afford a kind of molding compound, and to carry out practice-related tests on this material (refer to Sect. 6.1.2.3.7 for tests of emissions from phenolic resin bonded textile felts).

7.5.10
Plywood and Resorcinol Resin Adhesives

Plywood adhesives are subjected to a bonding test (production of a plywood sheet under precisely defined working conditions). The strength is determined in a cleavage test (DIN 69705). Adhesion tests that can be performed in the laboratory also exist for resorcinol resin adhesives.

7.5.11
Resins for Floral Foam

A simplified reactivity test can be performed with these resins. The resin is placed in an insulated paper cup and mixed with additives and pentane. The acidic curing agent is then stirred in, and a stopwatch simultaneously started. The stirrer is removed after 15 s, and the length of time the foam mix requires to reach the upper lip of the sample container measured.

References

1. ISO 10082 (1999) Plastics – phenolic resins – definitions and test methods
2. Bakelite AG (June 1997) Analysis of Rütaphen resins. Company Publication, Bakelite AG, Iserlohn, Germany
3. ISO 13320 (1996) Particle size analysis – guide to laser diffraction methods
4. Cilas (1998) Cilas-2 in 1. Company Publication, Cilas, Marcoussis, France
5. ISO 11401 (1993) Plastics – phenolic resins – separation by liquid chromatography. 1st edn, Dec
6. Gardziella A, Suren J (1997) Phenolic resins as impregnating agents for refractories – present state of development. UNITECR '97, New Orleans, USA, pp 975–996
7. VDG Test Methods (1997) Verein deutscher Giessereifachleute, Düsseldorf, Germany

CHAPTER 8

Industrial Safety and Ecological Questions (Raw Materials, Recycling, Environment)

Consideration and solution of environmental and workplace-related problems in production and processing of phenolic resins (Fig. 8.1) have been emphasized in continued development of phenolic resins for many years [1]. The diagram in Fig. 8.1 helps to illustrate the topics that are addressed under the above heading.

It must be expected that components classified under the heading *"Hazardous Materials"* (C) can develop in production (A) and processing (B) of phenolic resins and phenolic molding compounds. This means that raw materials (industrial materials) require hazardous substance markings (G) when they contain certain levels of such compounds. Compliance with occupational exposure limits must also be ensured when volatile substances can be liberated, and the composition and quantities of exhaust gases (E) that can enter the environment in the form of flue gas must be controlled. The emission limits must also be observed. Wastes (D) can be created during production (A) and processing (B), and require a decision as to whether they are to be recycled (F), otherwise utilized, or sent to a waste disposal site.

Figure 8.2 explains this process using a foundry facility where synthetic (for example phenolic) resins are used as mold and core binders for silica sand as an example. Various emissions (E_1, E_2, E_3) develop during mixing, in production of cores and molds, and during pouring, and from different sand wastes (A_1, A_2, A_3) during core production (residual core sand A_1 and core breakage A_2) and after pouroff (A_3). All "E" and "A" type areas can present problems that must be overcome.

Fig. 8.1. Effect of production and processing on the workplace and environment

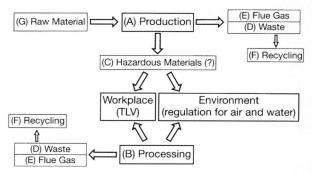

Fig. 8.2. Emissions and waste in a foundry (schematic)

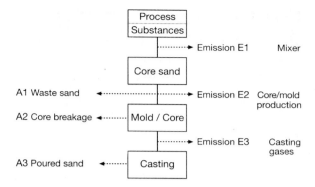

In every production and processing system, analytical questions involving the workplace, flue gas, and emissions into the environment must be solved, and – if necessary – technical activities initiated in the areas of occupational safety (for operating personnel), flue gas treatment, and reduction of environmental impact.

Reduction of wastes and emissions to an acceptable level, one that can be regularly monitored, is a general principle that also applies to production and processing of phenolic resins.

The question of resource conservation, specifically to recycle raw materials (G in Fig. 8.1) or to make use of renewable raw materials, also represents a part of this consideration.

8.1
Toxicological Properties, Hazards Labeling [2 – 5]

According to European guidelines for classification of hazardous materials and compositions, the acute toxicity assignments shown in Table 8.1 are presently valid. The LD_{50} is the mean lethal dose of a substance or composition that, when assimilated in the bodies of experimental animals of the same species following introduction into the stomach (oral) or application to the skin (dermal), causes death in half (50%) of the animals; it is expressed in milligrams per kilogram of body weight (mg/kg b.w.). Extensive toxicological studies of solid and liquid phenolic resins afford unequivocal information on their acute toxicity [6].

The mean lethal dose (LD_{50}) for solid phenolic resins (novolaks and resoles) by the oral route of administration, determined in the rat, is more than

Table 8.1. Acute toxicity (animal tests)

	Oral LD_{50}, rat mg/kg b.w.	Dermal LD_{50}, rat mg/kg b.w.
Very poisonous	< 25	< 50
Poisonous	25 – 200	50 – 400
Slightly poisonous	200 – 2000	400 – 2000

16,000 mg/kg body weight. Resins with relatively high levels of free phenol (solid resoles with levels up to about 8 wt%) have also been investigated. No effect due to dermal exposure could be determined at levels up to 3000 mg/kg body weight. This means that classification for hazards labeling according to the above criteria is unnecessary in the case of solid phenolic (novolak and resole) resins. Studies of the dermal and mucous membrane tolerance showed that the resins that were tested and are representative for the pertinent types of resins may not be classified as "dermal irritant" or "mucous membrane irritant." Although no sensitizing properties could be determined in animal studies, hypersensitivity reactions in susceptible persons cannot be excluded.

The acute oral toxicity of *liquid, aqueous phenolic resins* depends on the level of free phenol to a greater extent than in the case of the solid types of phenolic resin. Depending on the phenol level (up to about 10 wt%), the oral LD_{50} in rats is more than 16,000 to 6000 mg/kg body weight. The oral LD_{50} in rats is still above 2300 mg/kg b.w. even at (extremely high) phenol levels up to 24 wt% [7]. No mean lethal dose (LD_{50}) could be determined for dermal exposure. Thus, liquid phenolic resins similarly fall under none of the classification described in Table 8.1. The dermal and mucous membrane tolerance of liquid phenolic resins is affected both by their level of free phenol and by the alkali content of the resins. Although the primary skin irritancy results did not lead to classification as an "irritant," some types of phenolic resins can induce mucous membrane irritation in cases of lengthy, repeated exposure. Animal tests for a sensitizing effect similarly afforded negative results in this case, but hypersensitivity reactions cannot be excluded.

The inhalative effect resulting from the presence of solvents dominates the toxicological properties of *phenolic resin solutions*. Solvents exert only a minor effect on the toxicological properties of phenolic resins with respect to the acute oral and dermal toxicity as well as the skin and mucous membrane tolerance.

Notwithstanding the above, current practice for phenolic resins as expressed in presently valid European (EU) Guidelines results in a purely formal stipulation – not involving toxicological testing – that phenolic resins must be marked with a St. Andrew's cross ("X") when their level of free phenol is greater than or equal to 1% and less than 5%, and with a skull and crossbones when the level is greater than or equal to 5%. Although this very strict classification is patently unjustified on the basis of the described toxicological studies, it is nonetheless practiced. Similarly formal limits for purposes of hazards markings apply to the level of free formaldehyde.

8.2
Workplace, Exhaust Air, Low-Monomer Resins

Environmentally relevant substances – particularly phenol and substituted phenols, formaldehyde, and possible solvents – can be liberated during various operations, for example, those involving mixing, impregnation, and drying. The main emissions during curing of phenolic resins are of phenol and formaldehyde, ammonia also being emitted when novolak-hexamethylene tetramine systems are used. Table 8.2 surveys the international occupatio-

Table 8.2. Survey of occupational exposure limits measured in ppm (TWA = "time-weighted average"; STEL = "short-term exposure limit"; TRK = "technical guideline level for carcinogenic substances")

	Formaldehyde		Phenol		Ammonia		Ethanol		Benzene		Methanol		Furfuryl alkohol	
	TWA[a]	STEL[b]	TWA	STEL	TWA	STEL	TWA	STEL	TWA	STEL	TWA	STEL	TWA	STEL
Australia	1	2	5	–	25	35	1000	–	5	–	200	250	10	15
Belgium	1	2	5	–	25	35	1000	–	10	–	200	259	10	15
Czech Republic	0.4	0.8	5	10	29	58	540	270	3	6	78	390	–	–
Denmark	–	0.3	5	–	25	–	1000	–	5	–	200	–	5	–
Finland	–	1	5	10	25	40	1000	1.250	5	15	200	250	5	10
France	–	2	5	–	25	50	1000	5.000	5	–	200	1000	10	–
Germany	0.5	1	5	–	50	–	1000	–	1[c]	–	200	–	10	–
Hungary	–	0.5	1	2	26	39	540	1.720	–	1.6	39	78	–	2
Japan	0.5	–	5	–	25	–	–	–	10	25	200	–	5	–
Netherlands	1	2	5	–	25	–	1000	–	10	–	200	–	5	50
Poland	0.4	0.8	2.5	5	27	39	540	1.720	3	12	78	225	–	–
Sweden	0.5	1	1	2	25	50	1000	–	0.5	3	200	250	5	10
Switzerland	0.5	1	5	10	25	50	1000	–	5	–	200	400	10	–
UK	2	2	5	10	25	35	1000	–	5	–	200	250	5	15
USA[d]	0.75	2	5	–	25	35	1000	–	–	–	200	250	10	15

[a] TWA = Time-weighted average.
[b] STEL = Short-time exposure limit.
[c] TRK = Technical guideline level for carcinogenic substances.
[d] OSHA.

nal exposure limits [8], which are quite different in some cases, but nonetheless exhibit trends toward lower ranges. The occupational exposure limit ("MAK" in Germany) is the maximum permissible concentration of an industrial material in the form of a gas, vapor, or suspended matter in the air of the workplace that, based on present knowledge, does not generally impair the health of the workers or represent an excessive nuisance to them even in case of repeated and lengthy exposure (as a rule 8 h), with a 40 h limit on the mean weekly working hours. As a rule, the MAK represents the mean of the concentration integrated over periods of up to one working day or one shift. There are additional restrictions on peak concentrations. The modes of activity of the substances are primarily considered in establishing the MAK limits. The main emphasis in this work is on scientifically founded health protection criteria rather than on the technical and economical possibilities of realizing the limits in practice.

In Germany, the concentration limits are developed by the DFG or "Deutsche Forschungsgemeinschaft" (German Research Society) in Bonn. Table 8.2 provides no information on the duration and frequency of peak levels per shift. Details may be requested from the pertinent national institutions. In the USA there are two different assessments, those by OSHA and the ACGIH. The latter is a US governmental facility for industrial medicine and generally applies stricter criteria. Compulsory measurements at intervals that depend on the percentage by which the concentrations fall below the limits exist in Germany. In principle, analytical technology permits continuous monitoring, possibly in the form of base measurements of the "organic carbon" parameter.

A large number of work area analyses and control measurements that have been carried out in German plants demonstrate that in far more than 80% of all cases the determined concentrations are less than three-quarters of the MAK limit. In many cases, a reduction in the concentrations can be achieved by means of simple organizational or technical measures [9, 10] in the relatively small number of workplaces where employee exposures were above the indicated level. On the basis of many reliable exposure determinations, it may thus be noted that phenolic resins can today be used and handled without problems from a technical or industrial medical viewpoint. The types and levels of the phenolic resin-derived components in the process air depend to a marked extent on the production and processing methods. Due to the regular, simple structure of the phenolic resins, emissions from curing processes that proceed at temperatures up to a maximum of 250 °C are limited to the monomers that remain in the system – phenol and formaldehyde – and ammonia. When organic solvents are involved in the work, these can also be emitted [11].

Even years ago, the general, positive trend toward radical reduction of all types of environmental burdens led responsible phenolic resin manufacturers to minimize the levels of free monomers, odoriferous elimination products, and solvents in phenolic resins. As a result of this development work (Table 8.3), novolaks with a level of monomeric phenol below 0.2% and resoles with a level below 2% can presently be used in nearly all areas. It is even possible to reduce the level of free phenol in novolaks to values below 0.05% for special applications. Special resoles with a free phenol level approaching zero for applications such as coated abrasives meanwhile exist.

Table 8.3. Monomer (phenol, formaldehyde) levels in modern, environmentally friendly phenolic resins

	Free phenol %	Free formaldehyde %
Novolak, solid	<0.2 (<0.05)	≪0.01
Novolak, solutions	<0.1	≪0.01
Novolak, low molecular mass (m.p. <50 °C)	<0.15	≪0.01
Resole, solid	ca. 2	<0.5
Resole, aqueous	<1.5	<0.5
Resole, solutions	<1	<0.3

Novolaks are essentially free of formaldehyde, and the level of free formaldehyde in liquid resoles, assuming the application-related demands on the products manufactured from these permit this, can be reduced to 0.1–0.2% depending on the type of resin. For reasons of quality, resins with higher monomer levels must continue to be used in some areas. In these cases, particular attention must be devoted to occupational safety and the MAK limits must be met, or the emissions even better held considerably lower by appropriate technical measures [20, 21].

In addition to carbon monoxide, carbon dioxide, and phenols, aromatic and aliphatic hydrocarbons can arise in small amounts during thermal decomposition of uncured or cured phenolic resins in an atmosphere which is low in or free of oxygen, for example, during pyrolysis of phenolic resin bonded refractory bricks for steelmaking. For many years, measurement of "key" components such as phenol, benzene, formaldehyde, total organic carbon (TOC), and carbon monoxide has provided a satisfactory method for toxicological assessment of pyrolysis gases in practice. A branch of industry in which introduction of this key component concept was very important was the foundry industry [12], where phenolics, furan resins, urea resins, epoxies, and phenolic/polyurethane systems are used as sand binders for cores and molds.

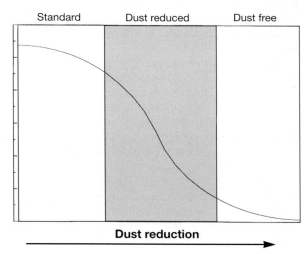

Fig. 8.3. Measurement of dust development (standard and dust-free)

As already described in Sect. 6.2.2.5, p. 327 the levels of dust produced when powder resins are processed have also been reduced. This particularly applies to the use of phenolic resins in production of brake lining compounds or in mixing processes encountered in the grinding wheel industry. As shown by the dust levels in Fig. 8.3, both low-dust and dust-free powder resins exist. These properties are achieved by particularly homogeneous distribution of specific active ingredients throughout the powder resins [13]. The dust measurements are performed using special instruments in which dusting is produced by tumbling the powder resin in a test drum, the dust carried off by a flow of air and trapped in a filter. In the case of dust-free powder resins, this method affords results that approach zero and meanwhile agree with experience in practice.

8.3
Flue Gas Treatment

Additional facilities for flue gas treatment are required if the officially permissible emission limits for phenolic resin processing cannot be met due to the process technology. Fundamentally applicable treatment systems are (1) biological flue gas treatment, (2) flue gas scrubbing, (3) thermal incineration, and (4) adsorption [14].

Biological flue gas treatment [15] was long regarded as a future-oriented process. It was presumed that this process offers a wealth of possibilities for variation. A relatively lengthy process of development and testing has meanwhile taken place, and trials have also been been conducted in the area of phenolic resin processing, specifically in the foundry industry. On the whole, these have shown that the biological initiation/activation surface assumes very large dimensions when large quantities of flue gas are involved, generally preventing successful introduction of the process.

Physical and chemical flue gas scrubbing shifts the focus from the flue gas to the scrubbing medium, and thus generally lead to an effluent disposal problem. On the face of it, flue gas scrubbing is a relatively economical process; however, it has the drawback that the inevitably required effluent treatment can be very costly in the case of phenolic resins, consequently eliminating cost advantages.

Thermal incineration is a process for reducing the levels of hazardous organic compounds in flue gas which has been in use in industry for many years. A combustion chamber temperature of 750–850 °C is required for incineration of phenolic components in flue gas. For economical reasons, thermal incineration is principally used for treatment of solvent-containing flue gas, i.e., gas incorporating a high level of energy, in combination with an efficient heat-recovery system. Direct regenerative combustion (Fig. 8.4) is a thermal flue gas treatment process [16] in which the pollutant-containing exhaust is passed over hot ceramic packing material. This completely incinerates the organic pollutants. The ceramic packing material functions as a regenerative heat exchanger. Up to 95 % of the heat of combustion is recovered; the system thus operates without a primary energy input even at low pollutant levels. This results in relatively low operating expenses and a high treatment effect.

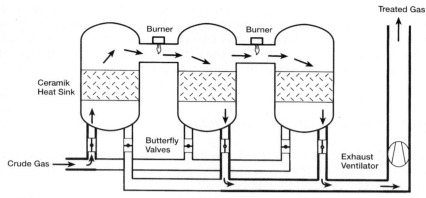

Fig. 8.4. Thermal exhaust treatment process (DRV) of Siemens AG

The *adsorption or active carbon process* (Fig. 8.5) is primarily used for solvent recovery and in treating large volumes of air containing low levels of pollutants. The porous structure of the active carbon [17, 18] provides internal surface areas of 400–1500 m^2/g. Physical adsorption (for example of phenols) is only possible at temperatures ranging up to the boiling point of the adsor-

Fig. 8.5. Active carbon process – fluid bed adsorber

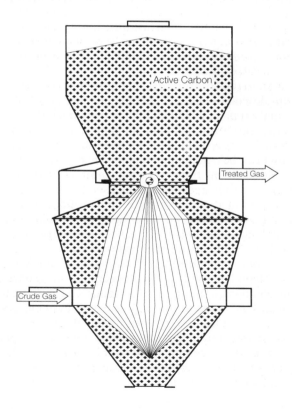

bate, and is particularly effective at lower temperatures. Chemisorption on pretreated active carbon (for example, of formaldehyde) demands a certain minimum temperature, and also proceeds at temperatures above the boiling point of the adsorbate. The active carbon process is oriented toward the pollutant levels to a greater extent than industrial incineration.

8.4
Analytical Determinations at the Workplace and in Flue Gas

Since it would exceed the scope of this publication to consider the details of hazardous materials testing at the workplace and in flue gas, only basic descriptions of the methods will be mentioned [19].

Tests to determine the gaseous substances to which a worker is exposed during a shift and to monitor compliance with the legal limits [20] as well as to determine the extent to which the concentrations exceed or fall below these limits can be carried out using various methods [21, 22]: (1) in the form of basic preliminary tests in the working area using analytical tubes, where the results are indicated by graduated color reactions; (2) using digitally indicating instruments that are also available for some substances, and can be used for preliminary measurements or control tests and represent an evolved form of the analytical tube method; (3) with personal testing devices during the work shift using either (a) diffusion badges that absorb pollutants from the flow of air that bears them or (b) portable instruments that are equipped with absorption fluids or solids and a pump; or (4) by stationary measurements in the workplace area.

Analytical tubes represent one of the classical methods of determination. In simplified terms, a gas analysis tube is a glass tube containing a chemical formulation that undergoes a color reaction with the substance to be determined. The tubes are fused shut on both ends; these are opened and air pumped through to perform the determination. The concentration at the time of sampling can be determined (colored response) from a printed scale.

Because of their wearing comfort, *pumpless long-term measurement systems* and badges are mainly used for personal measurements. However, the question of their accuracy is controversial. *Long-term analytical tubes* with pump systems can be used for both stationary and personal measurements over lengthy periods of time. They afford mean results, but do not detect brief peak levels or concentration variations.

In *indirect measurements*, sampling of the hazardous substance, for example, phenol or formaldehyde, is carried out by enrichment from the workplace air on appropriate collection media. The analytical evaluation is then carried out at a different time and place. Small, light battery-operated pumps attached to the individual are used for personal sampling. Regulation of the air flow is particularly important for proper sampling. The pertinent sampling probe for the specific type of measurement is attached to the carrying belt, the vacuum pump set to the desired flow rate using a calibration system, the pump similarly fastened to the belt, the pump connected to the sampling probe, and the

8.4 Analytical Determinations at the Workplace and in Flue Gas

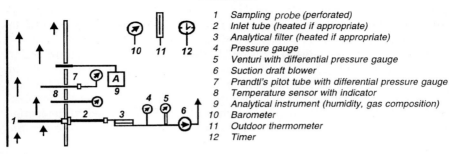

Schematic

1. Sampling probe (perforated)
2. Inlet tube (heated if appropriate)
3. Analytical filter (heated if appropriate)
4. Pressure gauge
5. Venturi with differential pressure gauge
6. Suction draft blower
7. Prandtl's pitot tube with differential pressure gauge
8. Temperature sensor with indicator
9. Analytical instrument (humidity, gas composition)
10. Barometer
11. Outdoor thermometer
12. Timer

Fig. 8.6. Diagram: determination of air pollutants (formaldehyde, phenol, ammonia, sulfur dioxide, organic carbon, various organic solvents, etc.) in flue gases

pump switched on. The sampling period generally extends over an 8 h shift. The evaluation is carried out at a different time and place using gravimetric, optical, atomic absorption, colorimetric, or gas chromatographic methods appropriate to the type of measurement. Stationary workplace measurements may also be carried out in a manner similar to that used for flue gas determinations (using solid and liquid collection media) and described below. Figure 8.6 shows a diagram of the test system [23].

The following *absorption* and *adsorption* media may be used:

1. AHMT (4-amino-3-hydrazino-5-mercapto-1,2,4-triazole) solution, 30 ml per washing bottle (formaldehyde determination by the AHMT procedure).
2. Distilled water, 30 ml per washing bottle (formaldehyde determination by the pararosaniline procedure).
3. Sodium hydroxide, 30 ml 2% solution per washing bottle (phenol determination).
4. Hydrochloric acid, 30 ml 1% per washing bottle (ammonia determination).
5. Silica gel adsorption tube (organic carbon or solvent determination).
6. Further appropriate absorption and/or adsorption media as required.

8.4.1
Example of a Measurement Procedure

A 20 mm opening is cut in the exhaust gas chimney or pipe to sample a partial flow for the measurement. The inner inlet probe is inserted into the measurement opening and connected to the heated inlet probe. The open cross section is sealed with a stopper. The absorption unit (three washing bottles, the first two charged with a certain amount of the appropriate absorption medium and the third as a reserve bottle) or adsorption unit (two adsorption tubes a + b connected in series) is connected to the end of the heated probe. If multiple pollutants are to be determined simultaneously, a glass distribution system may also be attached to the heated probe. In order to prevent formation of condensate in the glass distribution system, the latter is jacketed with insulating tubing. The absorption/adsorption units are then connected to the glass distribution system. Sampling is carried out for a period of 30 min (in special cases for longer or shorter periods) at a flow rate of 3 l/min. The contents of the collection units are then chemically analyzed.

The color reaction with pararosaniline was long favored for determination of *formaldehyde* [24]. However, hexamethylene tetramine interferes with the pararosaniline method [25]. Use of the AHMT procedure [26] is thus preferable, particularly for testing exhausts or emissions arising during novolak-HMTA curing. The color reaction with *p*-nitroaniline is used to determine *phenol* [27], and the reaction with Nessler's reagent to analyze for *ammonia* [28–30].

8.5
Waste, Recycling

The subject of flue gas treatment relates to compliance with pure air regulations. In addition to these, regulations pertaining to waste disposal sites and waste utilization as well as the general process of material cycles must be observed in production and further processing of phenolic resins (Fig. 8.7). The subject of utilization of phenolic plastic and phenolic bonded material wastes is quite varied and decisions must be made for each specific case [31, 32]. In principle, recycling, incineration, and site disposal of cured or uncured systems is possible.

In many cases, it is impossible to avoid creation of process-related, uncured, or partially cured residues when phenolic resins are used. If these residues cannot be subsequently cured, all such incompletely cured material represents hazardous waste. In principal, materials such as incompletely cured resins can be handled either by disposal at a "properly managed waste site" or by incineration, although in the case of incineration it should be noted that the gross calorific value of such waste is generally relatively high and the pertinent incineration facility must be capable of dealing with waste exhibiting such

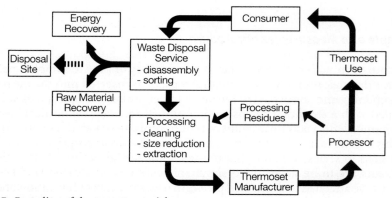

Fig. 8.7. Recycling of thermoset materials

high gross calorific values. If appropriate, such waste can be mixed with products exhibiting low gross calorific values. Depending on the quantities, volumes, and process opportunities, production waste composed of fully organic materials such as resin-containing fibers from used industrial (phenolic resin bonded) textile felts or waste from processing phenolic molding compounds containing organic fillers can be recycled [33–38].

However, the lack of disposal site capacity and hazardous waste incineration facilities as well as the rapid rise in disposal costs make it wise to consider waste avoidance or minimization as well as possible recycling of residues. At an early date, the manufacturer and processor of phenolic resins were thus required to deal with the question as to how the waste and used materials arising in connection with production and processing of these products could be disposed of in an environmentally acceptable manner. Nonetheless, the methods that can be used for this purpose depend on the specific circumstances.

Reworking of production residues is successfully practiced by notable thermoset processors. This requires specific organizational structures to ensure the required standard of quality. Filled or reinforced phenolic plastics can basically be reused by means of particulate recycling. Depending on the end use, however, the risks with respect to the achievable level of secondary material quality should not be underestimated. In view of these aspects, the question as to whether the effort involved in recycling is justifiable, or whether it may be economically and ecologically better to exploit the energy content of certain materials by thermal utilization rather than carry out particularly cost-intensive recycling, must be considered in each individual case.

On the other hand, case studies have demonstrated that residual materials from production of phenolic-impregnated filter papers and clutch linings as well as resin-impregnated textiles [39–41] from grinding wheel fabrication offer interesting possibilities of reuse. Despite this, greater utilization of such opportunities is frequently hindered by logistic or separation problems.

Phenolic molding compounds can be recovered by particulate recycling. This can be done without significant losses in quality both in the processing plant and during molding compound production. Available experience shows that such activities are not only ecologically, but also economically desirable. The cured or partly cured fabrication waste produced during thermoset processing can be directly milled and incorporated into new goods in the processing plant. Thorough investigations confirmed that this can be done without losses in quality [42, 43]. This recycling technique has since been successfully practiced by many molding compound processors.

However, recycling of particulate material to produce high-quality molding compounds assumes that this material satisfies the same demands of quality as all other raw materials. This includes the requirements that it be free of foreign materials such as metals and other plastics, as far as possible cleanly sorted and of defined composition. The latter principally refers to the type of binder and the main fillers. This material is recycled either into the original product, or into products specially developed for this use. The PF 2000 molding compound (Fig. 8.8) represents an example of this type of product. It contains

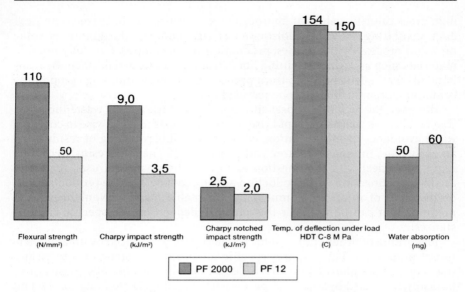

Fig. 8.8. Comparison of various material parameters of recycled material – containing molding compound PF 2000 and type PF 12 (source: Bakelite AG, Iserlohn, Germany)

50% recycled material, about 15% chopped glass fiber, and 35% phenolic resin including curing agent and miscellaneous additives. This formulation not only offers excellent processing properties, but also a range of material properties that in some cases exceeds the quality of the primary material. This enhancement is mainly due to the double impregnation effect.

Although fabrication wastes are generally obtained in straight (unmixed) form, and are free of foreign contaminants, the situation with respect to rejects (moldings) in areas such as the automotive industry is completely different. Two circumstances that render recycling considerably more difficult are encountered in such cases: (1) the materials are directly or indirectly bound to other products in complex construction elements, and cannot be simply separated from these, and (2) the components are exposed to a wide variety of environmental effects in their applications, and are accordingly contaminated. The parts to be recycled must thus be sorted by material types and cleaned before use.

Specific studies have similarly demonstrated that returns (used or defective parts) can also be recycled. Despite this, the risks with respect to the achievable material quality should not be underestimated. In view of this aspect, it is probably better to avoid undue effort and make reasonable and economical use of the energy content of such materials in individual cases by thermal utilization.

8.6
Renewable Raw Materials (Furfuryl Alcohol, Lignin, Tannin)

For ecological and economical reasons, renewable raw materials are presently the object of increasing attention. Furan resins offer a classical example of such utilization, since their basic raw material, furfuryl alcohol, is obtained from corn stalks (Fig. 8.9). Phenolic resins can similarly be modified with furfuryl alcohol, for example, in production of foundry binders and in acid-resistant construction [44].

Although only 4% of the total oil production is used in manufacture of plastics, the limited availability of petroleum provides a constant motivation to free completely or partially thermosetting materials from their dependence on oil wherever this is possible. The ability of phenolics to incorporate high levels of inorganic fillers and reinforcing agents that are available in unlimited quantities offers such a possibility. Possible filler levels in materials such as common phenolic molding compounds are 50–65%. In general, thermosets can incorporate up to 80% fillers, whereas the limit on use of fillers and reinforcing agents in thermoplastics, the bulk of which represent unfilled systems, is generally limited to a filler level of 30%.

In the search for alternatives to oil-derived binders, two naturally occurring raw materials in addition to furfuryl alcohol – tannin and lignin (see Sect. 6.1.1.6.2 and 6.1.1.6.3) – offer interesting possibilities. Lignin (Fig. 8.10) is obtained in large quantities in production of cellulose. In the plant world, it represents a widely occurring basic substance that can exhibit various structures depending on the type of plant it is derived from, but in contrast to other biopolymers it exhibits no regular structure. Wood contains about 20–30% lignin. Since several structural changes occur in the lignin during the digestion process, this process can exert a great influence on the quality of lignins. On the basis of its structure, lignin may be considered a kind of phenolic condensate in the broadest sense of the term, making it logical to use it in production of phenolic resins. Pertinent tests involving the use of lignin in various areas

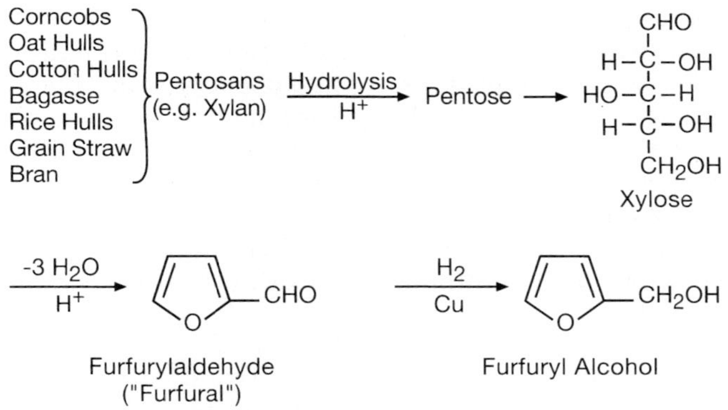

Fig. 8.9. Origin of furfuraldehyde and furfuryl alcohol

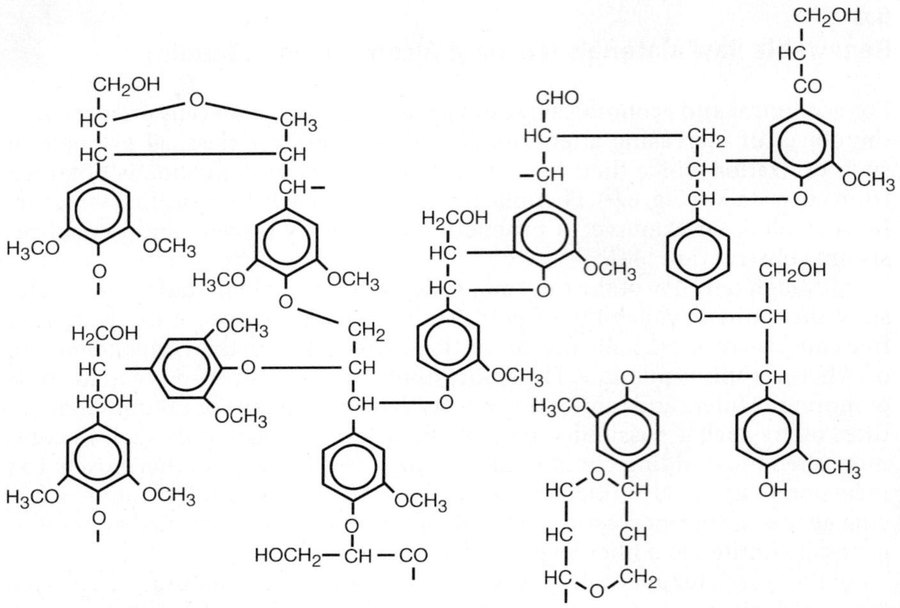

Fig. 8.10. Structure of lignin (segments)

have been carried out in the last few years, and some have been successful [45–49]. At present, it is necessary to wait and see whether the cellulose and paper industry can further develop sulfur-free digestion processes and bring these into large-scale production, since a basic requirement for the use of lignin is generally the availability of sulfur-free, low molecular mass lignins. Such materials are basically available by way of the Organosolv process. However, the Organosolv process, involving organic wood digestion, is currently only used to a limited extent on an industrial scale.

Tannin obtained from specific types of wood (used in large quantities for tanning) is already employed in production of chipboard at present, and is also under consideration for other applications since it can be easily transformed into biopolymers and incorporated into phenolic resin systems [50–54]. Thus, tannins have been industrially used as binders for chipboard and fiberboard for several years, not only in South America and South Africa, but also in Europe [55]. Tannins can also be used as curing accelerators or phenolic resins in the plywood industry (Fig. 8.11). The high reactivity of the tannins results from their polyphenolic nature. Depending on their chemical composition, polyphenols differ from one another with respect to their reactivity toward formaldehyde. The reaction with formaldehyde, which forms the basis for the use of these binders, affords high molecular mass polycondensates. Since crosslinking sites are only present to a limited extent, the formaldehyde demand is small, and the crosslinking density in the cured binders is relatively low (see Sect. 6.1.1.6.3, p. 148–9).

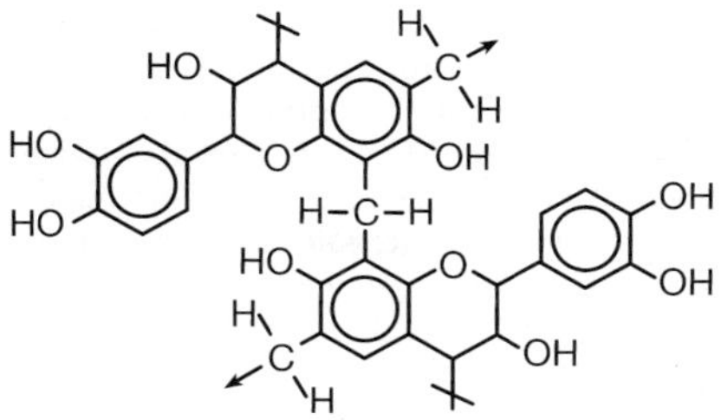

Fig. 8.11. Structure produced by reaction of tannin and formaldehyde

When used as binders for chipboard and fiberboard, tannin extracts thus offer several advantages such as the low formaldehyde demand during curing, the possibility of using chips with higher moisture levels, and the faster rate of cure. In addition, the polyphenolic substances in the extract undergo ideal curing under slightly alkaline conditions. Wooden materials bonded with these products thus contain very little alkali and only liberate very low levels of formaldehyde. In these respects, they largely correspond to chipboard made with straight phenolic resins, and furthermore exhibit a lower sorption capacity than straight phenolic chipboard [56, 57].

Whereas tannins are used alone and in combination with other materials as primary binders for wood materials, lignins – in this case not only those derived from the Organosolv process – can be employed as extenders for phenolic resins. Such materials may represent kraft lignins, various sulfite liquors, waste liquors from non-conventional wood digestion processes, or lignins from annual plants.

This discussion on the use of furfuryl alcohol, lignins, and tannins as renewable modifiers or partial replacements for phenolic resins is intended to show that these products do not merely represent "alibi substances" in the question of resource conservation, but have meanwhile even become irreplaceable raw materials in a number of areas.

References

1. Gardziella A, Müller F-J (1990) Umweltrelevanz und Anwendungsfunktionen der Phenolharze. Kunststoffe 4:510–514
2. Bakelite AG (1992) Toxikologie, Arbeitsschutz, Emissionen, Rutaphen-Phenolharze. Company Publication, Bakelite AG, Iserlohn, Germany, Sept, pp 15–17
3. Carl Hanser (1988) Chemie und Physik der duroplastischen Harze. In: Kunststoff-Handbuch, vol 10, Duroplaste. Carl Hanser, Munich pp 1–155
4. Galle F, Decker KH, Exner HD, Grünewald R, Kugler F, Meier B Toxizität, Gewerbehygiene, Vorschriften: Phenolharze. In: Kunststoff-Handbuch, vol 10, Duroplaste. Carl Hanser, Munich, pp 135–146
5. Deutscher Bundes-Verlag (1996) Verordnung über gefährliche Stoffe-Gefahrstoffverordnung (GefStoffV) vom 26.8.1986. In the Revision of July 1996, Deutscher Bundes-Verlag, Bonn
6. Industrieverband Giessereichemie (1980–1990) Untersuchungen toxikologischer Daten von Giessereibindemitteln, 1980–1990. Commissioned by the Industrieverband Giessereichemie, Frankfurt/Main
7. Bakelite AG (1980–1990) Untersuchungen toxikologischer Daten von Phenolharzen, 1980–1990. Commissioned by Bakelite AG, Iserlohn, Germany

8. Senate Commission (1989) Maximale Arbeitsplatzkonzentration und Biologische Arbeitsstofftoleranz-Arbeitsstofftoleranzwerte 1989. 33rd Communication of the Senatskommission zur Prüfung gesundheitsschädlicher Arbeitsstoffe (Senate Commission on Testing of Workplace Materials That are Detrimental to Health) of July 1997
9. Gardziella A (1981) Ursachen und Verringerung von Schadstoffen bei der Herstellung und Verwendung von Phenol- und Furanharzen. Giesserei 8:223–232
10. Gardziella A (1979) Reduzierung von Emissionen und Stäuben bei Herstellung und Anwendung von Phenolplasten, Humane Produktion – Humane Arbeitsplätze, Issue 4, Verlag für technische Literatur, Frankfurt
11. Bakelite AG (1986) Emissionen bei der Verarbeitung von Phenolharzen und Möglichkeiten zu ihrer Reduzierung. Company Publication, Bakelite AG, Iserlohn, Germany, Oct
12. Carl Heymanns (1997) Technische Regel für Gefahrstoffe (TRGS) 402, Ermittlung und Beurteilung der Konzentrationen gefährlicher Stoffe in der Luft in Arbeitsbereichen. Version of 1997, Carl Heymanns Verlag, Cologne
13. Verein Deutscher Giessereifachleute (1980) VDG Technical Information Publication R 311, Gase und Dämpfe beim Einsatz von Formstoff-Bindemitteln und Formüberzugsstoffen – Beurteilung und Praxisempfehlung. Verein Deutscher Giessereifachleute, Düsseldorf, Feb
14. Bakelite AG (1993) EP-B 0 595 003
15. Eisenmann Maschinenbau (1989) Leitfaden Umwelttechnik. 4th Revised and Expanded Edition, Company Publication, Eisenmann Maschinenbau GmbH, Böblingen, Jan
16. Arasin (1987) Biokatalytische Abluftreinigung. Company Publication, Arasin, Voerde
17. Siemens AG (1995) Direkte regenerative Verbrennung – DRV. Technical Information Brochure, Siemens AG, Energy Production Department, Erlangen, Germany
18. Müller F-J, Winkel J (1983) Aktivkohle im Einsatz bei phenol- und formaldehydhaltigen Abgasen. Chemie-Technik 12(2), Hüthing-Verlag, Heidelberg
19. Norit Adsorptions (1990) Aktivkohle-Eine Einführung. Company Publication, Norit Adsorptions GmbH, Düsseldorf
20. Hans Heger (1986) Erste Allgemeine Verwaltungsvorschrift zum Bundes-Immissionsschutzgesetz (Technische Anleitung zur Reinhaltung der Luft – TA-Luft). Verlag Dr. Hans Heger, Bonn
21. BHI (1980) Richtlinie für die Eignungsprüfung, den Einbau und die Wartung kontinuierlich arbeitender Emissionsmessgeräte. GMBl. 1980, 343, BHI Circular of July 21
22. Drägerwerk (1994) Dräger-Röhrchen-Handbuch, Boden-, Wasser- und Luftuntersuchungen sowie technische Gasanalyse, 9th edn, Drägerwerk AG, Lübeck
23. GSM (1992) Personenbezogenes Probenahme-Gerät, System zur Probenahme von Stäuben, Fasern, Dämpfen und Gasen am Arbeitsplatz. In: Messgeräte für den Arbeitsschutz, 3rd edn. GSM, Neuss, Germany, April
24. Bakelite AG (1995) Determination of air contaminants in exhaust gasses. Technical Information Brochure, Bakelite AG, Iserlohn, Germany, Aug
25. Bakelite AG (1994) Determination of formaldehyde with pararosaniline. Technical Information Brochure, Bakelite AG, Iserlohn, Germany, Sept
26. Gardziella A, Haub H-G, Mühlhauser S, Müller F-J (1988) Messung von Emissionen bei der Aushärtung von Phenolharzen – Probleme durch Querempfindlichkeiten bei der Bestimmung von Formaldehyd. Staub-Reinhaltung der Luft 48:145–149
27. Bakelite AG (1995) Determination of formaldehyde with 4-amino-3-hydrazino-5-mercapto-1,2,4-triazole (AHMT). Technical Information Brochure, Bakelite AG, Iserlohn, Germany, July
28. Bakelite AG (1995) Determination of phenols with p-nitraniline. Technical Information Brochure, Bakelite AG, Iserlohn, Germany, Aug
29. Bakelite AG (1994) Determination of ammonia with Nessler's reagent. Technical Information Brochure, Bakelite AG, Iserlohn, Germany, Sept
30. Bakelite AG (1995) Determination of organic carbon (silica gel). Technical Information Brochure, Bakelite AG, Iserlohn, Germany, Aug
31. Bakelite AG (1994) Determination of organic solvents with silica gel or active carbon. Technical Information Brochure, Bakelite AG, Iserlohn, Germany, Sept

32. Braun U (1995) Ökologische Aspekte beim Einsatz von Duroplasten. In: SKZ Duroplastische Werkstoffe aktuell. Symposium, Würzburg, April 26-27
33. Emminger H, Decker K-H (1988) Entsorgung von duroplastischen Abfällen. In: Kunststoff-Handbuch, vol 10, Duroplaste. Carl Hanser, Munich, pp 148-156
34. Eisele D, Gardziella A, Marutzky B, Müller F-J (1992) Duroplastisches Textilvliesmaterial: bewährt und verbessert. Kunststoffe 3:229-234
35. Braun U (1991) Lösungen für technische und umweltrelevante Fragen beim Einsatz duroplastischer Formmassen. SKZ Polymere Hochleistungswerkstoffe in der technischen Anwendung. Symposium, Würzburg, June
36. Oppermann M (1983) Entwicklung eines Recyclingverfahrens zur direkten Wiederverwertung von Duroplastabfällen. BMFT T 83-303
37. Fischer V (1983) Möglichkeiten und Grenzen der Wiederverwendung duroplastischer Formmassen-Abfälle. Seminar No 118,033 Aufbereitung und Wiederverwertung von Kunststoffabfällen, Wuppertal
38. Niemann K, Braun U (1992) Möglichkeiten einer gesicherten und sinnvollen Entsorgung. Plastverarbeiter 1
39. Adolphs P (1992) Rohstoff- und Energiebilanz bei compoundierten Duroplasten unter beproberer Berücksichtigung des Recyclings und der Rückgewinnung von Energie. SKZ Duroplaste im Automobil, Symposium, Würzburg, Sept 23-24, pp 27-42
40. Bakelite AG (1991) DE-C 4 142 251
41. Bakelite AG (1992) DE-C 4 211 446
42. Bakelite AG (1992) DE-C 4 211 446
43. Gardziella A (1988) Furanharze. In: Kunststoff-Handbuch, vol 10, Duroplaste. Carl Hanser, Munich Vienna, pp 70-86
44. Gardziella A, Schröter S (1994) Lignin – ein nachwachsender Rohstoff – Möglichkeiten und Voraussetzungen zum Einsatz als Rohstoff für Phenolharze. Unpublished in-house communication, Bakelite AG, Iserlohn, Germany
45. Rütgerswerke AG (1992) EP-B 0 540 835
46. Bakelite AG (1992) EP-B 0 540 836
47. Rütgerswerke AG (1992) EP-B 0 540 837
48. Roffael E, Dix B (1991) Lignin and lignin sulfate in non-conventional bonding. Holz als Roh- und Werkstoff 49:199-205
49. Roffael E (1995) Duroplastgebundene, umweltfreundliche Holzwerkstoffe. In: SKZ Duroplastische Werkstoffe – aktuell Symposium, Würzburg, April 26-27
50. Roffael E (1976) Über die Reaktivität von wässerigen Rindenextrakten gegenüber Formaldehyd. Adhäsion 20:297-298
51. Kulvic E (1976) Chestnut wood tannin extract in plywood adhesives. Adhesive Age 19:19-21
52. Kulvic E (1977) Chestnut wood tannin extract as a cure accelerator for phenol formaldehyde wood adhesives. Adhesive Age 20:33-34
53. Long RK (1991) Quebracho based polyphenols for use in wood-panel adhesive system. Holz als Roh- und Werkstoff 49:485-487
54. Rütgerswerke AG (1993) EP 0 639 608 A1
55. Rütgerswerke AG (1994) DE 4 402 341 A1
56. Bakelite AG (1994) EP-BO 648 807
57. Rütgerswerke AG (1993) DE 4 331 656 A1

CHAPTER 9

Conclusion: Guidelines for Future Developments in Phenolic Resins and Related Technologies

The impression sometimes arises that development, production, and consumption of phenolic resins and other thermosets – particularly amino and furan resins – are stagnant or in decline. However, this is not, in fact, the case. It is safe to assume that this group of products will always be required for many end-use applications due to its performance characteristics. The reader of this publication will recognize that phenolic resins still have a bright future as adhesives for industrial applications and thermosetting molding compounds. Pertinent examples are matrix resins for high performance fiber composites, applications in the transportation industry, components for carbon-based materials and refractories that afford a high carbon yield in the carbonization process, and glass fiber-reinforced phenolic molding compounds exhibiting high, constant thermal resistance.

The versatility of phenolic resins as exhibited by the six bonding functions and the range of applications may stimulate further modifications to resin, process, or final product.

Phenolic resins participate in both high volume commodity market areas as well as low volume specialty markets where product performance is crucial. Selected examples include photoresists for computer chips, and carbon carbon composites for aircraft brakes and nozzles for the U.S. space shuttle.

Aside from the technical performance parameters of phenolic-based materials and commercial aspects related to their use, future interest will be primarily concerned with developments that must offer an expanded range of properties. Future acceptance of phenolic-based materials will depend on the extent to which they are modified to meet more rigorous requirements.

Furthermore, shortened innovation cycles (laboratory to commercialization), challenges resulting from raw material shortages or changes in resources, more economical production methods, and demands relating to workplace safety and environmental aspects lead to certain guidelines for continued development of phenolic resin systems. The following proposals are presented to stimulate areas of technical research in expanding the utility of phenolic resins:

1. Development of production methods that reduce energy and increase yields. This refers to both continuous and discontinuous, computer controlled, modern processes for manufacture of resins and molding compounds. Computer simulation for desired resin preparation and/or formu-

lation, microwave, and E-beam methods of cure are additional areas of interest.
2. Development of phenolic adhesive systems in which alterations in the reactivity allow reduction of energy and time during cure. Pertinent examples of such goals are reduction of curing temperatures in phenolic-based coatings systems, reduction of the throughput time in production of coated abrasives, and reduction of the press time required for wood composites and all types of molding compounds, as well as a transition to core adhesives that cure at lower temperatures (e.g., using gassing methods) in foundries.
3. Enhancement of workplace safety and environmental protection during processing of phenolic resins by reduction of monomers and other pollutants in the original resin. Continued efforts in solvent free, aqueous, or melt processing (grinding wheel area) are recommended. Certain manufacturers have introduced „environmentally friendly" resins and more are expected. Thus, residual monomer (phenol, formaldehyde) levels in a wide variety of phenolic resins have been reduced to the limit of detection. Some processes involving changes in conventional resin synthesis will require some modification if the range of properties provided by the adhesive are to be maintained.
4. Reduction of exhaust emissions by development of new phenolic resins that liberate less pollutants to the environment during curing or pyrolysis. This is accomplished by systematically controlling the cleavage products produced during cure or pyrolysis of the phenolic resins with the objective to reduce emissions and simplify or eliminate the need for flue gas treatment.
5. Enhancement of the adhesive strength of phenolic resins by various means with the objective of reducing the resin level in phenolic bonded materials to an acceptable minimum. Reduction of the resin level is accompanied by a possible decrease in the levels of cleavage products liberated during abrasive or destructive utilization of materials such as grinding wheels, friction linings, or foundry molding materials.
6. Development of phenolic resins exhibiting higher levels of mechanical properties for use in materials such as fiber reinforced composites, that in addition should be non-flammable and exhibit reduced smoke density, for applications in airplane construction or mass transportation systems such as high-speed trains. Synergism of resin structure and char formation should be considered.
7. Optimization of phenolic resin and phenolic molding compound recycling capabilities to permit processing flash/waste and used articles to be returned to the molding material production process. Recycling of phenolic resin-based processing flash/waste by solid or particulate recycling is possible, and is already practiced today.

It is safe to assume that phenolic resins will maintain or even expand their market position relative to other plastics in the future, since their range of properties is unique in many areas. Process technologies that generally identify targets for products that compete with phenolic bonded materials should be optimized. On the other hand, even high-performance thermoplastics have their limits in both properties and processing technology. Appropriate modification – for example, with other resin systems – with the goal of further improving the processing technology or achieving a high-performance level of physical properties will become increasingly significant in phenolic resin systems.

Due to their relatively low energy demand, low environmental impact, high filler utility and compatibility, and low petroleum dependency, continued growth in phenolic-based materials is expected in the future. Furthermore, utilization of their typical range of thermal, electrical, and chemical properties will lead to additional expansion of opportunities for use of this versatile family of polymers.

Subject Index

Abietic acid 376
Ablation 104, 261
Ablative use 267
Ablation products 262
Abrasion 267
– Low 224
Abrasion resistance 266, 372, 373, 403
Abrasives 72, 103, 113, 122, 124, 125, 291, 314–353, 510
– Industry 115, 314
ABS (Automatic braking system) 76, 213, 215
Accelerators 135, 174, 190, 191, 193, 231, 284, 285, 307, 309, 321, 399, 400, 407
Acetal 296
– process 299, 312
– resins 18
Acetaldehyde 14, 21, 22
Acetic acid 46, 147, 221
Acetone 6, 8, 11, 13, 21, 47, 50, 73, 98, 285, 286, 391, 407, 409
Acetophenone 68
Acid Reactivity of Liquid Phenolic resins 500
Acid 58, 61, 76, 119, 224, 283, 287, 297, 298, 299, 312, 382, 437
– organic 160, 287
– weak 41, 46, 224
– catalyzed 48, 58, 160, 165, 302, 305, 307
– heat 297, 298
– inorganic 58, 160
Acid curing 305, 380
Acidic 70, 78, 453, 456, 457
Acetylated 98, 99, 148
ACGIH 518
Acid hydrolysis 147
Acoustical insulation 187
Acoustical properties 175, 179, 182
Acoustical responses 278
Acoustical tiles 155
Acrolein 22

Acrylate resins 112, 114, 297, 298, 313, 340, 344, 376, 380, 397, 405, 414
Acrylic fibers 172
Actinic radiation 394
Actinic light 385, 387, 397
Activation energies
Addition 28, 29, 33
– Reaction 78
Additives 64, 131, 133, 190, 249–251, 253, 254, 283, 285, 286, 288, 301, 303, 312, 344, 358, 365, 372, 375, 377, 384, 385, 405, 415, 435, 439, 460
Adhesion 52, 73, 125, 126, 152, 240, 254, 258, 264, 285, 287, 363, 373, 382, 383, 406
– Promoter 231, 337, 380, 405
Adhesives 12, 50, 51, 57, 124, 125, 127, 130, 131, 134, 135, 137, 140, 141, 143, 149, 152, 256, 274, 373
– cold setting 149, 150
– modified 142, 153
– room temperature 149, 150, 153
Advanced composites 27, 72, 76, 260, 261, 272
Aerodynamic matting 176, 177
Aerogels 168, 169
Aerospace 58, 65, 78, 104, 110, 261
Aging 57, 73, 179, 180
– Resistance 252, 254
Agriculture
– Products 130
– Waste 149, 153
Agriwaste 130, 149, 153
Air conditioners 174, 180, 219
– Filters 252
Air treatment processes 516–524
Aircraft 56, 78, 105, 115, 117, 124, 149, 232, 256, 259, 268, 269, 270, 276, 277, 413, 415
– Flooring panels 269, 270
– Commercial 261, 262, 266
– Manufacturers 105, 266
– Brakes 56, 66, 449
– Composites 57, 127

Subject Index

Aircraft
- interior 57, 73, 105, 127, 167, 262, 263, 270, 271, 280, 281

Alcohol 62, 128, 147, 224, 254, 285, 287
Aldehydes 3, 14, 21, 22, 48, 122, 394
Aliphatic polymer 387, 388, 397
Alkali 33, 89, 101, 224, 382, 451, 452
- resistance 282, 283, 376, 377

Alkaline 33, 78, 119, 157, 384, 394, 421, 451, 453
- Earth 89, 287
- Hydroxides 156, 159, 287

Alkoxylation 455
Alkylation 5, 10, 22, 60, 61
Alkyd 326, 341, 374, 375, 376, 380, 381, 383, 384
Alkyd resin/isocyanate
- No-bake 299

Alkylphenol resins 357, 363, 364, 370, 371, 373, 374, 375, 376, 379–381, 383, 385, 398–400, 402, 405–410, 415, 456, 457, 459

Alkylphenols 3, 5, 9–12, 363, 375–377, 384, 402, 464

Alloy 24, 75–77
Allyl 24, 51, 55, 78
- -ether 55

Alumina 103, 320, 425, 426, 433
Aluminum 267, 268, 292, 293, 302, 304, 307, 310, 311, 377, 387, 410, 414, 451
- Die cast 218, 225
- Melts 256, 258

Aluminum oxide 11, 339, 433
Al (OH) 3, 162, 174, 196, 248
Ambient temperature 282, 284, 287, 297, 298, 305

Amides 61, 69, 194
Amide/imide 69
Amine 59, 61, 69, 89, 99, 305, 307–310, 383, 420, 455, 460, 462, 464
- Primary 57, 99

Ammonia 21, 74, 75, 89, 99, 142, 174, 182, 200, 301, 331, 332, 363
- Free 195, 196, 221, 224, 442

Ammonium chloride 136
Ammonium nitrate 303
Ammonium polyphosphate 162
Anacardic acid 359
Analyses
- End group 92
- Fracture mechanics 77
- Wet chemical 147

Analytical
- methods 142, 145, 423
- probe 62
- techniques 62

Analytical masticator test 209
Anchor 213
Ancillary motor components 218
Andalusite 427, 432, 433, 434, 436, 437
Aniline 4, 5, 58, 59
Anisotropic 202
Animal
- Glue 344
- Oils/fats 224

Anthracene oil 325, 329, 338
Anthracene sulfonate 396
Anticorrosion primers 380–383
Antidusting agent 174, 327, 441
Antimony 256
Antimony trioxide 231, 248, 323
Antimony trisulfide 321, 359
Antioxidants/Stabilizers 460–468
Antioxidants 11, 22, 50, 406, 408, 433
Appendage 24, 51, 60, 101
- Ethyl acetate 101–103, 105
- Triethoxyethylene ether 102–103, 105

Application Technology Testing 507–513
Applications 122, 124–127, 133, 141, 153, 159, 169, 175, 179–181, 185–187, 192–194, 207, 212–224, 230–234, 238, 241, 248, 250, 251, 253, 254, 258–260, 269, 278, 283, 284, 286, 287, 290, 291, 294, 312, 314, 363, 380, 381, 398, 401, 403, 405, 409, 410, 415, 416, 425, 432, 446, 454, 455
- Critical 261, 262

Aramid 151, 256, 261, 264, 266–269, 272, 278, 289, 333, 358, 415

Arches 149, 278
Asbestos 358
- Fibers 192, 193, 195, 214, 288, 358

Ashtrays 215, 219, 221
Aspen 129, 131
Associated 97, 98, 99, 105,
ASTM 203, 204, 262
Audio equipment 238, 241, 243
Autoclaves 284, 434
Automatic 285, 368
- Machines 285
- Transmission 232, 258

Automation 281, 314
Automobiles 130, 142, 153, 180–182, 188, 414, 415

Automotive 110, 170, 176, 180, 218, 225, 232, 252, 256, 272, 280, 281, 301, 355, 383, 385, 405
- Areas 193, 213, 215, 277, 304
- Electronics 221, 224, 231,
- Engineering 175, 180, 212, 380
- Industry 185, 219, 295, 353, 383, 406, 415
- Construction 115, 126, 127, 410

- Tires   373, 398, 401, 402, 404
Autooxidation   461, 462, 467
Auxiliaries for petroleum and natural gas production   125, 370–373

B-stage time determination   157, 249, 253
B-staged intermediate   62, 265
Baekeland   109–112, 187, 197, 233, 314, 374, 489
Bagasse   127, 149, 153
Bakelite   110, 122, 123
Ballistics   265, 266–267
Balsamic resin   285, 286
Bar code   385, 460
Barium hydroxide   33, 89, 102, 156,
Barytes   282, 359, 408
Base   55, 56, 61, 280, 285, 286, 290
- Metal   285, 287
Basic   25, 70, 78, 150
Battery separators   232, 256
Batts   155, 156
Bauxite   426, 427, 432–434, 436, 437
Beams   145, 151, 276, 278
Bearing   221, 224, 251, 272, 291, 411, 446
- Shells   193, 251
Bench life   307, 308
Bending strength   179, 205–207, 239, 418
Benzaldehyde   14, 69
Benzene   5, 9, 12, 98, 101, 224, 423
Benzoic acid   8, 9
Benzotriazole   466–467
Benzoxazine   57–59, 67, 69, 78,
Benzyl   55, 78
Benzyl alcohol   37, 395
Benzylamines   67, 69
Beverages   377, 460, 463
BHA   463
BHT   463–465
BH Incandescent Rod Flammability Methods   207
Billets   143, 145
Bimodal   139, 390
Binder   89, 125, 126, 153, 156, 170, 171, 190, 191, 198, 214, 250, 267, 283, 285, 287, 292, 294, 296, 298, 299, 301, 302, 304, 305, 307, 310, 312–314, 340, 341, 348, 351, 354, 358, 361, 366, 370, 375–378, 385, 404, 415, 423, 426, 427, 431, 432, 441
- Inorganic   293, 294
- Organic   293, 294, 357
- Synthetic resins   296, 298, 299
Binders for the Adhesives Industry   405–415
Bisbenzoxazines   61, 72, 74

Bismaleimide   56, 58
Bismethylol   48, 72
Bisoxazolines   61, 72, 73
Bisphenol F   36–39, 51, 67, 111, 396
Bisphenol A   3–5, 13–14, 37, 38, 51, 56, 59, 111, 377, 385
- tetramethylol   72
Bisphenols   56, 385
Blend   24, 136, 285, 305
Blowing agents   159, 160, 162–164, 168
BMI   56, 58, 112
Bobbins   220, 224
Bolted connections   179, 219
Bond   152, 285, 292
Bonded Abrasives   316
Borates   309, 312
Boric acid   312
Boron   174, 363, 364, 420
Boron nitride   317, 321, 323
Boron carbide   321, 339
Bottle method   184–185
BR   398
Brake   353
- Disk   291, 357, 364
Braking systems   215, 258, 268
Brake linings   117, 122, 126, 200, 218, 291, 351, 373, 417
Branched   41, 97, 99, 105
Branching   41, 45, 46, 78, 97, 134, 149, 150
Brashness   155, 156
Brass   292, 359
Bricks   425, 426, 433, 435–437
- Acid resistant   282
- Magnesia   426, 434, 436
- Dolomite   426, 431, 434, 436, 437
- Manufacture   434–438
Bridge   41, 149, 151, 261, 278, 279, 380
Brittleness   24, 75, 162, 165, 281, 332, 374
Bronze   292, 302, 339
Brush cements   287–288
Brushes   281, 287
Bubbles   200, 285, 287, 349, 361, 363, 414
Building industry   115, 125, 167, 169, 405
Building trade   124, 163, 186
Buildings   145, 277
Bulk molding compound (BMC)   117
Bulk density   176, 179, 185, 205, 256, 286, 494
Burns   207, 266, 271
Burst strength   232, 321
Bus   105, 281,
Butanol   249, 376, 379, 381
Butyl rubber   400, 405, 414
Butylphenol resin-p-tert   407
Butylphenol p-tert   3, 11, 12, 95, 377, 406

Subject Index

Butyraldehyde 14, 21, 22
By-product 51, 78, 116, 147, 312

C-C Composites 261, 262, 267–278, 276, 288, 445, 447–449
– Brakes 261, 262, 268, 449
Calcium stearate 300, 301
Calcium carbonate 134, 135, 162, 196, 280, 343
Calcium hydroxidc 33, 89, 156
Calcium oxide 321, 325, 408, 432, 433, 437
Calcium silicate 162, 409
Calendar mills 197, 365
Calibration curves 44, 92
Calixarene 47, 48, 100–103, 105, 106, 389, 396
Campus data base 195
Camshaft 219, 302
Cannizzaro reaction 15, 16
Capacitors 169, 208, 242, 455
Caprolactame ε 4, 5
Carbon and Graphite Materials 444–455
– Phenolic resins 446–450
Carbon Forming Bonding 125, 126, 296
Carbon Forming Function 296, 355
Carbon 196, 444
– Black 65, 194, 231, 403, 404, 428, 432, 435, 444, 450
– Donor 117, 444, 445, 446, 447, 451
– Glassy 296, 310, 446, 452-5
– Materials 117, 125, 426, 490
– Polymeric 126, 417, 419, 420, 435, 445, 452–455
– Yield 119, 125, 126, 416–420, 441, 445, 450
Carbon Fibers 58, 73, 103, 104, 151, 261, 264, 265, 268, 272, 278, 280, 288, 358, 445, 446
Carbon fiber matrix system 267, 445
Carbon fiber/PF prepreg 262
Carbonates 61, 156
Carbonium ions 37, 400
Carbonization 126, 190, 261, 267, 355, 416, 418–420, 430, 432, 434, 439, 445, 446, 448
Carbonless Copy Paper 12, 125, 126, 385, 456–459
Carbon monoxide 12, 408, 423
Carburetors 219, 464
β Carotene 460
Carrier film 279, 280
Cashew 51, 61,
Cashew nut shell liquid (CNSL) 359, 363, 374, 403

Casting 126, 164, 287, 292, 293, 296, 299, 301, 302, 304, 310, 313, 438
– Centrifugal 382, 384, 452
Casting Resins 287–288
Catalysts 24, 33, 50, 60, 77, 109, 156, 159, 162, 164, 190, 191, 279, 280, 287, 298, 308, 382
– Acidic 20, 21
– Content 62, 160
Catechol 148
Cation 13, 27, 33, 103
CB coating 456–458
Ceiling/sidewall 154, 266, 270
Cell elements 140, 161
Cell
– Shape 168, 169
– Size 168, 169, 269
Cellulose 16, 103, 128, 147, 192, 194, 196, 257, 313, 415
– Fibers 190, 192, 195, 252, 253, 256, 358
– Paper 246, 248, 256
Cements 127, 282, 294, 426, 428
– Acid resistant 113, 125, 282
CEN 203, 204, 244, 489
Centrifugal force 213, 214
Ceramic 231, 314, 318, 321, 322, 351, 428, 444
CF layer 456–459
CFC (Blowing agent) 161, 168
CFC (Carbon Composite Carbon Fiber) 444–450
Char 57, 58, 104, 166, 261, 262
Charpy Impact Resistance 206
Charpy Impact Strength 197
Charpy Notched Impact Resistance 206, 221, 224
Chelate 46, 407, 408, 458, 459
Chemical, Physical and Application Technology Parameters 488–513
Chemical amplification (CA) 387, 395, 396
Chemical resistance 103, 119, 126, 127, 165, 225, 230, 251, 252, 254, 261, 262, 272, 278, 283, 287, 290, 373–378, 380, 382, 384, 385, 444
Chemical stress 283, 377
Chemically Reactive Bonding 355, 374, 385, 455–468
Chemically Resistant Putties & Chemical Equipment Construction 282–285
– Resins 282
Chemically resistant putties 282
Chip 52, 359, 386, 389, 397
Chipboard 89, 110, 186, 389, 397
Chlorination 9, 10
Chlorobenzene 9, 96

Subject Index 539

Chloroform 47, 98, 101, 105
Chromated copper arsenate (CCA) 152
Chromatography 91, 92, 169
- HPLC 504, 506
- TLC 505-508
- GPC 489, 504
Chromite 291, 294, 426
Circuit boards 238, 241, 243, 251,
Civil engineering 275, 276, 281
Cladding 176, 179, 281, 410, 413
Cladding Adhesives 413
Clay 64, 278, 294, 313, 426, 458, 459
Cleaning 200, 266
Closed cells 159, 161-163, 166
Clutches 256, 258, 291
Clutch linings 117, 291, 353-355, 361, 364, 365, 367, 368
Co-continuous 57, 76, 77
Co-cure 76, 78, 137, 174, 332, 336
CO_2 161, 165, 169, 297, 308-310, 391, 433
- gassing agent 297
- resole cold box 296, 297
- resole process 312
Coal 12, 16, 194, 446
- Tar 12, 450
Coat 283, 287
- Base 340, 344, 345, 347-349
- Sizer 340, 343, 344, 347-349
CoatedAbrasives 316, 340-353, 490
Coatings 11, 38, 50, 51, 113, 116, 122, 125, 137, 176, 231, 283, 284, 340-353, 373, 490
- Anticorrosion 374, 375, 381
- Coil 374, 378
- Electrical insulating 374, 375, 383-384
- Electrophoretic dip 374, 375, 381, 383
- Laminate 282, 283
- Oil 374, 380, 381
- Packaging 373, 374, 377, 379
- Photosensitive 374, 377, 384
- Powder 54, 224, 374, 377, 381, 384, 385
- Protective 284, 374, 377, 379, 382
- Resins 490, 511
Cobalt 8, 46, 310
Co-condense 136, 137, 156
Coconut shell flour 134, 135
Code approved 130, 133, 145
Coefficient of expansion 114, 230
Coefficient of friction 214, 218, 224, 354-356, 363
Cohesive strength 214, 336
Coil forms 212, 231
Coiled 97, 99, 105, 134
Coke 444, 446, 450, 451
Cold flow 114, 225
Cold pressing 328, 330, 338

Cold-box process 296
- MF 296
- Plus 296, 310
- PUR 296, 309-311
Collapse 161, 292, 296, 301, 309, 310
Color 137, 142, 193, 194, 195, 214, 215, 230, 287, 288
Columns 284, 444
Combustion 104, 288, 312, 419
Comfort behavior 357, 361
Commodity 51, 276
Commutators 212-214, 220, 221, 231, 455
- Molding compounds 214, 215, 230
Compaction 294, 306, 311
Compatible 296, 348, 407
Compatibility 355, 374, 376, 400, 401, 406, 416
Complementary bonding 125, 126, 296, 355, 373-415
Complexation 101-103, 389
Compression 133, 151, 157,
- Molds 204
- Moldings 179, 192, 194, 195, 197, 214, 233
- Method 199, 200, 209, 233
- Molding compounds 197, 198, 199, 357
Composite Compressive strength 267
Composite Materials 273, 284
Composites 51, 56, 65, 76, 113, 117, 125, 232, 241, 256, 257, 260, 281, 288, 351, 413
- Carbon fiber 65
- Carbon carbon 66
Composites Fabrication Processes 264-281
Composites Industry 259, 260
- Advanced composites 260
Compression molding 57, 73, 117, 201-203, 365, 446
Compression Strength 225, 278, 307, 308, 330, 336
Compressive stress 219
Computer 83, 237, 238, 241, 280, 385, 386
- Printer 460
- Simulation 29, 31, 392, 393, 464
Concrete Flow Promoters 287-278
Condensation 29, 33, 35, 36, 78, 198, 270
Conduction 158
Cone calorimeter 105, 167, 168, 169, 262
Conformations 47
- Bimolecular 100, 101, 105
- Cyclic 100, 101, 105
Construction 104, 106, 132, 141, 259, 261, 275, 288
- Design freedom 224, 225
- Elements 256, 258
- Materials 212, 250, 251

Subject Index

Consumer
- Articles   193, 243, 259
- Electronics   244
- Goods   290

Contact adhesives   406, 410

Continuous   37, 140, 254, 272, 275
- Phase   140, 445
- Process   87, 89, 154
- Production   37, 143

Convection   158

Conveyer   251, 403

Coolant pump   218, 219

Cooling   77, 286, 301, 336

Copolymerization   112, 392, 466

Copolymers   59, 93, 392
- Block   393, 397
- Random   397

Copper   8, 9, 46, 237, 242, 243, 302, 305, 358, 359, 387
- Adhesion   214, 221
- Alloys   293, 301
- Clad   232, 237, 239, 245
- Foils   233, 240, 242, 243, 414

Copper sections   213, 214

Co-reaction   61, 72, 74, 76, 78, 136, 160, 165, 299

Core   126, 269, 291, 293, 294, 299, 300, 301, 304, 305, 312

Corebox   293, 300

Coremaking
- Processes   295–297
- Shooters   293, 295

Core/Mold Processes   298, 299
- History   296
- Sands   296

Corrosion   114, 162, 163, 239, 359, 377, 380
- Problems   159, 160

Corrosion resistance   247, 251, 259, 272, 277, 278, 373, 375, 383, 444, 453

Corundum   317, 320, 323, 341, 351, 426

Cost   133, 150, 153, 175, 218, 225, 241, 253, 281, 292
- Low   136, 141, 199, 258, 280, 292, 307, 309
- Reduced   135, 157

Cost-Performance Relationship   224, 228

Cotton   170, 171, 174, 180, 248, 253, 254, 289, 333, 340, 404
- Fiber   190, 192, 194, 195, 221, 250, 253

Cotton cloth   232, 246, 249

Cotton fabric laminates   235, 245

Coupling agents   152, 157

CR   398, 400, 405

Cracking   203, 445

Crankcase   295, 304, 464

Creep   75, 151, 415

Cresol   338, 373, 376
- bismethylol   61, 393
- methylols   396
- novolak   46, 48, 72, 230, 231, 281, 385, 387–391, 393, 394, 395, 397
- trimer   72,

Cresol resols   249, 379, 381, 405, 409

Cresols   3, 8–11, 48, 67, 99, 325, 391, 423

Croning process   297, 300, 371

Crosslink   61, 62, 69, 109, 111, 230

Crosslink density   52, 59, 62, 65, 66, 70, 97, 105, 200, 294, 310, 361, 403, 417, 422, 423
- high   62, 225

Crosslinked networks   58, 70,

Crosslinked resin   56, 64, 70, 109, 188, 325, 383

Crosslinking   57, 62, 65, 66, 69, 70, 113, 264, 376, 385, 388, 395, 399, 400, 408, 412, 420
- agents   395, 396, 404
- reaction   161, 162

Crown ether   27, 28, 47, 48, 103, 105

Cryolite   321, 323

Crystal violet lactone   456, 459

CTBN   59

Cumene   5, 8
- hydroperoxide   7, 313
- process   4, 5, 6
- sulfonic acid   160

Cup test   209, 211

Cure   57, 61–75, 125, 133, 140, 143, 145, 184, 230,
- Chemical   63, 145
- Complete   62, 280
- Mechanical   63, 145
- Room temperature   135, 280
- Parameters   62, 63, 134, 138
- Optimum   62, 64
- Temperature   135, 265
- Rapid   45, 136, 141, 164, 307
- Thermal   61, 62
- Non-hexa   61, 72–74, 78
- Post   57, 61, 65, 69, 75, 78, 163, 265, 273, 278, 287, 355, 452
- Level of   61, 62, 74–75, 78
- Maximum   63, 74
- Rates   191, 277
- Residual   63, 74
- Monitor   66, 97, 105
- Behavior   72, 74, 134
- Agents   72, 73
- Time   73, 74, 143, 200–202
- Characteristics   145, 185, 236

Cured resin   96–97, 119,

Curing Characteristics, measurement 498–500
Curing 75, 76, 209, 253, 254, 273, 283, 286, 288, 296, 301, 313, 434
- Cold 283, 305, 313, 375
- Conditions 273, 368, 369, 490
- Process 200, 202, 270, 285, 312, 330
- Rapid 286, 304
- Rate 212, 341, 350
Curing agents 72, 190, 191, 230, 231, 250, 282–284, 298, 302, 304–307, 309, 334, 385, 398, 409, 437, 438
- Latent 303, 304, 348, 372
Curing curves-No Bake 308
Curtain coater 135, 143
Curves
- Shear modulus 225
- Stress/strain 226
CVD 267
CVI 267
Cyanate 24, 51, 78, 264
Cyanate esters 50, 56–57, 78, 104, 280
Cyclodextrin 27, 28
Cyclohexane 9, 101
Cyclohexanone 4, 9, 381
Cymene 10, 11

Dampening 322, 357
Damping 278
- Mechanical 77
Data systems engineering 234
Decking
- Tunnels/Mass transit 277
Decomposition 69, 70
Decorative laminates 124, 236, 414
Decorative materials 179, 185
Defects
- Casting 294, 310
- Expansion 294
Defibrillate 252, 253
Degreasing 377, 414
Delamination 152, 236, 267,
- Resistance 152
Deluge systems
- Wet/dry 274
Density 142, 153, 157, 158, 166, 168, 204, 282, 290, 293
- Determination 494
Design 103, 128, 161, 264, 357
- Engineer 114, 251
- Stresses 130, 133
Devises 52, 142, 397
DFG 518
Dialcohols 48, 49

Diameter 287, 288, 289
Diamond 317, 318
- Grinding wheels 224, 339
Diazonaphthoquinone 389
Dibenzoxazine 58, 59
Dicyanate ester 56, 57
Dicyclopentadiene 50, 60
Dielectric Analysis(DEA) 62
Dielectric Constant 56, 204
Dielectric Loss Factor 63, 208, 239, 247
Dielectric Number 208, 239
Dielectric Properties 238, 248
Dielectric Strength 204, 208
Diethers 48, 49
Differential scanning calorimeter (DSC) 62, 64, 145, 174, 191, 491, 492
Differential thermal analysis (DTA) 492
Difunctional 51, 56, 57, 58, 376
Diisocyanates 165, 297
Diisopropyl ether 13, 161
Dimensional stability 114, 125, 126, 127, 145, 188, 204, 214, 218, 225, 238, 248, 261, 267, 304
- Permanent 296, 355
Dimensional tolerances 225, 292
Dimer 31, 33, 38, 42, 43, 47, 66, 78, 92, 100, 101, 105, 150, 152, 157, 372, 392
Dimethylol Phenols 25
DIN 203, 204, 209, 220, 239–241, 245, 246, 247, 250, 489, 491
Diodes 52, 241, 455
Dioxane 46, 147
DIPAC 168, 169
Diphenolic compounds 56, 57
Diphenyl methylene 4, 4' diisocyanate (PUR) 307, 310, 455
Disc 209, 340
Disc flow test 209, 211
Dishwashers 180, 214
Disk brakes 218, 359, 364, 413
Disk brake pads 353, 355, 357, 365, 367
Dispersed 96, 164
Dispersions 286, 348
Dissipation factor 56
Dissolution inhibitor 389
Dissolution rate 388, 391, 394
Distortion 75, 117, 203, 214
Distributor rotors/caps 215
Disubstituted 41, 45
DMSO 162, 163
DNQ 389, 390, 395, 396, 397
Domes 278, 284
Doors 142, 258
Door skins 130, 133
Double belt 165, 166

Doughy   197, 365
Douglas fir   129, 143
DRAM   52, 386, 387
Drum brake linings   353, 355, 357, 359, 363, 365, 366, 413
Dry prepreg   264, 265
Drying   248, 252, 285
Drying stage   248, 253
Ductile   38, 293
Ducts   57, 272, 281
Ductility   24, 263
– factor   76, 77
Dumping   296, 301
Dust   156, 192, 196, 313, 338, 519, 520
– Low   198, 327, 363, 399, 442
Dust explosions   21
Dustmeter   327
Dye developers   127, 456–459
Dyes   194, 287, 457
Dynamic mechanical analysis (DMA)   52, 62–64, 74, 134, 145,
Dynamometer   267, 291

E1, E2, E3   141
E-zero   141
E-beam   387, 389
Ecological   170, 252, 295, 296
Economical   256, 278, 295, 296, 307, 312, 313, 315, 343, 380
Economics   62, 114, 140, 153, 272, 346
Efficiency   218, 295
Elasticity   119, 193, 252, 285, 286, 306, 321, 350
Elastomer modified   76, 355
Elastomeric Materials   78, 187, 373
Elastomers   111, 112, 126, 175, 199, 399, 402, 403, 405–407, 410, 412, 463, 464
Electric tools   215, 221
Electrical and Electronic Industry   232, 234, 237, 246
Electrical applications   74, 195
Electrical engineering   188, 212, 213, 221, 224, 232
Electrical conductivity   453, 490, 497
Electrical industry   115, 212
Electrical insulating power   208, 214, 225, 251
Electrical parameters   214, 230
Electrical laminates   11, 51, 56, 59, 110, 113, 124, 127, 207, 237–244, 248, 250, 254, 414, 490
Electrical properties   195, 196, 204, 212, 220, 230, 234, 238, 239, 240, 245, 248, 250
Electrical resistance   207, 208

Electricity meters   224
Electrode   207, 208, 209, 383, 444, 446, 450–452, 454
Electrolyte   209, 248
Electronics   78, 230, 386, 415
Electronic components   57, 115, 127, 231, 238, 241, 384
Electronic molding compounds   230–231
Electronic potting compounds   48
Electroplating   193, 240, 251
Electrostatic   380, 385
Elevated temperatures   230, 240, 280, 284, 297, 300, 382
Elongation   205, 226, 289
– At break   225, 403
Embedded   283, 284
Emergency running properties   251
EMI shielding   280
Emissions   133, 142, 159, 165, 170, 174, 182, 200, 250, 252, 295, 296, 304, 332, 333, 417, 423, 425, 430, 431, 437, 451, 514, 515, 519
Emulsions   175, 184, 341, 348, 382
Enamels   374, 375, 381, 383, 384
Encapsulation   52, 57, 230, 295, 345, 455
Energy   77, 78, 110, 115, 149, 185, 225, 267, 278, 299, 304, 309, 311, 355
Engine block   219, 292, 304
Engine compartment   185, 219, 254
Engineered lumber products   128, 130–145
Environmental   125, 127, 128, 307, 312, 313, 314, 357, 363, 514
Environmental impact   323, 427
Environmentally friendly   264, 277, 309, 375, 383, 428, 439, 519
Enzyme   24, 26, 50, 78
EPDM   398, 400, 401, 402
Epoxy cresol novolak (ECN)   51, 230, 231
Epoxy molding compounds   192, 193, 199, 214, 215, 225, 229, 230, 231
Epoxy hardeners   48
Epoxy co-cured   270, 271
Epoxy resins (EP)   4, 24, 38, 50, 51, 53, 56, 58, 76, 78, 110–120, 127, 151, 152, 174, 187, 190, 232, 233, 235, 236, 238, 239, 241–243, 245, 246, 248, 250, 252, 258, 261, 262, 270, 282, 287, 291, 297, 298, 321, 325, 326, 341, 363, 373–376, 378, 380, 381, 383, 385, 402, 411, 417, 447, 452
– Acrylate modifieds   313
– Flame retarded   76, 118, 271
Epoxy SO_2 process   296, 298, 299
Equipment   115, 142, 145, 192, 218, 220, 251, 258, 281, 443, 444, 454
Ester   297, 299, 305, 307, 310, 312, 437
Etched   240, 242, 384, 388

Ethanol   21, 50, 286
Etherification   374, 375, 376, 378
Ethoxylation   371, 455
Ethyl acetate   47, 98, 406–409
Ethylene glycol   18, 162, 437
Exhaust gas treatment   254, 310–312
Exhaust gases   296, 298, 302
Exothermic   63, 77, 78
– Heat of reaction   160
Expansion   58, 78, 137, 142, 160, 256, 269, 293, 294
Expensive   137, 278, 287, 296
Explosion chamber discs   251
Exposure   274, 285
Exposure levels
– limits   14
Extenders   64, 131, 147, 288
Extractibles   72, 73, 128
Extraction   466
– Cation   101, 102, 105
– Metal cation   47, 48
Extruders   365, 399, 415
– Continuous   197
– Mastication   193, 198
Extrusion   163, 165, 201

Fabrication   234, 280, 281, 285, 288, 301, 307, 310, 361, 363, 368
– Plants   387
– Processes   260, 264
Fabric   234, 246, 250, 251, 256, 268, 271, 288, 351
Fabric laminates   250, 251
Face sheets   265, 268
Fading   355, 363
Failure   137, 277
Fats   461, 462
Fatty acids   128, 194
Fax machine   385, 460
FDA   378, 463, 466
Felt   74, 170, 174, 288
Festoon   235, 345, 346, 350
Fibers   130, 153, 170, 172, 175, 176, 192, 196, 260, 263–265, 267, 288, 289, 355, 358, 365, 366, 441
– Developments   263, 264
– Inorganic   124, 153, 169
– Natural   288, 289
– Novolak based   288, 290
– Organic   124, 153, 169, 192, 288, 354
– Phenolic   125
– Synthetic   190, 288, 289
– Vulcanized   236, 340, 341, 344, 349, 351
Fiber glass   153, 258, 415

Fiber reinforced composites   58, 59, 73, 104, 119, 125, 127, 170, 259, 262, 270, 275, 280, 284
Fiber optic   57, 465
Fiber boards   132, 142, 353, 410
Fibrous materials   282, 283, 284
Filaments   155, 156, 272
Filament winding   57, 65, 236, 261, 263, 264, 265, 271, 272–275, 281
Fillers   52, 64, 65, 109, 111, 113, 134, 150, 162, 165, 185, 190, 192, 195, 197, 198, 200, 231, 248, 265, 278, 280, 323, 355, 358, 365, 366, 377, 385, 401, 403, 404, 408, 409, 415, 441
Films   179, 242, 270
Filters   125, 232, 253, 254, 255, 258,
Finishing operation   253, 254
Fire   91, 104–106, 158, 159, 163, 166, 174, 225, 256, 258, 261, 262, 273, 281, 288, 290, 425, 426
Fishing rods   125, 232, 256, 259
Fittings   224, 304
Flakes   140, 145, 153
Flame   105
– Barriers   266, 290
– Resistance   103, 117, 126, 160, 174, 180, 245, 247, 258, 277, 288, 289
– Retarding properties   38, 196, 232, 234, 246, 248
– Retardants   114, 117, 180, 231, 238, 248, 249, 254, 271, 280
Flammability Test   204, 207
Flammability   104, 161, 180, 204, 207, 224, 225, 240, 271
– Limit   8, 18
Flat bars   236, 237
Flatware   176, 180, 181
Flexibility   103, 155, 234, 252, 253, 287, 324, 341, 346, 361, 363, 374, 376, 379, 382, 385, 409
Flexible   248, 253, 283, 340, 363, 378, 399, 433
Flexural strength   271
Flexural stress   205
Flexure   133, 138
Floorings   38, 129, 142, 163, 256, 270, 282, 283
Floors   154, 186, 283, 408
Floral foam   125, 162–166, 169
Flour   192
Flow   138–140, 155, 202, 205, 248, 286, 377, 384
– Distance   286, 332, 361, 490
Flow Behavior of Phenolic Molding Compounds   209

Subject Index

Flow distance, determination of 497–498
Flower support 159, 163
Flue gas 87, 297, 446, 520–524
Fluorescence microscope 140
Foam 105, 113, 116, 124, 135, 153, 154, 159, 160–166, 170, 455
Foaming equipment 165
- batch 165, 166
- continuous 165, 166
Foaming operation 161, 163
Foods 377, 460, 463
Food packaging 117, 270, 377
Foot imprint 159, 164
Formaldehyde 14–21, 24, 25, 48, 57–60, 64, 65, 78, 85, 87, 93, 95, 98, 99, 111, 116, 122, 123, 142, 148, 150, 156, 159, 160, 182, 308, 309, 312, 391–394, 438
- Determination of free 502, 503
- Emission 64, 137, 141, 148, 185
- Free 49, 50, 131, 133, 136, 139, 157, 164, 184, 253, 254, 263, 273, 277, 281, 305, 306
Formic acid 15, 16, 309, 312
Foundry 61, 103, 110, 124,–126, 133, 292–313, 490
- Binders 113, 125, 291, 455, 456, 509
- Industry 115, 258, 295, 296, 299, 371
Foundry Processes
- Acetal 296
- Cold Box 296
- MF 296
- Plus 296
- PUR 296
Epoxy SO_2 296–298
Free radical 298, 461
Furan SO_2 296–298
- Hot box 296, 298
- Plus 296
- No Bake 296, 298, 300
- Ester curing 296, 297
- Furan 296, 298
- Phenolic 296, 297
- Shell Molding 296, 297
Fourier transform infrared spectroscopy (FTIR) 31, 92, 96, 147, 148
Fracture 77, 267, 371
Friability 159, 161–165
Friction 57, 59, 72, 193, 201, 291, 359
Friction linings 113, 124, 256, 258, 353–370, 405, 490, 509
- Adhesives 412, 413
- Composition 358–361
- Manufacture 365–370
Friedel-Crafts 10, 51, 60, 287
FRP 77, 103, 104, 106, 110, 117, 151–153, 263, 264, 278–281

FST 73, 76, 91, 104–106, 159, 162, 163, 165, 261, 264, 270–272, 280
Fuels 147, 225, 251, 252, 254, 446, 464
Functional groups 51, 65, 97, 128
Functionality 24, 25, 66, 76, 91, 96, 126
Furan resins(FF) 110–114, 119, 120, 282–284, 294, 296–300, 302, 304–308, 313, 374, 417, 421, 422, 427, 430, 437, 452
- No bake 296–299
Furan SO_2 296, 299
Furfural 14, 323, 329, 330, 338, 339, 378, 429, 439, 445, 527
Furfuryl alcohol 110, 111, 114, 115, 300, 305, 329, 338, 372, 437, 452, 527, 529
Furnance 155, 425, 427, 442, 449
Furniture 132, 141, 142, 186, 405, 406, 410, 415

Gap filling 139
Garnet 341, 351
Gas 17, 200, 224, 252, 273, 304, 307, 387, 452
- Curing process 295, 296, 298–300, 307, 308–313
Gas chromatography (GC) 22, 92
Gasoline 17, 213, 219
Gaskets 59, 256, 403, 405, 415
Gassing process 309, 310, 312, 313
Gating 201–203
Gears 170, 221, 224, 251
Gel time, determination of 87, 89
Gel time 141, 148, 174, 277, 361, 490
Gel permeation chromatography (GPC) 22, 42, 44, 46, 92, 97, 105, 138, 489, 504
Gelation 39, 45, 63, 275,
Generator 219, 384
Geostationary broadcast satellites 57
Glass 73, 103, 113, 155, 196, 246, 264, 268, 272, 285, 314, 321, 351, 428
- E glass 272
- Special glass 264, 265, 279
Glass cloth 134, 232, 238, 246
Glass fabric 232, 233, 238, 242, 243, 251, 283, 323, 333, 335, 336
Glass fibers 151, 154, 156, 157, 169, 190, 192, 194, 214, 215, 230, 243, 250, 253, 259, 261, 265, 281, 358, 413
- Chopped 193, 279, 280
Glass mats 232, 233, 238, 283
Glass reinforced composites 73, 77, 127, 128, 243, 264, 279, 280
Glass transition temperature 226, 230, 238
Glassy Carbon 416, 452–455
Gloves 290

Subject Index   545

Glow heat resistance   192
Glue bond   131, 133, 135, 404
Glue laminated timber beams   278
Gluing Process   133, 135, 345
Glueline   133, 135, 137, 140
Glulam   127, 149, 151–153, 278, 279, 281, 404
Glycols   287, 384
GNP   386
Granulated materials   201, 450
Graphite   162, 190, 193, 224, 251, 282, 284, 291, 409, 428, 435, 439, 444–446, 451, 453
Graphite crucibles   416, 432, 440, 441
Graphite materials   119, 125, 416, 443
Gratings   272, 276, 277
Gray iron   292, 293, 302
Green strength   317, 323, 445, 446
Grinding   125, 259, 291, 314, 315, 413
Grinding materials   314–353
– Coated   340–353
– Wheels   110, 113, 117, 122, 232, 291, 317–339, 373, 490
Grouts   120, 282, 442
Growth rate   259, 281
Guide rails   224, 251
HALS   461–463
Handles   123, 214, 221
Hand lay up   260, 262, 264, 265, 280, 281
Hardboard   136, 142, 143
Hardener   62, 135, 298
Hardness   192, 224, 317, 332, 336, 361, 373, 382, 403
Hardox process   313
Hardwood   136, 142, 143
Hatrack   180, 182
Hazardous materials   514
– labeling   515, 516
Hazardous Occurrences   77, 78
HCFC   161, 168
Headers   143, 145
Heat   62, 63, 65, 73, 109, 117, 131, 158, 160, 168, 280
– Shield   104, 110
Heat conductivity   224, 277
Heat exchangers   447, 450
Heat/Moisture balance   179, 180
Heat release   167, 168, 261, 271,
– Low   73, 261
– Total   262
Heat resistance   62, 76, 136, 137, 165, 220, 252, 261, 274, 285, 321, 376, 400, 403, 405, 409
Heat distortion temperature (HDT)   103, 206
Heat treatment   200, 226, 227

Heat transfer   201, 274
Helmets   267
Hemiformals   15, 38, 39, 46, 64, 92
Heptamer   101–103
Hexamer   43, 47, 100, 101, 105
Hexamethoxymethylmelamine   398, 399, 403, 404
Hexamethylenetetramine (Hexa, HMTA)   21, 60–62, 66, 67, 69, 70, 72, 74, 76, 87, 97, 148, 157, 174, 184, 190, 191, 193, 198, 200, 253, 254, 285, 286, 297, 299, 300, 301, 304, 325, 331, 353, 361, 372, 398, 403, 404, 418, 420, 422, 429, 435, 437, 438, 439, 441, 446
– Content, determination   503
– ^{13}C and ^{15}N 66
Hexane   161, 407
Hg-Xe lamp   387
Hide glue   341, 351
High chemical resistance   284, 288, 314
High Compressive strength   268, 372
High Density   56, 318
High Density Fiberboard(HDF)   132, 142, 143
High peel strength   270, 340, 350
High Performance Composites   266
– Components   272
High performance & Advanced Composites   259–281
– Market segments   260–263
High performance liquid chromatography (HPLC)   31–34, 92, 504, 506
High reactivity   286, 346
High speed   56, 267, 272
High strength   267, 270, 278, 296, 304, 306, 355, 363, 442, 445, 454
High temperature   285, 296, 314, 355
High strength temperature resistance   49, 291, 446
High thermal resistance   283, 285, 286, 322, 354, 361
High voltages   208, 248, 250, 251
High voltage resistance   250, 251
Homogenized   198, 201, 282, 285
Honeycomb Core Sandwich Construction   268–272
– Process   269–270
Honeycomb   232, 256, 257, 264, 265, 268, 270, 271, 413
Hood   284, 290
Hot Box   296, 302–304
– Plus   296, 304
Hot curing   300, 313
Hot/Warm Box   299
Hot melts   57, 254, 405, 415
Hot pressing   137, 138, 143, 317, 337, 452

Hot storage 207, 231
Hot tensile strength 296, 297
Hot water resistance 119, 120
Household appliances 170, 174, 180, 186, 188, 212, 214, 218
Housing 129, 142, 221,
Housings 446
- Electric meters 212
Huggins constant 97, 98
Humid-environment 134, 137, 157
Hybrid 136, 165, 278, 397
Hydraulic components 218, 251, 252
Hydraulic presses 433, 434
Hydrocarbon 10, 12, 161, 310, 423, 427, 456, 464, 465
- Chlorinated 224, 284
- Resins 374, 402, 415
Hydrochloric acid 46, 85, 160, 444
Hydrodynamic volume 91, 105
Hydrofluoric acid 290, 444
Hydrogen 15, 16, 418, 423
Hydrogen bonds 58, 101
- Intramolecular 41, 47, 70, 101
- Intermolecular 47, 101
Hydrogen donating 462
Hydrophilic 128, 384
Hydrophobic 128, 248, 254, 340, 384, 389
Hydroxyl end group 76
Hydroxyl group 51, 64, 147
Hydroxylamine 467
Hydroxymethyl 64, 65, 72, 392
Hydroxymethylation 60, 61

IEC 204, 207-209, 227, 239, 245
IFWI 77
Ignitibility 167, 261, 262,
Ignition 207, 215, 221, 224
Image 385, 388, 394, 395, 457, 459
Imaging 140, 242, 385-397
Imidization 51, 67
IMO 274
Impact 266
- Charpy 76,
- Damage resistance 59
- Resistance 103, 192, 195, 204, 206, 273, 378, 385
Impregnated carrier materials 236, 237
Impregnating 38, 122, 198, 235, 249, 254, 258, 340, 367, 373, 375, 383, 432, 442, 448, 489
- Agents 238, 256, 404, 427, 444-446, 450
- Resins 248, 249, 250, 252, 450, 510
Impregnation of Paper and Fabric 231-259

Incandescent 207, 224, 285
Incineration 185, 418
Industrial Filter Inserts 251-254
Industrial filters 253, 254
Industrial laminates 120, 233
Industrial Safety and Ecological Questions 514-529
Infrastructures 145, 281
Inhibitor 388, 394
Injection compression 202, 203
Injection embossing 199, 203
Injection molding 110, 197, 199, 201, 203, 452, 455
- Machines 201-203, 209, 210
Inorganic compounds 128, 154, 164
Inorganic Fibers 153-158, 231
- Resins 156-158
In situ 39, 59, 62, 65, 297
Insulation 103, 153-186, 207, 213, 214, 248, 455
- Acoustical 113, 124, 127, 153, 158, 170
- Thermal 113, 124, 127, 153, 158, 170
Insulating materials 124, 125, 127, 247
Insulators 123, 208, 209
Integrated circuits 52, 242
Interface 128, 152, 161, 260, 264, 281, 336, 401, 402, 412
Interior parts 133, 258
Interlaminar adhesion 264, 271
Intermediate and Carbon Forming Bonding 416-455
Intermediate handling 126, 296
Intermediates 37, 65-67, 69, 70, 75, 109, 264
Internal bond 132, 139, 140, 142
Interpenetrating polymer networks (IPN) 24, 56, 75-77
Intractibility 64, 66
Intramolecular 34, 41, 52
Ion beam 387
Ion exchange resins 96, 169
Iron 292, 293, 301, 307, 314, 359, 427, 428, 433
- Casting 304, 351
Irons 123
ISO standards 195, 203-212, 220, 222, 223, 245, 249, 253, 426, 489, 491
Isocyanates 152, 161
Isophorone 11, 310
Isopropanol 46, 50, 286, 351
Isostatic pressing 429, 432, 433, 438-441, 446

JIS 204
Jute 127, 130

Subject Index

K-factor 154, 159, 166, 168, 169
Karl Fischer method 490, 504
Ketones 224, 310
Kitchen appliances 214, 221
Kynol 288, 289, 291

Ladders 272, 276
Ladle 425, 428, 431, 438
Lamda value λ 158, 159, 162, 163, 166, 168, 169
Laminate coatings 282, 283
Laminates 38, 113, 122, 124, 125, 127, 163, 179, 232, 233, 242, 243, 250, 254, 262, 270, 271, 282, 490
- Fabric 234, 248
- Glass 235, 236
- High voltage resistant 248, 249
- Markets 238, 243, 250
Lamp 221, 285, 286
Lamp sockets 231
Lamp Base Cements (Socket Putties) 285–287
- Resins 286
Latex 193, 340, 348, 350, 404, 415
Lead 46, 310
Lederer-Manasse reaction 25
Lethal dose LD_{50} 515, 516
Lewis acid 55, 400
LIFT 167, 169
Light color 116, 285, 286
Light fastness 116, 117
Lighter 274, 278
Lightweight 267, 278
Lightweight construction 256, 259
Lignin sulfonate 147, 157, 288, 426
Lignin 12, 51, 61, 127, 128, 131, 146, 147, 148, 153, 157, 174, 313, 527, 528
Limiting oxygen index (LOI) 288
Line widths 387, 389, 397
Linear 46, 97, 98, 101, 111, 376
- Expansion 142
Liner 265, 290
Linings 258, 284
Linkages (o,o';o,p';p,p') 33, 34, 41, 42, 389
Linseed oil 380, 383
Liquid Chromatography for Separation of Phenolic Resins 504, 505
Lithium
- Hydroxide 33
- Cation 95, 103
Lithography 387, 391
Load 205, 207
Load bearing 127, 149
Long term resistance 127, 151

Long term 206, 224
Longitudinal tensile force 205
Low density 234, 250, 251, 267
Low profile additive 117, 279, 280
Low smoke 73, 258
Low smoke emissions 274
Low wear 267, 291, 355
Lubricants 190, 193, 194, 358, 359, 372
Lubricating oils 254, 464
Lumber 127, 130, 143, 152
LVL 127, 130, 131, 133, 143, 144, 153

Machine 215, 220, 285, 288, 301, 353, 364, 380
Magnesia 416, 426, 427, 432, 433, 435, 436
Magnesium 8, 46, 292
Magnesium hydroxide 33, 248
Magnesium oxide 11, 321, 406–408, 433
MAK-value 518, 519
MALDI 92–96, 105
Mandrel 236, 237, 265, 272
Marble flour 285, 286
Marine 105, 133, 259, 262, 272, 275, 277, 280, 281, 380
Mark-Houwink-Sakurada expression (MHS) 97–99, 105, 134
Markets 78, 143, 154, 159, 243, 259, 279, 288, 314, 463
Mask 290, 388
Mass transportation 250, 262, 272, 280, 281
Master curve 75
Masticators 197, 284, 365, 399, 409, 415
Matrix 59, 66, 117, 152, 232, 243, 259, 264, 266
- Resin 52, 57, 104, 233, 260, 261, 272, 394, 395, 397
Mats 155, 281, 333
- Mineral fiber 157, 158
- Uniform 146, 153
Maximum Application Temperature 206
Measurements 63, 77, 128, 208,
Methylene glycol 20, 37
MDF 127, 128, 130–132, 137, 142, 145, 148, 149, 153
MDI 4, 127, 128, 130, 131, 133, 138, 140, 141, 143, 145, 149, 310
Mechanical engineering 115, 212, 232, 233, 251, 380, 415
Mechanical Properties 57, 58, 74, 78, 91, 103, 105, 118, 132, 135, 138, 142, 143, 180, 182, 183, 188, 195, 196, 234, 239, 245, 256, 264, 271,
Mechanical strength 192, 194, 214, 221, 246, 250, 254, 256, 271, 306, 361, 377, 441

548 Subject Index

Mechanical stress   125, 221, 238, 251, 354, 377, 400
Mechanism   24, 58, 64, 66, 77, 136, 158, 267, 392, 400, 401
Mechanistic   61, 72, 78, 96
Mechanization   295, 314
Melamine   60, 111, 137, 142, 395
Melamine formaldehyde resin (MF)   62, 110–117, 127, 130, 133, 134, 141, 142, 187, 190, 201, 209, 214, 215, 232, 233, 235, 248–250, 288, 296, 297, 363, 374, 403, 404
Melamine phenolic resins (MPF)   187
– Molding compounds   215, 228, 229, 230
Melamine urea formaldehyde (MUF)   62, 134, 141, 149
Melt   62, 190, 292, 294
Melt flow process   197, 198
Melting range   87, 184, 191, 236, 492
Melting point   301, 402, 407, 408
Memory capacity   387
Metal dithiocarbamates   460
Metal dithiophosphates   460, 464
Metal powders   193, 435
Metallic fibers   218, 251, 354
Metals   73, 113, 203, 205, 224, 225, 251, 258, 276, 285, 287, 292, 301, 351, 358, 359, 373, 410, 425
Metal salt   24, 26, 46, 56, 393
Methane   18, 160, 418
Methanol   14–18, 98, 142, 248–250, 253, 254, 286, 287, 312, 391
Methacrylate resins   112, 113
Methyl formate   297, 299, 308, 309, 311, 312
Methylal   308, 309, 312
Methylene   41, 97, 150
– Bridge   33, 46, 64, 65, 67, 69, 70, 456
– Linkage   39, 44, 45, 46, 70, 92
Methylene (p,p';o,p';o,o')   41, 65, 69, 92
Methylene ether   47, 64, 399, 400
Methylene diisocyanate   18, 127
Methylol   65, 150, 374, 376, 380, 392, 399, 400, 407, 411
– Phenols   25–27, 39, 41, 46, 150, 157, 396
Methylolated   148, 396
– Monomers   29, 31
– Phenols   31, 33, 78
Methylstyrene   6, 8
Methyl tert-butyl ether (MTBE)   12, 17
Mica flake   190, 192–196, 214, 215, 220, 230, 359
Microcapsules   456–458
Microcomponents   230
Microdielectric spectroscopy   62
Microelectronics   50, 386
Microstructure   389, 392

– Phenolic   76
– Resole   99
Microwave   56, 143, 145, 201, 230, 279
Military   65, 78, 261, 262, 268, 272
Mimosa   148, 149
Mine/tunnel foam   159, 164–166, 169
Mineral oil   12, 224, 251
Minerals   128, 194, 196, 415
Mineral wool   89, 103, 124, 153, 155, 157, 169, 441
Mineral fibers   154, 155, 194, 195
Miniaturization   230, 386
Mix   284, 296, 313
Mixer   198, 301, 305
Mixture   284, 285
Model compounds   66, 137
Modulus   57, 62, 63, 73, 74, 205, 266, 289
– Low   52, 56
– High   52, 62, 63, 75, 76
– Elasticity (MOE)   132, 133, 138, 142, 357, 400, 403
– Rupture (MOR)   138, 142
Moisture   52, 135, 289, 304, 309
– Content   62, 64, 133, 135, 137, 138, 140, 141, 145, 146,
– Detection   143, 490
– Level   205, 346, 437
– Resistance   62, 127, 136, 141, 142, 157, 230, 261, 341, 441, 459
Molar ratio   62, 78, 131, 150, 156, 159, 287
Mold   165, 192, 193, 199, 201–203, 209, 224, 225, 279, 280, 281, 284, 287, 291–294, 296, 299, 305, 310, 311, 313
Molded products   59, 66, 127, 130
Molded laminates   232, 233, 234–238, 241, 242, 246, 250, 251
– Standardization   244–247
Molding   38, 75, 125–127, 176, 180, 181, 191, 200, 250, 293
– Compression   188, 189
– Injection   188, 189
– Transfer   188, 189
Molding compounds   74, 109, 112, 113, 117, 122, 124, 125, 127, 200–203, 205, 207–209, 214, 456
– Composition of   188–194
– Production of   197, 198
– Standardization of   194–198, 220–224
Molding Materials   72, 74, 75, 103, 260, 291, 293–296, 312
Molecular Configuration   91, 97
Molydenum sulfide   251, 360
Monitor   62–64, 77, 78
Monodisperse   390, 397
Monomer   66, 91, 109, 112, 113, 161, 392

Monomer levels   430
  Reduced   254, 346, 349, 379
Monomeric   56, 58, 78
Montmorillonite   458
Morpholines   162, 163
Morphology   53, 57, 77
Motor   213, 215, 220, 221, 383, 384
– Adjustment   220, 221
Motorcycle clutches   258
Motor racing   126, 363
Motor vehicles   124
Mounting parts   221, 224
Multifunctional cyanate esters   56, 57
MW   97, 134, 139, 150, 390, 391
MWD   15, 42, 92, 139, 375, 389, 390, 391

Naphthalene   423, 424, 431
– sulfonates   288
Napped   253, 254
Natural gas   16, 272, 370
Natural resins   146–149, 285, 286, 314, 374, 376
Natural products   51, 61
NBS smoke chamber   117, 167
Negative resists   388, 389, 396
NEMA   238–241, 245, 247
Net shape   280, 281
Nickel   46
NIST   133, 167
Nitric acid   290, 453
Nitrile-butudiene rubber (NBR)   76, 398, 401, 403–405, 409, 412, 413, 415
Nitrile-Rubber Adhesives   408, 409
Nitrogen content   70, 306, 420
NMP   162, 163
NMR   15, 37, 39, 64–66, 69, 96, 97, 105, 138, 139, 391
– ^{13}C   22, 31, 33, 39, 40, 41, 44, 45, 46, 65, 69, 78, 92, 96, 97, 105, 147, 148, 150,
– 1H   44, 45, 92, 96, 97, 105, 147, 148
– ^{13}C CP/MAS   64, 66
– T_2 65, 97, 105
– $T_{1\rho H}$ 65, 97, 105
– ^{15}N   66, 69, 96, 97,
– ^{31}P   96, 97, 147
– Solid state   53, 64, 65, 69, 96, 105
No Bake Process   296, 305–307
– Ester curing   296, 307
– Furan   296
– Isocyanate   307, 308
– Phenolic   296
Noise   127, 158, 180, 357
Non volatile components   282, 490
  determination   500, 501

Non-wood materials   130, 149
Non-wovens   265, 268, 290, 349
Nonferrous metals   292, 293, 314, 351, 428
Nonvolatile   89, 249, 253, 341
Nonyl phenol   3, 12, 402
Novolak/furfural   169
Novolak   24, 25, 36–50, 59, 61, 72–76, 78, 85–89, 91–93, 95–97, 103, 105, 122, 159, 165, 191, 193, 197, 214, 230, 232, 253, 254, 304, 334, 354, 359, 372, 375, 377, 380, 395, 398, 403–405, 409, 411, 412, 414, 419, 420, 422, 423, 429, 432, 437, 452, 456, 458
– All ortho   101–103, 105, 106
– Diazo   384, 389
– Fibers   290
– Foundry   296, 297, 300
– Low free phenol   250
– Low melting   250, 325, 446
– Solution   122, 252, 253, 254, 432, 435, 436, 437, 440–443, 450
– Hexa   63, 74, 76, 77, 300
– Random   39–46, 78, 87, 97, 98
– Resole   302, 444
– High ortho   46–48, 78, 87, 97, 98
– Conventional   46, 49, 67, 70
– Modified   48–51, 78, 325, 405, 439
– Reactions   51–61
– Hardener   51–53
– Powdered   62, 63, 122, 124, 439
– Preparation   78, 394
– Cure   66–72, 78
– Non-hexa cure   72–74, 78
– Epoxidized   192, 230, 455,
Nozzles   201, 202, 203, 438
Nuclear reactor   446, 450

Occupational exposure   517
Octyl phenol   3, 11, 12, 400, 402
Odor   74, 170, 174, 184, 252, 254, 306, 310, 374, 437, 442
Offshore   262, 273, 274, 281
– Installations   263, 272
– Oil platforms   274, 276, 281
Oil recovery   370, 373
Oilfield   12, 261, 263, 272, 273, 281
Oils   382, 461, 462
Oligomer   31, 34, 38, 42, 43, 57, 58, 70, 91, 93, 95, 101, 102, 109, 112, 113, 157, 160
– Ortho linear   47, 48
– Silanol/Siloxane   76
Oligomerization
– Novolak   39–50
– Resole   27–36
Olivine   291, 294

Opaque 288, 394
Open cells 159, 161, 163, 169
Organosolv pulping 147, 148, 528, 529
Oriented strand board (OSB) 62, 63,
 127–131, 136–141, 145–147, 152, 153
– Resins 138, 140
Orifice Flow Test 209, 212
Ortho 22, 34, 41, 71, 309
– Linkage 34, 389, 394
– Site 67, 70, 78
Ortho-linked oligomers 393, 394, 397
Orthopedic foot impressions 159, 164
Orthopedic foam 159, 164, 165, 166, 169
OSHA 518
OSU 57, 73, 105, 167, 262
Oven 254, 272, 286
Oxalic acid 11, 37–39, 41, 87, 391
Oxazine 51, 58, 264
Oxidation 8, 9, 64, 69, 403, 462, 464
– Mechanism 461
Oxygen 297, 288, 289, 382, 420–423, 431, 432, 453, 460
Oxygen index 105, 167, 262

Paints 115, 124
Panel 127, 138, 142, 256, 258, 266
– Manufacture 130, 140, 145, 153
Panel properties 133, 135, 137, 146, 149
Paper 103, 147, 232–234, 236, 238, 239, 246, 248, 249, 251–254, 256, 265, 268, 340, 351, 415, 441, 456
– Web 248, 252–254
Paper Impregnation 232–250, 258
Paper Laminates 103, 235, 239, 240, 243–247, 249, 250, 251, 353
– Electrical 233, 239, 240, 373
– Epoxy 235, 239–241
– Markets 244–245
– Phenolic 235, 239, 240, 242, 244, 247, 248, 251
– Raw materials 248–250
Papermaking machine 252, 253
Para position 67, 70, 71, 78
Paraform 27, 46, 148, 150
Paraformaldehyde 16, 20, 38, 13
Parallam 127, 130, 131, 133, 143–145, 153, 279
Particle size 139, 172,
– Distribution 490
Particle boards 98, 103, 127, 130–132, 136, 141, 142, 145, 146, 149, 153
Parting agent 174, 193, 194, 231, 301
Partitions 127, 163, 186
Passengers 105, 256, 262, 281

Paste 282, 285–287
Patent literature 270, 288
Pattern making 272, 300, 390, 449
PBOX 72, 73
Pecan nut 148, 149
Peel strength 271, 272
Pellets 198, 199
Penetration 128, 134, 139, 140, 145
Pentamer 101, 102
Percolation theory 391
Perforated 141, 208
Permanent Bonding 72, 125, 126–291, 296
Permeability 252, 372, 446, 450
Peroxides 190, 313, 399, 462
Petroleum 12, 111, 370, 372
– Products 455, 460, 464
PG 463
pH 33, 34, 46, 62, 70, 77, 78, 139, 149, 157, 305, 496
Phase 57, 62, 66, 76, 140, 208, 263, 408
Phenol 3–12, 89, 98, 111, 123, 147, 148, 150, 156, 159, 160, 286, 391, 392
– Emissions 185, 301
– Free 66, 67, 131, 133, 157, 164, 174, 184, 230, 253, 254, 263, 264, 273, 277, 281, 302, 305, 306, 310, 328, 333, 363, 435
– Residual 62, 78, 159, 480, 501, 502
Phenolate 25, 27,
Phenol sulfonic acid 160
Phenolic compression molding compounds 193
– Glass fiber-reinforced 193, 195
Phenolic Foam 153, 154, 158–169
– Resins 159, 160
– Blowing agents 160, 161
– Equipment 165, 166
– Applications 162–165
– Foam Testing 166–168
Phenolic functionality 51, 374
Phenolic groups 60, 375
Phenolic intermediates 62, 423
Phenolic microstructure 76, 270, 271
Phenolic Molding Compounds 187–231
– Applications 212–224
– Composition 188–194
– Glass reinforced 188, 189, 218, 226–228
– Production 187, 188
– Properties compared to other materials 224–230
– Special purpose 221–223
– Test methods 203–212
Phenolic novolak 56, 57, 191, 192, 301, 302
Phenolic prepregs 264
Phenolic/reinforced fiber systems 260–263, 353

Phenolic oligomers 66, 182
Phenolic rings 69, 170, 171
Phenolic Resins as Additives in the Rubber Industry 398–405
Phenolic Resins as Binders for Carbon and Graphite Engineering Materials 446–455
Phenolic resins as Binders for Refractories 425–444
Phenolic resins (PF)
- Acid curing 299
- Amino resin modified 258
- Core making grades 294
- Elastomeric 270, 363, 364
- Epoxy modified 258, 333, 337, 339, 373
- Highly alkaline 307
- Hydroxyl 41, 47, 65, 73, 78
- Liquid 282, 303, 319, 323
- Matrix 105, 261
- Modified 127, 232, 234, 243, 253, 256, 258, 259, 264, 266, 267, 271, 275, 283, 323, 361, 381
- Network 160
- Polysiloxane modified 273, 274
- Powdered 137, 172, 285, 319, 323, 444
- Reactions 297
- Rubber modified 337
- Toughened 76, 77, 78
Phenolic Resin Fibers 288–291
- Products 290
Phenolic Resins-Bonded Textile Felts 170–186
- Processes 176–178
- Resins 172–176
Phenols 122, 182, 423, 424, 431, 438
Phenoxy 326, 363
- Methylene ethers 64
Phenylphenol 66, 408, 459
Phosphatizing 377, 380
Phosphites 461, 462, 464–466
Phosphoric acid 5, 41, 87, 160, 290, 306, 310, 380
Phosphorous 174, 363, 364, 420, 461, 462
Photoresists 11, 46, 48, 72, 242, 375, 381, 385–397
Photoacid generator 395, 396
Photolithographic 384, 386, 389
Photolithography 387, 397
Photosensitive 384, 388
PHS 385, 387, 388, 394, 395, 396, 397
Physical properties 130, 195, 245, 267, 277, 373, 494
Pigments 117, 190, 194, 231, 248, 280, 287, 377, 384, 385, 409
Pine 131, 148, 149

Pipe 263, 272–274
- Filament wound 76, 261, 273, 281
Piston 201, 215, 221
Pitch 282, 423, 424, 426, 428, 430, 431, 444, 451
Plastics 193, 204, 225, 256, 262, 314, 357, 399, 433
Plasticizers 259, 372, 375, 380, 383, 405, 409, 415, 442
Plastisols 254, 414
Platens 140, 142
Platforms 261, 277
PLB 76, 77, 118, 270, 271, 281
Pleated 253–255
Plywood 62, 63, 98, 103, 127, 129, 130–138, 140, 147, 148, 152, 404, 512
- Exterior 131, 134, 136
- Hardwood 137, 142
PMF 137
PMR-15 58, 104, 280
Polar solvents 46, 162, 407
Pollutants 302, 305
Polyacetal resins 467
Polyacrylonitrile 256
Polyaddition 113, 119,
Polyamide 340, 349, 404, 415, 467
- Glass fiber reinforced 225, 228
Polyamide Fibers 172
Polyamide-imide 383
Polyaryl ethers 76, 77, 104,
- phenolic terminated 76
Polybenzoxazine 58, 104
Polybutadiene resins 463
Polycarbonate 4, 13, 38, 467
- Glass fiber reinforced 226
Polychloroprene 406–408, 410–412
Polychlal 289
Polycondensation 112, 119,
Polycyclic aromatics 423, 427, 430, 451
Polyester amide resin (PEAR) 73
Polyester (UP)/glass 276
Polyesters 171, 193, 340, 341, 358, 383–385, 404, 467
Polyether 55
- imide(PEI) 57, 76, 384
- sulfone(PSF) 57, 76, 104
Polyethylene 56, 179, 190
Polyethylene terephthalate 226, 228
Polyimides 57, 60, 73, 243, 384, 452
Polyisocyanates 305, 308, 310, 374, 408
Polymeric Materials 464, 465
Polymerization 50, 58, 112, 117, 119
Polymethylmethacrylates 115
Polyols 111, 165
Polyphenolic 147, 312

Polypropylene 275, 465, 466
Polysiloxane 251
Polystyrene 42, 44–46, 56, 92, 110, 154, 155, 168, 179
Polysulfone 57, 467
Polyurethane foam 155, 179, 180
Polyurethanes 112, 165, 307, 309, 321, 326, 341, 374, 384, 385, 405, 409, 410, 445, 455
– Adhesives 254
– Thermosetting 114, 119, 120
Polyvinyl alcohol 16
Polyvinylbutyral 267, 325, 326, 339, 372, 380, 381, 413, 414
Polyvinyl fluoride 266
Positive resists 388, 397
Positron annihilation 53
Pores 200, 337, 450, 453
Porosity 168, 445
Porous network structure 128
Potassium hydroxide 33, 307, 312
Potential, 207–209, 241
Potentiometer 248, 249
Poured 287, 288, 292, 294, 296, 336
Powder resin 87, 170, 437, 441, 442, 446, 490
Power brake actuator 215, 221
Power sockets 221
PPO 59
PPS 104, 227, 228
Precipitation process 297, 308, 309
Precure 176, 200, 236, 303
Preheating 200, 201, 254, 255, 286
Preimpregnation resin 249, 250
Prepreg 57, 112, 232, 256, 258, 259, 264, 265–272, 333, 335
Prepreg face sheets
– Epoxy/carbon fiber 269, 270
– Epoxy/glass 269, 270
Presses 131, 133, 135, 145, 153, 201, 236, 250
Press temperature 131, 133, 135, 176
Press plates 138, 142, 145, 146
Pressure 75, 109, 131, 283, 284
Pressure sensitive adhesives
– recording papers 459
PRF 131, 134
Price level 114, 115, 117
Primers 383
– Wash 375, 380
– Shop 375, 380
Printed circuit boards 127, 233, 237–244, 251, 384, 414
Printing 256, 384
Printing calculators 460
Printing inks 12, 242, 374, 375, 381, 384
Process 114, 140, 143, 193, 201, 225, 254, 281
– Conditions 131, 133

Processability 76, 264
Production of phenolic resins 83–90
Productivity 200, 252, 309, 361, 363
Profiles 233, 236, 237, 246, 278, 279, 281
Propagation 77, 164
Propene 12, 13
Proppants 340, 371, 372
Propylene 5, 10, 21
Protective clothing 290, 455
Pultruded 110, 128, 143, 145, 151, 232, 261, 263, 264, 265, 275–279
Pumice 341
Pump housing 218, 446
Punch 209, 377
Punching properties 239, 246
– Cold 245, 247, 248
PUR 113, 115, 125, 161, 297, 299, 308
– Cold box 299, 307, 313, 455
– No bake 299, 305, 307, 308
– Rubbery phase 76
PUR/PIR 154, 167, 168
Putties 285, 286
– Furan 282, 283
– Highly filled 283
– Phenolic 282, 283
– Thermosetting 285
Putty 285
Putty powder 282, 283, 286
PVA 348, 405
PVC 262, 340, 409
– Plastisols 254
Pyrite 321, 323
Pyrolysis 126, 169, 304, 416, 421, 430, 431, 441, 454, 455
– Conditions 296, 355
– Controlled 262
Pyrolysis of Phenols 417–425
– Controlled 261
Pyrolytic applications 117
Pyromelletic anhydride 60

Quality control 140, 203, 209, 488, 489
Quality assurance 244, 489
Quartz flour 230, 231, 282
Quartz 293, 294
– Resin bridge 294
– Sand 293, 294, 300, 301, 303, 305, 310, 313, 408
Quebracho 148, 149
Quinolines 464
Quinone methides 400

R curves 77
„R" value 153, 154, 156,

Subject Index

Radiation 158, 159
Radical scavengers 462
Radome 56, 57
Rail 105, 256, 281, 353, 355, 414
Ramming mixes 282, 310, 311, 425, 429, 432, 437
Random 41, 393
Raw materials 111, 115, 123, 125, 127, 130, 149, 150, 176, 190, 199, 250, 260, 293, 313, 370, 373, 376, 433, 437, 446, 451, 455, 514
- renewable 170, 313
Reactors 12, 13, 37, 38, 83, 87, 89
Recycle 179, 184, 185, 190, 196, 230, 313, 335, 514
Recycled products 170
Recycling methods 186
Refractive index 84, 89, 157,
Refractories 113, 115, 117, 119, 122, 124, 126, 328, 416, 425-444, 452, 490, 509
Reinforced materials 214, 279
Reinforcing Resins 398, 403-405
Reinforcing Agents 109, 113, 190, 192, 193, 195, 198, 214, 224, 230
Regulatory agencies 105
Relative humidity 64, 65, 182
Relay sockets 224
Release agents 280, 441
Renewable Raw Materials 527-529
Resin transfer molding (RTM) 57, 73, 280, 281
Resin 133, 264, 284
- Advancement 65
- Chemical reactivity 297
- Core 138
- Cover sheet 248
- Dispersion 140
- Liquid 302
- Low monomer 254
- Matrix 209, 234 321
- Microstructure 41
- Phases 270, 271,
- Spray dried 131, 138, 139
Resin content 195, 200
Resin viscosity curves 270
Resistance 75, 208, 242, 251, 256,
Resist line 387, 379
Resists 384, 388, 394
Resite 200, 403, 417, 420, 424
Resole Resin 21, 24-36, 74-78, 89, 91-93, 97, 103, 105, 122, 131, 133, 142, 161, 162, 164, 165, 191, 198, 265, 302, 304, 305, 323, 324, 331, 333, 334, 349, 354, 359, 372, 376, 379, 381, 400, 405, 409, 411, 412, 415, 432, 438, 441, 443, 452
- Acetylated 98

- Alkaline 311-313, 323, 333
- Amine catalyzed 98-100
- Aqueous 89, 122, 124, 232, 249, 250, 258, 340, 342, 348, 361, 380, 420, 422, 439, 442, 444, 446, 450
- Boron containing 309
- Cresol 249
- Cure 62-66, 78,
- Foundry 296, 300
- High ortho 89, 165, 296, 297, 299, 305, 307-309, 455
- Highly water soluble 156, 248,
- KOH containing 296, 305, 309, 312
- Liquid 65, 89, 92, 122, 131, 138, 152, 159, 435, 436, 440, 446
- Low monomer levels 254, 258
- Mixed with novolak 253, 254
- Modified 253, 282, 296, 300, 302, 375, 412, 414, 415, 429, 440
- Neutral 297, 323
- Powdered 131, 138, 140,
- Solid 61, 72, 74, 85, 89, 122, 124, 152, 420
- Solutions 122, 124, 232, 436
- Spray dried 131, 138
- Urea extended 156
- Water based 263, 265, 277, 281,
Resorcinol 12, 13, 20, 149, 150, 160, 248, 404
- Adhesives 150, 151, 512
- Hydroxymethylated (HMR) 152, 153
- Resins 153, 296, 312, 349, 405,
- Novolak 98, 160, 404
Resorcinol-formaldehyde (RF) 12, 62, 63, 134, 135, 149-153, 169
Retention power 251, 253
Rheological 63, 209
Rice 130
- Straw 149, 153
Rigidity 62, 63, 134, 136, 157, 275
Ring 214, 285
- Formation 57-59, 78, 417
- Opening 58, 73
- Strander 146
- Structures 70, 264
Rock flour 190, 192, 193
Rock wool 155, 358
Rocket 104,
- Motor cases 65, 262, 272
- Nozzle 65, 261, 262, 267
- Technology 446, 449
Rods 233, 246, 250, 251
Rolls 221, 233, 251, 255
Roller mills 198, 399, 409
Rollers 236, 248, 251, 284, 314, 403
Roof 129, 159, 180, 181

554   Subject Index

Rosin   48, 286, 402
- acids   110, 376
Rotary shooting machines   301, 303
RPF   150, 151, 152, 153
RSST   78
Rubber   12, 59, 76, 125, 190, 193, 314, 321, 325, 341, 354, 361, 363, 365, 373, 398–401, 404, 405, 409, 410, 414, 434, 460, 461, 464, 512
Runaway reactions   77, 78
Rupture   161, 205, 266, 385, 457
Rutapox process   313

S Glass   266, 267
SO_2-Process   297
Safety precautions   161, 286,
Salicylic acid   385, 459
Sand   103, 291, 294, 296–299, 301, 302, 304, 305, 307, 311, 313, 370, 441
- Coated   300, 371
- Coremaking   299, 310
- Reclaimed   306, 307
Sandblast   377, 413
Sandwich construction   268, 269, 270
Saponifies   312
SBR   398, 401, 404
Scaling   168, 169
Scanning electron microscopy (SEM)   294, 320, 418, 421, 422, 445, 457, 458
Scratch resistance   116
Screen analysis   286, 490, 492–494
Screen printing   242, 413
Screening   251, 300
Screw   142, 201, 202, 203, 221
Sealants   373, 405, 414, 415
Sealed   285, 295, 297, 309
Self-extinguishing   224
Seismic   278
Semiconductor technology   241, 385, 386, 387, 397
Semifinished materials   250, 251
Sensitivity   77, 395, 396
Separation   91
- Efficiency   254
- Microphase   53
Separators   232, 256
Service life   218, 254, 266, 285, 291, 355, 357, 412, 427, 464
Shakeout   292
Shaped Products   432–434
Shape   52, 109, 126, 127, 176, 291, 293, 301, 340, 425, 426, 439
Shaping   200, 233, 236, 303, 307, 311, 365, 452

Shear modulus   226
Shear modulus curves   226
Shear strength   271
Sheathing   129, 138
Sheet goods   126, 232
Sheets   249, 250, 256, 281, 340
Shelf life   131, 148,
Shells   300, 301
Shell molding process   296, 297, 299–302, 313, 371
Shellac   286, 321, 326, 374
Shipbuilding   258, 380
Ships   149, 274
Shock   193, 378
Shoe   403, 405–407
- Custom   164
- Industry   410
- Orthopedic foam   164
Shooting   301, 310, 311,
Shredding   176, 177, 185
Shrinkage   119, 132, 285, 343, 434, 449, 452
- Behavior   116, 214
- Low   73, 280
- Post   117, 192, 215
- Processing   117
- Secondary   192, 201, 204, 205, 214, 230
- Zero   58
Shuttle reentry surfaces   104
Sidewall   401
Silanes   76, 152, 155, 230, 337, 372
Silica   52, 168, 293, 326, 408, 425, 433
- Sands   282, 291
Silicate/CO_2 (and ester)   299, 309
Silicon   241, 263, 387, 460
Silicon wafer   387–389
Silicon carbide   320, 321, 339, 341, 351, 425, 432
Silicon resins   112, 114, 245, 285
Silicones   76, 78, 372
Siloxanes   52, 53, 55, 76
Sink marks   202, 225
Sintering   292
Six Bonding Functions   122–126
Size exclusion chromatography (SEC)   92
Slide gate plates   432, 440, 441, 443
Sliding/guide elements   224
Slip plates   251
Slip properties   224
Slip rings   193, 446
Slot die   163, 166
SMC   264, 265, 279–280
- Compositions   280, 281
Smoke   104, 105, 117, 261, 383, 404
- Emissions   274
- Values   262

Smoke density   105, 117, 232, 262, 271, 288, 289,
Smoke obscuration   167
SO_2 308, 309, 313
Socket Putties   285–287
Sodium carbonate   3, 89
Sodium cation   101, 103
Sodium hydroxide   3, 12, 13, 15, 27, 29, 30, 33, 34, 62, 89, 98, 133, 150, 159, 297,
Sodium silicate   309, 426
Softening point   42, 226, 253
Softwood   129, 131, 134, 153, 192
Sol gel   168, 169
Solder bath behavior   414
Solder crack resistance   52, 240
Soldered   52, 243, 285
Solenoid switch covers   221
Solid moldings   282, 284
Solid polymer   158, 159
Solidification processes   297, 298
Solidifies   286, 292
Solubility parameter   128
Solution properties   91, 97–101
Solvent   50, 62, 97, 250, 254, 263, 283, 301, 307, 310, 365, 377, 382, 387
Solvent resistance   76, 119, 120, 248, 252, 282, 287, 290, 374, 375, 379, 385, 403
Southern pine   129, 143, 152
Soybean peroxidase   50
Space shuttle   65, 267
Spark extinguishing chambers   251
Sparkplugs   215, 231
Specifications Performance test   133
Specific Optical Density (Ds)   117
Specific Surface Resistance   207
Specific Volume Resistance   204, 208, 239
Specific gravity   132, 139, 142, 164, 289, 452
Speeds   176
- High testing   77
- Impact type   77
- Line   146, 153
Spinning   251, 288, 441
Sports products   259, 272
Spray cans   377, 378
Spray dried resins   131, 138, 139
Spray methods   135, 140
Sprayed   155, 310
Sprueless injection molding   199, 202
Stainless steel   16, 274
Standard Chemical and Physical Tests   490–507
- Chemical methods   491, 504–507
- Physical methods   491–503
Standards
- International   195, 203, 244

- National   195, 197, 203
Starches   128
Statistical   41, 292
Steam injection pressing (SIP)   63, 140, 145, 146, 153
Stearates   194
Steel   115, 218, 276, 278, 292, 293, 301, 307, 314, 339, 351, 380, 404, 427, 428, 441, 451
Steel making   444
Steiner tunnel test   105, 167
Stiffness   114, 145, 151, 176, 224, 269, 278, 403
Stoichiometry   24, 34
Stones   314, 321, 351
Storage   126, 131, 157, 160, 174, 254, 270, 284, 296, 300, 301, 310, 327, 380, 464
- Period   207, 285
Storage batteries   232
Storage life   89, 282, 372, 385
Strands   129, 131, 133
Straw   127, 130
Strawboard   130
Strength   76, 126, 149, 151, 153, 175, 200, 262, 266, 269, 278, 285
- Bending   118, 204
- Bond   72, 145, 146,
- Compressive   103, 138, 204
- Flexural   76, 103, 132, 197
- Fracture   77, 192, 195
- High   76, 114, 117, 224
- Impact   38, 72, 73, 76
- Interlaminar shear   76, 118
- Mechanical   74, 157
- Peel   76, 118
- Shear   151, 152
- Ultimate bonding   72, 73
Stress   52, 103, 146, 152, 188, 193, 213, 282, 285,
Stress cracking resistance   224
Stress-strain curves   225
Stripping plate systems   198
Stripping times   306–308
Strong acid resistance   283
Structural
- Analyses   65
- Assignments   150
- Characteristics   64, 65
- Components   275
- Composites   276
- Identity   64
- Integrity   166, 225, 272, 277
- Materials   284
- Oligomers   39
- Properties   259
Structural Wood Gluing   149–152

Structure (Methods of Analysis)   91–106, 304
- Cellular   159, 161
- Network   160
- Secondary   263, 272
- Tertiary   263, 272
- Compositional   96, 97
- Ring   24
- Infusible   24
- Laminar   192
- Polymeric   73
Styrene   111, 116, 117
Sugar   128, 420
Sulfamic acid   282
Sulfonation   9, 11
Sulfonic acids   160, 163, 281, 304, 306
- Copper, aluminum salts   304
Sulfuric acid   11, 13, 20, 41, 84, 95, 160, 290, 297, 298, 306, 308–310, 312, 313
Supercritical Fluid Chromatography (SFC)   92, 391
Supercritical drying   169
Supersizer   340
Surface flammability   105, 262
Surface   152, 161, 182, 192, 193, 208, 209, 231, 248, 264, 281, 310, 401
- Characterization   128, 182
- Hardness   114, 225
- Mount configuration   52
- Resistivity   197, 204
Surfactants   160–165, 256,
Swelling   75, 132, 143, 145, 182, 190
Switch gear cross bars   221
Switches   212, 251
Synthetic resins   190, 288, 298, 321, 341, 379, 426, 427, 430, 431, 435, 447, 457

T_2 (spin-spin relaxation time)   65
$T_{1\rho H}$ (proton spin-lattice rotating frame relaxation time)   65, 66
Tack-free   355
Tackifying resins   398, 401, 402, 405, 407, 415
Tan delta curve   63, 74
Tanks   261, 272, 273, 380
Tannin   51, 61, 110, 127, 128, 146, 148, 149, 153, 313, 528
- Extracts   148
- Formaldehyde   62
- Resoles   148
Taphole mixes   426, 427, 430, 432, 440, 442
TBHQ   463

Techniques
- Chromatographic   91, 105
- Soft ionization   92
Teflon   52, 387
Telecommunications   221, 224, 247
Telephone equipment   238
Television sets   237, 238, 241, 243
TEM   76
Temperature   64, 69, 70, 74, 75, 205, 221, 274, 289
Temperature of deflection under load   197
Temperature Index(TI)   227, 228
Temporary Bonding   125, 126, 291–373
Tennis rackets   272
Tensile strength   197, 205, 227, 278, 289, 403
- Wet   313
Tensile stress   205, 403
Tensides   370, 371, 373, 455, 456
Tension   151
Tensioning pulleys   219, 221
Terminal boards   221, 224
Terpene   128
Terpene-phenolic resins   402, 407
Test Methods (Application of ISO Standards)   203–212
Test pieces   204–208, 227
Tests   133, 262, 488
- Acoustical   180
- Boiling   135, 137
- Chamber   182, 184
- Emissions   182
- Fracture toughness   77
- Fracture mechanics type   77
- Heat durability   135
- Impact   77, 266
- Lap shear   145
- Loss on ignition (LOI)   156
- Method   288
- Odor   182
- Round robin   142
- Scarf and finger joint   135
- Shear   137
- User-friendly type   77
- Vacuum pressure   135
Tetrahydrofuran (THF)   98,
Tetramer   43, 47, 100, 101, 105
Textile   113, 115, 170, 446,
Textile cord   404
Textile felts   72, 74, 174, 175, 179–182, 186, 512
Textile fibers   193, 233, 288
Textile flakes   190, 192, 193, 195, 224
Textile mats   124, 126, 127

Tg (glass transition temperature)   56–59, 62, 65, 74, 103, 146, 390, 391
- High   52, 74–76, 261, 273, 278, 281, 391
- Low   52
- Predicting   75
TGA/FTIR   142
Thermal coefficient of expansion (TEC)   52
Thermal behavior   117, 190, 226
Thermal conditions   61
Thermal conductivity   154, 158, 159, 168, 251, 279, 290, 446, 452
Thermal decomposition   209
Thermal deformation resistance   192, 195, 200, 224
Thermal expansion   138
Thermal insulation   153, 158, 159, 162, 168, 169, 289, 290, 425
Thermal insulators   289, 290
Thermal Mechanical Analysis (TMA)   62, 65, 138, 226
Thermal properties   240, 304
Thermal post treatment   214
Thermal resistance   114, 168, 179, 180, 194–196, 204, 214, 215, 218, 225, 227, 230, 251, 280, 282, 283, 452
Thermal shock   240, 291, 339, 413, 429, 435, 454
Thermal stability   57, 103, 227, 267, 270, 349, 406, 407, 408, 415
Thermal stress   52, 192, 213, 219, 238, 251, 294
Thermal resistivity   153, 156
Thermal/Viscoelastic   62–64
Thermography   385, 459, 460
Thermogravimetric analysis (TGA)   58, 418, 490
Thermoplastic resins   77, 78, 104, 111–113, 117, 125, 126, 151, 175, 190, 195, 199, 203, 205, 225–227, 285, 460, 463, 464
- High performance/Engineering   467
Thermostat housing   219
Thermoset molding compounds   187, 190, 199, 245
Thermosets   109–120, 126, 195, 225
Thermosetting cure   286
Thermosetting resins   56, 58, 104, 125, 151, 185, 187, 259, 261, 279, 284, 373, 417
Thickening agents   279
Thickness   132–135, 142, 143, 153–156, 167, 168, 182, 208, 209, 253, 301, 304, 378
- Wall   192, 201, 202, 208, 225
Thickness swelling   132, 137, 139

Thin layer chromatography   505–508
Thioesters   464
Thixotropic   165, 265, 343, 371
Three D (3D) structure   128, 388, 417
Tile flooring   120, 283
Tiles   282, 283
Timber   127, 313
Timberstrand   127, 130, 131, 133
Time
- Clamping   200, 201, 209
- Cycle   200, 201, 309
- Ejection   200
- Injection   201
- Period   262, 285, 291
Time curves   75
Time-temperature   62, 75
Tire   122, 186,
Tire cord adhesive   150
Titanium dioxide   248, 286, 409
TLV   302
Tocopherol γ 460, 463
Toluene   9, 10, 21, 101, 249, 406–409, 424, 431
- Sulfonic acid   12, 49, 160, 282, 306, 312, 378
- Sulfonyl chloride   282
Tooling   281
Torsion.braid analysis (TBA)   136
Toughened   57, 76
Toughness   73, 75–77, 117, 206, 224, 225, 232, 263, 321, 332
- Fracture   59, 77
Towpreg   272
Toxic gases   104
Toxicity   105, 123, 288, 427
Toys   122, 123, 238, 241, 243
Tracking resistance   204, 209, 215, 230, 239, 248
Trains   256, 258, 259, 268, 281
Transfer compression molding   201
Transformers   247, 384
Transfer molding   52, 199, 201, 225, 230, 452
Transistors   52, 241, 386, 455
Transmissions   56, 258, 268
Transportation   104–106, 110, 115, 117, 259, 296
- Industry   117, 119
- Mass   163
- Vehicles   133
Trees   128, 129, 153
Triazine   69, 467
Trimethylol phenol   25, 33
Trimer   20, 31, 42, 47, 100, 101, 105, 150, 152, 157, 393

Trioxane 20, 21, 286,
Triphenyl phosphine 72, 73
Triphenylsulfonium hexafluoroantimoniate 395
Troughs 432, 437, 441, 442
Truck 180, 181, 301, 353, 363
Trunk 127, 182
Tube inserts 263, 272
Tubes 233, 236, 237, 246, 250, 251, 254, 255, 273, 285, 286, 375, 377, 378, 382, 443, 446, 454
Tundish liners 427, 438, 442
Tung Oil 248, 249, 373, 374, 376, 380
Tungsten carbide 321
Tunnel test 105, 262
Turbomix process 198, 199
Turnkey plant operations 149
Twin screw extruder 163, 166

UHMWPE 264, 267
UKOOA 274
Unclad material 232, 237, 247, 249
Underground stations 281
Underwriters Laboratories (UL) 204, 207, 220, 221, 224, 240, 247
Unidirectional reinforced composites 275
Uniformity 66, 97, 105, 165
Unsaturated products 376, 377, 380, 461
Unsaturated polyesters (UP) 110, 112–117, 119, 120, 151, 187, 190, 209, 233, 243, 245, 258, 261, 279, 280, 282, 321, 326, 373, 380, 417, 477
UP molding compounds 198, 199, 215, 228–230
Urea 11, 18, 111, 149, 150, 155–157, 189
Urea-formaldehyde resins (UF) 110–117, 127, 130, 131, 133, 134, 136, 137, 141, 142, 145, 148, 149, 165, 190, 201, 296–298, 300, 341, 351
Urea- furan resins 296, 302, 303
Urea phenol resins 136
Urea resin molding compounds 187, 188
User friendly 75, 264, 277
UV 31, 92, 96, 290, 387, 389, 461, 466, 467
UV-spectra
– Absorbers 280
– Extreme 387, 389
– Deep 387–389, 394, 397
– Light 388

Vacuum 208, 280, 296, 305
Vacuum cleaner 180, 215, 220

Vapor phase solder bath 52
Vapor pressure 146, 200, 450
Vapor solder reflow 52
Varnish 110, 122, 374, 375, 377, 378, 380, 381, 383
VDE 204
Veining 294
Veneers 129, 131, 133, 134, 137, 140, 143
– Dryer 135
– High moisture 135
– Low porosity 135
Venting 200, 203, 286
Vermiculite 154,
Vessels 272, 273, 283, 284
Vibrating 310, 311
Vicat Test 204
Video camera 140
Vinyl ester 119, 120, 151, 261, 262
– Glass 276
Vinyl benzyl halide 55
Viscoelastic 62, 349
– Properties 75
Viscosity 131, 139, 141, 157, 159–161, 164, 212, 248, 249, 253, 265, 282, 306, 307, 310, 341, 489, 490
– High 313
– Low 38, 51, 62, 273, 277, 281, 283
– Solution 38
– Determination of 494
– Melt 38, 87, 361
– Modifiers 162
– Intrinsic 97
– Regulate 287
VOC emissions 142
Void 164, 285, 336, 453
– Content 51, 264, 281
Volatile 160, 355
– By-products 75
– Components 248, 253, 336, 338
– Emissions 163
– Reaction products 214
Volatilize 62, 161
Vulcanization 398, 399–400, 403, 405, 410, 415
Vulcanized fiber abrasives 316

Waferboard 127, 129, 130, 137, 139, 141, 142, 147, 148, 152
Wafers 129, 137
Walls 154, 225, 283
– Interior/exterior 163
– Panel 181
– Partition 251
– Nozzle 267

Wall paneling   163
Warm Box   296, 298–300, 304, 305, 311
– Vacuum   296, 300, 305
Warp   135, 246
Washing machines   180, 213, 215
Wash out   140, 145
Waste Recycling   524–527
Water   62, 75, 78, 135, 157, 160, 161, 163, 200, 251, 274, 282, 297, 301, 418, 423, 437, 438
– By-product   264
– Flowing   274
– Reduced   288, 306
Water absorption   52, 53, 132, 143, 163, 182, 192, 194, 197, 207, 240
– Low   247, 254
Water content   62, 160
Water dilutability   131, 138, 141, 157
– High   157
Waterglass   294, 312
Water miscibility   490, 497
Water resistant   131, 133, 141, 151, 248, 380, 381
– Soaking   152
Wavelengths   387, 389, 394, 395
Wax   140, 174, 194, 372, 442
Wear   126, 192, 203, 281, 314, 354, 357, 361, 363
– Gradual   291
– Resistance   125, 428, 429, 442
Weathering   117, 373
– Resistance   273, 382
Weather conditions   131
Webs   176
Weight loss   65, 74
Weight savings   225
Wet conditioning process   197
Wet scrubber   297
Wet strength   253, 337
Wettability   128
Wetting   134, 152, 388
Wetting agent   162, 164, 325, 329, 330, 338
Wheat   130
Wheat straw   130, 149, 153
Wheat flour   64, 134
Whiskers   421, 422
Winding   233, 236
– Bodies   251
– Dry   272
– Machines   236
– Process   272, 368
– Wet   272
Windings   214, 375, 383, 384, 446
Windshield   213, 215, 221

Wiper rings   275
Wire enamels   374, 375, 381, 383, 384
Wire terminals   214
Wires   285
Wood I-joists   143
Wood Composites   127–153
Wood   103, 113, 147, 185, 196, 252, 314, 351, 353
– Adhesives   12, 62–65, 85, 98, 127, 133, 146–148, 150
– Binder   62
– Composites   62, 63, 127, 129
– Fibers   22, 131, 252
– Flour   150, 190, 192–195, 339
– FRP composites   278
– Glulams   281
– Laminates   233
– Lesser quality   279
– Materials   110, 113, 115, 116, 124, 125, 127, 149, 187
– Pore   128
– Reinforce   278
– Sources   128, 129, 148, 152
– Species   128
– Waste   131, 141,
– Moisture   131, 145
– Failure   137, 140
– Strands   137, 138, 143, 279
– Flake   140, 145, 146
– Coupling agent   153
– Surface   128, 137, 140, 152
Wool   172, 289, 359
Work place   286, 307, 313, 314, 357, 363, 430
Work Place Safety   514–529
– Emissions   301
Woven fabric   258, 264, 265, 267, 289
Woven preform   280

X-ray   387, 389
X-Ray microtomography   128
X-Y plotters   460
XPS   128
Xylene   38, 381, 423, 431
– Meta   95
– Sulfonic   160
Xylenol   3, 10, 11, 373, 376, 379, 391, 396
– 2, 4   66, 67
– 2, 6   66, 67
– Cresols   248, 249
– Novolaks   405, 415
Xylenol resols   379, 381, 396, 405, 409
Xylok resin   49, 286, 287

Yield strength 77

Zinc 46, 459
- Acetate 87, 150
- Alloys 293
- Borate 278
- Bromide 11
- Oxide 406, 408, 409
- Silicate 425
- Sulfide 321, 323
Zircon 291, 294
Zirconia 168
- Corundum 321, 341

L.A. Pilato, M.J. Michno

Advanced Composite Materials

1994. XIII, 208 pp. 50 figs., 49 tabs.
Hardcover DM 164*
£ 63 / FF 618 / Lit. 181.120
ISBN 3-540-57563-4

Advanced composite materials or high performance polymer composites are an unusual class of materials that possess a combination of high strength and modulus and are substantially superior to structural metals and alloys on an equal weight basis. The book provides an overview of the key components that are considered in the design of a composite. Prime advanced composite materials application areas encompass ballistics/armor/ordnance, commercial/military air craft, sports/leisure, tooling, aerospace, automotive and other market areas that require these materials on a cost/performance basis.

Please order from
Springer-Verlag
P.O. Box 14 02 01
D-14302 Berlin, Germany
Fax: +49 30 827 87 301
e-mail: orders@springer.de
or through your bookseller

* This price applies in Germany/Austria/Switzerland and is a recommended retail price. Prices and other details are subject to change without notice. In EU countries the local VAT is effective. d&p · 65517 SF · Gha

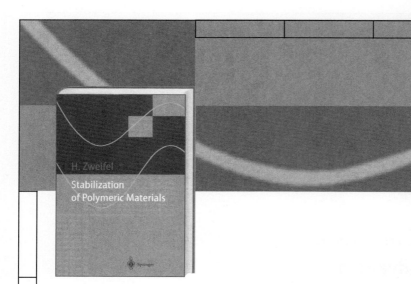

H. Zweifel

Stabilization of Polymeric Materials

1998. XII, 219 pp. 84 figs.
(Macromolecular Systems -
Materials Approach)
Hardcover DM 219*
£ 84 / FF 825 / Lit. 241.870
ISBN 3-540-61690-X

This book provides a concise and comprehensive overview of the basic mechanisms of plastic degradation processes caused by heat and light. At its core is a detailed description of the stabilization of different polymers, including an explanation of stabilization mechanisms and the influence of commonly used additives such as fillers, flame retardants and pigments on the stability of plastic. Every polymer scientist, material technologist, or application engineer dealing with the design of the properties of plastics will benefit from this new overview.

Please order from
Springer-Verlag
P.O. Box 14 02 01
D-14302 Berlin, Germany
Fax: +49 30 827 87 301
e-mail: orders@springer.de
or through your bookseller

* This price applies in Germany/Austria/Switzerland and is a recommended retail price. Prices and other details are subject to change without notice.
In EU countries the local VAT is effective. d&p · 65487/2 SF · Gha

Printing (computer to plate): Mercedes-Druck, Berlin
Binding: Stürtz AG, Würzburg